Sir William Rowan Hamilton

William Rowan Hamilton

Sir William Rowan Hamilton

THOMAS L. HANKINS

The Johns Hopkins University Press
Baltimore and London

Library of Congress Cataloging in Publication Data

Hankins, Thomas L
Sir William Rowan Hamilton.

Bibliography: p.
Includes index.
1. Hamilton, William Rowan, Sir, 1805-1865.
2. Mathematicians—Ireland—Biography.
QA29.H2H36   510′.92′4[B]   80-10627
ISBN 0-8018-2203-3

For my parents

Contents

Illustrations

Photographs

Figures

Acknowledgments

This book has been ten years in the making and owes its completion to the work of many individuals besides myself. If I fail to mention any of the persons who have helped me, the omission should be ascribed to my notorious absent-mindedness and not to any lack of gratitude. The manuscript has been read in its entirety by Fred J. Levy, Robert H. Kargon, and James B. Gerhardt. They have suggested many improvements and caught many errors in the text, and I am grateful for all of their valuable advice. Separate chapters have benefited from the criticisms of my colleagues in the History Research Group and from the students in my graduate seminar. Carl Allendoerfer helped me with the algebra, John Wheeler and Rudolf Peierls advised me on quantum mechanics, Michael J. Westwater helped with the dynamics, Paul Dietrichson straightened out my misunderstandings of Kant, P. J. Federico helped with the history of graph theory, June Z. Fullmer showed me how to decipher Hamilton's shorthand, and Meyer H. Abrams, Thomas McFarland, and Frederick Pottle all gave valuable advice on Coleridge. For this generous assistance I am extremely grateful.

The task of organizing Hamilton's papers was made much easier by Virginia Barto, who spent many hours at the microfilm reader creating order from the confusion of Hamilton's letterbooks. Her insight often helped me to understand the psychological aspects of Hamilton's life. I am also grateful to Susan Frey for her conscientious labor as research assistant, to Joan Scott for the line drawings that appear in this book, and to John S. Barss and Mott Greene, who helped check the final proof.

I owe a great debt of gratitude to Hamilton's heirs, Mrs. Phoebe Alice Abbott O'Regan and Hamilton's great-grandsons, John W. H. O'Regan and Michael L.V.R.H. O'Regan, for allowing me to read and quote from the large collection of letters that they discovered at their home in Marlborough. They have also allowed me to take copies of family photographs and have assisted me in every possible way over the past ten years. I also wish to thank Michael Crowley for his listing of the Hamilton-O'Regan papers, and Patrick Wayman, director of Dunsink Observatory, for photographs and for allowing me to use the observatory records.

I wish to thank Trinity College Dublin for allowing me to use the manuscripts and for providing photographs, and I am especially grateful to the Keeper of the Manuscripts William O'Sullivan and to Stuart Ó Seanóir for their help while I was working at Trinity College. I also wish to thank the National Library of Ireland, the Royal Irish Academy, the Archive for History of Quantum Physics, and the American Philosophical Society for allowing me to use their manuscript collections. For access to the Hamilton-Dunraven correspondence I wish to thank the Seventh Earl of Dunraven, A.P.W. Malcomson, and the Public Record Office of Northern Ireland. I am grateful to Virginia Rowe for letters from Elinor De Vere, to Christina Colvin for letters and photographs of the Edgeworth family, and to A. J. McConnell, Stella Mills, Desmond MacHale, J. G. O'Hara, and Aldon Bell for their helpful advice. I also wish to thank Christine L. Reid and the Libraries of Columbia University for photographs.

Research for this book has been supported by the National Science Foundation and by the Graduate School Research Fund of the University of Washington. Their generous assistance has given me time to write and has made possible three indispensable trips to Ireland.

I wish to thank The Cornell University Press, The Ohio State University Press, and *Isis* for permission to use previously published material. Material in Chapter 4 was published originally in "How to Get from Hamilton to Schrödinger with the Least Possible Action: Comments on the Optical-mechanical Analogy," by Thomas L. Hankins, in *The Analytic Spirit: Essays in the History of Science in Honor of Henry Guerlac*, edited by Harry Woolf, Copyright © 1980, The Cornell University Press, All rights reserved. Material in Chapters 6 and 7 was published originally in "Algebra as Pure Time: William Rowan Hamilton and the Foundations of Algebra," by Thomas L. Hankins, in *Motion and Time, Space and Matter: Interrelations in the History of Philosophy and Science*, edited by Peter K. Machamer and Robert G. Turnbull, Copyright © 1978 by The Ohio State University Press, All rights reserved, and in "Triplets and Triads: Sir William Rowan Hamilton on the Metaphysics of Mathematics," by Thomas L. Hankins in *Isis* 68 (1977): 175-93.

Finally, I would like to thank my family for its support during the years that this book was written.

Introduction

On June 13, 1865, only three months before his death, Sir William Rowan Hamilton received a letter from the American astronomer Benjamin Gould informing him that the newly created National Academy of Sciences had elected him first on its list of Foreign Associates. The fifteen Foreign Associates were, in the opinions of the members of the academy, the most prominent scientists in the world outside of the United States. Gould told Hamilton that the debate in the academy had turned not on the choice of the associates, but on the order in which their names were to be inscribed on the rolls. Hamilton was voted to first place by a substantial majority, signifying that the academy considered him the greatest living scientist.[1] Since Hamilton's name is not recognized immediately by most historians today, or by most scientists, for that matter, one may well ask why the Astronomer Royal of Ireland, who was not a particularly successful astronomer himself, qualified for such an extraordinary honor. The answer is that Hamilton was one of the most imaginative mathematicians of the nineteenth century. In 1827 he devised a completely new theory of optics that allowed him to describe any optical system in a very general way, and in 1832 he predicted the phenomenon of conical refraction in biaxial crystals, a prediction that was dramatically confirmed by his colleague Humphrey Lloyd at Trinity College Dublin. Augustus De Morgan wrote that "opticians had no more imagined the possibility of such a thing, than astronomers had imagined the planet Neptune, which Leverrier and Adams calculated into existence. These two things deserve to rank together as, perhaps, the two most remarkable of verified scientific predictions."[2] Then in 1834 Hamilton created the most general method known for describing the motion of a system of particles.

In addition to his work in theoretical physics, Hamilton made major contributions to algebra. In 1835 he showed that by writing complex numbers as number pairs one could dispense with "imaginary" numbers in algebra. His most dramatic discovery, the one that he considered most important, and the one that won him first place on the roster of Foreign

Associates of the National Academy, was his discovery in 1843 of quaternions. Quaternions are hypercomplex numbers composed of one real part and three complex parts. For years mathematicians had been trying to find systems of numbers, other than real numbers and complex numbers, that would follow all the rules of ordinary algebra. Quaternions obey all those rules except the commutative law—the product is not independent of the order of multiplication of the factors. Hamilton's achievement was not just one of finding a new algebra. More important was his realization that one could sacrifice one or more of the rules of ordinary algebra and still have an algebra that was meaningful. The discovery of quaternions opened the way to much of modern algebra.

All of these achievements are important, but there is more to be learned from Hamilton's career than just his mathematics. His life tells us a great deal about the social context of Victorian science. He was educated at a time when the mathematical and scientific curriculum at British universities was undergoing a major revolution. He was one of the original members of the British Association and took an active interest in its affairs, and he was president of the Royal Irish Academy in Dublin. His contacts with other scientists help to clarify the organization of science during the Victorian era. In addition to the formal organization of science, Hamilton's life tells much about the broader literary and intellectual world in which that science grew. He had important connections with Maria Edgeworth, William Wordsworth, Samuel Taylor Coleridge, and Aubrey De Vere, with whom he discussed the scientific and the aesthetic views of nature, their possible conflicts, and their connection to religious belief.

It is also informative to study the subsequent fate of Hamilton's ideas. Although he was famous in his time for his mathematical prowess, much of his work was so abstract that it had little immediate application. The practical value of his mechanics was not realized until the advent of quantum mechanics in the twentieth century, and his celebrated quaternions became important for physicists primarily through the contribution they made to vector analysis. Thus the real impact of Hamilton's mathematics was delayed, a delay that tells much about the course of science in the intervening years.

For many historians the most important aspect of Hamilton's work is its connection with German metaphysical idealism. There is a noticeable difference between scientific theories in the eighteenth century and those in the nineteenth. Eighteenth-century theories are largely mechanistic. They attempt to explain physical phenomena in terms of the movement and interaction of particles—either atoms or some other material corpuscles that might or might not have forces between them. Mechanics was the science par excellence in the eighteenth century, because it gave the rules for motion of all these parts.

In the nineteenth century new concepts appeared, such as the electromagnetic field, the wave theory of light, and the concept of energy, that could not be explained in mechanical terms. Whether these new nonmechanistic theories appeared because they were demanded by an advancing experimental science, or whether they were the result of new philosophical views of the cosmos, is one of those questions that will never be settled conclusively. In fact it is certainly erroneous to ask the question in such an either-or fashion. The stimulus to scientific discovery must come from more directions than we commonly realize. The question will continue to defy resolution also because it is not subject to historical proof. No one can say for certain why an original thought pops into a person's head at a particular place and at a particular time, and any historical explanation of its origin must remain conjectural, at least in part. If, however, the new direction of science in the early nineteenth century did have an origin that was philosophical, the most obvious candidate was that current of German metaphysical idealism starting with Kant and passing through Fichte into Schelling's *naturphilosophie.* It was an explicitly antimechanistic philosophy that emphasized the dynamic elements of "force" and "power" against the changeless atoms of mechanism.

Hamilton is an especially important figure in this regard, because he was one of the earliest British scientists (or philosophers, for that matter) to read Kant and to take him seriously. He was also a great supporter of Coleridge, whose philosophical writings did much to introduce Schelling's views to the British. Where the writings of other British scientists may contain an occasional tantalizing reference to German philosophy, indicating that they knew of its existence, Hamilton's manuscripts are full of information about what he read, when he read it, and how he used Coleridge and Kant to build his own philosophy. Hamilton was certainly atypical in this regard, but because he was one of the major scientists of the first half of the nineteenth century, his philosophical connections and the uses he made of them are important.

If Hamilton had been more of a physicist and less of a mathematician and metaphysician the connection between his science and his philosophy would undoubtedly have been much closer. He denied the existence of material atoms and believed the universe could best be explained by "power" acting in space and time, but he never created a physical model that illustrated his metaphysical ideas in a concrete way. Instead he remained almost entirely in the realm of mathematical abstraction. His famous papers in optics and mechanics were all highly analytical. Rational mechanics in the eighteenth century had been regarded as a branch of mathematics and had been treated as such. Hamilton's work was very much in that tradition. His starting place was Lagrange and Laplace, not Newton. A curious example of this appears in a manuscript notebook from the summer of 1826, when he was preparing for his examinations and at the same time thinking out some ideas for his optics and mechanics. At one point he carefully copied out Newton's Laws in order to learn them for the examination; apparently he did not know them very well. And yet the same manuscript contains notes outlining a system of mechanics and optics more general and more abstract than anything in existence at the time.[3] Hamilton was able to achieve such brilliant successes in mechanics precisely because his work was so mathematical. His contemporaries understood that. While Hamilton was writing his "General Method in Dynamics," he wrote to William Whewell at Cambridge to tell him that he thought he had "made a revolution" in mechanics. Whewell, who had been trying to clarify the fundamental concepts of the science, replied: "I am glad to hear you have been turning your thoughts to mechanics, and have no doubt you will make a hole quite through them with your long analytical borer, and, for aught I know, bring up purer waters from greater depths than we have yet known. . . . In the meantime I, who have been long muddling at the bottom of the well, have persuaded myself that I have got the mud to subside, and have been trying to distinguish how much of the stuff comes from the clear spring of intellect, and how much is taken up from the base mud of the material world."[4] Any philosophy of nature has of necessity to deal with the material mud of the physical world, but Hamilton always preferred to stay with the clear spring of intellect.

As it was, Hamilton's metaphysics did not lead him to new physical interpretations, as happened in the case of Oersted, Ritter, Mayer, and other *naturphilosophen*. Instead, his metaphysics operated at a much higher level. It led him to seek the greatest possible generality and abstraction in all of his work. His success in uniting the sciences of optics and mechanics did not involve any new physical concept. It was a formal unity describing both sciences by the same mathematical system. But the fact that it was only a mathematical unity made it no less meaningful for Hamilton. This was true also in his work in algebra, where his study of Kant convinced him that hypercomplex numbers, such as quaternions, *had* to exist, or rather had to be possible of construction. In this way his metaphysics led him to seek a much greater generality in the concept of "number."

Biographers of mathematicians usually describe the accomplishments of their subject interspersed with biographical information of an anecdotal sort. There seems to be little connection between the man's life (his religious beliefs, his social position, his character, and so on) and the mathematical work that he does. The reason is that mathematics is the most "internalist" of all the sciences. It has a logical coherence independent of physical events or their chronology, and it is therefore not surprising that the history of mathematics has an independence that separates it from the histories of other scientific disciplines. Of course the separation is not complete, because the natural sciences present problems that require new mathematical methods for their solution, and in turn, many kinds of mathematics that at first seem to be entirely "pure" eventually find an application in the natural sciences. And yet these connections are often difficult to find and even harder to demonstrate conclusively.[5] The same is true for the relations between mathematics and philosophy. One cannot prove, for instance, that Hamilton would not have discovered his method of representing complex numbers by number couples if he had not followed Kant's idea of constructing number from the pure intuition of time. He *did* follow Kant, and he *did* describe his system in a paper entitled "Preliminary and Elementary Essay on Algebra as the Science of Pure Time," and he *did* say that his own philosophy greatly affected the course of his mathematical research, but one cannot prove it, because the number couples are not *logically* dependent on the concept of time. One can argue and it has been so argued that the proper representation of complex numbers is an entirely algebraic problem, and that the metaphysical trappings in which Hamilton presents his ideas are superfluous adornments that actually obscure and interfere with the important mathematical content.[6] But beyond the doubt that a mathematician can compartmentalize his mind with such complete exclusiveness, there is the realization of the biographer that the process of invention in mathematics, or in any science, for that matter, is very different from the logically coherent exposition that appears in the final published papers. Earlier biographers of scientists mistakenly saw their purpose as one of assigning priorities for "discoveries" made. It was a matter of granting credits—the awarding of literary "prizes" to the leaders of science. We now recognize that such a procedure obscures the history of science. It is more important to understand how scientific concepts are created and how they develop than it is to attribute them to certain individuals. And if we are going to try to understand how concepts develop, it is important to also understand the intellectual context in which they appear.

The biographer has an advantage when it comes to explaining the intellectual context of scientific ideas if he will only take it. The relationship between scientific currents and other intellectual currents is notoriously difficult to discover and to describe. The biographer at least has the certainty that the ideas and events he has to recount were all the property of a single individual, and were integrated by that individual into his own life. To this extent the biographer is on more solid ground than the historian of intellectual "movements," who is constantly frustrated by seeing his generalizations dissolve as he brings them down to particular instances. But in order to take advantage of his opportunity the biographer must attempt to integrate all the varied aspects of his subject's life into a single whole. It will not do to write the life of a mathematician excluding his mathematics. Equally unhelpful is a description of the mathematics that ignores the subject's philosophy or his education. Biographies that present only one facet of the subject's life can be entertaining, but they are of little value to the historian seeking the origin and development of scientific ideas.

This biography is an attempt, as all biographies must be, to bring together the many aspects of an individual's life and work into the expression of a single personality. If that attempt at unification is successful, it will knot together in one man and at one point in historical time several strands of history that may previously have been at loose ends.

The reader will find some chapters that contain mathematics and some that contain philosophical exposition. These chapters concentrate on the fundamental ideas in Hamilton's writings and do not attempt to cover the entire range of his interests. There is, for instance, a great deal in the *Elements of Quaternions* that is not even mentioned in this book, and the mathematical reader will want to consult the three volumes of *Mathematical Papers* for comments on the more technical aspects of Hamilton's work. Hamilton's presentation of his optics, mechanics, and algebra is quite different from that usually given in modern textbooks. Also, because he was a poor expositor of his own ideas, it has seemed important to present his arguments in the clearest possible fashion, and this I have attempted to do.

Anyone reading Hamilton's manuscripts will conclude that he must have spent most of his life with pen in hand. He wrote incessantly, usually in notebooks of all sizes and shapes, but also on pieces of loose paper, particularly if he was drafting an article or a lecture.[7] He wrote on walks, in carriages, during meetings of the Royal Irish Academy, on his fingernails if no paper was handy, and, according to his son, even on his egg at breakfast.[8] Occasionally he attempted to achieve some order in this mass of papers, but never with success. The papers flowed over the tables, onto the floor, and under the beds. On one occasion, when he was giving a large formal dinner, he had to clear the library. It took two days of solid effort, and even then the task was accomplished only by resorting to bags and baskets to contain all the papers.[9] Tradition has it that when Hamilton's son sorted this mass after his father's death he found plates of desiccated chops interleaved with the manuscripts, creating a true archeological midden for his literary executor.[10]

When Hamilton died, his friend Robert Perceval Graves received all of these papers and immediately added to their bulk by writing to Hamilton's friends and associates requesting any letters they might have received from him. Graves was very thorough, and since he knew most of Hamilton's correspondents himself, he was able to collect a large number of letters. His labor resulted in a massive "life and letters," published in three volumes containing 2090 pages. The last volume appeared in 1889, twenty-four years after Hamilton's death. Graves had been working more or less steadily the entire time. He ruefully admitted that "the immense mass of papers" left by Hamilton and the "regard paid by him habitually to small things, as well as great, may probably have had an injurious effect upon his biographer."[11] Graves won an honorary degree from Trinity College for his efforts, and no doubt he welcomed the opportunity to write about Ireland's greatest scientist, but he must have thought occasionally how convenient it would have been if some of those papers had been accidentally mislaid.

When he had finished his biography, he deposited the manuscript notebooks, a large quantity of loose papers, and a substantial part of the scientific correspondence at the Library of Trinity College. He returned most of the letters that he had borrowed to their original owners. During the past century many of these letters have found their way back into the manuscript collections at Trinity College and at the National Library of Ireland. In 1968 and again in 1974 Hamilton's great-grandsons, Michael and John O'Regan, found trunks and boxes of the family letters and papers in the attic of the O'Regan home in Marlborough, England. These manuscripts have more than doubled the number of letters extant. Graves made notes on the manuscripts as he read them, and the ubiquity of his blue pencil is a constant testimony to his thoroughness. Few papers have come to light that did not pass under his scrutiny. I have read through ten boxes of loose papers, approximately 250 manuscript notebooks (some very large), and more than 6000 letters. We know from his letter lists that Graves had considerably more than that.

The enormous amount of documentation permits one to follow Hamilton almost week by week through his life. Anyone writing a modern biography is tempted to follow Graves's lead. He was thorough, fair in his judgments, and bold enough to distress some of Hamilton's friends. Nevertheless, there were many things he would not or could not say in his biography. He was extremely cautious about references to persons still living, and he barely hinted at some aspects of Hamilton's most personal life. Such hints had meaning for those who were privy to Hamilton's closest thoughts, but they hid from the general reader the events that most influenced Hamilton's life. Because he lacked the necessary training, Graves was not able to do justice to Hamilton's mathematical work, nor was he sufficiently distant from events to be able to put that work into accurate historical perspective. I have made no attempt to include all the details that Graves was able to pack into his three massive volumes. Therefore a reader interested in any particular period or aspect of Hamilton's life will still do well to consult Graves's biography. Instead, I have tried to bring forward those aspects of Hamilton's life and work that I believe to be most important. In some cases this has meant including information already known, in some cases it has meant changing interpretations, and in some cases it has meant presenting entirely new material. The result, I hope, will reveal to the reader the truly remarkable character that I discovered in my research.

I

Birth and Education

1
Early Years

In July of 1806 the Irish patriot Archibald Rowan was returning to his estate at Killyleagh after eleven years of exile. He was one of the few surviving members of the United Irishmen and a former friend of Wolfe Tone. Convicted and imprisoned in 1794 for "endeavouring by tumult and by force to make alterations in the constitution and the government and overturn them both," he had tricked his jailer into taking him to his Dublin house from which he had made his escape into exile, first to France, then to the United States, Germany, and England.[1] Now, with the uprisings of 1798 sufficiently far in the past, he was at last coming home to Ireland. His agent, Archibald Hamilton, had arranged an extravagant procession suitable for Rowan, who enjoyed pomp and lavish display. The road from Donaghadee to Killyleagh was crowded with excited peasantry, who removed the horses from Rowan's coach and attempted to draw it by hand. His terrified wife insisted that she and Rowan get out and walk rather than trust themselves to their overly enthusiastic supporters. At the castle an enormous bonfire "blazed to the skies"; the houses and cottages for miles around were illuminated with candles; rockets and crackers blasted away until midnight, and ale and whiskey were distributed liberally. There were many burned hands and singed whiskers, and the Rowans' agent, Hamilton, who wrote this account to his wife the next day, complained that his burns made it difficult for him to hold the pen.[2]

Archibald Rowan and Archibald Hamilton were respectively godfather and father of Sir William Rowan Hamilton, the subject of this biography, and the connection between these two Archibalds is one of the mysteries of William Rowan Hamilton's parentage and early life. Archibald Hamilton had become agent to the Rowans in 1800, when Mrs. Rowan had left Killyleagh Castle to join her husband in Germany. Before leaving she had hired Hamilton to look after the estate. It had not been an easy or lucrative undertaking. Because Rowan was a fugitive from justice he was not permitted to draw from his main income and had to live on whatever rents his agent could extract from his tenants at Killyleagh. Mrs. Rowan complained constantly that they were "starving in a foreign land" and Rowan himself admitted: "The fact is I have always lived in too high a style and my family do not like to give up indulgence."[3] Rowan was heir to substantial property in Ireland if only he could lay his hands on it. His father continued to live on at Killyleagh Castle with a mistress and an illegitimate son, and Rowan was afraid that when his father died (an event that seemed imminent in 1804) the mistress would take as much as she could. Rowan warned his agent in Dublin, who locked the family valuables in the muniments room of the castle, and later, when Rowan's father died in April 1805, Archibald Hamilton called out the Killyleagh Volunteers, who were still loyal to Rowan, evicted the mistress and son amid great howls of misery, and began to negotiate for his patron's return to Ireland.[4] Hamilton was still in the north of Ireland when his own son was born at Dominick Street in Dublin at midnight between the third and the fourth of August, 1805. It was the same street from which Rowan had escaped nine years previously. In gratitude Rowan agreed to serve as sponsor, and it was from him that the child took his middle name.

Rowan's gratitude, however, was not long lived. In response to the Rowans' constant importuning for money, Archibald Hamilton had borrowed at a high rate of interest, and when Rowan refused to pay (arguing that his agent's services had been acts of friendship), Hamilton went bankrupt. Money problems were already apparent in June of 1807, when Hamilton's sister Sydney wrote that "Mr. R. had better try and return to London elegantly caparisoned for surely he will be ashamed to show his face here."[5] His creditors sued and obtained some compensation, but his business had been ruined. He managed to recover somewhat, but was in financial difficulty throughout his life. To his wife he admitted: "I can manage anything but my own money concerns," which would indicate that Rowan was not completely to blame for all his troubles.[6]

Archibald Hamilton died in 1820, but Rowan lived on until 1834 and was frequently in Dublin cutting the rather ridiculous figure of an outmoded rebel.[7] Young William Rowan Hamilton, his godson, never saw him, nor did he mention Rowan in his letters or use that portion of his own name. The signature *William Rowan Hamilton* first appears in 1826, when the family's indignation had cooled and William had left his uncle's tutelage for the more liberal atmosphere of Trinity College.[8] In 1835 William was knighted by the lord lieutenant and became Sir William Rowan Hamilton at the young age of thirty. He received a long letter from Rowan, full of rather belated sponsorial advice, urging him never to forget his religious duties, and to make his name and that of Ireland great in the scientific world. There is no record of any answer from William.[9]

William lived with his parents at Dominick Street until he was almost three years old. Probably because of the family's worsening financial condition, in 1808 he and his sisters Grace and Eliza went to live with relatives. William was sent to be raised and educated by his uncle James Hamilton, who ran the diocesan school at Trim, about forty miles northwest of Dublin in County Meath.[10] Unlike William's father, Uncle James was a graduate of Trinity College Dublin and a scholar of the classics. His letters are full of classical allusions and flowery Ciceronian passages, usually written in a hasty scrawl. James was curate of Trim, a subordinate position in the Church of Ireland under the vicar, who was the Reverend Richard Butler. The Anglican Church of Ireland was the established church, and therefore appointments in the church were in large part a measure of one's access to patronage and family influence. James remained a lowly curate until his death in 1847, in spite of all his own efforts and those of his nephew in later years.[11]

At Trim, William lived in the midst of spectacular ruins. The diocesan school was located in Talbot's Castle, a fortified manor house built in 1425 by Lord Talbot, the lord lieutenant of Ireland.[12] Part of the house contained ruins of the much older St. Mary's Abbey, and the Hamiltons referred to the house both as Talbot's Castle and as St. Mary's Abbey. The castle served both as home and school. James's mother had bought it in 1800 and had lived there with James at least since 1802.[13] Talbot's Castle is on the edge of the town of Trim. One side of the house drops precipitously to the River Boyne. Across the river are the remains of St. John's Castle. This is a "real" medieval castle—huge, with immense stone walls, moat, keep, and all the other accoutrements of a medieval fortification. On the other side of the house is a field containing the Yellow Steeple of Trim. It stands a few steps from the door to Talbot's Castle and is a piece of the old tower of St. Mary's Abbey, built in 1368. This remnant of the old abbey is quite tall and overlooks the town. William used it as a giant sundial, marking the hours by the position of its shadow in the field, and also as a station for a semaphore system that he invented with his school friend Tommy Fitzpatrick. Half a mile down the Boyne was another large ruined abbey, the Abbey of St. Peter and St. Paul, and there were numerous other ruins further along the river. William was a great rambler who had all these monuments as his playground. He also liked swimming and went into the river whenever his Uncle James would allow it.

Under the tutelage of James, William prepared for entrance to Trinity College for what was expected to be an illustrious career. In addition to his academic studies, he learned about Irish politics. They were not always

Trim, County Meath. The large medieval fortress to the right of center is King John's Castle; the ruined tower to the left is the Yellow Steeple; and the large building between the two is Talbot's Castle, the Diocesan School of Trim and the boyhood home of William Rowan Hamilton. The river Boyne flows between Talbot's Castle and King John's Castle.

the same lessons as the ones his own father would have taught him. He later told Augustus De Morgan:

Understand, first, that I don't pretend to be an unprejudiced man. Deeply prejudiced I know myself to be; not thereby admitting that I am wrong. From childhood I have had political leanings, and always to the *il*-liberal side. My father (Archibald Hamilton, of Dominick-street, Dublin) was a liberal, almost a rebel; he assisted Hamilton Rowan to escape from prison, and deeply involved himself by other efforts in his favour. On the other hand, his brother, my uncle, the Rev. James Hamilton, who lived for forty years the Curate of Trim, and died as such, was a Tory to the back-bone, and doubtless taught me Toryism along with Church of Englandism, Hebrew and Sanskrit .... My father used to enjoy provoking me into some political or other argument, in which I always took my uncle's side.[14]

In his younger days Hamilton was not as conservative as this later statement would indicate. He favored parliamentary reform (much to the disgust of his friend William Wordsworth) and expressed occasional sympathy for the cause of Catholic emancipation, but as he grew older and

Irish politics changed, he joined the Conservative Society, and stood to the right as a staunch supporter of the Union.

Hamilton was born and lived his entire life under the Union, which made Ireland part of the United Kingdom, subject with England, Scotland, and Wales to the same king and governed by the same Parliament in London. Legally the Irish were to be treated the same as any other subjects, but what was desirable in theory was impossible in practice. Differences between England and Ireland were too great to make any smooth amalgamation possible, and centuries of conflict, rebellion, and suppression left long memories that refused to fade. During Hamilton's lifetime the great political watchwords in Ireland were *emancipation* and *repeal*, both sounded most vigorously by the "Great Liberator," Daniel O'Connell. *Emancipation* meant emancipation of the Roman Catholics, who were denied access to most public offices until 1829, when O'Connell finally forced the Catholic Emancipation Bill through Parliament. *Repeal* meant repeal of the Union between England and Ireland, and a return of an Irish Parliament. Repeal turned out to be a much tougher issue than emancipation and one that has not been resolved to this day.

Hamilton was a member of the Protestant ascendancy, the group that had essentially controlled Ireland for three centuries. The Protestants were the major landlords, and they controlled most manufacturing and industry as well as most commerce. They alone had the right to sit in Parliament, and their church, the Church of Ireland, was the established church of the land. Hamilton's Uncle James received a large proportion of his income from parish tithes. These tithes were collected from all persons residing in the parish whether they were members of the Church of Ireland or not. In 1800 only one-tenth of the inhabitants of Ireland were members of that church. Around Dublin the Protestant population was greater, but in some parts of Ireland there were parishes that contained no Protestant parishioners at all! The tithe was the most hated of all the tax burdens, and in the 1830s there was a tithe war centered in County Meath that made the parish clergy almost destitute. The tithe war was accompanied by coercion acts and an enormous increase in agrarian crime that made Uncle James's position extremely uncomfortable.

During Hamilton's lifetime the Protestant ascendancy was gradually losing its control over the country, and this tended to make Protestants like Hamilton and his Uncle James increasingly conservative in their politics. The late eighteenth century had been the high point of prosperity for the Protestants. At that time Dublin was the second largest city in the British Isles. Many of the monumental buildings of Dublin, such as the Custom House, the Four Courts, the Bank of Ireland, and the elaborate Georgian residences, date from that period. These were symbols of Protestant control. When Hamilton visited friends in Dublin he went to Fitzwilliam Square, Merrion Square, Upper Baggot Street—all fine residences on broad avenues with parks and coach houses. But prosperity for the Protestants was not necessarily prosperity for all, and the rebellion of 1798 had required some more permanent solution to the eternal "Irish question."

The Act of Union in 1800, whatever its intent, did not unify all offices in Ireland and England, and it certainly did not unify the people. The administration of Ireland continued to be through the lord lieutenant, his chief secretary, and an under-secretary, with the administrative offices located in Dublin Castle, the traditional center of Protestant authority. The lord lieutenant maintained an elaborate vice-regal court in Phoenix Park with levees, balls, and dinners that Hamilton attended from time to time. Phoenix Park was close to Dunsink Observatory, where Hamilton spent most of his life. It is symbolic of his social position that when he became Astronomer Royal in 1827 he took the two sons of Anglesey, the lord lieutenant, into his house to provide for their education. In 1835 Hamilton was knighted in full panoply by the lord lieutenant, who preserved the royal prerogative of conferring knighthood.

The Act of Union was not a real union at all, but an effort to preserve the Protestant ascendancy and a recognition by England that Ireland required a separate administration and needed to be treated in a manner different from the rest of the United Kingdom. One result of the Union was an increase in the sectarian spirit of the country. During the eighteenth century there had been "Irish" questions in which Protestants and Catholics stood together against England, but within twenty years after the Act of Union, the Protestants as a body recognized the Union as their defense against the threat of rising Catholicism. They tended to become more "English" and less "Irish." Hamilton was always conscious of his Irishness; at meetings of the British Association he often acted as spokesman for the "Irish delegation," and he even wrote sentimental poetry when he first viewed the Irish coast on his return across the channel, but his politics were conservative and his devotion to the monarchy was total.

There were other problems with the Union. It always seemed that the English accepted the Irish as equals when it was in their interest to do so, but when the Irish ran into serious difficulties, they were expected to take care of their own problems. The biggest problem in Ireland was economic. As the century advanced Irish prosperity declined. Industry, which was rapidly expanding in England, failed to take root in Ireland except in the north, in the region around Belfast, and even there its growth was not as rapid as it was in England. Ireland retained its linen industry, again in the north, but all other industry declined in the face of competition from England. The cost of the Napoleonic Wars was a tremendous burden. The Act of Union specified that Ireland was to pay two-seventeenths of the total expenditures of the United Kingdom. In 1800 this was not an obviously unfair levy, but as England prospered, Ireland declined, and the total budget was determined by the needs of Great Britain and by Great Britain's ability to pay. England could afford the Napoleonic Wars; Ireland could not. Ireland's share rose from £4¹/₂ million in 1800 to over £13 million in 1816, at a time when the country was increasingly depressed economically.[15] The influx of capital that might have provided an industry to pay the debt never came. The English were reluctant to invest in Ireland at a time when expanding British industry provided a safer and more profitable haven for their capital.

While the other major cities in the British Isles rapidly grew, Dublin was left behind. In 1821 Dublin was still next to London in population, but by 1831 it had been passed by Manchester, and by 1841 it ranked fifth in the United Kingdom. The character of the city changed, too. Before the Union it had been the seat of an independent parliament, and the home of many peers. After the Union the nobility moved in large numbers to England, the Parliament Building became the central office for the Bank of Ireland, and the city took on a less dignified and more commercial character.[16] It also became increasingly Roman Catholic. In 1800 Dublin was already 70 percent Catholic, but its leadership was dominated by Protestants. In 1816 the Catholic procathedral was built to rival the centers of Anglicanism, St. Patrick's Cathedral and Christ Church Cathedral. The Catholic population grew, and with the passage of the Catholic Emancipation Bill in 1829, the Protestant ascendancy began to lose its political control.

Although Dublin did not grow appreciably, the total population of Ireland increased at a rapid rate after 1800. The first reliable census was in 1841, and it set the population at 8,175,124, the highest ever recorded. By 1845, when the Great Famine struck Ireland, the population must have exceeded 8,500,000. Comparing these figures to the estimates taken from tax records for 1785 of under 3 million, one recognizes how remarkable the growth in population was.[17] Ireland became one of the most densely populated countries in Europe, while at the same time it remained almost entirely agricultural.

This immense increase in population was supported largely by the lowly potato. It was the main staple and, often, the only item in the diet of half the population. At times of normal harvest an adult Irishman ate between ten and twelve pounds of potatoes a day, which he could grow on an extremely small garden plot. Marriages occurred early and families were large. A father divided his meager farm into tiny plots for his sons, who then subdivided them again for the next generation.[18] Even the fantastic productivity of the potato could not sustain this kind of population pressure, and by the early 1840s almost one-fifth of the population had no work or adequate land and was destitute for at least part of the year. When the potato crop failed in 1845 distress became widespread. When it failed again in 1846 all over Ireland, the famine became extremely serious. One million people died, and another million emigrated, so that by 1851 the population was down to 6,552,000. Throughout the nineteenth century the population continued to decline; in 1900 it reached about 4 million and has remained constant ever since.

The areas with the most dense population were also those that were the most remote. A map of population for the 1840s shows a high density in the city of Dublin proper, but just outside the city the density drops to the lowest in the entire island. Moving westward from Dublin, the population density gradually grows until one reaches the west coast, where the density often exceeds 400 people per square mile. County Meath, containing the town of Trim where Hamilton grew up, was one of the least populated counties, having a density of fewer than 250 per square mile; it also had a high proportion of large farms used primarily as grazing land. A contemporary traveler's account is as laudatory about County Meath as it is horrified by much of the rest of Ireland: "There is indeed no part of Ireland where the Englishman will find himself so completely at home."[19] Hamilton grew up, then, in an area that escaped the worst population pressure, although dispossessed laborers from other regions of Ireland swarmed to Meath at harvest time, and the slums of Dublin continued to swell with the poor.

Even so, Hamilton was not entirely shielded from the distress of the country. Even in County Meath much of the population lived on the edge of famine during his youth. Hamilton's Aunt Sydney, in her letters to his mother, interspersed her comments about "Sweet William" with occasional descriptions of conditions around Trim. In 1809 she hoped the harvest would be a good one because "the poor labourers were literally starving for want of employment."[20] And again in 1812:

There is nothing new here but the melancholy scarcity of provisions and distress of the Poor who are starving and I am convinced that if something is not done to prevent the Farmers storing up potatoes, meal etc. there will be a Rebellion. Drum said to me the other day 'sure ma'am we may as well lose our lives fighting as be starv'd to death.' . . . The meal is 4-4d a stone, potatoes 13d but as to the last they are scarcely to be had and the worst of it is the poor not being able to pay the price required for *seed* cannot set them so that they will of course be equally scarce next year. Mr. Fox as a great *compliment* has promised to sell us a sack at 13d per stone our being out as well as the rest of our neighbors. It is really scandalous that the distillation from oats is not stopped.[21]

In 1822 Hamilton wrote to his sister Eliza in Ballinderry in the North: "Does any degree of that alarming scarcity of provisions, which is so great in the South and West exist near you? How much honour the English have done themselves by their prompt liberality."[22] These recurring crises came during the so-called "meal months" before the potato harvest, when the previous year's supplies were gone and the people had to live on meal if they could afford to buy it.

The state of incipient famine in Ireland and the frequent rapacity of the landlords meant that agrarian revolt was common. The poor organized into societies called "Whiteboys" or "Ribbonmen." They attacked the landlords' holdings and took revenge on any persons taking up a lease on land from which a former tenant had been evicted. Insurrection in the countryside was a constant fear. In 1823, when Hamilton's sister was on a trip to the north, he wrote that his "Cousin Sarah did not wish her going to Ballinderry spoken of, for fear an ambush should be laid on the road. But somehow or other, there is always someone to blab out these wonderful secrets."[23] And in writing to his Cousin Arthur the previous month he mentioned that he wished to send a letter to his sister Grace, but did "not chuse putting a letter with her direction upon it in a country Post Office like this."[24] And again, in 1829, when Hamilton was expecting a visit at the observatory from William Wordsworth, he received word from Wordsworth, who had never been in Ireland, expressing real concern about traveling through the countryside during his visit.[25]

And yet Hamilton's youth was comfortable and relatively secure. In addition to his Uncle James and Aunt Elizabeth, his father's sister Sydney also lived at Talbot's Castle. Aunt Sydney kept Hamilton's mother appraised of all his antics and intellectual successes, and her letters are full of good humor. Hamilton's earliest education was as much in her hands as it was in the hands of her brother; we read of her hearing William's lessons, and her letters and some early manuscripts indicate that she had a good command of Hebrew and Latin. At Trim were also the children of Uncle James and the boarders at the school. Aunt Elizabeth, James's wife, was from the La Touche family of Bellevue in County Wicklow, a place that Hamilton visited occasionally. Another close relative was "Cousin Arthur" Hamilton, a Dublin barrister and graduate of Trinity College who was almost the same age as James.[26] William had relatively little contact with him until he was old enough to travel to Dublin. Uncle James disliked Dublin, and William seldom saw his parents or his Cousin Arthur until he was about twelve years old. Then trips to Dublin became more frequent, and Cousin Arthur's house on South Cumberland Street became a second home where William and his sisters would meet, especially at holiday time. As William neared the age for entrance into Trinity College, Cousin Arthur took a greater interest in his career and figured more prominently in his life.

Hamilton's mother was from the Dublin family of Huttons. Her sister Susan married Joseph Willey, a minister in the Moravian Brethren, who lived in a Moravian settlement named Gracehill near Ballinderry in Northern Ireland. Hamilton saw the Willeys infrequently, but his sisters lived with them at different times and there were many comings and goings between Trim, Dublin, and Ballinderry. Joseph Willey was an amateur astronomer and astrologer, and he kept up a regular correspondence with Hamilton on astronomical subjects. Hamilton always showed extraordinary kindness and patience for his Uncle Joseph. Even when he became Royal Astronomer and had more elevated occupations to take up his time, he was willing to undertake long calculations and to send lessons of instruction to Joseph. The Moravians were strongly Calvinistic, and it is probably from the Willeys that Hamilton's favorite sister, Eliza, drew the strain of narrow piety that characterized her religious convictions.

Hamilton had four sisters who survived infancy, and no brothers. Grace was the eldest, three years older than Hamilton, and she was the sister who ran the household when they were all together at the observatory. She had been raised by the Moravians, first in Fairfield, England, and then at Gracehill, when the Willeys moved there in 1817. Eliza was two years younger than Hamilton and the sister closest to him in age and temperament. She was also the one for which he had the greatest attachment. She had a literary bent, and Hamilton always referred to her as his "poet sister." She lived at Gracehill part of the time, too, but in 1823 we find her at the school of the Misses Hincks in North Great George Street, Dublin.[27] She seemed the most promising of all the girls, and extra effort and money were spent on her education. Sydney was five years younger than Hamilton and the most adventurous of all the sisters. In 1828 she was working as an assistant at the school of Mrs. Swanwick at Rhodens, near Belfast. She later emigrated with one of Hamilton's sons, William Edwin, to Nicaragua, where she stood the climate and deprivation much better than he did. She later returned to Ireland and emigrated again to New Zealand, where she died in 1888 at the age of 78. Archianna, born in 1815, was the youngest of the sisters. Although she lived in Dublin until her death in 1860, there is little mention of her in the family correspondence. She must have led a very quiet and retired life to be so inconspicuous in the mass of letters that Hamilton left behind. None of the four sisters ever married. None of them had any close connections with their parents. They lived at Dominick Street occasionally, but Uncle James, Uncle Willey, and Cousin Arthur took the major responsibility for their upbringing.

When William first arrived at Trim he had not yet reached his third birthday. He had been taken there on a visit immediately after his christening, and the decision to educate him at Trim had been made before he was one year old.[28] Even though William was still an infant, Uncle James realized that he had a nephew of extraordinary ability. Education at that time, both elementary and at the university, consisted largely of science and the classics. James's strong point was classics, with an added interest in oriental languages, and these subjects he fed to William, who showed an astonishing ability to digest them all. William was reading the Bible in English soon after he arrived. Uncle James set about teaching him to spell. James was strong on method, and he taught William to spell following a system that he had devised. First he combed dictionaries and spelling books for all monosyllabic words in which the letter *a* occurs. When William had mastered these he went to *b*, and so on through the alphabet. After covering all the monosyllabic words he moved on to words of two syllables.[29] From the beginning William was working on obscure words that most adults had never heard or seen. His extraordinary precocity must have been the only thing that saved him from total confusion.

Classical languages came next, and again, following the Hamilton method, William was first introduced to Hebrew. He was little beyond the toddler stage and his muscle coordination was still undeveloped. The earliest manuscript notebooks contain Hebrew characters scrawled by the hand of a three-year-old.[30] After Hebrew, William absorbed Latin and Greek. What most astonished learned visitors to Trim was such extraordinary learning attached to childish behavior. When a visiting curate was given a Greek Homer from which to examine William, he said, " 'Oh this book has contractions, . . . of course the child cannot read it.' 'Try him, sir,' said James. To *his amazement* Willy went on with the greatest ease. Mr. Montgomery dropped the book and paced the room; but every now and then he would come and stare at Willy. . . . He would not, he said, have thought so much of it had he been a grave quiet child: but to see him the whole evening acting in the most infantine manner and then reading all these things, astonished him more than he could express."[31]

James continued to pursue his relentless course almost to the point of absurdity. William's father was pleased, of course, by the favorable accounts of his son, and concluded that William's skill in languages would suit him for a career in the foreign service or with the East India Company. When William was ten, Archibald boasted to a friend that his son knew Hebrew, Persian, Arabic, Sanscrit, Chaldee, Syriac, Hindoostanee, Malay, Mahratta, and Bengali, in addition to Latin, Greek, and the modern European languages. He was hoping for Chinese also, but was having difficulty getting Chinese books in Dublin.[32]

For a while it was hoped that William's sisters might follow the same course, and Archibald wrote to them: "Now my dear children, Grace and Eliza, only look to this and be encouraged. Boys are supposed to be idle, girls are supposed to be industrious; but your young brother is determined not to relax a moment his pursuits. Providence is very gracious in giving me such a son, and you such a brother. Now, my dear children, as life is uncertain, and I may be called away, value as you ought such a brother, and prove yourselves by your industry and attention deserving of his support and countenance." Five months later he concluded another letter to

Grace with the admonition "Go thou and do likewise," but it appears that William had monopolized most of the genius in the family.[33]

William must have occasionally become obnoxious. Like most precocious children he was not averse to displaying his talents. Instead of complaining that it was too early to get up, he argued that "though Diana had long withdrawn her pale light, yet that Aurora had scarce unbarred her gates, and therefore he begged to be allowed to lie still."[34] He fought against his Saturday bath by proclaiming that since he had begun studying Hebrew he was observing both the Christian and the Jewish Sabbath. And when he ripped loose some railings near the hall door of the castle and threw them into the courtyard it was "to shew in a metaphorical sense [the] horribleness" of having them in a state of disrepair.[35] William also liked to organize his sisters into games reenacting the Trojan War, about which he had all the details at his command. By age eight he was wont to burst forth in long Latin declamations that he then translated for any adult companions that happened to be nearby—in other words, he enjoyed showing off. He was not particular about his audience. One day when Reilly, the carman, who usually was tolerant of William's prattle, was in too great a hurry, William "attached himself to Fotterell the smith, . . . who, though one of the most savage men in the county of Meath, sat for a quarter of an hour listening to him reading a poem, and seemed quite delighted."[36] Aunt Sydney put up with William's antics with good spirits, and she exercised a valuable maternal influence on him. James was obviously proud of his protégé, but resisted displaying him too widely, at least in his earliest years, because, as Aunt Sydney reported, "He is a most sensible little creature, but at the same time has a great deal of roguery about him. James does not let him much out for fear of his being spoiled by praise for he says he thinks that is the reason so few clever children grow up clever."[37]

In later life Hamilton looked back on his early days at Trim with real fondness, but recognized that Uncle James did not always act for the best. When he later met one of his cousins, Bessie Hamilton, after a long separation, he was pleased that she had grown into such a happy and intelligent woman, because he could remember the brutality with which her father drove her to her studies.[38]

William's first public display of his talents was not in classics, but in arithmetic. James was not in a position to push mathematics at William quite as fast as he pushed the languages, but before William was six Aunt Sydney reported that "You would find it difficult to puzzle him in addition or multiplication; but even in that he must go some strange way unless he is fought with."[39] Apparently William did not appreciate traditional methods of computation and devised his own. There is little mention of his studying mathematics before age thirteen. At that time he read

Clairaut's *Algebra* and composed for himself "A Compendious Treatise of Algebra" summarizing his new learning.[40]

His real skill, or the skill that Uncle James recognized, was in computation. In 1818 Zerah Colburn, the famous American Calculating Boy, arrived in Dublin to display his talents. He was about a year older than William and had the ability to perform extremely difficult calculations in his head without knowing how he obtained the answers. For instance, he gave almost immediately the number of minutes in 1,811 years. It took him about twenty seconds to find the two prime factors of 4,294,967,297. Zerah was the son of a poor farmer in Vermont, who first exhibited him in the United States and then brought him to England in 1812. He lived in France for eighteen months and studied at the Lycée Napoléon, but France was not particularly interested in calculating boys and his father brought him to London and Dublin, where he was much more successful. The Earl of Bristol patronized him and for three years he studied at Westminster School. Unfortunately, he turned out not to be particularly intelligent, nor did he have a real aptitude for mathematics beyond the power of lightning calculation. William was put in competition against him in 1818 and lost consistently. This impressed William, who was not used to meeting children who could beat him in any contest of intellect.[41]

In 1820 Zerah was back in Dublin and William had several chances to talk with him at Cousin Arthur's house on South Cumberland Street. Zerah's father had quarrelled with the Earl of Bristol, the Colburns were as poor as ever, and Zerah, still accompanied by his father, had decided to take up acting for a career, much to William's disgust. The visit in 1820 was not a test of skill but a reunion and a chance to discuss methods of computation. Zerah explained to William all his tricks for extracting roots and finding factors.[42] By 1820 William was less interested in rapid calculation than he was in finding out why Zerah's methods worked, and he wrote up a sheet of "Remarks" showing their limitations. William's skill at computation never left him. Throughout his life he liked to test his ability by performing long calculations in his head. He also tried to recall poetry and would boast in letters that he had written out sonnets composed years before without looking for a copy. Unlike most theoreticians he seemed almost to enjoy calculation, and he checked his theories by solving complicated problems that required sheets of computation. He liked to calculate anything, from the volume of the Egyptian tomb at Edfu using Röber's method, to the velocity of Christ's ascension into the heavens.[43]

In addition to his visit with Zerah Colburn, William had come into Dublin to listen to the Fellowship Examinations at Trinity College. In 1820 William was fifteen years old and we find him making more frequent trips to the city, where he usually stayed with his Cousin Arthur. His new independence had come in part as a result of his renewed acquaintance with his father. From the time that William went to Trim until his mother's death in May 1817, he had almost no contact with his parents. In the following August his father invited him to accompany him on a business trip to Derry. In describing the trip forty-four years later, Hamilton wrote: "I remember well my going with him in the year 1817, when I was twelve years old, almost 'en prince' in a luxurious postchaise, or what then appeared to me such, scattering half-pence or 'bawbees' to poor people (a very unwise thing, as I have since come to think), to the north of Ireland."[44] Archibald enjoyed traveling in a grand manner. He also tended to be verbose and pompous when writing. His letters to William read like legal drafts, and those to his wife are full of his oratorical triumphs before judge and jury. In some cases he frankly admitted that his clients were guilty, but that he won their cases by the power of his eloquence. One gets the impression of a rather imprudent but congenial man who loved to talk not only with his colleagues at the bar, but also with innkeepers, farmers, or anyone he encountered in his travels.[45]

William was fascinated by his father's business and by that of his Cousin Arthur, who frequently traveled on circuit. The Assizes at Trim had always been exciting moments for him. William attended, followed the trials carefully, and lugubriously recounted the numbers assigned to the gallows.[46] In March, 1819, he received the first letter that his father ever wrote to him. It was long, wordy, and full of fatherly advice, but seemed written more for a client than for a son. Archibald feared that his son's education might lead him from the path of true Christianity, and he closed his letter with the admonition: "Recollect how great will be your responsibility if after all the advantages you have enjoyed this gracious Word of God should not be treasured up in your mind so as to prove a corresponding source of good fruits. . . . Finally let me guard you from any indifference to the Bible from the extensive range of classical and other literary pursuits to which you may be called."[47] William responded in kind. His letters to his father during this year were unnaturally formal and stilted, but they probably did not reflect his true feelings. He wrote to his sister at the same time, "My letters to my Father are, of course, corrected etc. by my Uncle before they are sent."[48] Why they needed correction is not clear. Uncle James certainly was no stylist, although he may have thought that he was.

During the summer of 1819 Archibald was still recovering from the death of his wife, and continued to write long letters advising his son about his future career. He went through all the professions open to William. He warned him away from the Church, obviously with Uncle James's fate in mind. Divinity was honorable, but it was also an illiberal profession: "Success beyond the fate of a poor curate seldom falls to the lot of the greatest gifts in the ministry unless accompanied by very great and singular interest with the government and this connection generally lowers the independence of mind every minister of Christ ought to possess." Archibald would have liked to see his son become a great statesman, and urged him to acquire the gentlemanly polish requisite for that position. The study of history, government, and political economy were other necessities, but above all he wanted his son to pursue the study of languages: "In fact you should stop at no satisfaction in this particular short of being able to converse in a diplomatic capacity with all the crowned heads of Europe and their respective Plenipotentiaries or resident ministers at their own courts."[49] Archibald had grandiose dreams.

At the end of May he wrote a more personal letter admitting that he was lonely and asking his son to come and visit him.[50] Uncle James gave his approval, and William spent more than two months with his father and sisters at Booterstown, where his father was then staying. It was an exciting time for William because his father included him in adult society for the first time. He went to dinner at his father's friends' houses around Dublin; he went to the theatre (there was a debate as to whether it was proper); he visited the observatory with two of his father's assistants; and he helped his father calculate costs for some law business, all the while continuing his studies as best he could without James's direction. He also studied shorthand and penmanship with a Mr. Jones. The shorthand stood William in good stead, and throughout the rest of his life he made frequent use of it in drafting letters and in copying extracts from articles and sermons.[51]

The three oldest Hamilton sisters were also staying with their father at Booterstown, and William organized them into a government entitled the Honourable Society of Four. He himself was peer, Eliza and Sydney were commoners, Grace was lady lieutenant, and their father was king, with veto power over all legislative acts (crimes were punishable by slaps on the back). The statutes of the society were long and elaborate and signed by William, who now began to attach a flourish to his signature.[52] Good government was a major enthusiasm of his at the time; when he returned to Trim he organized the school into a regular senate similar to the Society of Four.[53]

One of the first acts of the Society of Four had been to draw up a petition to King Archibald requesting that William be allowed to stay in Dublin longer than the period to which Uncle James had originally agreed.[54] William was evidently enjoying his father's company. But Archibald must have found living with a genius a bit trying at times. He had expected William to help attend to the needs of his sisters after they went back to join the Willeys at Gracehill. He advised the eldest sister, Grace, that she would have to write William "peremptorily, . . . for he is very full of his own weighty concerns amongst the rest making a new Code of Laws for your mysterious Society . . . let your Aunt write him also . . . of her Court of Laws since he will attend to the orders of the female Lieutenant tho' not to the King's commands."[55] William could not be counted on to shop for his sisters. Archibald wrote again in August: "I am greatly annoyed at William's and everyone's want of attention to your commissions. . . . I blame William very much, but he is as wild and thoughtless on common subjects as possible."[56]

During his visit William became aware of his father's intention to remarry. Archibald had developed an attachment to Mrs. Barlow, a widow who lived in the neighborhood of Booterstown. William had a chance to meet her before she left for London in search of medical advice for her son. In October Archibald joined her in London, where they were married. These two months in 1819 were almost the only time that William and his father spent together. On the basis of this brief contact Archibald claimed that he had raised his son with care. In a long, unctuous, and egotistical letter to a friend he described his relationship with his son in the following terms:

He has had, what I never had, the advantage of a father's care, advice as of a companion, and expostulation without austerity. He has had the advantage of the free communication of a father's experience in every changing scene of life, from youth upwards; he has had every sunk rock, upon which the youthful mariner may make shipwreck, accurately traced on the chart of his voyage; and what an advantage that is can be conceived by those who recollect the bulges of their own vessel sustained for want of such a chart, or for want of looking to it with attention. The absolute advantages, I trust, he may prove; be that as it may, I am already rewarded in the success that has hitherto attended my parental affection and care, and by the consciousness of having so far discharged one of the greatest moral duties, as well as by the reflection that I have left my son in that state of mature initiation in every principle of honour and justice, that, with his own talent, unless abused, must ensure his own success, and render him a honour to himself and to his country, and a comfort and a blessing to his family and friends.[57]

Archibald was never one to end a sentence prematurely.

William saw their relationship a little differently. The following February he wrote to his sister Grace: "It astonishes me to reflect how little conversation I have had with you and with some of those with whom it may be or might have been both profitable and pleasing, my father for instance . . . . Nearly all our intercourse has been by letters, but of these I have received and retain many long and valuable, and therefore the less regret our having so seldom conversed, I mean seriously and alone."[58]

In November William was back in Dublin, apparently with his uncle's blessing, in an attempt to gain an introduction to Mirza Abul Hassan Khan, the Persian ambassador, who was causing great excitement in the city. Persian was one of William's languages, and he arrived at the Bilton Hotel, where the ambassador was staying, armed with a letter of his own composition. It was his first opportunity to pursue his father's wish that he be able to "converse in a diplomatic capacity with all the crowned heads of

Europe and their respective plenipotentiaries." The ambassador was packing and had "a headache," but William did get to see his secretary, who told him that his letter was superior to another Persian letter received from another Dublin gentleman wishing an introduction to the ambassador. William had to be satisfied with this limited success.[59]

In December Archibald died suddenly, only two months after his marriage and three months after the long visit from his son. William felt strongly his responsibilities as the only son. To Eliza he wrote that he was not allowed to weep at his father's funeral.[60] He suddenly was required to be much more adult. The combination of Mr. Jones's lessons, the influence of his father's letters, and the solemnity of his new position made his letters pompous and formal. Eliza, who received more of these letters than any of his other sisters, was informed by her fifteen-year-old brother: "I have heard from various quarters, [that] your natural indolence . . . grows upon you—and you may be sure that if you allow it to do so—if you allow the habit to take possession of you, of not bending your whole energy of thought and mental power to whatever pursuit or study you are engaged in, those energies, however great, will as it were grow rusty and decay."[61] Eliza did not like being bossed around by her brother and let him know it, but it had little effect.[62] For the rest of that year William continued to take himself very seriously. He was also serious about preparing himself for entrance into Trinity College. As he studied, however, he discovered that he did not completely share his Uncle James's passion for the classics. In 1823, just before entering Trinity, he told his sister Eliza: "One thing only have I to regret in the direction of my studies, that they should be diverted—or rather, rudely forced—by the College Course from their natural bent and favourite channel. That bent, you know, is Science— Science in its most exalted heights, in its most secret recesses. It has so captivated me—so seized on . . . my affections—that my attention to Classical studies is an effort, and an irksome one."[63]

His future was being formed.

2

Trinity College and
Dunsink Observatory

In LOOKING BACK over his earliest work in mathematics, Hamilton remarked that whatever study he was pursuing at the moment seemed the most interesting. "I believe it was seeing Zerah Colburn that first gave me an interest in those things. For a long time afterwards I liked to perform long operations in Arithmetic in my mind; extracting the square and cube root, and everything that related to the properties of numbers."[1] When his uncle introduced him to Euclid he discovered that there were other joys in mathematics besides rapid computation. After Euclid he somewhat reluctantly took up introductory algebra. In September, 1822, he wrote that if one added Stack's *Optics* and a little popular astronomy to his Euclid and introductory algebra one would "have the whole of my acquirements in Science, at the beginning of last year."[2] These modest and traditional achievements in mathematics were all that William was allowed until 1821, the bulk of his energies being directed to languages and to the classics. But in August, 1821, when he went to Dublin to witness the visit of King George IV to Ireland, James sent along Bartholomew Lloyd's *Analytical Geometry,* a book required for the course of study at Trinity College. It opened William's eyes. "Ill omened gift! it was the commencement of my present course of mathematical reading, which has in so great a degree withdrawn my attention, I may say my affection, from the Classics."[3]

Child prodigies often turn to mathematics, music, and languages, because these are the subjects that reveal their talents most dramatically. William was no exception. Although he had no aptitude at all for music, languages had come easily to him, and he found that he could read mathematical treatises as one reads a novel. But more important than his ability to absorb the mathematical work of others was his realization that he could discourse in his mind on equal terms with the authors he read. From the very beginning he regarded "science" (by which he meant mathematics pure and applied) as an aesthetic creation, akin to poetry, with its own mysteries and moments of profound revelation. In his first study of analytical geometry he was struck by what he called the "demonstrable mysteries" of the subject. The fact that a hyperbola can continue to approach its asymptote without ever meeting it he found to be demonstrable by "rigidly mathematical proof," and yet "it is difficult if not impossible to conceive how [it] can be true . . . If, therefore, within the very domain of that science which is most within the grasp of human reason, which rests upon the firm pillars of demonstration, and is totally removed from doubt or dispute, there be truths which we cannot comprehend, why should we suppose that we can understand everything connected with the nature and attributes of an Infinite Being?"[4] It is an old argument, but one that William discovered for himself.

Beginning in August of 1821 William made rapid progress, even though his mathematical study was pursued "only at stolen intervals" from his classical studies with Uncle James.[5] James objected to William's mathematics not because he disliked the subject, but because William did not always read the books prescribed for the course of study at Trinity College, and James was preparing him to overwhelm all rivals in the Examination Hall. In the spring of 1822 William was seriously ill with the whooping cough. His cousin Arthur wanted him to come to Dublin in April, but he replied in a matter-of-fact way, "I cannot quit Kate [his cousin, James's daughter], of whom there is now no hope. She will not (in all probability) outlive this day. . . . My cough is heavy but I have not yet hooped (sic)."[6] A week later he reported that he had begun to whoop. The doctor bled him; his cousin Sarah recommended Iceland-moss, but William concluded: "As every one recommends something, I like an ass between two bundles of hay, take nothing at all."[7]

After a month of illness the doctor recommended that he take a change of air and accept Cousin Arthur's invitation to Dublin. He went, over James's vigorous objection. It was not only the state of William's health that bothered James. William admitted to Cousin Arthur that "Uncle has, you know, an old prejudice against Dublin and considers it another Pandora's box."[8] The Pandora's Box was mathematics. Once out from under his uncle's thumb William again began to taste the forbidden fruit—this time the *Leçons de calcul différentiel* of Jean-Guillaume Garnier and the *Mécanique céleste* of Laplace. In fact he tasted it so vigorously that by the end of the month he had found a mistake in Laplace's work.[9] It was not a particularly crucial mistake, but it was surprising that a schoolboy just beginning his study of mathematics would be reading Laplace at all. A friend of the family took William's corrections to John Brinkley, the Astronomer Royal of Ireland and one of the leading mathematicians of Dublin, who was sufficiently impressed that he wanted to see William. The visit never materialized, because William was back in Trim by the time Brinkley's message reached him.[10] When he did finally visit the observatory later in the year, he went armed with other papers on the theory of curved surfaces that were more impressive than his criticism of Laplace.

Hamilton's trips to Dublin in the summer of 1822 were fortunate because they brought him into contact with the new mathematical curriculum at Trinity College. His introduction was through Charles Boyton, a family friend, who had become a Fellow of Trinity just the preceding year.[11] In July of 1822 Hamilton wrote to his sister: "I called on Charles Boyton, the Fellow, last week. He was trying to solve a problem in Analytic Geometry, which he showed me, and I had the pleasure of solving it before him; for, two days after, when I brought the solution, I found that he had not succeeded. Charles Boyton is eminent as a mathematician in College. He will be my tutor. He has lent me several French books."[12]

The "French books" that Boyton lent Hamilton represented a dramatic change in the mathematics curriculum at Trinity. Ten years earlier Bartholomew Lloyd had been appointed Erasmus Smith Professor of Mathematics, and had immediately set about revising the course of mathematical study. The previous course had been limited to "English" mathematics, centered on the writings of Newton. These included Newton's *Arithmetic, Optics,* and selections from the *Principia* plus MacLaurin's *Fluxions,* a necessary supplement, because Newton never published a treatise on his own fluxional calculus.[13] In addition to Newton, fellowship candidates studied Greek geometry—the first six books of Euclid and Conic Sections—and a selection from Helsham's *Lectures on Natural Philosophy.* The great continental mathematicians—Euler, the Bernoullis, d'Alembert, Lagrange, Laplace, Fourier, and Monge—were ignored completely. The same condition existed in England. It was only in the first decade of the nineteenth century that the British recognized their error. At Cambridge, George Peacock, John Herschel, and Charles Babbage led the revolt by translating S. F. Lacroix's *Differential and Integral Calculus* (1816). At Trinity College Dublin, Lloyd led a similar mathematical revolution. He wrote a book on *Mechanical Philosophy* for the undergraduate course and it was his *Treatise on Analytical Geometry* (1819) that had first stimulated Hamilton's interest in mathematics. Lloyd also persuaded the younger fellows to write still other books for the college course. Henry Hickman Harte translated portions of Laplace's *Mécanique céleste* and *Système du monde* as well as Poisson's *Mécanique.* T. Romney Robinson, later Astronomer at Armagh Observatory and a close friend of Hamilton's, wrote a *Treatise on Mechanical Philosophy.* Thomas Luby wrote a *Treatise on Physical Astronomy* and *Elements of Trigonometry.* Dionysius Lardner began his extraordinary career as a scientific writer just as William prepared to enter the college. Lardner wrote a *System of Algebraic Geometry* in 1823, followed by an *Elementary Treatise of the Differential and Integral Calculus* (1825). When Hamilton graduated and moved to the observatory, Lardner went to London University to take the chair of Natural Philosophy. A continuous stream of textbooks and popular scientific works came from his pen as well as the *Cabinet Cyclopedia* that he edited and that finally reached 133 volumes. In addition to the Fellows of Trinity College, John Brinkley at the observatory was well versed in the new curriculum and wrote an *Elements of Astronomy* for the undergraduates.[14]

By the year 1822, when Hamilton began his mathematical study under the guidance of Charles Boyton, the undergraduate science medal course at Trinity consisted of Woodhouse's *Trigonometry,* Lardner's *Algebraic Geometry,* Lacroix's *Traité du calcul différentiel et du calcul intégral,* Lloyd's *Mechanical Philosophy,* Poisson's *Mécanique,* and selections from Newton's *Principia* and Laplace's *Mécanique céleste.* This was an enormous change from the curriculum of just ten years before. Of course Hamilton followed the specified course of study, first for entrance, and then for the different special premiums and medals awarded by the college. But Boyton also introduced him to French texts that were not part of the college course. In 1821 Hamilton had been reading books from the old course of study—Helsham, Hamilton's *Lectures,* and the portions of Newton's *Principia*—prescribed for examination. By 1822 he was reading not only the books prescribed for the science medal course outlined above, but also Garnier, Lagrange, and Puissant. Later he added Francoeur, Monge, Cousin, and Malus. These "extras" that Boyton had given him were texts written for the Ecole Polytechnique in Paris, in particular Jean-Guillaume Garnier's *Cours d'analyse algébrique, à l'usage des élèves de l'école Polytechnique* (1803) and Louis-Benjamin Francoeur's *Cours complet de mathématiques pures: Ouvrage destiné aux élèves des écoles normales et polytechniques, et aux candidats qui préparent à y être admis* (1809). By mastering these texts as well as Lagrange's *Théorie des fonctions analytiques,* and *Mécanique analytique,* and Laplace's *Mécanique céleste,* Hamiton had a thorough grounding in the mathematics of the Ecole Polytechnique. The Polytechnique was the most important center in the world for the study of pure and applied mathematics, so Hamilton was preparing himself well.

It is surprising that with this introduction he did not follow continental mathematics more closely in later life. Instead he went more or less his own way, studying the works of his continental contemporaries only when his attention was called to them, or when they worked on the particular problem that interested him at the time. But this was probably a reflection of his own character and method of study. Hamilton was an independent thinker who struck out in new directions for himself without much awareness of what was going on around him. In later years he had little connection with his own mathematical colleagues at Trinity College, who were building an important mathematical school in Dublin.[15]

From the beginning of his mathematical study Hamilton was not content merely to absorb what he found in textbooks. He always added his own speculations. His discovery of Laplace's error and his solution of a problem that his future tutor could not solve were only two of the many indications in the manuscripts that testify to his ability as a creative mathematician. The manuscripts from 1822 also contain "Preliminary Remarks on Division," "Example of an Osculating Circle Determined without Any Consideration Repugnant to the Utmost Rigour of Analysis" (November 14), "Osculating Parabola to Curves of Double Curvature" (no date), "Essay on Consecutive Values" (no date), "On Contacts between Algebraic Curves and Surfaces" (December), and a paper on "Developments." When he finally visited Brinkley at the observatory it was these papers that he carried with him.[16]

Even more important was an "Essay on Equations Representing Systems of Right Lines in a Given Plane," also written in 1822. It was a preliminary draft of his article "On Caustics" that he submitted to the Royal Irish Academy in 1824 and later became the basis for his "Theory of Systems of Rays."[17] In May of 1823 Hamilton wrote to his cousin Arthur: "The time I have given to science has been very small indeed; for I fear becoming again infatuated with it, and prefer giving my leisure even to less valuable reading, if it can be connected in any way with Classical literature. I find, however, that I have not lost much ground. In Optics I have made a very curious discovery—at least it seems so to me."[18] This is the earliest indication that Hamilton was extending his thoughts about systems of straight lines to the problems of optics, a subject that would make his reputation during the next ten years. As early as September, 1822, he had written to his sister Eliza:

I have some curious discoveries—at least they are so to me—to show Charles Boyton when we next meet. ... No lady reads a novel with more anxious interest than a mathematician investigates a problem, particularly if in any new or untried field of research. All the energies of the mind are called forth, all his faculties are on the stretch for the discovery. Sometimes an unexpected difficulty starts up, and he almost despairs of success. Often, if he be as inexperienced as I am, he will detect mistakes of his own, which throw him back. But when all have been rectified, when the happy clue has been found and followed up, when the difficulties, perhaps unusually great, have been completely overcome, what is his rapture! Such in kind, though not in degree, as Newton's, when he found the one simple and pervading principle which governs the motions of the universe, from the fall of an apple to the orbit of the stars.[19]

All this sounds extravagant coming from a seventeen-year-old boy, but Hamilton's enthusiasm was genuine enough. He had a fiery ambition to become a great mathematician. Uncle James's constant prodding only reinforced his natural competitiveness, and when he encountered "modern" mathematics and realized that in this field he could contribute something entirely new, he turned to it hungrily.

Classics provided no such opportunities. As a classicist he might acquire a reputation for great learning, but he would always be reworking other men's thoughts. All he could hope to do would be to refine and polish the existing body of scholarship. Science and mathematics were a different matter:

Who would not rather have the fame of Archimedes than that of his conqueror Marcellus, or than any of those learned commentators on Classics, whose highest ambition was to be familiar with the thoughts of other men? If indeed I could hope to become myself a Classic, or even to approach in any degree to those great masters of ancient poetry, I would ask no more; but since I have not the presumption to think so, I must enter on that field which is open for me.

Mighty minds in all ages have combined to rear upon a lofty eminence the vast and beautiful temple of Science, and inscribed their names upon it in imperishable characters; but the edifice is not completed: it is not yet too late to add another pillar or another ornament. I have yet scarcely arrived at its foot, but I may aspire one day to reach its summit.[20]

William did not add that he felt he could scramble to the top in a rather short time. A reputation as a classicist would take much longer to build.

There was one other area in which William thought he might excel; that was poetry, and at times he even considered leaving science for it. Fortunately he was steered into a more profitable course by no less an authority than William Wordsworth. It was good advice, because Hamilton's poetry never matched the beauty of his mathematics. For the care of his own soul, however, he believed that poetry was an essential avocation. His poetry was romantic in the extreme, and in the same spirit he tended to romanticize his science, even when it consisted largely of differential equations. He saw science and poetry as two aspects of the same creative spirit. When, in the preface to his most famous paper, "On a General Method of Dynamics" (1834), he called Lagrange's *Mécanique analytique* a "scientific poem," he meant what he said. The imagination and the intellect were, for him, closely related faculties uniting the creative acts of mathematics and poetry into a unity. Hamilton held these beliefs even before his serious study of Coleridge and Kant began in the 1830s, and they led him to a position of strong idealism. Take, for instance, these lines written to his sister Eliza in October, 1824:

> Yet 'twas the hour the Poet loves
> Alone to wander through the groves,
> Unheeded, uncontrolled, to pour
> His spirit forth in verse; to soar
> Up to the heaven of heavens, to climb
> Above the bounds of space and time;
>
> To call ideal worlds to view,
> His own creation bright and new.
> And I, although I dare not claim
> That lofty meed, the Poet's name,
> Enjoy in Solitude like this
> A portion of the Poet's bliss.[21]

Those "ideal worlds," those "creations bright and new . . . above the bounds of space and time," were the airy atmospheres in which he believed poetry and mathematics to reside. Hamilton had asked Arabella Lawrence, a friend of Coleridge and the Edgeworths, to judge his poetry. She found it obscure and unrevealing. He accepted her judgment, but added that while poetry could never be his profession, he needed it to keep his whole being alive.

It is the very passionateness of my love for Science which makes me fear its unlimited indulgence. I would preserve some other taste, some rival principle; I would cherish the fondness for classical and for elegant literature which was early infused into me by the uncle to whom I owe my education—not in the vain hope of eminence, not in the idle affectation of universal genius, but to expand and liberalise my mind, to multiply and vary its resources, to guard not against the name but against the reality of being a mere mathematician.

The pursuit of science and poetry had, for Hamilton, a strongly religious flavor. Mathematics presents

some of the sublimest objects of human contemplation; . . . its results are eternal and immutable verities; . . . it seems to penetrate the counsels of Creation, and soar above the weakness of humanity. For it sits enthroned in its sphere of isolated intellect, undisturbed by passion, unclouded by doubt. And I have thought that, in the infinity of Creation, there may be an order of beings of pure and passionless intellect, to whom Science in all its fulness [sic] of beauty is unveiled, and to whom our noblest discoveries appear but as the elements of knowledge. . . . And as we read that the mystery of our redemption affords a theme which angels desire to look into, so I think there may be angelic existences admitted to behold the whole of that vast connexion which binds together the material universe of God.

But Hamilton recognized that his was not a "pure and passionless intellect": "Man is not a creature of intellect alone, nor is he at liberty to bestow upon *it* an isolated cultivation. His heart is even more important than his mind; he was made to be a social creature, and his second duty is love to man. Now I think that poetry is eminently qualified to strengthen and refine the links which bind man to his kind."[22]

Hamilton said things in his poetry that he could not or would not put into his correspondence or into his occasional journals. His poems do not so much describe events or philosophical positions held as they express emotions and romantic musings. Sometimes they are banal; usually they cling too much to convention; and Arabella Lawrence was correct in her criticism that he did not reveal himself enough to make his poetry really powerful. And yet his poems express as much as he was willing or able to express. Occasionally in moments of real distress his feelings break through the conventional forms, but not often.

Hamilton's poetry was nourished by his relationship with his sister Eliza. A twentieth-century writer of psychological biography would find traces of latent incest in their relationship, but in the nineteenth century it would have been considered entirely conventional. Eliza was William's much-loved "poet sister," with whom he found the greatest spiritual and intellectual affinity. Uncle James had seen signs of genius in her as well as in William, and had announced that if her mind were properly nourished she would be "a wonder," but he never had the opportunity to apply his method to her education, because she was raised by the Willeys at Gracehill, where she stayed until 1822, when she entered a school in Dublin.[23] William must have come to know his sister much better after 1822 when they began to meet frequently in Dublin, usually at Cousin Arthur's house on South Cumberland Street. In the letter to his sister Grace in which he mentioned how little contact he had had with his father, William added: "Eliza is perhaps the person to whom I have talked most and most freely."[24]

In 1822 William was still capable of incredible pomposity, and he continued to scold Eliza for her excess of "natural indolence." As he passed through college Eliza followed every triumph, proud of his accomplishments, but fearful that he might be heading for a terrible fall some day. She was extremely religious, and most of her poetry dealt with religious themes. Robert Graves characterized her religion as a rather narrow piety, and mentioned that as early as 1819 William was evasive in describing his visits to the Dublin theatre (sanctioned by his father), because Eliza might find the subject offensive.[25] Her poetry had more force than her brother's—this was Wordsworth's assessment—but it dwelt overly much on sin and retribution. Her poem "The Boys' School," which William sent to Wordsworth, begins with a description of the vigor and optimism of youth, and ends with disillusionment, disaster, and a final appeal "to the perfect peace of God."[26] It is one of Eliza's favorite themes, and its treatment was overdone in her poetry.

Her first discussions of poetry with William were about the poems of

James Montgomery, who was also of Irish parentage and who also grew up in a Moravian establishment in England. Eliza's piety did not dampen her romantic spirit—she was an extreme romantic, even more so than her brother. They began to send original verses to each other in July 1822, just as William began his serious pursuit of mathematics. William initiated the exchange by sending his poetic musings on a dream: "This you will perhaps say is great nonsense—and I believe it is. Aunt Mary saw it, and asked me whether I did not live on vegetables, as I was a believer in the transmigration of souls. Sydney says that you are very fond of poetry, and that in the nightly visitations of your muse, you are so 'raised to fury, rapt, inspired,' that you do not allow anyone to sleep. Why, then, do you not favour me with a few of your compositions, in return for the many foolish ones I have sent you?"[27] Throughout 1822 Hamilton's letters were filled with descriptions of pastoral scenes observed on his frequent walks about Trim. Beattie's poem "The Minstrel" was his constant companion, and he wrote poems entitled "On the Scenery and Associations of Trim" and "Address to the Evening Star"—both full of rapture for the beauty of nature.[28]

While Hamilton's mathematical and poetic genius was growing he also began to take a renewed interest in astronomy. Cousin Arthur bought a telescope for him, which he used both in Dublin and at Trim.[29] But after a year of enthusiastic observing, Uncle James, Cousin Arthur, and the whooping cough forced him back into the track of preparation for the entrance examinations. William reported to Cousin Arthur:

We have been getting up before five for several mornings, that is, my uncle and I; he pulls a string which goes through the wall and is fastened to my shirt at night. The Constellations visible in the mornings are those that appear later in the winter in the evening (Orion). The Planets are Jupiter, Venus, and Saturn; but you can scarcely conceive how little I care now about making astronomical observations; my telescope lies untouched in a corner of my desk, and my coughs forget to trouble me. This is all your fault, for you broke me of the habit of star-gazing.[30]

In 1823, however, he was back observing, and this time it provided a subject for romantic poetry. In January and July of that year there were total eclipses of the moon. William calculated the time for the appearance of the first eclipse, walked out into the garden at the calculated time, and found the moon in total shadow, from which it gradually emerged. He described the scene to Eliza:

That Sunday night when the rest of the family had retired to rest, I remained for a good while admiring the effect of the snow in the moonlight. The fields were smiling in one dazzling and unbroken whiteness, except a few spots from which the snow had been drifted away. The borders of the river were covered with thin sheets of ice, but in the main channel, where the frost had no power, the small waves were all tipped with silver: while the ruins of the castle, which slept in shadow, formed a striking contrast by their dark and frowning majesty. You perceive that in writing to you I unite in some degree the poet with the astronomer: but it was such a scene as I could have wished you to have witnessed along with me. We should have '... felt how the best charms of nature improve when we see them reflected from looks that we love.'[31]

Eliza replied that while reading his description of the eclipse, "I almost fancy that I am with you and hear you speaking and long to tell you that I would have felt exactly so, and that the very same recollections and associations the same dreams of memory and hope would have endeared it to me."[32] When another eclipse of the moon occurred in July, William wrote an "Ode to the Moon under Total Eclipse," a poem that he always considered one of his best works. He and Eliza continued to work together on their poetry, William's becoming more a reflection of his ambition, Eliza's becoming more religious. Her poetry could sometimes contain a stern warning. The following, composed in 1832, is expressive of one side of her relationship with her brother.

To _____
"Though thou exalt thyself as the eagle and though thou make thy
nest among the stars, thence will I bring thee down saith the Lord." Obadiah

> Yes! thou indeed art as an eagle cleaving,
> High Solitudes profound.
> Thought's mountain summits far beneath thee leaving
> And who of earth shall bring thee to the ground?
> Thy wings of Intellect are dazzling bright,
> Oh! earliest loved! I know not where they soar,
> I veil mine eyes before the splendid sight,
> I only know that this must once be o'er.
>
> For take thy flight which hath a glorious seeming,
> Upward and upward wandering through light,
> Smile in thy heart at faith's prophetic dreaming,
> That aught shall pluck thee from thy sovereign height:
> Go to thy throne amid the Stars of Heaven,
> Where Death itself shall never touch thy crown!
> One dwelleth there: with Him if those have striven,
> Shall He not cast thee like the weakest, down?
>
> Is there around the lofty habitation,
> Of thy bright spirit, any guard from Him?
> Canst thou defy the inward desolation
> With which his wrath all brilliant thoughts can dim?
> Hast thou a heart that would not much be wounded
> Should burning arrows fall on it like rain?
> Should love be crushed and deepest trust confounded,
> And memory's self become unsleeping pain?[33]

Eliza knew her brother's heart. She also knew that by 1832 he had already felt more than his share of those "burning arrows."

William took the examination for entrance into Trinity College on July 7, 1823. He had intended to enter college the previous October, but James had been taken with a bad case of cold feet and had concluded, after a conference with Cousin Arthur, that a delay would give William a chance to prepare himself better and might allow him to compete against a weaker field of candidates.[34] William had his own doubts, which he confessed to Eliza. He was worried because he had never competed against a boy better than himself, and expected that college would be a lot different from the diocesan school at Trim.[35] Whatever the reasons for delay, they were exaggerated. William was first out of a hundred students sitting for the examination, and the next day, before he even began the college course, he was awarded a premium for an examination in Hebrew.[36] During his four years at Trinity College, William demolished all competition in examination after examination. Each class was divided into groups of about thirty students chosen by their order on the class lists. Each group had its own examiners, and a student competed only against other students in his group. Those students who were unfortunate enough to have names close to Hamilton's on the class list never stood a chance. The grades awarded on examinations were *valde bene, bene, satis, mediocriter,* and *vix medi.* The record of Hamilton's examinations is a monotonous series of *valde in omnibus,* meaning that he received the highest grade on every subject. In addition to the grades, there were "premiums" awarded in science and classics to the best student in each section. The premiums consisted of credit at the bookstore for buying college books. After winning one premium, a student could not get a second until the following year, but if he had the highest standing in a subsequent examination during the same year he was granted a certificate stating the fact. As expected William regularly won the premiums and certificates in all examinations plus two Chancellor's Prizes for poetry.[37]

His most striking accomplishment came in the Trinity examination of 1824 when he was awarded an *optime* for the examination on Homer. An *optime* was a grade off the scale, so to speak. It meant that the student had a complete mastery of the subject and that he was not in a position to be examined. No *optime* had been awarded in the past twenty years, according to his tutor Boyton, and none had ever been given to a freshman. Hamilton's *optime* was granted by Dr. Thomas Elrington, who had been responsible for a reform in the classics course comparable to Lloyd's reform of the science course.[38] The *optime* came as a surprise, and it gave Hamilton the thought that he might attempt an entirely unparalleled feat of intellectual prowess. At the final examinations of the senior year, members of the class competed as a whole for two gold medals, one in classics, and one in science. Each examination for a medal had a special

Trinity College Dublin—West Front

"medal course" of reading and preparation and a student would compete for one medal or the other, depending on his intellectual inclination. The competition was intense, and consequently the medals were much in the minds of the students during their final year. At Trinity the word for a "grind" or compulsive student was *medal*, a term commonly used by Hamilton and his friends.[39] Hamilton set as his goal obtaining both medals for his class, which meant following both courses simultaneously. He never got the opportunity to compete, but much of 1826 and 1827 were spent in preparation.

When Hamilton entered Trinity there were four ranks of scholars, each with its own privileges. These were *Nobilis* and *Filius nobilis*, fellow commoner, pensioner, and sizar. Each rank had its own distinct gown; the *Filius nobilis* wore gold and silver tassels while the lowly sizar dressed in plain black. The great body of the students were pensioners, which meant that they paid a substantial fee and ate by themselves. The fellow commoners paid a higher fee and had the privilege of dining at the fellows' table, while the sizars, who paid no annual fee at all, waited on the fellows and ate the leavings from their table. Hamilton entered Trinity as a pensioner, but there is some question about his status, because his tutor referred to him at one point as a sizar.[40]

Students at Trinity worked hard as a whole, but they also had a reputation for riotous behavior. A typical disturbance was described by a scholar to his father in 1789: "My dear Father, Commons on Trinity Sunday were very pleasant; geese thrown, trenchers broke, and everything torn and broke etc."[41] A sousing under the college pump was the usual treatment for anyone unfortunate enough to incur the wrath of the students.

Many students carried arms, and duels occurred, although they were declining in fashion during the beginning of the nineteenth century. Hamilton's manuscripts contain an exchange of letters between two duellists. Unfortunately none of them are dated, so it is difficult to tell if the affair occurred while Hamilton was in college or afterwards. Whatever the date, he was probably acting as a mediary and not as a participant, although he later confessed that he had once challenged a member of the Royal Irish Academy. On that occasion concerned friends were able to settle the quarrel before it came to an actual duel.[42]

During his college years Hamilton was much in demand in Dublin society, but we read little about student life within the college. He lived most of the time at South Cumberland Street with Cousin Arthur, or at Trim. Since the course of study for each examination was laid out in advance, he could have prepared himself without regular attendance at lectures.

Hamilton's college success lacks interest just because of the sheer inevitability of it all. In some ways the course of study was a disadvantage for him, because it meant that he had to suppress his own creative talents in order to prepare for set examinations. There were a few moments of suspense during the first year until William's record was well established. Eliza told her Aunt Willey in February, 1824, that Cousin Arthur had waited among a crowd of anxious friends outside the college to hear who had won the premiums. "At last there was a distant cry of 'both—both premiums,' 'Oh!', then said Cousin Arthur, 'I am at home—none could have both but himself.' "[43]

The term and gold medal examinations were all *viva voce* and in Latin. William trained for them as if he were entering an athletic contest. In May, 1824, he was "vaulting tables for exercise," and in October of the next year he asks: "Are my Cap and Gown ready? A skipping rope would complete my equipment; ... Bessy [his cousin] has got a very fine long rope. I am taking a good deal of exercise and of air." In June of the same year he wrote to his sister Grace: "Send me word at what hour the tide will serve on the 23d and 24th that I may be able to conjecture whether I can bathe in the sea between the hours of examination."[44]

After completing the college course, a graduate of Trinity College could compete in the fellowship examinations. A fellowship was William's real goal; at least that had been his goal before he entered Trinity. He told his cousin Arthur: "My life as a student has always to me been divided into two principal parts—preparation for Entrance; preparation for Fellowship."[45] Winning a fellowship gave one a secure position at the college as a teacher and scholar, as well as an opportunity for advancement at the college or through the church. The examinations were given annually on the Wednesday before Trinity Sunday and lasted four days—two hours in the morning and two hours in the afternoon. They were conducted in the Examination Hall by the provost and the senior fellows.[46] The examinations were public, and potential candidates would listen eagerly to the questioning in order to prepare themselves for the day when they would themselves be struggling for the answers.

William began attending the examinations four years before he entered college in order to gain experience.[47] It was possible to take the examination any number of times, and year after year candidates would come back for another try after honing their skills on the set course of preparation. It could be very stultifying for the men involved, because the examination did not test creative capacity or provide an opportunity for the candidates to strike out on their own. Hamilton was more fortunate than most, because he found that he was able to compete successfully in the examinations and pursue his own interests at the same time. In spite of the clamorings of Uncle James and his own guilty conscience driving him back to his classics, it was during his college years that Hamilton began almost all of the highly original work that was later to make him famous.

College also led Hamilton into a much wider circle of friends. Once his reputation had been established, he was in demand at all sorts of social occasions. Trim had been a very quiet place, and Hamilton, no matter how much he loved the surroundings, would gladly leave it for the excitement of Dublin. Equally important were visits to country houses in the vicinity of Trim. In August of 1824 Hamilton was introduced to the Edgeworths of Edgeworthstown in County Longford, about forty miles west of Trim. It was his first opportunity to make the acquaintance of a major literary figure. Maria Edgeworth was fifty-six years old when Hamilton met her. At that time she was one of the most famous novelists in the British Isles, probably second only to Walter Scott in popularity.[48] Jane Austen's works have since cast Maria's into the shade, but in the 1820s she was a sensation, particularly because she succeeded at a time when "female authorship" was regarded with great suspicion.

Maria's first and most famous novel was *Rackrent House,* published in 1800. It was set in Ireland and depicted the character of the country gentry and the peasantry in a way that was both amusing and harshly critical. Sir Walter Scott saw in *Rackrent House* and *The Absentee,* which followed it, a new direction for his Waverly novels, and he consciously pursued Maria's style, trying to depict the Scots in the way that Maria had depicted the Irish.[49]

Before *Rackrent House,* she had written a series of books on education

Edgeworthstown House

with her father—books that have in recent years been recognized as early attempts to improve the position of women. Maria had traveled widely, and by the time Hamilton met her, she was acquainted with most of the leading literary and scientific figures in France and in the British Isles. Her conversation and her letters were full of wit, and she and Hamilton seemed to take to each other immediately. Maria wrote to her sister Honora: "Mr. Butler came with young Mr. Hamilton, an 'Admirable Crichton' of eighteen; a real prodigy of talents, *who, Dr. Brinkley says, may be a second Newton*—quite gentle and simple."[50] And Hamilton on his side described Maria to his sister Grace as follows:

She far surpasses all that I had heard or expected of her, though I confess that, at first sight, I was disappointed by her personal appearance; and though she said at once, "Mr. Hamilton, I am sure," I was not at all prepared to say, "Miss Edgeworth, I am sure." Yet even in beauty she seemed to improve, as if that of her mind cast reflected graces upon her person. In her conversation she is brilliant, and full of imagery to a degree which would in writing be a fault. Accordingly, if you would study and admire her as she deserves, you must see her at home, and hear her talk.

She knows an infinite number of anecdotes about interesting places and persons, which she tells extremely well, and never except when they arise naturally out of the subject. She has, too, a great talent for drawing people out, and making them talk on whatever they are best acquainted with. To crown her merits, she appeared to take a prodigious fancy to me, and promised to be at home, and made me promise to be at Edgeworthstown, for a fortnight, some time in the next long vacation.[51]

The Mr. Butler who brought Hamilton to Edgeworthstown and introduced him to Maria was the Reverend Richard Butler, vicar of Trim, and Uncle James's immediate superior. Two years later Richard Butler married Maria's sister Harriet, which tied the families at Trim and Edgeworthstown more closely together.

In 1824 Maria was the dominant figure at Edgeworthstown, but much of the character of the place, and her own character, for that matter, was the work of her father, Richard Lovell Edgeworth, who had assisted Maria with all her writing until his death in 1817. Richard had been an Irish prodigy of the practical sort. He had been a prominent member of the Lunar Society at Lichfield, where he knew James Watt, Matthew Boulton, and Josiah Wedgewood, as well as Erasmus Darwin and, later, Joseph Priestley. Richard Lovell was an indefatigable inventor. He designed a carriage on a new design, tracked vehicles that would climb over walls, a telegraph link between Galway and Dublin to warn of an expected French invasion. At Lyons he directed an effort to divert the Rhône River; in Paris he wrote treatises on the construction of mills. He had a deep interest in education and had brought up his eldest son following the method of Rousseau's *Emile,* and even took his son to Rousseau to display the results of his efforts. Back at Edgeworthstown he served on the board of education and set up projects to drain and reclaim bogland. He designed a new steeple for the parish church following his own ingenious plan; Hamilton, on a visit to Edgeworthstown, was proudly led on a tour to see this product of Richard Lovell's mechanical skill.

Maria's brilliant social gifts and her father's desire for "improvement" have more the character of the eighteenth century than the nineteenth. It was in country houses such as Edgeworthstown that the eighteenth-century manner lasted the longest. During the 1830s and 1840s Maria began to be left behind by changing tastes and changing society. Her novels lost their immense popularity, members of her family died or married, and the political turmoil of emancipation and repeal raised a barrier between her and her tenants that had not been there before. When Hamilton met her, however, he was able to savor some of the lingering brilliance of the social life of the preceding century.[52]

In 1824 Maria had a large family around her. She never married herself, but her father had had four wives and numerous children by each. Maria, the eldest child still living in 1824, was born in 1768. She was one year older than her stepmother. Michael Pakenham, the youngest, was born in 1812. In between there was a house full of "children" of all ages. Maria lived in the midst of constant commotion, writing her books in the drawing room with the rest of the family about her.

In addition to Richard Butler, the Edgeworths had other family connections that were later important for Hamilton. Francis Beaufort had been a close friend of Richard Lovell. His sister became Richard Lovell's fourth wife, and Francis himself married one of Richard Lovell's daughters by his third wife. Francis Beaufort was hydrographer at the Admiralty in London, a member of the Royal Society, and an important contact for Hamilton.[53] Another Edgeworth sister married T. Romney Robinson, astronomer at Armagh and another of Hamilton's scientific friends.

Hamilton formed attachments among Maria's younger brothers and sisters, too. On the first of September, just a few days after he had been introduced at Edgeworthstown, Eliza told him that his sisters had come to the conclusion that "the *beautiful* Miss Edgeworth is at Edgeworthstown" and having heard that she had appeared "like a vision at a Coach office I can not be without my fears."[54] Grace even hinted that he might no longer be a threat to the other fellowship candidates, since the Fellows of Trinity were celibate. William did not much enjoy being teased, and only replied that he was still free to offer himself as a candidate.[55] In 1828 the teasing resumed. Cousin Arthur mentioned that Edgeworthstown was apparently "more than ordinarily attractive," and Maria, in writing to invite William for a visit added: "It is but honest to tell you that Miss Edgeworth is not at home."[56] In July Uncle James told him that "Mr. Butler seemed rather shy of answering on the subject of Edgeworthstown. He now has confessed that the intention formally impending of an union with one of that family is likely to be soon realized."[57] The nineteenth-century habit of not mentioning names in letters makes it difficult to determine just who was the recipient of Hamilton's affections. Maria had several sisters of marriageable age. The most likely candidate in 1824 was Lucy, born the same year as Hamilton and later married to T. Romney Robinson. He might have been interested in Fanny. She was Maria's favorite sister, six years older than Hamilton, and an enthusiastic astronomer. Sophy and Harriet are also possibilities, but Sophy married in 1824, soon after Hamilton first visited Edgeworthstown, and Harriet was the subject of Richard Butler's affections. More important than Hamilton's temporary attachments to Maria's sisters was his friendship with Francis Edgeworth, who was four years junior to Hamilton. He was a poetic young man, a follower of Platonic philosophy, and an extreme romantic. It was with Francis more than anyone else that Hamilton discussed his theories of aesthetics, the relationship between science and poetry, and the relative domains of "truth" and "beauty."

In the 1820s Maria made frequent tours to England, Scotland, and France. Her father's death in 1817 had been a severe blow to her and she had not been sure that she would be able to continue writing without his criticism and assistance. Editing his memoirs had been an act of devotion for her, and when some of the reviewers published less than favorable comments (usually not about her, but about her father), she had become defensive and had turned to writing children's stories about which there could be little controversy. Gradually she discovered that she could manage quite well by herself and became more assertive in society. In her father's shadow she had tended to be timid. On her own, she was witty and forceful—too forceful, according to some accounts.[58] In 1820 she took Fanny and Harriet with her on a Continental tour, and in 1821 they spent a full winter season in London. In 1823 Harriet and Sophy accompanied her on a tour of Scotland, where they stayed with the philosopher Dugald Stewart and spent two weeks at Abbotsford with Sir Walter Scott. Maria had never met Scott, but had corresponded with him for some time. It was an important meeting, and in 1825 Scott came to Ireland and took Maria with him on a trip to Killarney.

It was in the midst of these social and literary successes that Hamilton first met Maria, and learned from her the excitement of the drawing rooms and literary salons of the time. In 1826, when Hamilton began to study philosophy on his own, it was the works of Dugald Stewart, an old Edgeworth friend, that he read, and from the copy of Stewart's *Philosophical Essays* in the Edgeworth library. He particularly valued the book because it had marginal comments by the Edgeworths, and he noted in his journal: "I like reading books that have been marked by persons that I know and care for."[59] Maria's judgment always had great weight for Hamilton. She advised him when he had to decide whether or not to accept the position of Astronomer Royal in 1827, and again when he was elected president of the Royal Irish Academy in 1837. He carried on a long correspondence with her, asking her what she believed would be the best organization and mode of operation for the academy. Maria took it upon herself to instruct him on the subject.

At almost the same time that he first visited the Edgeworths, Uncle James introduced Hamilton to another family that would be even more important for his future life. This was the Disney family at Summerhill. Thomas Disney was estate agent for Lord Langford; he resided at Rock Lodge, although some of Hamilton's letters seem to indicate that the Disneys may have been living in the main house of Summerhill in the 1820s.[60] Thomas Disney had five sons—all close to Hamilton in age—and at least four daughters. All the sons went through Trinity College and those in college with William became his closest friends.[61] Hamilton was extremely grateful to his Uncle James and wrote to thank him for introducing him to a group of such congenial friends. What he did not tell his uncle was that he had fallen head over heels in love with their sister Catherine the first time he set eyes on her. That meeting occurred on August 17, 1824, a date that Hamilton never forgot to the very end of his life. Apparently he committed all kinds of social blunders on that first visit; he ignored Mrs. Disney, whom he should have led into dinner, and took Catherine's arm instead, and completely monopolized her during the whole evening. It was a typical case of love at first sight. [62]

Hamilton's infatuation for Catherine must have been obvious to those around him, but he was not yet in a position to do much about it except to cultivate her company and that of the rest of the Disneys. The Disneys were wealthy, and William still had to finish three years of college and find a secure position before he could think of marriage. Of the sons, Edward was his favorite: "Ardent in ambition, and in friendship, of a pure and lofty mind, tempering by his piety and modesty the lustre of his talents which I consider as of the first order. Had I been allowed to select for myself a companion in the race which I have run, what character could I have chosen more congenial than that which I have described."[63] As for Catherine, he wrote to James: "It is absolutely necessary that I should no longer defer speaking of Miss Disney. Beautiful as she is, the stranger only can observe her beauty; her mind and her heart, with those who know her, are the objects which engage their attention and secure their love."[64] Though Hamilton could not hope to marry Catherine in 1824 he was not entirely without hope, because, as he wrote to his uncle: "Mr. and Mrs. Disney have shown a desire to cultivate our society. Mr. Disney called on Cousin Arthur, and Mrs. Disney has paid us a still more welcome and delicate attention, by making a visit to my sisters, who are now with me."[65] These signs of parental approval were a good omen.

Lambert Disney had been studying for entrance to Trinity with Uncle James at Trim, but either he became homesick or Uncle James's "method" was too much for him and he left Trim to study with a tutor in Dublin. Hamilton tactfully refers to Lambert's "almost too finely affectionate disposition [that] had lost in melancholy the power of adequate exertion."[66] Beginning in 1824 Hamilton was frequently in Dublin and so were his sisters. Together with the Disneys they formed a group in 1826 called the "Stanley Society," which met once a week for breakfast at the house of James Stanley, another college classmate. Hamilton's contributions to this literary society were numerous; most were reflections on their common life: "On College Ambition," "On the Long Vacation," "On a Visit to Belfast" (where he met Sir Humphry Davy), "The Epanados" (a humorous essay on rhetoric), "On Learned Ladies," and a more serious essay, "On the Connexions between Science and Poetry."[67]

The friendships between the Hamiltons and the Disneys became close. Eliza and Anne Disney in particular were extremely intimate. Hamilton did not mention it at the time, but in later correspondence with Louisa Disney he mentioned Eliza's "almost *romantic* fondness" for Anne, whom they nicknamed "Araminta." Their affection was *"quizzed* [in its meaning to ridicule or to scoff], shocking to say, at the time, by some of us, though *not by me*," and in other letters he referred to their "extreme intimacy."[68] Hamilton was too shy or too constrained by social convention to express his love for Catherine directly, but he did confide in Eliza. Eliza, however, was prevented by "the instinctive and right reserve of her sex" from ever relaying back any information about Catherine's feelings, and one gets the impression that Hamilton did not have a good idea of where he stood.[69]

In February 1825 Catherine's mother told Hamilton that Catherine was to be married in May to a clergyman by the name of Barlow, who had been hovering in the sidelines, more or less unnoticed.[70] It came as a terrible shock to Hamilton, who had never been told that this marriage had long been contemplated by the elder Disneys. Barlow was about fifteen years older than Hamilton and was wealthy and well connected.[71] Catherine pleaded desperately against the marriage, or so Hamilton said later, but her father had an "iron will" and Barlow was too proud to let his prize be taken by a "mere boy." "That boy, her lover, had not heard that his love was returned, but had heard that she was engaged, from her mother, whose anguish of manner, whether arising from compassion for me, whose love she no doubt divined, or through pity for her daughter, at the moment of her speaking those words, which nearly killed me at the time, and have coloured my whole subsequent life, I never can forget." Catherine's relatives told her it would be a sin to break the marriage agreement. The family's honor was mentioned and she was "led as a victim to the altar."[72]

Such was Hamilton's description of the event thirty years after it happened. Probably at the time it was less dramatic; his feelings and Catherine's were kept under control, but they were all the more intense for being unexpressed. Catherine was married on May 25 and Hamilton became ill.[73] It was in large part an emotional illness that brought him to the verge of suicide. Shortly after he heard of Catherine's engagement he was walking to the observatory to attend a party given by Brinkley, when he "experienced, in all but its last fatal force, the suicidal impulse." When he later moved to the observatory he had occasion to pass the same spot often, each time being a reminder of his earlier agony. What saved him was his ambition: "I wish that I could add that it was religion—the Christianity of the Anglican Church, in which I have been baptised and confirmed and to which I adhere—or even generally my belief in the Bible, which protected me. My recollection has always been that it was simply a feeling of personal *courage*, which revolted against the imagined act, as one of cowardice. I would not *leave my post*; I felt that I had *something to do*."[74] There were enough fragments of his shattered ego left intact for him to survive the experience, but it left a mark that proved to be permanent.

Hamilton's love for Catherine Disney and the way in which he recovered from his loss set the theme for the rest of his life. He went on to love other women, even to marry and to make many successful friendships, but the remembrance of Catherine runs through them all. His correspondence and poetry leave the impression of a man making the best of things, living on the surface while underneath powerful emotions are ready to break through. Occasionally they did. The most honest lines Hamilton ever wrote occur in a letter to Aubrey DeVere in 1848, when, in a moment of real crisis, he confessed: "The same remembrance has run like a river through my life, hidden seemingly for intervals, but breaking forth again with an occasional power which terrifies me—a really frightful degree of force and vividness."[75] In a negative way, his love for Catherine helped Hamilton. It was a focus for his life, a foil for his ambition, (since it caused him to turn even more ferociously to his studies), and an inspiration for his poetic feeling.

Hamilton said that he could never express his agony in prose. He did, however, express it in verse. Many of his poems contain concealed references to Catherine, but only one of them, "The Enthusiast," is a direct account of his feelings.

> He was a young Enthusiast. He would gaze
> For hours upon the face of the night-heaven,
> To watch the silent stars, or the bright moon
> Moving in her unearthly loveliness;
> And dream of worlds of bliss for pure souls hid
> In their far orbs. At other times he loved
> To listen to the mountain torrents roar,
> To look on Nature in her many forms,
> And sympathise with all: to hold sweet converse
> In secret with the genius of the stream,
> The fountain or the forest, and to pour
> His rapture forth in some fond gush of song;
> For the bright gift of Poetry was his;
> And in lone walks and sweetly pensive musings
> He would create new worlds and people them
> With fond hearts and sweet sounds and sights of Beauty.
> He had been gifted, too, with sterner powers.
> Even while a child he laid his daring hand
> On Science' golden key; and ere the tastes
> Or sports of boyhood yet had passed away
> Oft would he hold communion with the mind
> Of Newton, and with awed enthusiasm learn
> The Eternal Laws which bind the Universe,
> And which the stars obey. As years rolled on,
> Those high aspirings visited his soul,
> Which Genius ever breathes. He longed to leave
> Some great memorial of himself, which might
> Win for him an imperishable name.
> Fame was around him early, and his path

Was bright with honour, and he had a home,
And hearts that loved him and could sympathise
In all his joys; he was perchance *too* happy;
For love had not yet swept with fiery hand
Over his chords of feeling, calling forth
For one short moment all their melody,
Then leaving them for ever mute and broken.

It was an August evening, and the youth
Had numbered nineteen summers, when, a guest,
He came within an old romantic mansion,
With dark woods round. He found a brilliant circle
And, holier charm! a happy family.
But oh! how soon and how entirely faded
All else when his enthusiastic gaze
Had fallen upon a form of youth and beauty,
A maiden in her simple loveliness,
With locks of gold and soft blue eyes, and cheeks
All rich with artless smiles and natural bloom!
He sat beside her at the board, and still
He saw her only, thought of her alone;
But now it was on other charms he dwelt,
Her thoughts, her tastes, her feelings, and these were
So full of mind, of gracefulness and nature,
Blended with such retiring timidness,
They riveted the chain her beauty wove.
    They met again, too often for his peace;
For what had he, but Genius, Hope, and Love?
Her image became twined into his being;
His musings were of her, of her his dreams;
She was the star of his idolatry,
But like a star he deemed her all too high
To bow to love for him. Yet he hoped on.
Who hath not felt how heavenly Hope can live
And freshen even amid what should be death,
Like to the self-renewing bird of Araby
Which springs to life from its own funeral pyre!
    One eve she woke the harp. The fond enthusiast,
O'erpowered by feeling, sate him down apart,
And hid his face; he could not look and listen!
And then she sang a sweet and simple air;
Her voice aroused him, and with altered mood
In silent trance of pleasure he hung o'er her.
But these were moments all too exquisite,
Too richly fraught with transport, to last long;
The dream was to be broken, the chain sundered.
    He had not talked of Love. His happiest hours
Were those he passed with her; yet then his words
Breathed only such respectful tenderness

As if he were addressing a dear sister:
And she—she thought of him but as a brother.
He knew himself in fortune her inferior,
And therefore would not seek to win her heart;
But he did *not* know that her troth was plighted,
And a few months must bring her bridal day.
The tidings when they burst upon him crushed
Awhile to earth his energies of soul;
Or left them but to add new stings to agony,
New power of pain to torturing remembrance.
At length his bitter anguish passed away,
But left him darkly changed. His mind awoke;
Its powers were unimpaired, and the affection
Of his fond friends could warm his bosom still;
And he seemed happy; but his heart was chilled,
And he was the enthusiast no more.[76]

In 1830 Hamilton published this poem in the *Dublin Literary Gazette and National Magazine,* and throughout his life he sent copies of it to new acquaintances who he believed would receive his confession with sympathy. His romance with Catherine, therefore, became semipublic property. It was a personal experience for him, but, like Coleridge's ancient mariner, he felt the need to retell the tale constantly—to take others into his confidence and to make them witnesses to his distress.

The spring of 1825 was not a good time for Hamilton. No sooner was he devastated by Catherine's engagement than he ran into trouble on an examination. The Reverend J. Kennedy, the examiner in classics, gave Hamilton a *bene* in place of his customary *valde bene*, and stopped the classical certificate altogether, indicating that none of the students had answered satisfactorily. Mr. Kennedy had a reputation for asking complicated and often irrelevant questions on minute points of detail; he had also expressed displeasure at Hamilton's having been awarded an *optime* by another Fellow for his answering on Homer, especially since Homer was Mr. Kennedy's own specialty. But since *no one* received the premium it was hardly a mark of disgrace. Uncle James was not pleased, however, and concluded that William had made a "radical mistake" in assuming that he had a "surplus of time for extraneous pursuits," by which he meant helping Edward Disney and working on his own mathematical research.[77]

A third disappointment came in his mathematical work. In 1824 Hamilton completed a long paper, "On Caustics," that he submitted to the Royal Irish Academy on December 13. At age 20 it was his first venture of a professional kind, and, of course, he expected it to be received enthusiastically. The committee assigned to judge the memoir finally returned their verdict on June 13, 1825. They found it "novel and highly interesting" and observed that "considerable analytical skill" had gone into its preparation, but the arguments presented were "of a nature so very abstract, and the formulae so general as to require that the reasoning by which some of the conclusions have been obtained should be more fully developed, and that the analytical process by which some of the formulae have been obtained should be distinctly specified."[78] In other words, the examiners could not understand it, a not altogether unusual situation, since Hamilton was never a good expositor of his own ideas. Uncle James was miffed, this time more with the committee than with his nephew, but Hamilton himself had confidence in his own abilities. "Caustics" had been only a mathematical preliminary. He had "dreams of Discovery and Fame which Hope had interwoven in [his] mind with the renewed prosecution of [his] Caustics," and Charles Boyton, his tutor, who had helped him in the preparation of his paper, must have strengthened his confidence to resume his researches.[79] In April, 1826, Boyton was the examiner in science, and he gave Hamilton an *optime*. It was the first time that an undergraduate had been awarded an *optime* both in classics and in science. Hamilton went back to work on his caustics, and on April 23, 1827, he presented to the Royal Irish Academy his "Theory of Systems of Rays," a paper that made his reputation and set the course of his research for the next eight years.

3

Life at the Observatory

AFTER AWARDING Hamilton an *optime* in the spring of 1826, Boyton began to work on obtaining an appropriate position for Hamilton after graduation. The examinations in October were merely a formality. Boyton had again been given Hamilton's division to examine. He tried to find another examiner, even offering the division to Hamilton's old foe, Mr. Kennedy, because, as he said, his mind was already made up, and there was no point in holding the examination. No other fellow wished to accept the offer. Examining Hamilton was not a task to be taken lightly. He received the certificate as usual.[1] But there was more excitement afoot than the examination. In announcing the results to Uncle James, he added: "Among the rumors flying, I have heard it said, on the one hand that Dr. Brinkley is to keep the Observatory; on the other hand [that I] ought to be appointed to succeed him."[2] Boyton was almost certainly the source of the rumor. Brinkley did, indeed, resign his position at the observatory to accept the Bishopric of Cloyne, but it was not until the following year that any action was taken by the Board of Fellows towards finding a successor. Rumors continued to fly, however, and before the end of the year Hamilton received an invitation from T. Romney Robinson, astronomer at Armagh, to come visit him and gain experience in practical astronomy. Hamilton refused, begging the need to prepare for his examinations, which struck Robinson as being a bit too disingenuous. Robinson renewed the invitation anyway, apparently convinced that Hamilton was destined for the observatory at Dunsink. The more others mentioned his name for the position, the more silent Hamilton became. Uncle James continued to drive him unmercifully, urging him to finish the rest of his "Theory of Systems of Rays" and to prepare with total thoroughness for his onslaught on both gold medals.[3]

The appointment of an Astronomer Royal had to be made by the middle of June, 1827. Candidates had begun to present their credentials as early as January. Among the names submitted were several of the fellows of the college. MacDonnell, Harte, Luby, and Humphrey Lloyd all applied. Dionysius Lardner offered himself with a spectacular collection of testimonials and the boast that his book on the calculus had received the largest sum ever given for a mathematical text.[4] The most formidable applicant was George Biddell Airy, Senior Wrangler from Cambridge and later Astronomer Royal at Greenwich, who came over to Dublin with strong recommendations from John Herschel and Francis Beaufort. Beaufort had written to Brinkley earlier from London to ask why there had been such a delay in the appointment and to tell him that he was urging Airy to apply. When Airy arrived in Dublin, however, he found that the salary was only £300, and wrote to the board that he could not even apply unless the salary were raised to £500 in addition to the house and grounds.[5]

The salary was not the deciding factor, however, because on June 8, 1827, just a week before the decision had to be made, Boyton wrote to Hamilton in a letter marked "private and confidential" that the Board of Fellows would make the salary of the Astronomical Professor

certainly £500, probably £600, possibly £700, and that you would upon proper application obtain it. I have turned it carefully in my mind, and I have not the least hesitation in saying that, looking to *character,* and to *Pecuniary* advancement, it is your very wisest course to try to obtain it. I would be glad to see you provided for, and I would be glad to produce to the world, *one* creditable act of the Board.

Don't on any account *consult* Brinkley till you see me. You had better come to town at once and take the advice of your friends. The application needs to go in before Thursday next.[6]

In supporting Hamilton's candidacy before the board, Boyton had already stated that Hamilton had "been in the habit of staying days and nights at the Observatory employed at the instruments and [had] made himself acquainted with their use in detail," a claim that stretched the truth a good deal.[7] Hamilton did have some experience in observing with his own telescope, but he kept that telescope in his desk drawer. The great meridian circle at Dunsink was the most accurate instrument of its kind in the world. Operating it successfully required a skill gained only by experience and much practice.

On June 16, 1827, the board unanimously elected Hamilton to the position of Andrews Professor of Astronomy, an appointment that automatically included the title of Astronomer Royal and the directorship of Dunsink Observatory. He still had two quarterly examinations to take before he could receive his degree.[8] Brinkley was unhappy with the appointment. He had opposed Hamilton before his election and continued to express his doubts afterwards. To Boyton and the other fellows he had argued that although it would be hard to find a young person more prepared to engage in the management of an observatory, Hamilton had not "actually made himself acquainted with the detail and business to be done," and when he had done so he might find himself unwilling to give up his time to it.[9] He had written to Hamilton in much the same vein. Hamilton should wait for a fellowship, rather than commit himself to the detailed work of a practical astronomer. The income of the Astronomer Royal might look enticing now, but it was a fixed income, and, he added, "I struggled sixteen years with a family on the late small income of the Observatory, and my after changes of circumstances could not have been reasonably reckoned on. I say this that you and your friends might well consider."[10] Even when Hamilton arrived at Cloyne as the fellows had ordered, Brinkley's mind was unchanged. As Hamilton recounted Brinkley's arguments:

He expressed his fear that it had been an imprudent act on my part the accepting the Observatory. He said that I ought not to depend upon the Board, for they had acted shabbily to him. He too had begun very early (before he was twenty-four), and was told that he would certainly get some preferment soon; but he was left for many years without anything more than the small salary of the office. . . . If I were a Fellow, I might have got a dispensation, enabling me not to take pupils; and I would have been gradually gaining standing at least, if not income. My friends ought to have decided the thing for me, and not have left it to myself. To all this I could only reply, that so decidedly did I prefer the Observatory to Fellowship in point of liking, that I would have accepted it if it had been offered to me without any money at all; that as a Fellow, on the present system, I would either have had no time for pursuing Science, or must have made that time by exertions at extra hours and to the injury of health; that, in short, my tastes were strongly for the thing, and that my friends thought that prudence was for it also.[11]

Hamilton had given the same argument to Boyton and the fellows in making his application; in fact it was an argument that he had probably heard from Boyton and was repeating back for the benefit of the fellows. The appointment to the professorship would be "the gratification of my dearest wishes, and the object of my highest ambition; my most earnest desire being to be enabled to devote myself exclusively, and without obstruction from other avocations, to those pursuits of Science, to attain distinction in which has ever been the object nearest my heart."[12]

While Brinkley had urged him to hold out for a fellowship, T. Romney Robinson said that he had made a wise choice and that his talents were of "far too high an order to be thrown away on the drudgery of tuition, or what are called the Learned Professions."[13] But, like his colleague at Dunsink, Robinson suggested that Hamilton come up to Armagh and get in a little practice with the instruments. He tactfully mentioned that it might prove embarrassing if Thompson, the assistant at Dunsink, who had put in years at the transit and circle, should find that his new director was inept at observing.

Hamilton could almost certainly have obtained a fellowship soon after graduation. But a fellowship would have involved him in tutoring students and administration of the college. It would also have required him to become a clergyman, and, in 1827, it would have required him to remain celibate. The celibacy statute was removed in 1840, and before 1811 it had been ignored by the fellows, many of whom had married. But in 1811 a stringent statute had been instituted that again forced celibacy upon the fellows.[14] As far as Hamilton could tell in 1827, becoming a fellow would prevent him from ever marrying. At the observatory he would have greater independence, both in his academic pursuits and in his private life. The tutorial responsibility of the fellows was very demanding. T. Romney Robinson wrote in the preface of his book on mechanics (1820): "The Fellows of Trinity College can scarcely be expected to devote themselves to any work of research, or even of compilation; constantly employed in the duties of tuition, which harass the mind more than the most abstract studies. ... In the present case the author happened to be less occupied than most of his brethren, yet he was engaged from seven to eight hours daily in academical duties, for the year during which he composed this work."[15] Hamilton must have realized that the duties of the Astronomer Royal would be less time consuming than those of a fellow. At the observatory he would also be able to have his sisters come live with him, something that he desired very much.

Hamilton was never a success as a practicing astronomer. Observational astronomy in the early nineteenth century meant measuring the positions of the heavenly bodies with the greatest possible accuracy. The transit instrument measured the precise moment when a star passed the meridian of the observatory, and the meridian circle also measured the elevation of the star. The work was tedious in the extreme—night after night of measurements to be reduced during the day to star positions. Despite the impression given by the occasional mention of colds and fatigue brought on by observing late at night, Hamilton spent little time in the meridian room with the instruments. The record book of the observatory belies any argument that he was a successful or enthusiastic observer. Observations by Hamilton are entered with his initial H; those of his assistant Thompson are marked by T. His sisters Sydney and Grace also made entries (marked S and G), as did his noble pupil, Lord Adare (marked A). In August, 1827, Hamilton began observing regularly, as witnessed by a long string of H's in the record, but by 1828 the H's thin out and there are more and more T's.[16] Sydney and Grace showed considerable stamina, particularly in 1831, when they were doing much of the observing. In that year Hamilton had an opportunity to exchange the professorship of astronomy for that of mathematics. He confessed to Robinson: "My tastes, as you know, are decidedly mathematical rather than physical, and I dislike observing; which circumstance makes me *rather*

Dunsink Observatory

unfit for holding an Observatory as a contemporary and compatriot of you."[17] The Board of Fellows opposed the change of professorships and nothing came of it. Hamilton spent the rest of his life at Dunsink doing much mathematics, but little practical astronomy. After 1831 scarcely an *H* appears on the record.

Before settling down at the observatory, Hamilton accepted Robinson's invitation to get in some practice with the instruments at Armagh. He arrived towards the end of July and set to work learning the peculiar problems of making transit observations. His sister Sydney was leaving Dublin at about the same time to go to her teaching job at Rhodens near Belfast, and Robinson kindly invited her to spend a few days at Armagh on the way. Hamilton unfortunately forgot about his other sister, Archianna, who was to travel with Sydney (such lapses caused frequent annoyances for his sisters), but the Robinsons gladly took them both in when they arrived.[18]

What was to be a long visit at Armagh was suddenly terminated by the arrival of Alexander Nimmo, engineer for the Western District of Ireland, who was at that time actively involved in the great ordnance survey. Nimmo, who was extremely vigorous and adventurous, had little difficulty in persuading Hamilton to go with him to Killarney, where the survey was in progress. Robinson objected that he had much yet to learn about the instruments, but Hamilton promised to return after his brief excursion with Nimmo and take up where he had left off.[19] What was expected to be a brief week or two at Killarney turned into a regular Odyssey about Ireland, England, and Scotland. Hamilton had traveled little before 1827.

Except for two trips to the north (Derry and Belfast), he seems never to have strayed farther from Dublin than Edgeworthstown to the west and Wicklow to the south. In Nimmo's carriage they made the trip from Armagh to Limerick in a single day. Nimmo was directing the construction of the Wellesley Bridge and the docks at Limerick, and he showed Hamilton the construction works and even sent him to the bottom of the Shannon in a diving bell, where he witnessed the compression of the air at depth "in a very painful manner."[20] From Limerick they went on to Killarney, where Hamilton saw the famous lakes for the first time. Letters home mention dancing with the beautiful sister of a Limerick gentlemen at whose house he had twice dined, and another beautiful girl, this one the daughter of a Baroness whom he met on the steam packet to England.[21] Traveling with Nimmo apparently included frequent entertainments and social events.

To Cousin Arthur he wrote to check on the progress of remodeling at the observatory and to tell his assistant Thompson that he could let his cow graze on the lawn as long as he did not touch the hay. Thompson was taking advantage of the change of directors to acquire all of the benefits he could. The Board of Fellows had voted to provide him with a new house, and Hamilton felt the need to assert himself. He wrote: "I told [Thompson] that I only allowed [his cow on my lawn] during my absence, and that I would not continue to do so. I did not consider him as having been at all diligent while I was away [at Armagh], but thought that there was no use in taking notice of it until I return to take active cognisance of his proceedings,—*then* I shall keep him to his sharps, because however good he may be as a computer and observer, he will not do for me unless he be also industrious. In the mean time I shall let him alone."[22]

From Killarney Hamilton accompanied Nimmo along the south coast of Ireland, where they visited some of Nimmo's earlier engineering triumphs, until they reached Dunmore and from there rowed out to intercept the steam packet from Waterford bound for Bristol.[23] Hamilton saw for the first time the growing industry of the English midlands. From the battlements of Dudley Castle they looked down at night on the "vast masses of flames" from the iron smelters and from piles of burning waste coal. The next day they descended into the Dudley Caverns without any adequate guide and were lucky to make it to the workface of the mine without disaster.[24] A small boy started them on their way with only burning rope-ends for light, and they traveled too far before discovering that the remaining rope would not be enough to light their way back. They had to keep going, trusting that they would reach the miners before they were left in darkness. Hamilton found it all very romantic and poetical, and wished that Eliza could have been with him.

At Birmingham, Hamilton and Nimmo parted company, with Hamilton going on to Liverpool, where he visited the Misses Lawrence and one of his Hutton relatives. He wrote to Nimmo to say that their travels had instilled in him a new interest in practical things, an interest that might even compete with mathematics, but he concluded: "This, however, is an effect of which I need scarcely entertain any very serious apprehensions. My mathematical tastes are too deeply rooted and too solidly founded to be in danger from the rivalry of more elegant perhaps, but surely less fascinating pursuits—."[25]

Hamilton rejoined Nimmo in the Lake Country, where they called on William Wordsworth after first making a dramatic ascent of Helvellyn. Hamilton and his party had met Wordsworth the previous evening at the home of the Reverend John Harrison, and Wordsworth had invited them all to tea. After tea Wordsworth walked back with them to their lodgings. But when they arrived at the hotel, Wordsworth and Hamilton were so deep in conversation that Hamilton offered to return again with Wordsworth to his house. Even this trip did not exhaust their legs or inspiration and they set off again for Ambleside "without any companion except the stars and [their] burning thoughts and words."[26] Considering Hamilton's youth and Wordsworth's eminence it is surprising that they would be as strongly attracted to each other as they were and in such a personal way. After this first meeting, Hamilton became a poetic disciple of Wordsworth, and Wordsworth was moved to write to Hamilton, "Seldom have I parted—never, I was going to say—with one whom, after so short an acquaintance, I lost sight of with more regret. I trust we shall meet again."[27] Their letters exhibit a surprising familiarity from the beginning.

Hamilton was nothing if not enthusiastic about the lakes and their poetic associations. Wordsworth gave him a letter of introduction to Robert Southey that he employed the next day. Southey described the visit to Thomas Digges La Touche:

We had some of your countrymen here in the latter part of the season ... [including] Hamilton, the young professor of astronomy, who is so fond of the stars and so full of life and spirits that I dare say if the kites had been ready, and Mac had not been willing to undertake an ascent, he would. [Southey was referring to an imaginary project of designing a kite to pull carriages up the mountain of Skiddaw.] Nay, I believe that for the sake of making a tour among the stars, he would willingly be fastened on to a comet's tail. Nimmo the engineer was with them, but he indeed is a Scotchman.[28]

William did not tell his sister the subject of his long evening conversation with Wordsworth, but he did send her a poem he claimed to have "picked up at Ambleside." The poem was called "It Haunts Me Yet," and it was obviously written by Hamilton himself, because he sent the same poem to Wordsworth and to Arabella Lawrence, and to them he did not hide the fact that it was his work.[29] Beneath all his displays of exuberant vitality and good spirits the remembrance of Catherine still bothered him. The poem begins:

> It haunts me yet, that dream of early Love!
> Though Passion's waters toss me now no more;

"Passions waters" had not been completely stilled, however, for the third stanza concludes:

> Days of Emotion, ye are *not* forgot!
> The thought of you is twined with whatsoe'er
> Of more than common happiness my lot,
> Or more than common grief, to this thrill'd breast had brought.

After describing what the memory of Catherine had meant, the poem addresses Wordsworth:

> And THOU too, mighty Spirit! whom to name
> Seems all too daring for this lowly line;
> Thou who didst climb the pinnacle to Fame,
> And left'st a memory almost divine! ...
>
> All reverently though I deem of thee,
> Though scarce of earth the homage that I pay,
> Forgive, if 'mid this fond idolatry
> A voice of human sympathy find way;
> And whisper that while Truth's and Science' ray
> With such serene effulgence o'er thee shone,
> There yet were moments when thy mortal day
> Was dark with clouds by secret sorrow thrown,
> Some lingering dream of youth—some lost beloved one.
>
> If then thy history I read aright,
> O be my great example! and though above,
> Immeasurably above, my feeble flight,
> The steep ascent up which thy pinions strove,
> Yet in their track my strength let me too prove;
> And if I cannot, quite, past thoughts undo,
> Yet let no memory of unhappy love
> Have power my fixèd purpose to o'erthrow,
> Or Duty's onward course e'er tempt me to forego!

These last two stanzas and the invitation to intimacy contained in them caught Wordsworth's attention. As an objective measure of their impact he confessed: "They affected me much, even to the dimming of my eye and faltering of my voice while I was reading them aloud." He then proceeded to demolish the poem line by line, exposing every fault. Wordsworth said the verses were "animated with true poetic spirit" and were "evidently the product of strong feeling," but emotion alone could never create good poetry. "The logical faculty has infinitely more to do with poetry than the young and the inexperienced, whether writer or critic, ever dreams of. Indeed, as the materials upon which that faculty is exercised in poetry are so subtle, so plastic, so complex, the application of it requires an adroitness which can proceed from nothing but practice; a discernment, which emotion is so far from bestowing, that at first it is ever in the way of it." Hamilton had asked for Wordsworth's criticism of Eliza's poetry as well. Wordsworth liked her poems better than William's, but saw in them the same faults. Her poetry might come to something, but to William he warned: "You especially have not leisure to allow of your being tempted to turn aside from the right course by deceitful lights."[30]

Wordsworth's advice was harsh, but valuable. On the trip to the Lake Country Hamilton had recovered some of his poetic ambitions; Wordsworth laid them quickly to rest. As the time approached for him to return to the observatory and begin his career William was dwelling much upon his destiny. His poem to Wordsworth closed:

> There is a monitor within my heart,
>   A secret voice that passeth not away;
> A burning Finger that will not depart
>   But urges onward still and chides delay;
> Summoning to excellence's onward way;
>   And though yet feeble, I will follow still,
> Till every cloud be lost in perfect day,
>   And I have reached the summit of that hill
>   Where more than earthly light my strengthened gaze shall fill!

Arabella Lawrence reacted to these verses in a different way. Before Hamilton had left on his journey she had warned him not to become too excited by his new honors.[31] Now upon reading his poem she saw signs of a dangerous morbidity, and warned him against "indulging too often this train of thought," even though his abstract studies provided a "never failing antidote against its influence."[32] Arabella did not wish him to be less sensitive, but she saw dangers in his continued brooding over Catherine. Not the least was the possibility that he might attempt to assuage his pain by a marriage that would later prove to be unwise. She wrote: "I suppose I must expect soon to hear that you too are about to [lose] your liberty, but I hope as you are so fortunate of having sisters for friends you will not *rashly* commit yourself—but do cautiously that which cannot be undone."[33]

Hamilton returned to Dublin after a short tour of Edinburgh and Glasgow. The work on the observatory was nearly complete, and he wrote in one of his many notebooks: "It was on Saturday, October 13th, 1827, that we came to the observatory to reside."[34] Grace and Eliza came to live with him. Sydney and Archianna were still at Rhodens, where Sydney was teaching, but Hamilton hoped to have them at the observatory, too. He had always expected to have his sisters around him. It seemed the natural thing to do, and everyone expected it. When he accepted the astronomical professorship, Charles Boyton told him: "Now, you will go and settle quietly there with your sisters," and his Aunt Susanna Willey, who had been primarily responsible for raising his sisters at Gracehill, wrote that since he now had a home for them it was desireable that they go to it.[35] Hamilton hoped that his sisters would join enthusiastically in his work and that he could turn their interests to astronomy. Eliza in particular was the subject of his hopes. Together they would find fame among the stars, as had Caroline and William Herschel. Although Hamilton shared the Victorian prejudice that women were incapable of real intellectual attainment, he was less prejudiced than most. In the Stanley Papers he had written against those "absurd prejudices . . . by which 'Learned Ladies' were once looked upon as a sort of wild beasts, to be treated with a mixture of fear and aversion." "The times are gone," said Hamilton, "when working in tapestry was one of the highest accomplishments of princesses, and when women of inferior rank were not allowed to aspire much farther than the making of a shirt, or of a gooseberry pie." Hamilton was not even sure that "domestic excellence" was the greatest female skill, although he admitted that in spite of the new liberality towards women (he was writing in 1826), domesticity was still regarded as being their most important talent. He went on to argue: "I am not quite sure that in anything valuable the minds of men are really superior to those of the other sex. In taste, in imagination, in feeling, in affection, in piety, in the enduring of pain, and the charming away of distress, women have, in general, almost an allowed superiority; and . . . there are some recorded instances in behalf of the fairer sex which may perhaps excite a suspicion that if there have not been more, the cause has been the want of opportunity rather than the want of ability."[36] Hamilton hoped that his sisters might prove the truth of this statement.

Hamilton's sisters would have been more eager to join him in his work if he had been a better instructor. He thought of himself as a good lecturer to popular audiences and to beginning students. He did have a flair for oratory, but none for elementary instruction in mathematics. Every such attempt began with explanations so elementary and obvious that they were embarrassing to his listeners, and then suddenly would leap into material totally incomprehensible to them. He liked to try out new ideas on an audience, and later, when his sons were studying science and mathematics, he would gather them in front of the blackboard along with Thompson and mystify them after the first few sentences. "It was very amusing to watch Thompson—whose ideas moved slowly, and who could only go a very small way in the subject—standing by, spectacled and owlish, and chiming in with an occasional 'I see.' Sometimes, however, [Hamilton] threw a Parthian dart at him, such as 'just recapitulate the last six equations'; when it generally happened that Spica Virginis or Lyrae required immediate attention.[37]" Nor was he any better at written instruction. His papers constantly digress. A good example is his *Elements of Quaternions*, originally intended as a small *Manual*, because the previous *Lec-*

tures on Quaternions had been too long and too advanced for students. The *Elements* grew to over 800 pages. Adding to these handicaps the fact that all his work of substance was abstract and highly mathematical, one must conclude that his plans to take his sisters with him on his mathematical journey was doomed from the beginning.

Hamilton was also confused as to whom he wanted for female company. Somehow he imagined that if he were to marry, his sisters would remain at the observatory and they would all be one happy family. Mr. and Mrs. Wordsworth had Wordsworth's poet sister, Dorothy, in their house, and the house at Edgeworthstown contained a great number of Edgeworths related in various ways. Why should he not do the same thing? But Eliza recognized that her brother's marriage, if it occurred, would seriously complicate their relationship. It was on his tour to England that he began to contemplate the idea of helping Eliza in her studies. He wrote to her just after emerging from the Dudley Caverns, when he found himself in a particularly romantic mood: "I am beginning to long for home, and often think of the pleasure I shall enjoy when we shall set out together on our journey from earth to heaven. Would that it were such in every sense! We should then indeed be happy; but I only meant our journey along the beauteous and glorious path by which we shall mount almost to that unearthly eminence where unembodied spirits look abroad upon the wonderful spectacle of the Universe. To mount this path, however, we must needs begin at the bottom." He then outlined a course of mathematical study and astronomical observation that would allow them to "see the planets in their mystic dance still looking to their glorious central fire, and circling round its ever-burning altar. The comets too. . . . We shall see them rushing with a lover's joy to the presence of their beloved sun, but slackening their pace and lingering as they withdraw," and so on, for several pages. From this constant contemplation of God's works "we shall have our minds more and more ennobled, our hearts more and more softened and purified, and our souls more framed and fitted to modesty, piety, and virtue."[38] Eliza was frightened by this romantic effusion, and Hamilton had to reassure her that he would not force her to travel faster or further than she wished, but he had been serious, because in replying to her letter he concluded: "To speak more plainly, if you will not despair of becoming in ten or twelve years an accomplished astronomer should we both live so long, or of being *fit* to succeed me when I die, I will always let you stop in your lessons whenever you feel yourself even beginning to be tired."[39]

For a while it looked as if Hamilton might have realized the mistake of trying to pressure his sister into astronomy, but by the end of the month he was back at it again. This time he tried a different tack and replaced the romantic vision with pompous persuasion that reminds one of his father, Archibald, at his worst.

Feeling thus deeply, then, the almost insuperable difficulties of that enterprise in which, nevertheless, I have long determined to engage, I would not willingly augment those difficulties by neglecting to arm myself with the aid of friendly and female sympathy. And, therefore (to speak at present of you only), though I do not expect, and scarcely even wish, that you should ever pursue Science to the same extent that I shall . . . I yet indulge the thought that we may not wholly fail in uniting our pursuits, and blending our tastes together; that so we may not stand, as it were, aloof, in rival and opposite stations, but each be able, though with inferior skill, to sympathise with, encourage and even assist the other.[40]

These lines could be read almost as a threat, and Eliza must have read them that way, even though Hamilton's concluding invitation to follow him "hand in hand" breathes brotherly affection.

When Hamilton first settled at the observatory he set about observing with a vengeance, exposing himself to the cold and sometimes staying up all night, so that he soon made himself sick enough to require a vacation. For a change of scene he went to Edgeworthstown in the middle of March for an extended visit. At Edgeworthstown he found Maria's brother William and her sister Fanny busy at astronomy, which roused again his desire to have a sister for an astronomer. To his sister Sydney he frankly admitted that among Maria's sisters "The one that I like best is Fanny, who has a very strong taste for Science, and is a great assistant to her brother William in observing and calculating, as I hope that you will be to me at some future time, unless you should be otherwise disposed of." Sydney was still teaching at Rhodens and caring for Archianna. Although he would welcome her at the observatory he hints that because of his small income it might be better for her to retain her present employment for awhile.[41] Hamilton was enjoying Fanny's company, giving her lessons in astronomy, and receiving lessons in turn on botany. The lessons must have been at least a partial success, because he sent her his copy of the second volume of Laplace's *Mécanique céleste* as soon as he returned to the observatory.[42]

While at Edgeworthstown he again urged Eliza to join him in the study of astronomy, and this letter brought a sharp rejoinder from Eliza. Eliza's letter has been lost, but Hamilton's reply gives a good indication of its contents. In his biography of Hamilton, Graves states that Eliza was offended by Hamilton's continued insistence and apparent distrust of her devotion to him. Graves decided not to print Hamilton's rejoinder because "it is throughout too private and personal for publication," implying that Eliza's indignation was caused only by her brother's overbearing tone.[43] But the letter shows that Eliza had other concerns. Hamilton admitted that he had scarcely written his first letter when he began to think better of it. He had been dazzled by Fanny's scientific talent, "yet even in the moments during which I felt most strongly the influence of sympathy with my master-passion, I would not have exchanged for her any one of my own sisters, or any one of the Disney family."[44] The member of the Disney family now attracting his attention was Catherine's sister Louisa. Eliza had accused him of forgetting his own sisters and of not mentioning them in his letter, to which he replied that he wrote the letter because he loved his sisters so much that he could not let them go while he pursued his mathematical studies. "I know too well your independence of spirit to doubt that if I could be so ingenerous as to wish to separate you from me, it would require but a slight hint to carry that wish into execution. But believe me that the shadow of such a wish has never crossed my mind, and that ardently as I desire an union with Louisa this desire would be greatly damped if I looked forward to such a consequence."

Hamilton's emotions were in a confused state. He had announced his devotion to Louisa and was now wavering towards Fanny, in a fickle manner. Eliza was on Louisa's side and she scolded her brother in no uncertain terms for his inconstancy. Hamilton's letter reveals the reason for his vacillation. Fanny had liked his lectures, while Louisa had been bored by them. It was in this fickleness that Eliza saw a threat. Hamilton was prepared to measure his devotion by the amount of time his sisters and female friends were willing to devote to science. Hamilton went on: "If then I said to you that my affection for Louisa would have been deeply weaken'd, had I been sure that indifference about the Lectures was altogether real, it was *not* because I expected *her* to be a mathematician or astronomer, but because a *total* indifference to this attempt of mine to interest her in my pursuits, would have argued (as I thought to me) that total indifference to *me* (of which I was perhaps too sensitively apprehensive), and which would certainly have weakened my own affection as well as made it unwise to indulge it."[45] Hamilton did not marry Fanny or Louisa. He probably had no chance with Fanny anyway. Since 1819 she had wanted to marry Lestock Wilson, brother of Captain Francis Beaufort's first wife, Alicia, and had been prevented from doing so by Maria, who thought she could attract a more cultured suitor.[46] Hamilton may have been a possible candidate; he was relatively poor, but sufficiently prominent for Maria's aspirations. However, his devotion to Eliza and to the Disneys kept him from Fanny and she married Lestock Wilson the next year.

Towards the end of his life Hamilton met Louisa Disney again (now married and with the name of Reid) and corresponded with her in 1861. At that time he wrote: "I am sure that a stranger would say you were (not 7 but) 17 years younger than myself; but you were not *actually* a *baby*, when I first saw you, at Summerhill, though you were *very* young. Can *you* remember it *at all*?"[47] From the perspective of thirty-three years, Hamilton could evaluate his feelings for Louisa objectively. He wrote to Lady Wilde (Oscar Wilde's mother and one of his confidantes in later life): "You may conclude how intimate we once were, when I mention that

I found an opportunity for saying to her (*lately*) that I had *once* wished to marry her—or rather had *thought* that I so wished: but found that I had sought in vain to transfer my feelings from her (by me) lost sister and that I had mistaken affection to the family, for love to the individual, in short, had confounded recollections with hopes. ... She sweetly assured me, that she *knew*, she *understood* it all."[48]

Hamilton's attachment to Louisa had come in the summer of 1827, when he was only twenty-one and Louisa was, by the evidence of his later letter, only fourteen or fifteen. It is not surprising that she found his mathematical lectures boring. In his correspondence of 1861 Hamilton confessed that one of his poems written in the summer of 1827 had been inspired by a visit of Louisa's to the observatory.[49] After she left Hamilton found some flowers that she had forgotten and left by the great circle in the Meridian Room. They were the occasion of a poetic reverie, which indicated the choices that he believed he would have to make.

> And is it here, ye lovely ones,
>   That ye have chosen to fade?
> A bright but fragile offering
>   On Science' altar laid!
> Alas, too oft, 'mid scenes like these,
>   Must Feeling, too, decay;
> And in this air, serene but cold,
>   Her sweetness waste away!
>
> For Science on her votaries lays
>   A stern and deep control;
> Entire dominion she demands,
>   And empire o'er the soul ...
>
> Yet perish not, loved flowers,
>   So soon, so suddenly;
> Though parted from your native soil,
>   Yet bloom awhile with me:
> And be to me an emblem
>   Of hopes that change and fade,
> And of the heart's young sweetness
>   On Science' altar laid.[50]

A heart laid on the altar of science apparently had no room for Louisa.

Eliza never openly refused to accept Hamilton's tutelage, but he soon discovered that she was not going to be the scientific companion he had sought. By November, 1828, he had fixed his hopes on Sydney. He wrote to her: "You know I have set my heart on having *one* of my sisters an astronomer, and I cannot expect either Grace or Eliza to become one, as they are too much occupied with the care of the house and of my pupils [Hamilton had taken on the instruction of the Lord Lieutenant's sons], while Archianna will not for many years be ready (not to mention that she seems likely to prefer the lyre to the telescope). I have no resource but *you*, and I hope you will not disappoint me."[51] For over a year he had been sending Sydney mathematical lessons (she kept having trouble with logarithms) and he urged her on with accounts of famous women mathematicians and astronomers. He urged her cooperation partly for his own advantage, and partly because of his "zeal for the honour of womankind." Sydney had committed herself to stay at Rhodens throughout the year, but agreed to come in June of the next year. Hamilton was overjoyed and stepped up the tempo of his instruction.[52]

II
Ray Optics

4

The "Theory of Systems of Rays"

HAMILTON'S RAPID ELEVATION to Royal Astronomer was brought about not by any great achievements in astronomy, but by the success of his highly original theory of geometrical optics. He had been interested in optics ever since his first introduction to analytical geometry, and by the time he came to the observatory his optical ideas were just taking their final form. The most important aspect of the theory, which he called his "Theory of Systems of Rays," is its great generality. By his method it is possible to include all the properties of any optical system in a single equation that, when solved, will give a "characteristic function" completely describing the optical system. In other words, given any ray of light entering the optical system (telescope, microscope, or any system of lenses and mirrors), the characteristic function describes how the ray emerges at the other end. Finding the characteristic function is not always a simple matter; in fact it can be so difficult that it is often easier to attack a problem in optics by some less elegant but more practical method. Still, the method does find important application in the design of optical instruments, and the fact that such an extraordinarily general and powerful method exists at all is exciting to the mathematician.

For Hamilton the task of generalizing and abstracting was always more important than the search for practical results. He stated his position in a letter to Samuel Taylor Coleridge:

My aim has been, not to discover new phenomena, nor to improve the construction of optical instruments, but with the help of the Differential or Fluxional Calculus to remold the geometry of Light by establishing one uniform method for the solution of all problems in that science, deduced from the contemplation of one central or characteristic relation ... my chief desire and direct aim being to introduce harmony and unity into the contemplations and reasonings of Optics, considered as a portion of pure Science. It has not even been necessary, for the formation of my general method, that I should adopt any particular opinion respecting the nature of light. [1]

Leaving aside for the moment the question of what Hamilton meant by "pure science," his goal is obvious enough, to reduce all of geometrical optics to the most economical and most general mathematical statement possible. As early as 1826 he realized that his method would work not only for the science of light, but also for the science of mechanics. In 1834 he successfully made the extension to mechanics, and here again his purpose was more aesthetic than practical: "Even if it should be thought that no practical facility is gained, yet an intellectual pleasure may result from the reduction of the most·complex ... researches respecting the forces and motions of body, to the study of one characteristic function, the unfolding of one central relation." [2] The unification and generalization of mechanics, not practical results, were Hamilton's goal.

Considering Hamilton's love for the abstract and the ideal, it is not surprising that he directed his efforts towards creating an abstract and highly mathematical description of physical phenomena. The theory was his attempt to emulate those "pure and passionless intellects," mentioned in his letter to Arabella Lawrence, that are capable of comprehending the universe at a single glance. [3]

Hamilton's philosophy requires a more detailed description, and this may be found in other chapters of this book, but it is important to notice here how closely it parallels his mathematical work. By the time he was writing his papers on "A General Method of Dynamics" he was well into Kant, and he included in his "Introductory Remarks" his belief that "the science of force, or of power acting by law in space and time, has undergone already another revolution, and has become already more dynamic, by having almost dismissed the conceptions of solidity and cohesion." [4] Three months later he elaborated his position in a letter, in which he argued that "Power, acting by law in Space and Time, is the ideal base of an ideal world, into which it is the problem of physical science to refine the phenomenal world, that so we may behold as one and under the forms of our own understanding, what had seemed to be manifold and foreign." Hamilton wished to rid natural philosophy entirely of the concept of matter and replace it by a dynamic principle that he called variously "power," "force," or "unific energy."

His "powers" make an interesting comparison to Faraday's fields of electric and magnetic force. The two men met and discussed their ideas in June, 1834, just when Hamilton was writing the statements quoted above. But while Faraday's conception of the electromagnetic field was at the very heart of his understanding, and in his hands immediately productive of new experiments and new electrical phenomena, Hamilton never pursued his doctrine of "powers" in any way that was productive for physics. It was at the higher level of mathematical abstraction that Hamilton sought the principle of unity that his metaphysics persuaded him must exist. His attempts at a dynamical theory of light and electromagnetism were unsuccessful, because he could not bring himself to *study* experimental physics. His preference was always for metaphysical and mathematical subjects. His *mathematical* insight was uncanny. He could usually find a way if the physical problem could be reduced to one of mathematical analysis. More importantly, he was able to see among the various mathematical descriptions of nature a unity at a higher level of generality.

Of course the danger of pursuing such a highly abstract course is that when all is said and done, a theory whose elegance pleases the mathematician may have little to do with the actual physical constitution of light and matter. This is just the criticism that Heinrich Hertz brought against Hamilton's theory later in the nineteenth century. According to Hertz, Hamilton did not "unite" optics and mechanics at all. His contribution was entirely formal; that is, his mathematical technique could be used to describe and solve problems both in optics and mechanics (and most other branches of physics as well), but it did not "unite" the two sciences any more than Euclidean geometry united them.[5] Geometry can be used to solve problems in both optics and mechanics; so can the theory of the characteristic function. Hamilton's method is powerful, but it is still a mathematical method and not a physical theory.

Nineteenth-century writers on the sciences of optics and mechanics mentioned Hamilton's method with respect, but found little advantage in employing it. Hertz's criticism was certainly valid, but Hertz wrote before the advent of quantum theory. The creators of quantum mechanics found new value in Hamilton's method. Arnold Sommerfeld spoke most highly of it in his famous book *Atomic Structure and Spectral Lines,* from which most physicists learned their quantum mechanics in the 1920s.

Up to a few years ago it was possible to consider that the method of mechanics of Hamilton and Jacobi could be dispensed with for physics and to regard it as serving only the requirements of the calculus of astronomic perturbations and mathematics ... [but] since the appearance of the papers [of Epstein, Schwarzschild, and Sommerfeld] in the year 1916, which ... link up the quantum conditions with the partial differential equations of mechanics, it seems almost as if Hamilton's method were expressly created for treating the most important problems of physical [quantum] mechanics.[6]

Sommerfeld was describing the importance of Hamilton's methods for the "old" quantum theory. In 1926 the appearance of wave and matrix mechanics inaugurated a "new" quantum theory, also closely related to Hamilton's work. Erwin Schrödinger, who, along with Louis de Broglie, created wave mechanics, described Hamilton's importance for his new theory in the following terms:

The Hamiltonian principle has become the cornerstone of modern physics, the thing with which a physicist expects *every* physical phenomenon to be in conformity. . . .

The modern development of physics is continually enhancing Hamilton's name. His famous analogy between mechanics and optics virtually anticipated wave-mechanics, which did not have to add much to his ideas, [but] only had to take them seriously—a little more seriously than he was able to take them, with the experimental knowledge of a century ago. The central conception of all modern theory in physics is "the Hamiltonian." If you wish to apply modern theory to any particular problem, you must start with putting the problem "in Hamiltonian form."

Thus Hamilton is one of the greatest men of science the world has produced.[7]

Granting a certain amount of hyperbole to Schrödinger's statement, this still is a remarkable tribute.[8] The extreme generality and abstraction of Hamilton's methods, which made them impractical in the nineteenth century, were the precise qualities that made them transferable to quantum mechanics with a minimum of alteration.

Hamilton's first mathematical studies in 1822 were on osculating circles and parabolas, curves of double curvature, contacts between algebraic curves and surfaces, and developments. All of these topics are in analytical and differential geometry, a field that had recently undergone a revolution at the hands of the great French mathematician Gaspard Monge.

Analytical geometry as it had originally been conceived by Descartes was a method of using algebra to solve difficult problems in geometry. By the use of coordinates Descartes had shown how figures in space can be described by algebraic equations, and the manipulation of these equations can often solve problems that are intractable in pure geometry. But as Descartes had conceived his subject, it was "an *application* of algebra to geometry." In other words, it was a tool for solving problems in geometry, but not really a branch of mathematics in its own right. Even through the eighteenth century mathematicians such as Euler, who contributed a great deal to the subject, continued to ignore equations of the first degree, because these equations represented lines and planes in space—figures that could be handled by traditional geometrical methods and did not require new techniques.

It was Gaspard Monge who created analytical geometry as an independent branch of mathematics; in fact, it was he and his disciples who gave it the name. (Earlier mathematicians had always used Descartes's expression, "the application of algebra to geometry.") S.-F. Lacroix, a student of Monge and the chief expositor of his ideas, explained this new mathematical science in the following way: "There exists a way of looking at geometry that can be called *Analytical Geometry*, and which consists of deducing the properties of extension from the fewest principles by purely analytical methods, as Lagrange has done for the properties of equilibrium and movement in his *Mécanique*. . . . Monge is, I believe, the first to consider presenting the appliçation of Algebra to Geometry in this way."[9] Lacroix went on to argue that while this new science was closely related to traditional geometry, it was preferable to separate it completely from the synthetic methods of the ancients so that the subject could be pursued without diagrams or constructions.[10]

For Hamilton, whose greatest powers were in analysis rather than in geometry, the creation of this new science came at just the right time. The French writers whom he read, Lacroix, Garnier, Bourcharlat, and Puissant, were the disseminators of the new ideas.[11] Monge himself never wrote a proper textbook, which probably explains why his name does not appear more frequently in Hamilton's early manuscripts, but it is certain that Hamilton read him by July, 1823, even though his works were not required reading for the course of study at Trinity College.[12] In a much later letter to John Herschel, Hamilton sent an old poem of his own with the following reminiscence: "As I well remember the delight with which, at the same epoch of my life, I was reading and endeavouring to extend the profound and beautiful researches of Monge respecting families of surfaces, it would really seem to have been at one time a toss-up, whether I should turn out a rhymer or an analyst."[13] The poem referred to was written in the middle of July, 1823, which was the time that Hamilton was laboring on the caustics.[14] The previous May he had mentioned in a letter to Cousin Arthur, "In Optics I have made a very curious discovery—at least it seems so to me."[15] This is the earliest hint by Hamilton that he was on to some new trend of research in optics. It must have been a reference to his paper "On Caustics," which he presented to the academy the following year.[16]

"On Caustics" contains frequent references to Monge's famous work *Application de l'analyse à la géométrie.*[17] Obviously Monge had been the inspiration for much of the work. "On Caustics" was a study of what are now called "rectilinear congruences." A rectilinear congruence is a system of straight lines filling space. An example of a rectilinear congruence would be all the lines perpendicular to the surface of a given sphere. The connection to optics is obvious, because light rays, at least as they are conceived geometrically, are straight lines and they totally fill space.[18] In fact, the congruence of lines perpendicular to a sphere immediately describes the rays of light emanating from a point source.

A caustic surface is created by light rays reflected from a curved mirror or refracted by a lens. In most cases the light rays will not all come together in a perfect focus after reflection or refraction. Instead they will overlap in such a way as to create a surface called a "caustic," and it was the mathematical properties of this surface that Hamilton set out to investigate. The caustic curve is the envelope of the straight rays (see figure 4.1). It is found from two properties: first, the property that every point on the caustic is also a point on one of the reflected rays, and second, the property that all the reflected rays are tangents to the caustic. A caustic curve can be seen in a morning cup of coffee. Bright sunlight striking the inside edge of the cup is reflected onto the surface of the coffee so as to create the caustic curve. Although the caustic surface is defined in optics, its mathematics is part of the general theory of envelopes. In his paper Hamilton goes well beyond the caustic itself, to investigate the whole theory of rectilinear congruences, including the singular points.

In his *Application de l'analyse* Monge had investigated a wide variety of problems treating envelopes of families of planes, evolutes of curves of double curvature, and developable surfaces, but he had not taken up the problems related to rectilinear congruences. Hamilton believed that in his work he was opening an entirely new field. It was a great disappointment when Boyton showed him a work by Monge's pupil Etienne Malus entitled "Traité d'optique," which investigated rectilinear congruences in some detail.[19] Although Hamilton did not know it, Monge had already done a considerable amount of work on rectilinear congruences in an earlier paper that Hamilton had not seen.[20] Malus had taken up the whole problem as it related to optics and had developed it into a substantial treatise without mentioning Monge at all.[21] Thus Hamilton, in his paper "On Caustics," pays tribute primarily to Malus, although he is careful to state that he completed his paper before he was aware of Malus's treatise. In the body of the work, however, the influence of Monge is obvious, and the fact that Hamilton emphasizes the general theory of congruences rather

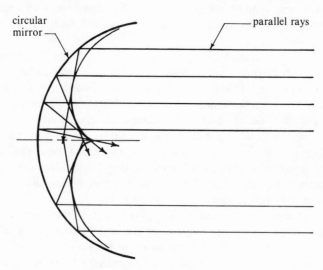

Fig. 4.1.   The Caustic Curve.

than the application to optics makes his work closer to that of Monge than Malus. There are no discussions of reflection or refraction; it is entirely a study of the general linear congruence and its focal properties and singularities. In the later "Theory of Systems of Rays" Hamilton does, of course, turn his attention directly to the problems of optics. But even there the influence of Monge predominates. It is interesting to note that several of the early studies and drafts of the "Theory of Systems of Rays" bear the title "Application of Analysis to Optics," indicating that Hamilton's original intention had been to write a work analogous to Monge's *Application de l'analyse à la géométrie.*

The generalized study of rectilinear congruences, which Hamilton began in "On Caustics" and continued in the "Theory of Systems of Rays," was later developed by E. E. Kummer, who, in his "Allgemeine Theorie der gradlinigen Strahlensysteme," credited Hamilton with having initiated the whole subject.[22] Thus we can conclude that while Hamilton's idealism convinced him of the worth of highly abstract and general theories in physics, it was the recent mathematical revolution in France and the availability of French mathematics at Trinity College that set him on the track of his theory of the characteristic function.[23]

Hamilton wrote the "Theory of Systems of Rays" in three parts; the first part was a study of reflected rays (mirrors), the second a study of refracted rays (lenses), and the third a study of extraordinary systems (double refraction) and systems of rays in general. He published only the first part, on reflected rays. The second part was published for the first time in 1931 in his *Mathematical Papers.*[24] The third part, which the editors of the *Mathematical Papers* believed was lost, has recently come to light among Hamilton's loose papers at Trinity College.[25] It is certain that Hamilton had completed all three parts before the first part was published, because he included in this first publication a detailed table of contents for the entire work that agrees paragraph by paragraph with the manuscript versions. He probably decided not to publish the second and third parts because they had become obsolete by the time the Royal Irish Academy had room for them in their *Transactions.* Instead he published two "supplements" in 1830, showing the new directions his research was taking him, and a third supplement, in 1833, which was a complete treatise in itself and the most advanced version of his optical theory. The third supplement contained the startling prediction of conical refraction in biaxial crystals, a discovery that did more than all his other mathematical labors to bring him to the attention of the scientific community.

The "Theory of Systems of Rays" is difficult to read, largely because Hamilton practiced great economy of language and compressed his most important ideas into the smallest possible compass. There are no diagrams or examples to help the reader, and to make matters worse, Hamilton expands less interesting aspects of his theory to the extent that they almost bury the important nuggets. Sir Joseph Larmor, another Irishman and a great fan of Hamilton's, looking back on the optical papers from the perspective of the 1880s, found them "very dishevelled in form," a weakness that Larmor attributed to "the distraction of [Hamilton's] wide range of philosophical and poetic interests."[26]

Hamilton begins his paper with a discussion of reflected rays. This is the simplest case because the rays are always traveling through the same medium before and after reflection and will therefore travel with the same constant speed. The time of travel will always be proportional to the distance of travel and the rays will always be straight lines. In the case of refraction at a surface (as with a lens or prism), the rays are still straight, but they travel at different velocities in different media so that the time of travel from initial to final point is not necessarily proportional to the total distance of travel. If the media are nonhomogeneous (an example would be light traveling through the earth's atmosphere), the rays will be curved. And a further generalization is added when the medium is allowed to be anisotropic, meaning that the velocity of the ray is dependent not only on the *place* in the medium through which it is passing, but also on the *direction* of its travel. This last generality is necessary to include the case of light passing through double refracting crystals. As Hamilton made his theory more general in successive supplements to the "Theory of Systems of Rays," he also constructed a mathematical formalism that came closer and closer to satisfying the needs of mechanics. The transition from the "Third Supplement" in 1833 to the "General System of Dynamics" in 1834 was relatively easy. The greatest amount of mathematical labor went into the optics—not the dynamics.

In the first part of the "Theory of Systems of Rays," Hamilton's theory depended on the proof of a theorem called the "Theorem of Malus." This theorem states that if light rays emanate from a point source (or from a surface perpendicular to the rays), it will always be possible to construct a surface perpendicular to all the rays after any number of reflections or refractions. As mentioned earlier, a system of lines such as those defined by rays of light from a point source is called by mathematicians a rectilinear congruence. If through every point a surface can be constructed perpendicular to all the lines, the congruence is called a "normal congruence." Light rays from a point source form a normal congruence.[27] The Theorem of Malus states, then, that a normal congruence will remain normal after any number of reflections or refractions. Hamilton was able to prove further that the rays of light, as they are propagated through an optical system of lenses and mirrors, always reach these perpendicular surfaces at the same *time*, just as the rays emanating from a point source of light reach spheres concentric about the point source at the same time.

If one assumed that light is composed of particles, and if it were possible to photograph a flash of light particles as they moved through any system of mirrors and lenses, the particles would always be found on a surface perpendicular to the direction of their motion. Or if one assumed that light is composed of waves, the moving wave front would describe the surfaces perpendicular to all the rays.[28]

The Theorem of Malus showed at once an important advantage of Hamilton's system. His view of geometrical optics did not require any decision about the physical nature of light. If one believed light to be composed of particles, the straight rays described their trajectories. If one believed light to be composed of waves, the surfaces described the advancing wave front and the rays gave the direction of the propagation. Hamilton mentions this "remarkable analogy" in the early pages of his treatise.[29] Later Hamilton proved that the Theorem of Malus holds even when the rays are curved. Malus had believed that the theorem would hold for a single refraction or reflection, but would *not* hold after a second refraction or reflection. Malus's error had been corrected by Dupin, Quetelet, and Gergonne before Hamilton published his "Theory of Systems of Rays," but Hamilton knew nothing about their work and believed that his correction of Malus's theorem was new.[30]

His proof was simplicity itself. It drew upon a principle that lay at the foundation of both his optics and his mechanics—the Principle of Least Action or Principle of Least Time, first introduced by Fermat for the case of optics, and extended (in a less than rigorous fashion) to mechanics by Maupertuis.

The Principle of Least Time states that a ray of light, in passing between a given initial and a given final point, will follow the path that takes the least time. The analogous Principle of Least Action states that a particle, in passing between two given points, will take the path requiring the least expenditure of "action," a quantity measured by the product of the mass, speed, and distance traversed by the body. Hamilton always referred to this principle as the Principle of Least Action, although Least Time would have been more appropriate for optics, because the quantity called "action" is properly defined only for mechanics.

To use the principle in either of its forms it is necessary to write the integral expressing the total action or time consumed in going from the initial to the final point. One then varies the equation of the path of the light to consider the time for the light to traverse a "neighboring" path joining the same two end points. By the theory of maxima and minima the difference between the time for these two paths will be zero if the original path is a minimum, that is, if it is the actual path as specified by the Principle of Least Time. Therefore to find the correct path one takes the variation of the integral, sets it equal to zero, and solves for the path.

Hamilton regularly writes the principle as

$$\delta \int v\,ds = 0$$

In the mechanical case of "least action," v stands for the speed of the particle, the mass being assumed constant. In the optical case, v stands for the index of refraction. Then

$$\delta \int v\,ds = \delta \int \left[\frac{\text{velocity of light in vacuum } (c)}{\text{velocity of light in medium } (ds/dt)} \right] ds = c\delta \int dt$$

and the integral takes the form of Fermat's Principle of Least Time. This method assures that the path will be an extremum, but does not guarantee that it is a true minimum; it could be a maximum. For this reason Hamilton renamed the principle the Principle of Stationary Action, meaning that the variation of the action integral is equal to zero.

Hamilton proves the Theorem of Malus as in figure 4.2, using the Principle of Least Time. Consider a ray of light AP perpendicular to the surface AB. The ray impinges on the mirror at P and is reflected through point A'. A "neighboring" ray leaves the same surface at B, is reflected at Q and passes through a point B' so that the two rays reach points A' and B' at the same time. We need to demonstrate that the surface $A'B'$ is perpendicular to the rays $A'P$ and $B'Q$, for if this can be shown to be the case, then a congruence that is normal before reflection will remain normal after reflection, and, indeed, after any number of reflections.

By Fermat's Principle, the time for a ray of light to follow the two paths APA' and AQA' must be the same (the paths must also be of the same length in this case), because the path APA' is given as the true path of the ray between points A and A', and according to the principle any slight variation of the path (by moving the point of incidence on the mirror an

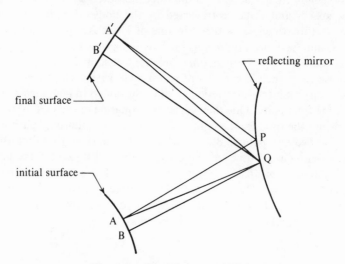

Fig. 4.2. The Theorem of Malus.

infinitesimal amount) does not change the time required for the light to travel between A and A'. Therefore

$$APA' = AQA'.$$

Because the ray BQ is given as perpendicular to the surface AB and because the distance AB is infinitesimal compared to BQ we can conclude that $AQ = BQ$ to the first order of infinitesimal differences. Then

$$BQA' = AQA' = APA'.$$

We chose point B' so that $BQB' = APA'$ and therefore

$$QB' = QA'.$$

Using the reverse of the previous argument, since $QB' = QA'$ to the first order of infinitesimal differences, the surface $A'B'$ must be perpendicular to the ray QB'. The entire argument may be repeated to prove that the surface $A'B'$ is also perpendicular to the ray PA'. Therefore at any time after any number of reflections the points reached by the light will lie on a surface perpendicular to the rays.

A completely analogous argument proves the theorem for the case of refraction. Hamilton's proof is geometrical. The same theorem can be proved analytically, but it is tricky, and as late as 1834 Hamilton was still working to enlighten able mathematicians who could not get it right.

Hamilton named these surfaces perpendicular to the rays "surfaces of constant action," because particles leaving the source at the same time will all reach the orthogonal surfaces at the same time and will all have expended the same amount of action. Because he thought of the light as particles for the purposes of analogy, he used the term *surfaces of constant action*, although for the optical case it would have been more proper to have called them *surfaces of constant time*.

At the beginning of his "Theory of Systems of Rays," Hamilton asks how to find a mirror $F(x, y, z)$ that will reflect rays of any given congruence into a focus at A'. The path length from a point A on the incident ray to A' is $\rho + \rho'$ (see figure 4.3)

where $$\rho^2 = (x - x')^2 + (y - y')^2 + (z - z')^2 \tag{4.1}$$

and $$\rho'^2 = (x'' - x)^2 + (y'' - y)^2 + (z'' - z)^2. \tag{4.2}$$

If we vary the point of incidence on $F(x, y, z)$ we get

$$\rho\delta\rho = (x - x')\delta x + (y - y')\delta y + (z - z')\delta z \tag{4.3}$$

$$\rho'\delta\rho' = -(x'' - x)\delta x - (y'' - y)\delta y - (z'' - z)\delta z. \tag{4.4}$$

In order to simplify these equations we note that

$$(x - x') = \rho\alpha$$

$$(y - y') = \rho\beta$$

$$(z - z') = \rho\gamma$$

where α, β, γ are the direction cosines of the incident ray. Equation 4.3 may then be written

$$\delta\rho = \alpha\delta x + \beta\delta y + \gamma\delta z. \tag{4.5}$$

The same simplification can be made for equation 4.4

$$\delta\rho' = -\alpha'\delta x - \beta'\delta y - \gamma'\delta z, \tag{4.6}$$

and adding these we get

$$\delta\rho + \delta\rho' = (\alpha - \alpha')\delta x + (\beta - \beta')\delta y + (\gamma - \gamma')\delta z.$$

By the principle of least action $\delta\rho + \delta\rho' = 0$ and therefore

$$(\alpha - \alpha')\delta x + (\beta - \beta')\delta y + (\gamma - \gamma')\delta z = 0 \tag{4.7}$$

This equation gives a family of surfaces $F(x, y, z) = $ constant, all of which would reflect the ray passing through A into a focus at A'.

It is not clear, however, that equation 4.7 is always integrable, or that

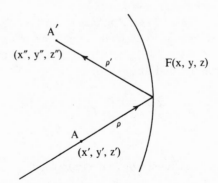

Fig. 4.3. The surface $F(x, y, z)$ reflects the rays to a focus at A'.

there is a family of focal reflectors for every congruence of incident rays. In seeking the conditions under which equation 4.7 is integrable, Hamilton points out that the equation contains only the direction cosines of the initial and reflected rays as functions of the coordinates of position of the mirror. We need not consider the initial rays as coming from a single point, but can regard them as a congruence of rays defined by the direction cosines of the rays at the surface $F(x, y, z)$. The reflected rays, however, all pass through the given focus A'. For this reason Hamilton says we can see immediately that the reflected portion of the path represented by equation 4.6 is integrable and an exact differential, because its integral is the path length from the point of incidence to the focus A'. We can also see that the reflected rays form a normal congruence, because concentric spheres about the focus A' are always orthogonal to the reflected rays.

But it remains to be proved that the equation for the entire path is integrable. The condition for integrability is the following:

$$(\alpha - \alpha') \left(\frac{\partial(\beta - \beta')}{\partial z} - \frac{\partial(\gamma - \gamma')}{\partial y} \right)$$

$$+ (\beta - \beta') \left(\frac{\partial(\gamma - \gamma')}{\partial x} - \frac{\partial(\alpha - \alpha')}{\partial z} \right)$$

$$+ (\gamma - \gamma') \left(\frac{\partial(\alpha - \alpha')}{\partial y} - \frac{\partial(\beta - \beta')}{\partial x} \right) = 0.$$

This can be rewritten as

$$(\alpha - \alpha') \left[\left(\frac{\partial\beta}{\partial z} - \frac{\partial\gamma}{\partial y} \right) - \left(\frac{\partial\beta'}{\partial z} - \frac{\partial\gamma'}{\partial y} \right) \right]$$

$$+ (\beta - \beta') \left[\left(\frac{\partial\gamma}{\partial x} - \frac{\partial\alpha}{\partial z} \right) - \left(\frac{\partial\gamma'}{\partial x} - \frac{\partial\alpha'}{\partial z} \right) \right]$$

$$+ (\gamma - \gamma') \left[\left(\frac{\partial\alpha}{\partial y} - \frac{\partial\beta}{\partial x} \right) - \left(\frac{\partial\alpha'}{\partial y} - \frac{\partial\beta'}{\partial x} \right) \right] = 0. \quad (4.8)$$

Because we know that equation 4.6 is an exact differential, then

$$\frac{\partial\beta'}{\partial z} - \frac{\partial\gamma'}{\partial y} = 0, \qquad \frac{\partial\gamma'}{\partial x} - \frac{\partial\alpha'}{\partial z} = 0, \qquad \frac{\partial\alpha'}{\partial y} - \frac{\partial\beta'}{\partial x} = 0,$$

and equation 4.8 reduces to

$$(\alpha - \alpha') \left(\frac{\partial \beta}{\partial z} - \frac{\partial \gamma}{\partial y} \right) + (\beta - \beta') \left(\frac{\partial \gamma}{\partial x} - \frac{\partial \alpha}{\partial z} \right)$$

$$+ (\gamma - \gamma') \left(\frac{\partial \alpha}{\partial y} - \frac{\partial \beta}{\partial x} \right) = 0. \quad (4.9)$$

Hamilton then goes on to prove that the quantities

$$\left(\frac{\partial \beta}{\partial z} - \frac{\partial \gamma}{\partial y} \right), \qquad \left(\frac{\partial \gamma}{\partial x} - \frac{\partial \alpha}{\partial z} \right), \qquad \left(\frac{\partial \alpha}{\partial y} - \frac{\partial \beta}{\partial x} \right),$$

if they are not equal to zero, then they must be proportional to α, β, γ, respectively. The only way that equation 4.9 can be satisfied for any surface other than a direct reflection of the rays back on themselves is to set these quantities equal to zero. This means that in order for equation 4.7 to be integrable, it must be an exact differential. It also means that in order to solve the problem of finding the family of focal surfaces $F(x, y, z)$, the initial rays must form a normal congruence, where the orthogonal surfaces cutting the rays are given by

$$\int (\alpha dx + \beta dy + \gamma dz) = \text{constant.}[31]$$

This result is what Hamilton calls his *Principle of Constant Action*. The integral $\int v d\rho = $ constant defines the family of surfaces that Hamilton calls the *surfaces of constant action*. He gives them this name because the rays all reach each surface at the same time. His *Principle* of Constant Action is the statement that these surfaces must be orthogonal to the congruence of rays.

We can now see how these rays and families of orthogonal surfaces are related (figure 4.4). The Law of Reflection is contained in equation 4.7, because it states that the vector normal to the mirror ($\alpha - \alpha'$, $\beta - \beta'$, $\gamma - \gamma'$) is the sum of the unit vector in the direction of the incident ray (α, β, γ) and the negative of the unit vector for the reflected ray (α', β', γ'). (Hamilton chose directions for both rays towards the mirror which makes his presentation confusing. I have chosen directions following the actual path of light.) This is equivalent to saying that the angle of incidence is equal to the angle of reflection. We did not use the Law of Reflection in deriving equation 4.7, but it is an immediate consequence of the Principle of Least Action, which we did use. To give physical meaning to figure 4.4, we can regard the "surfaces of constant action" as wave crests reflected off the mirror, although we must be careful to remember that these surfaces do not move along the ray as a wave crest would.

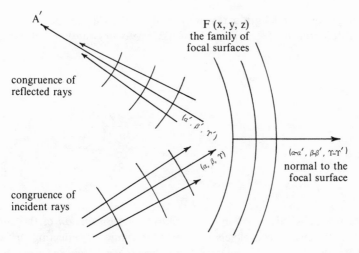

Fig. 4.4. *The family of focal surfaces F(x, y, z) reflects a congruence of incident rays to a focus at A'.*

Hamilton's next step is one of major consequence, since it leads directly to the characteristic function. If we now return to our example of a congruence of reflected rays (figure 4.3), but this time we think of the mirror as fixed and the end points as varying, we get the following expressions:

$$\rho\delta\rho = -(x - x')\delta x' - (y - y')\delta y' - (z - z')\delta z'$$

$$\rho'\delta\rho' = (x'' - x)\delta x'' + (y'' - y)\delta y'' + (z'' - z)\delta z''$$

$$\delta\rho = -\alpha\delta x' - \beta\delta y' - \gamma\delta z'$$

$$\delta\rho' = \alpha'\delta x'' + \beta'\delta y'' + \gamma'\delta z''$$

$$\delta\rho' + \delta\rho = \alpha'\delta x'' + \beta'\delta y'' + \gamma'\delta z'' - \alpha\delta x' - \beta\delta y' - \gamma\delta z'. \quad (4.10)$$

Because we know there exist surfaces perpendicular to the congruence at every point, this differential form is exact and therefore there must exist some function $V(x'', y'', z'', x', y', z')$ whose differential is

$$\delta V = \frac{\partial V}{\partial x''}\delta x'' + \frac{\partial V}{\partial y''}\delta y'' + \frac{\partial V}{\partial z''}\delta z''$$

$$+ \frac{\partial V}{\partial x'}\delta x' + \frac{\partial V}{\partial y'}\delta y' + \frac{\partial V}{\partial z'}\delta z'$$

V will be the path length between the end points. Equating coefficients in the differential to the coefficients of like terms in equation 4.10, we get

$$\alpha' = \frac{\partial V}{\partial x''} \qquad -\alpha = \frac{\partial V}{\partial x'}$$

$$\beta' = \frac{\partial V}{\partial y''} \qquad -\beta = \frac{\partial V}{\partial y'} \qquad (4.10a)$$

$$\gamma' = \frac{\partial V}{\partial z''} \qquad -\gamma = \frac{\partial V}{\partial z'}.$$

Hamilton calls this function V the *characteristic function* of the optical system, because it completely describes the family of reflecting surfaces, and because it gives the relationship between the initial and final points for any ray passing through the system—exactly what we need to know in order to describe the properties of any optical system.[32] Because of the identity relation of the direction cosines

$$\alpha^2 + \beta^2 + \gamma^2 = 1, \qquad \alpha'^2 + \beta'^2 + \gamma'^2 = 1$$

and substituting from equation 4.10a, we get

$$\left(\frac{\partial V}{\partial x''}\right)^2 + \left(\frac{\partial V}{\partial y''}\right)^2 + \left(\frac{\partial V}{\partial z''}\right)^2 = 1$$

$$\left(\frac{\partial V}{\partial x'}\right)^2 + \left(\frac{\partial V}{\partial y'}\right)^2 + \left(\frac{\partial V}{\partial z'}\right)^2 = 1. \qquad (4.11)$$

The simultaneous solution of these two differential equations of the first order and second degree will give the characteristic function for the optical system.

Thus far we have treated only the case of a single reflection, but the method can be easily extended to a system with any number of reflections and refractions. In the second part of his "Theory of Systems of Rays," which Hamilton never published, he derived comparable equations for the case of a single refraction. In the case of refraction, however, the path of least time is not the shortest path, because light moves at different speeds in different media. Therefore instead of minimizing the path length, we minimize the "optical path," which is the product of the path length ρ and the index of refraction ν, and is proportional to the time of propagation of the light. We should note that this represents a change in the definition of the characteristic function V, which makes it more general. Equation 4.7 written for a single refraction becomes

$$(v\alpha - v'\alpha')\delta x + (v\beta - v'\beta')\delta y + (v\gamma - v'\gamma')\delta z = 0. \qquad (4.12)$$

Equation 4.10 becomes

$$\delta V = v'(\alpha'\delta x'' + \beta'\delta y'' + \gamma'\delta z'') - v(\alpha\delta x' + \beta\delta y' + \gamma\delta z'), \qquad (4.13)$$

and equations 4.11 become

$$\left(\frac{\partial V}{\partial x''}\right)^2 + \left(\frac{\partial V}{\partial y''}\right)^2 + \left(\frac{\partial V}{\partial z''}\right)^2 = v'^2$$

$$\left(\frac{\partial V}{\partial x'}\right)^2 + \left(\frac{\partial V}{\partial y'}\right)^2 + \left(\frac{\partial V}{\partial z'}\right)^2 = v^2.$$

$$(4.14)$$

Next we consider the case of rays passing through an optical system consisting of any number of different media and surfaces of reflection and refraction (see figure 4.5). In order to find the characteristic function V of the optical system we vary the end points of the ray $AP \ldots P_n A'$ to the points B and B'. The ray $BQ \ldots Q_n B'$ is the "neighboring" ray joining B and B'. Both of these are actual paths of the rays. We now consider the path $BP \ldots P_n B'$. This is *not* an actual path, but it represents a variation of the actual path $BQ \ldots Q_n B'$ between fixed end points B and B'. Both paths between B and B' have the same optical length by Fermat's Principle. But the path $BP \ldots P_n B'$ follows the same path as $AP \ldots P_n A'$ except in the initial and final media. Therefore the difference in total optical length caused by varying the end points involves only the variation of the optical length in the initial and final media, which means that we can

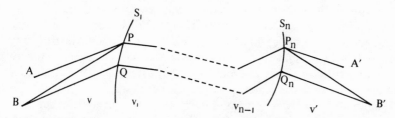

Fig. 4.5.   *Varying the end points of a ray passing through an optical system composed of several media separated by several reflecting and refracting surfaces.*

write the equations for the entire optical system as they are written for a single refraction in equations 4.13 and 4.14. By the Theorem of Malus, we know that the surfaces of constant action remain orthogonal to the rays throughout the entire optical system, and therefore we know that the equations are integrable.

The method is not quite as magical as it first appears, however. If we are given the initial position and direction of the ray, we still have to follow it through the entire optical system in order to find the direction cosines of the final ray. In order to do this we write the optical length of any ray between A and A' as a function of the different surfaces and indices of refraction. Then we apply Fermat's Principle of Least Time to find the actual ray between A and A'. Once we have this function for the optical length between two specific end points, we can vary those end points to obtain equations 4.13 and 4.14 for the characteristic function, and they then give the relationship between the initial and final positions of *all* rays passing through the system.

The characteristic function completely describes the optical system and is in turn completely determined by the optical system. Hamilton called it the *characteristic* of the system, because from it all the properties of the optical system could be deduced. Therefore it "characterizes" the system completely. He probably took the name from Monge, who had called the lines of contact between families of surfaces depending on a single parameter the *characteristics* of that family. These lines determine the way in which the surfaces generate the envelope to the family, and therefore they "characterize" the family of surfaces.[33]

In the first two parts of the "Theory of Systems of Rays" Hamilton develops the theory of the characteristic function for the reflection and refraction of straight rays. He then goes on to use the theory to investigate the focal properties of rays (pencils, foci, caustics, lines of reflections, images, and so on), optical aberrations, and optical density. At the end of this second part Hamilton introduces for the first time systems of curved rays. Thus all of the published first part of the "Theory of Systems of Rays" and almost all of the unpublished second part dealt solely with systems of straight lines. Hamilton's exclusiveness is justified not only because of the mathematical interest inherent in his study of rectilinear congruences, but also because the practical design of optical instruments deals only with straight rays. And while he did not construct his theory specifically to solve practical problems, the theory of lenses and mirrors had always been at the center of geometrical optics.

Hamilton's "Theory of Systems of Rays" is entirely theoretical, containing no examples or discussion of practical application, but he did work out in great detail the practical application of his theory to the symmetrical optical system.[34] Unfortunately it was not published until J. L.

Synge and A. W. Conway included it in their collection of Hamilton's *Mathematical Papers*. These papers are revealing of Hamilton's methods of work. The published papers seem to be the work of a pure mathematician who would not deign to waste time on any practical considerations. In reality, as the manuscripts prove, Hamilton regularly tested his theories on practical problems, some of which required laborious computation. In this respect he provided a striking contrast to his famous colleague at Trinity College, James MacCullagh, who would spend hours looking at a problem, trying to discover some shortcut that would allow him to circumvent the chore of computation. Hamilton often preferred to attack the problem head-on. Sometimes he seemed almost to glory in the task. J. L. Synge wrote out a list and description of Hamilton's manuscript notebooks in preparation for his own edition of the Hamilton papers. Large sections of the notebooks are described by the single entry of "heavy computation."

In his study of curved rays, Hamilton defined the characteristic function in its final integral form for nonhomogeneous but still isotropic media.[35] He writes it as

$$V(x'', y'', z'', x', y', z') = \int v d\rho$$

where v is the index of refraction and $d\rho$ the element of the path. The coordinates x', y', and z' refer to the initial point and x'', y'', and z'' to the final point of the ray. The integral is proportional to the time of travel of the light along a curved path. If Hamilton can prove that his Principle of Constant Action applies also to curved rays, he will have direct access to the characteristic function, because the existence of surfaces orthogonal to the rays at every point will lead to the same equations as before. Thus his first purpose is to show that the surfaces of constant action (or constant optical length) are orthogonal to the curved rays. Fermat's Principle states that the actual path followed by the ray will be the one that takes the least time, or, as Hamilton correctly said, the path along which the time is "stationary." This means that if the path is varied slightly to a neighboring path (it is not necessarily the actual path of a ray) connecting the same end points, the time will remain the same and the "variation" of the integral will be zero.

$$\delta \int v d\rho = 0$$

Hamilton expands this equation to obtain

$$\delta \int_{(x',y',z')}^{(x'',y'',z'')} v d\rho = v'(\alpha'\delta x'' + \beta'\delta y'' + \gamma'\delta z'') - v(\alpha\delta x' + \beta\delta y' + \gamma\delta z')$$

$$+ \int_{x'}^{x''} \left(\frac{dv}{dx} \cdot d\rho - d \cdot v \frac{dx}{d\rho} \right) \delta x$$

$$+ \int_{y'}^{y''} \left(\frac{dv}{dy} \cdot d\rho - d \cdot v \frac{dy}{d\rho} \right) \delta y$$

$$+ \int_{z'}^{z''} \left(\frac{dv}{dz} \cdot d\rho - d \cdot v \frac{dz}{d\rho} \right) \delta z = 0 \qquad (4.15)$$

The first two expressions, $v'(\alpha'\delta x'' + \beta'\delta y'' + \gamma'\delta z'')$ and $v(\alpha\delta x' + \beta\delta y' + \gamma\delta z')$, are due to the variations of the end points (x'', y'', z'') and (x', y', z'). Because Fermat's Principle requires the variation to be between *fixed* end points, these two terms are zero. Then because the variations δx, δy, δz are completely arbitrary, the sum of the integrals can be zero only if the integrands individually are zero, and Hamilton obtains the following equations of the ray:

$$\frac{dv}{dx} \cdot d\rho - d \cdot v \frac{dx}{d\rho} = 0$$

$$\frac{dv}{dy} \cdot d\rho - d \cdot v \frac{dy}{d\rho} = 0 \qquad (4.16)$$

$$\frac{dv}{dz} \cdot d\rho - d \cdot v \frac{dz}{d\rho} = 0.$$

These are the Euler equations for the variation of the integral. They are the optical equivalent of Lagrange's equations in mechanics.

So far this is a solution following Fermat's Principle and contains nothing new. It has limited value, because, as Lagrange and Poisson had already pointed out, it only gives the path after two *fixed points* have been chosen. What is needed is a method of determining the path as a function of the initial and final points. This is precisely what the characteristic function provides.

If in equation 4.15 we vary the end points (x'', y'', z'') and (x', y', z'), the first two terms in parentheses in equation 4.15 may not be zero, because the variations $\delta x''$, $\delta y''$, $\delta z''$, $\delta x'$, $\delta y'$, $\delta z'$ are no longer zero. Because the three integrals in the right member of equation 4.15 must follow Fermat's Principle for *any* given end points, equations 4.16 will still hold and we can write equation 4.15 as

$$\delta \int v d\rho = v'(\alpha' \delta x'' + \beta' \delta y'' + \gamma' \delta z'') - v(\alpha \delta x' + \beta \delta y' + \gamma \delta z').$$

The variation of the action integral (actually it is the integral of the optical length) is then determined entirely by the variation of the end points of the ray. Also the integral $\int v d\rho =$ constant gives the family of surfaces that Hamilton calls surfaces of constant action, but we still have to prove that these surfaces are normal to the curved rays at every point. Let us restrict the variation of each end point to one of the surfaces of constant action. Then, because the action integral equals some constant, its variation will be equal to zero and

$$\delta \int v d\rho = v'(\alpha' \delta x'' + \beta' \delta y'' + \gamma' \delta z'') - v(\alpha \delta x' + \beta \delta y' + \gamma \delta z') = 0.$$
$$(4.17)$$

Because the differential is equal to zero, it is exact (more precisely, "locally derived"), and we have proved that the surfaces of constant action are perpendicular to the rays in the case of curved rays of light as well as in the case of straight rays. Equation 4.17 is the same as equation 4.13, and therefore equation 4.14 follows as before.

The two crucial aspects of Hamilton's "Theory of Systems of Rays" are his Principle of Constant Action, which states that the surfaces of constant action are always perpendicular to the rays, and his characteristic function, which is a consequence of the Principle of Constant Action. The theory of the characteristic function is a *variational* principle, like Maupertuis's Principle of Least Action, but it has the important advantage over Maupertuis's principle of allowing the end points of the ray to vary. Using the characteristic function, Hamilton reduced the problem of describing any optical system to the solution of two simultaneous partial differential equations of the first order and second degree (equations 4.14).

The extension of his optical theory to curved rays and his definition of the characteristic function as the time integral point the way directly to a system of mechanics analogous to his system of optics. Since the theory of least action for mechanics is mathematically equivalent to the Principle of Least Time in optics, it should be possible to use it directly. The immediate analogue to a curved light ray is the path of a single particle following a curved path through space.

5

Creating the "Theory of Systems of Rays"

HAMILTON LEFT an enormous mass of optical papers from which it should be possible to reconstruct the development of his ideas. Unfortunately many of these manuscripts are undated, and it is impossible to arrange them in chronological order with complete security. Hamilton almost always dated entries in his notebooks, whether on mathematical researches or memoranda on other subjects, but often he did not date drafts of works to be published or notes on "scratch paper" that he had lying about him. Many of the optical papers fall into this latter category—drafts for his "Systems of Rays" and loose papers.

One must also approach with skepticism the published dates on Hamilton's papers. Part one of "Theory of Systems of Rays" bears the presentation date of December 13, 1824, but we know that the paper presented on this date was his "On Caustics," a paper completely different from the final "Theory of Systems of Rays." Hamilton signed his introduction to the "Theory of Systems of Rays, Part I" in June, 1827.[1] Some version of the paper had certainly been read to the Royal Irish Academy in April, 1827,[2] but in February, 1828, he was still working on it[3] and found it necessary to apologize to the president of the academy for holding up the publication of their *Transactions* while he finished his researches on optical density![4]

The confusion is multiplied by the fact that Hamilton wrote and rewrote the "Theory of Systems of Rays" many times, going back each time to incorporate new ideas into the earlier versions. Hamilton mentions committing many of the early drafts of his optical papers to the flames, so we can safely assume that there were many early papers that did not survive in any form.

The greatest problem of chronology is determining when Hamilton invented the characteristic function. There is, for instance, an undated draft entitled "Application of Analysis to Optics," part of which appears verbatim in the "Theory of Systems of Rays," parts one and two.[5] This manuscript employs Fermat's Principle from the beginning, and treats reflection, refraction, and the general case of curved rays in a heterogeneous nonisotropic medium. Therefore it appears to be a relatively late draft of the "Theory," *but* it contains no mention of the characteristic function at all. Although Hamilton employed the characteristic function from the beginning of his *published* "Theory," the first unambiguous reference to it in the manuscripts is in conjunction with the most general case of rays, those that are "any number of times reflected and refracted, by any given combination of crystalline or uncrystallized mediums, disposed and bounded in any manner whatsoever."[6] The characteristic function was apparently a late invention that was then employed by Hamilton to revise his "Theory" from the very beginning.

Any reconstruction of Hamilton's thought must of necessity be conjectural, but the most likely chronology seems to be the following: In his work "On Caustics" (1824) Hamilton concentrated on the focal properties of a normal congruence. There is no discussion of reflection or refraction, and so the paper is not really about optics at all. Charles Boyton pointed out to Hamilton the work of Malus on the subject, but Malus's famous theorem does not appear in "On Caustics." When "On Caustics" was refused publication by the Royal Irish Academy in June, 1825, Hamilton decided to expand it and to apply his ideas more specifically to optics. Throughout 1825 he was forced to devote most of his energy to college work, but early in 1826 he returned to his paper, which he now called "My Essay on the Theory of Systems of Rays,"[7] and on June 24, 1826, he sent to Brinkley a description of his new theory.[8]

Beginning in April, 1826, Hamilton kept a journal recording his preparations for the college examinations. Aside from a substantial amount of reading in the classics, most of his study was devoted to Brinkley's *Astronomy* and several works in mechanics. The reason becomes clear on June 16, when there is an entry, "Let me consider how I may best prepare myself in Physics and Astronomy for tomorrow," indicating that the examination was to be held on the following day.[9] After the June examination there is a gap that indicates a relaxed summer vacation spent in visits to Edgeworthstown, Bellevue in Wicklow (the home of the La Touche family), and Belfast, where he met Humphry Davy.

But on September 17 the journal resumes after Hamilton returned to Trim to begin preparing for examinations in October. This time he was having trouble, because his mind kept wandering off classics and onto optics. Sometimes he compromised by doing optical problems in Latin, but usually he chastised himself ineffectually: "Now don't you think I had better, till after the next Examinations, abstain from all further investigations of my own?"[10] What was occupying his mind was his "Principle of Constant Action," which he must have proved by this time for straight rays.

The Principle of Constant Action was the subject of one of his first conversations with his uncle when he arrived at Trim.[11] He wrote to Cousin Arthur two days later that while riding in the coach he had "discovered a still further generalization of [his] principle respecting the surfaces of constant action, and a simple demonstration which [included] all particular cases."[12] His impatience is manifest: ". . . very anxious to calculate the equation of the surfaces of constant action. . . . It was then dark and I remained in my room, till Tea, thinking of my Principle of Constant Action."[13] On September 18, five days after his arrival at Trim, he noted: ". . . began to think of applying my principle of Constant Action to Astronomical Refraction." Immediately following his attack on the problem of atmospheric refraction he applied his principle to "the case of molecules shot off with given velocity . . . from a given point" or perpendicular to a given surface. (There is a scrap among the loose papers containing Hamilton's calculations on atmospheric refraction.) His results were exciting, but on the twentieth he cautioned himself:

I must endeavour to abstain a little from any new investigations before the Examinations and confine myself to readying for the Examinations and to perfecting what I have already done. Went out into the garden, and after walking for some time, came to the conclusions that I was right in my general theorem of yesterday. After dinner, took a walk with Eliza and the children and read the Percy Ballads . . . and after tea was engaged in Biot and my own investigations (manque all my resolutions) till ten. I will now go to bed, and try to read more classics tomorrow.[14]

The next day Hamilton dutifully began reading Livy, but before he knew it he was "engaged once more in my Principle of Constant Action." He returned to the problem of parabolic trajectories for particles shot off perpendicular to a given surface and moving in a gravitational field. He finds that the variation of the action integral is the now-familiar expression $v(\alpha\delta x + \beta\delta y + \gamma\delta z)$, "which shews that the Trajectories are cut perpendicularly by the Surfaces of Constant Action."[15] Two days later, on the twenty-third of September, he wrote: "I began to consider how I could develope the whole of optics from the principle [of constant action]."[16]

It was during these days in September of 1826, when Hamilton began to search for the surfaces orthogonal to systems of curved lines, that he saw the great generality of his new system of optics. It is significant that he tested his Principle of Constant Action first on an optical case (that of atmospheric refraction), and immediately afterwards on a mechanical case (that of the parabolic paths of projectiles). It proves that at least from September, 1826, he was aware of the optical-mechanical analogy for curved rays and trajectories. It is also significant that in all these discussions of the Principle of Constant Action there is no mention of the characteristic function. (He writes the variation of the action integral as δA or δI, but does not identify it as the characteristic function.) Either he coined the term sometime after September, 1826, and then rewrote the earlier parts of the "Theory of Systems of Rays" to contain it, or (and this seems more likely) his earlier use of the characteristic function had been limited to the simple case of linear rays, and in the month of September he had not yet applied it to the case of curved rays.[17]

He decided to add his new discoveries regarding the properties of curved rays in a brief appendix to the second part of the "Theory of Systems of Rays." Finally, just before publication of his complete table of contents, he added a note regarding the Principle of Constant Action for anisotropic media as well as the statement that he had "an analogous principle respecting the motion of a system of bodies"—all this hidden away in the table of contents.[18]

In 1830 Hamilton added two supplements to his "Theory of Systems of Rays." Both papers were condensed from longer treatises that he had worked out in manuscript. The first supplement, which treats the most general case of rays in a heterogeneous anisotropic medium, replaced the unpublished third part of the "Systems of Rays." Hamilton found that the characteristic function (V) as a function of initial and final coordinates was often less convenient than a function of the initial coordinates and the final *direction cosines* of the ray. This new function he denoted by W and employed it in both the first and second supplements to investigate crystalline media and the theory of thin lenses.[19]

He continued to work extremely hard on optics through 1831 and 1832. Ten years later, while engaged in one of his periodic efforts to arrange his papers in some kind of order, he suddenly uncovered all his labors from the year 1831: "And now I must confess that a kind of awe fell upon me, somewhat like to that of him who in the morning saw with terror some precipice which in the night he had, unfearing, passed. So great was the mass of paper! representative of so much labour undergone! I felt even a *compassion* for my own old self, ... I am absolutely startled by the amount of calculation, respecting Optics, which I committed to paper in the same year."[20] The year 1832 was no less toilsome. His hopes of marriage were again dashed and he responded by an almost fanatical devotion to work. He told Wordsworth that he had become "most studious and hermit-like," and wrote to the Countess of Dunraven (the mother of his pupil Lord Adare) that he was "sitting up and getting up later than ever, and grown so much of a hermit that unless I find a pair of garden shears in some of my few visits to the garden, my beard, which already defies razors, will rival the chins of the old philosophers."[21] And to Aubrey De Vere he called his work a "mathematical trance": "I am writing a Third Supplement to my *Theory of Systems of Rays,* and have been engaged in it for the last few days to a most unearthly and Egerian degree: a structure of piled equations rising like an exhalation to my view."[22]

The product of all this labor was a series of investigations into the aberrations of optical systems of revolution (which was published for the first time in Hamilton's *Mathematical Papers*) and a third supplement to his "Theory of Systems of Rays."[23] The third supplement is not really a supplement at all, but a complete treatise in its own right. Starting with the definition of the characteristic function as the action (or optical length) of the path, it develops the theory in its greatest generality. By September, 1832, he had almost finished it, and reported to Lord Adare:

Great masses of my manuscripts I have, after examining their contents, and sucking out their marrow condemned to the flames: and have written out for the press, in a form which I really think I will let stand with perhaps verbal alterations, a large part of the tenth or twentieth copy of my Third Supplement. The various delays and interruptions have made this Supplement more complete, by giving me time to render the subject more familiar to myself and more of a whole: many old and new separate investigations have gradually arranged themselves better in subordination to my general view.[24]

He read portions of the third supplement to the Royal Irish Academy on January 23 and October 22, 1832. Certainly the great bulk of the paper was not finished until the second date. He signed the introduction June, 1833, indicating another eight-month delay before its final appearance in the number for the year 1833.[25]

In the third supplement Hamilton again extended the generality of his system and introduced another "auxiliary" function, the T function, which characterizes the optical system by the directions of both the incident and final rays. The method that Hamilton worked out in his third supplement is still the most general method available for analyzing the properties of optical instruments. Uncle James had warned Hamilton in 1827, when he was getting ready to publish for the first time, that publishing the "Systems of Rays" in the *Transactions* of the Royal Irish Academy was little better than committing it to "a tomb."[26] For once Uncle James knew what he was talking about. When Hamilton became president of the Royal Irish Academy he was astounded to find that not even the French Académie des sciences received their *Transactions*, nor did most of the other major academies on the Continent. Hamilton advertised his third supplement by sending copies directly to a large number of scientists, but these were soon filed away, so that after Hamilton's death in 1865 the optical papers were again unknown or certainly little known. He had learned his lesson by 1834, when he published his papers on mechanics. These appeared in the *Philosophical Transactions* of the Royal Society and were widely distributed.

An anomolous result of this uneven exposure occurred sixty years later, when H. Bruns discovered his famous "Eikonal" method of geometrical optics. The Eikonal was Hamilton's characteristic function resuscitated, but not from Hamilton's optical papers, which Bruns had never seen. Bruns had essentially worked backwards from Hamilton's theory of the characteristic function in mechanics to reach an optical analogy, without realizing that Hamilton had traversed the same analogy in the opposite direction nearly a century before![27] In his desire to bring luster to his country and to its leading scientific institution, Hamilton committed his ideas to an unwarranted obscurity that lasted until the end of the century. It was equally unfortunate that the great value that he placed on abstract reasoning prevented him from recognizing the value of his practical studies of instrument design. These he never published at all.

6

Conical Refraction

THE THIRD SUPPLEMENT was not all wasted effort, in spite of its obscurity, because in applying his theory to the wave theory of light, Hamilton had made a startling discovery. In certain crystals that have two axes of symmetry, the wave front following the theory of Fresnel has four conical cusps, or dimples, located symmetrically relative to the optical axes. Hamilton calculated that these cusps are such that if the wave surface makes contact with a plane, the cusps will touch the plane in a circle "somewhat as a plum can be laid on a table so as to touch and rest on the table in a whole circle and has in the interior of the circular space, a sort of conical cusp."[1]

This wave surface that Hamilton describes is defined as the position of the wave front a short time after it leaves a point source of light. In a homogeneous isotropic medium, the wave surface is a sphere with the light source at its center, but in a crystalline medium the light will divide into two surfaces, one spherical and one ellipsoidal. The sphere is the surface of the "ordinary rays," while the ellipsoid is the surface of the "extraordinary" rays. In a crystal with a single optic axis (the optic axis is the direction in which the ordinary and extraordinary rays move with the same velocity), the two surfaces are a sphere and a spheroid with one surface inscribed within the other so that the inner surface touches the outer surface at two points. (Iceland spar and quartz are examples of uniaxial crystals.) These two points lie on the optic axis. In 1818 David Brewster discovered that there are crystals (such as aragonite) that have *two* optic axes. In these "biaxial" crystals the wave surface is much more complex. It is again composed of two surfaces, one inside the other, except that at four points the surfaces pass through each other forming the conoidal cusps that Hamilton describes.[2]

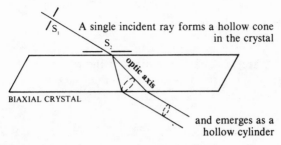

Fig. 6.1. Internal conical refraction. S_1 and S_2 are movable pinholes.

Augustin Cauchy had suggested that in biaxial crystals a light ray might break up into three rays, rather than the usual two rays of double refraction, but he predicted that the third ray would probably be too faint to be observed.[3] Hamilton found that he could better understand the properties of the Fresnel surface by relating it to another surface, already discovered by Cauchy (unknown to Hamilton), which he called the *surface of components.*[4] It was in studying the reciprocal relationship between these two surfaces that Hamilton discovered the fact that at each of the four cusps, the wave surface touches the tangent plane in a full circle. This means that if a biaxial crystal is cut with two parallel faces and a ray strikes the upper surface at the correct angle, it will break up into a hollow cone of rays within the crystal and emerge from the lower face of the crystal in a hollow cylinder.[5] Hamilton called this *internal conical refraction* because the cone is formed inside the crystal (figure 6.1). If, instead of a single ray of light, a cone of rays is made incident on the crystal at the correct angle (the incident cone can be created by a simple lens), then the cone of rays will become a single ray inside the crystal traveling in the direction of the center of the cusp, and will emerge from the lower face as a hollow cone. Hamilton called this *external conical refraction,* because the cone is formed outside the crystal (figure 6.2).[6]

Although Hamilton's "Theory of Systems of Rays" is a theory of geometrical optics, and therefore not of physical wave optics, we have seen that it gives directly the surfaces normal to the rays, which in the undulatory theory represent the successive positions of the wave front. In the case of double refracting crystals, however, the velocity of a light ray depends not on its *position,* but on its *direction.* Under these circumstances the congruence of rays is no longer normal and the wave front is not perpendicular to the rays. The Principle of Constant Action no longer holds, but Hamilton, treating the most general case in his third supplement, came up with a principle equivalent to his Principle of Constant Action that holds for the anisotropic medium of biaxial crystals.

$$\delta V = \delta \int v ds = \frac{\partial v}{\partial \alpha}\, \delta x + \frac{\partial v}{\partial \beta}\, \delta y + \frac{\partial v}{\partial \gamma}\, \delta z,$$

where α, β, γ, are the direction cosines of the ray at point x, y, z, and $v(x,\, y,\, z,\, \alpha,\, \beta,\, \gamma)$ is the index of refraction—a function of both the position and direction of the ray. Because

$$\delta V = \frac{\partial V}{\partial x}\, \delta x + \frac{\partial V}{\partial y}\, \delta y + \frac{\partial V}{\partial z}\, \delta z,$$

we can set

$$\frac{\partial V}{\partial x} = \frac{\partial v}{\partial \alpha} = \sigma \qquad \frac{\partial V}{\partial y} = \frac{\partial v}{\partial \beta} = \tau$$

$$\frac{\partial V}{\partial z} = \frac{\partial v}{\partial \gamma} = v.$$

V is again the characteristic function, and the σ, τ, v, are Hamilton's abbreviated symbols, which he calls *components of normal slowness*. They are called components of normal slowness because they are *reciprocals* of the components of velocity in the moving wave front.[7]

At the beginning of the third supplement, Hamilton showed that by the identity relation between the direction cosines, it is possible to write a single partial differential equation $\Omega(\sigma,\, \tau,\, v) = 0$ completely describing the most general optical system.[8] Hamilton takes this equation to represent a surface that he calls the surface of components. It was an important surface for him to investigate because it came directly from his optical theory. Because the components are reciprocal to the ray velocity,

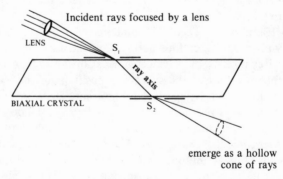

Fig. 6.2. *External conical refraction. S_1 and S_2 are movable pinholes.*

the surface Ω is *not* the wave surface. But the two surfaces are related by simple properties that allow one surface to be easily constructed from the other. The surface of components also has cusps, and by the reciprocity of the two surfaces, Hamilton was able to see that the existence of a cusp on the surface of components meant that there was necessarily a circle of plane contact on the wave surface. Cauchy had already used Hamilton's surface of components, but he had not seen the particular relationship between the two surfaces that Hamilton found.[9]

Hamilton realized that if his prediction of conical refraction could be verified experimentally, it would be a major triumph. Nothing like it had ever been observed in the long history of experimental optics, and the fact that he had deduced it mathematically without any hints from experiment made his prediction all the more dramatic. He gave an oral report of his discovery to an evening meeting of the Royal Irish Academy on October 22, 1832, and the next day he asked Humphrey Lloyd if he would try to verify the prediction experimentally.[10] Humphrey Lloyd was son of Bartholomew Lloyd, the creator of the new mathematics curriculum at Trinity College. He had succeeded his father as professor of natural and experimental philosophy in 1831 when Bartholomew became provost. Humphrey Lloyd was the author of a *Treatise on Optics* and had done a substantial amount of research himself, so he was the person to whom Hamilton would naturally turn for confirmation of his theory. During the next week Hamilton continued to refine his prediction by giving the angles and shape of the cone for the crystal aragonite. He and Lloyd decided to employ aragonite because it was more highly refractive than many other crystals and because the angles of the optical axes had been carefully determined by Rudberg. Unfortunately Lloyd had only a thin plate of the crystal (about one-twentieth of an inch thick), and he could not get a large enough separation between the ordinary and extraordinary rays to make reliable measurements.[11]

Hamilton was chafing against the delay, and in a letter of October twenty-fifth to George Biddell Airy he dropped an obscure remark about having deduced some interesting results from Fresnel's wave theory.[12] When Lloyd's further efforts failed for want of an adequate crystal, Hamilton became impatient and told Lloyd that Airy was asking for information. He said that "if you should, as you seemed to think likely, be prevented by want of apparatus or of leisure from making soon any decisive experiment on the point, I believe it will be well to mention the theoretical result to Airy."[13] Lloyd replied that he had now discovered that his sheet of aragonite not only was too thin, it was a "macled" crystal, composed of several crystals crossing each other, and therefore entirely unsuitable for the experiment, and he agreed that Hamilton should write to Airy and let him attempt the verification.[14]

Lloyd did not give up, however, and by December he had obtained a good crystal of aragonite from Dollond in London. On December 14 he wrote to Hamilton that with this new crystal he had found the cone.[15] Lloyd had observed *external* conical refraction, in which a cone of rays incident on the crystal passes through as a single ray and emerges as a hollow cone at the other face. Four days later Hamilton wrote to Airy and Herschel giving all the particulars, and on the same day Lloyd wrote to say that he had now obtained a striking demonstration of the cone by projecting it on a screen and had obtained a circle up to two inches in diameter.[16] Because the light ray used in the experiment is not a mathematical ray, but must have some angular aperture, the circle observed in conical refraction is actually *two* concentric circles with a fine dark ring between them. In Lloyd's less precise experiment these two circles were blurred together, and he probably observed a single thick ring of light.[17] It was a moment of great triumph for Hamilton and Lloyd and also a time for them to quickly establish their priority and publish their results.

The next day, December 19, 1832, Hamilton wrote to Lloyd that he would apply to the Royal Irish Academy "for leave, and am sure the leave will be given, to publish a notice in the Annals of Philosophy or some such place, without waiting for the slow appearance of our volume."[18] Two disturbing circumstances made haste important. A letter soon arrived from Airy, stating that he had long known of the existence of the conical cusps on the wave surface, but that he had not realized that there were also circles of plane contact at the cusps. Airy added the additional information that the light rays forming the cone should be polarized in a peculiar fashion. Rays on opposite sides of the cone should be polarized at right angles, which means that as one moves around the cone through 360 degrees, the plane of polarization of the rays moves through 180 degrees or exactly *half* as fast (figure 6.3). It was this peculiarity of "conical polarization" that puzzled Airy, because it seemed to him that it made the mathematical problem indeterminate. The ray passing through the crystal could be resolved into any pair of rays or into any number of rays when it emerged from the other side. "The whole construction for determining a ray's direction fails as I con-

Fig. 6.3. Planes of polarization for rays undergoing internal conical refraction.

ceive. ... I think that the wave surface must be abandoned, and that the expressions for velocity alone must be retained."[19]

Lloyd had observed some strange chromatic effects associated with conical refraction, but he had said nothing about the polarization effects in his letter to Hamilton of December 18. In his published account, Lloyd did describe the polarization phenomenon, but he made no mention of Airy's prediction stating that he had discovered it experimentally by viewing the cone through a tourmaline plate.[20] There is no reason to believe he was not telling the truth, but it had been a close call. Airy had nearly walked off with at least a portion of the prize. Hamilton then proceeded to bear down hard on the analytical problem. He was already immersed in a detailed study of Cauchy's theory, and announced on January 2, 1833, that he had calculated the planes of polarization for the rays forming the cone.[21] Lloyd found a mistake in the calculations, which threw Hamilton into a bit of a tizzy, but he soon discovered that his slip had come only in the final calculation and that the theory as a whole remained sound.[22]

At this point Airy responded with a disheartening blast:

Now that, among the optical oddities of crystals, there may be such a thing [as conical refraction] is (for aught that I can see) very possible: and if it is real its laws would be a most interesting subject of experiment and speculation. But it has no connection with your theory. This I say without the least doubt, and you will see it with the slightest consideration. You will see that a single ray in the crystal can produce but one linear ray in the air. So that I am now puzzled more than ever about the phenomena which Prof. Lloyd has found.[23]

But by this time Hamilton had the theory under control and was no longer worried about unpleasant surprises from England. He soon set Airy straight, and received the admission that his theory was "all comprehensible and all true." Airy said that if he had "not been very dull, [he] might perhaps have guessed at some of it before."[24] By this time Lloyd had also verified *internal* conical refraction, and Hamilton's triumph was complete.[25]

Airy was not the only competitor close on the track of conical refraction. James MacCullagh, Hamilton's and Lloyd's colleague at Trinity College, had given an elegant geometrical construction of Fresnel's wave surface in 1830. His paper had been reviewed by none other than William R. Hamilton himself![26] MacCullagh also came up with interesting results that might have put him ahead of Hamilton in the research on the wave surface. In December, 1832, Lloyd told Hamilton that MacCullagh had found that in order to obtain the full cone of rays it was necessary that the ray passing through the crystal be unpolarized. Hamilton hastily responded on New Years Day that this was "exactly the same with the conclusion which I had formed in October, and I distinctly remember mentioning it to you in our interview on the 23rd of that month."[27] Three days later, in writing to Airy, he mentioned that "MacCullagh (another of our young Fellows, a paper by whom I once showed you) has deduced the same results by his geometrical methods, having however previously heard of my theory of conical refraction."[28] Hamilton was obviously worried about establishing his priority.

MacCullagh did not entirely agree with Hamilton's recollection of events, and the July issue of the *Philosophical Magazine* contained the following note from him:

When Professor Hamilton announced his discovery of Conical Refraction, he does not seem to have been aware that it is an obvious and immediate consequence of the theorems published by me, three years ago. ... The indeterminate cases of my own theorems, which, optically interpreted, mean conical refraction, of course occurred to me at the time; but they had nothing to do with the subject of that paper; and the full examination of them, along with the experiments they might suggest, was reserved for a subsequent essay, which I expressed my intention of writing. Business of a different nature, however, prevented me from following up the inquiry.[29]

Hamilton saw in this note the implication that MacCullagh had recognized the possibility of conical refraction and had publicly expressed his intention of pursuing the subject, but that Hamilton had rushed into print ahead of him. Hamilton was understandably miffed, and he prepared a rejoinder "in a somewhat satirical vein," which he showed to Lloyd. Lloyd responded in alarm and quickly acted to head off an unpleasant controversy. MacCullagh was induced to write another note to the *Philosophical Magazine* that was essentially a retraction. He admitted that he had not noticed the circles of plane contact until Hamilton pointed them out to him.[30] Even so it had been another close call. Hamilton did not realize just *how* close it had been until after MacCullagh's death in 1847. While going through MacCullagh's papers he confessed to George Salmon that MacCullagh "seems to have been very near finding the theory [of conical refraction] for himself."[31]

After this first brief clash with MacCullagh Hamilton became obsessively cautious about questions of priority. When his third supplement finally appeared in 1837, he was careful to give credit where he believed it was due, but even then he derived the formulas for polarization of the rays forming the cone without mentioning that Lloyd had first discovered them experimentally.[32] Of course Lloyd had already published his discovery, and there was no reason for Hamilton to credit him with the solution of the analytical problem.

Modern physicists are unlikely to get excited about conical refraction. It is a curiosity, comparable to the many other optical phenomena discovered in the early nineteenth century that are tucked away in the corners of modern textbooks. Hamilton himself called it a "subordinate and secondary result" in writing to Coleridge, and George Stokes pointed out that it did not even prove the correctness of Fresnel's theory, because one can construct other surfaces approximating Fresnel's that predict the same results.[33] At the time it was announced, however, conical refraction electrified the scientific community. Airy called it "perhaps the most remarkable prediction that has ever been made," and Whewell claimed that "in the way of such prophecies, few things have been more remarkable than the prediction of [conical refraction]." In a general study of wave surfaces, the German mathematician Plücker wrote, "No experiment of physics ever made such an impression on me ... it was a thing unheard of and completely without analogy." He added that he would never have expected a theory as abstract and complicated as Hamilton's to be confirmed experimentally in such a precise manner.[34] Herschel immediately responded in the same way. As a prediction it was almost without parallel, and it was commonly compared to the discovery of Neptune by Adams and Leverrier later in the century. In 1835 it won for Hamilton the Royal Medal of the Royal Society (Faraday was the other recipient of that year) and a knighthood from the lord lieutenant of Ireland.[35]

III
Marriage

7

Philosophical Romance

DURING THE 1830s Hamilton led an extremely active life. These were his most productive years, for he worked on optics, dynamics, and algebra all at the same time. It was also during this decade that he made his most thorough study of philosophy. He seemed to have almost boundless physical energy. We read of him walking the parapet of the observatory roof, climbing the stack of a steam packet while crossing to England, jumping in the water fully clothed in wild abandon at Adare Manor, rolling hoops in the observatory garden with Robert Graves, and riding his horse, Planet, in wide galloping orbits about the meadows.[1] Hamilton retained the habit of taking long walks, always with a book in his pocket, or sometimes even with a sack of books. At night the books accompanied him to his bedroom, where he slept among them and would wake and read at any hour.[2]

In all of his pursuits, whether in science, philosophy, or poetry, one can see his great love for the abstract and the ideal. In his poetry, for instance, he followed the style of the English romantics, but because he always remained in the domain of the abstract and almost never employed concrete images, his poetry lacked the force that characterized poems by his heroes, Wordsworth and Coleridge. His optics and mechanics were also highly abstract, and he searched for the foundations of algebra in Immanuel Kant's metaphysics.

The same love for the abstract and the ideal can be seen in his personal relationships, especially in his relationships with women. There is an aura of artificiality about all of his courtships. The emotions they produced were honest enough, but Hamilton always expressed his love in a theoretical way. Love existed for him in some unearthly realm apart from this world, and that is probably why his love for Catherine Disney remained so vivid throughout his life. As the years went by that love became increasingly idealized in his own mind to the point that it stood alone, pure, inviolate, and invulnerable to the circumstances of daily life.

During the 1830s Hamilton also made close friendships with three young men, Francis Beaufort Edgeworth, Viscount Adare, and Aubrey De Vere. They were all younger than he and they were all "metaphysically" inclined. Edgeworth and De Vere also had sisters who attracted Hamilton's attention at one time or another. Francis Edgeworth was the stepbrother of Maria Edgeworth. In 1828 Maria wrote to Hamilton asking if he would accept Francis as a pupil at the observatory. Francis had been studying at Trinity College Cambridge, but had found the atmosphere there thoroughly distasteful.[3] His inclinations were towards aesthetics and metaphysics, while his tutors had wished to cram him full of mathematics. Hamilton would seem to have been a singularly inappropriate tutor for anyone revolted by mathematics, although Maria may have hoped that Hamilton would be able to cure Francis of that particular distaste. In any case Hamilton had to refuse Maria's request, because he had already agreed to accept as pupils the two sons of the Marquis of Anglesey, lord lieutenant of Ireland. Since the sons were only ten and twelve years old, it is not clear why the lord lieutenant settled on Hamilton, who had no experience teaching children and could hardly have been expected to introduce them to advanced mathematics. He reluctantly agreed to accept the

Francis Beaufort Edgeworth

Aubrey De Vere

two boys, recognizing that they would take much of his precious time, but recognizing also that he could not refuse.[4] Although Hamilton said that he expected no favors from Anglesey's patronage, he must have recognized the advantage of having the lord lieutenant's support at the beginning of his career. The only favor he actually seems to have received was a basket of pheasant at Christmas time.[5]

Hamilton's decision not to accept Francis as a pupil did not, however, damage their friendship. Without the onus of having to prepare for examinations, they soon became close friends. It was in his arguments with Francis that Hamilton worked out his ideas on the relationship between science and poetry. Francis was a thoroughgoing Platonist, even rejecting Christianity and modern science in favor of the idealistic vision of the *Timaeus*. He argued that truth and beauty could be found only in the unity of an ideal world, and not in the inductive process of modern science, which focuses on bits of observed nature but never on the whole.[6] Francis had the same criticism to make of mathematics, because he regarded mathematics as merely a tool for the pursuit of inductive science. On this point, of course, Hamilton was eager to set him right. He concluded that any disciple of Plato would soon take a different view of mathematics after a brief exposure to geometry.[7] However, Euclid failed to arouse the passion in Francis that Hamilton had expected, and he finally had to admit that although he found it "difficult, certainly, to conceive a mind so different from my own as to feel no beauty in mathematics after it has begun to invent and create," he agreed that if other interests gave Francis no time to gain the facility necessary to appreciate mathematics, he could not blame him for abandoning the subject.[8]

Though he failed to convert Francis, Hamilton continued to share Francis's idealism and felt strongly the need to resolve the apparent conflict between his own two great enthusiasms—science and poetry. In 1829 William Wordsworth decided to visit Ireland for the first time. Hamilton had sent Wordsworth some of Francis's poems and had described him as a young man possessing an "amiable but uncommon mind." In a postscript to his letter he urged Wordsworth to come to the observatory.[9] Hamilton also forwarded an invitation from the Edgeworths asking Wordsworth to stay at Edgeworthstown during his tour. Wordsworth accepted the invitation, but almost did not come because of reports of serious disturbances in the Irish countryside. As the time for Wordsworth's arrival neared, Hamilton and his sisters reached a high pitch of anticipation and anxiety. At the last moment a storm struck Dublin, ripping away the door from the dome of the observatory and soaking the house below.[10] "We conclude Wordsworth will not come to Ireland in this weather," Eliza reported to Cousin Arthur, "but we must be prepared I suppose. —*William* has not I think given up hopes of him yet."[11] And then suddenly Wordsworth appeared. Eliza, who kept the record of visitors to the observatory, wrote an account of his stay. He and her brother took long walks along the Tolka and through the observatory grounds, discussing poetry and philosophy.

Hamilton was eager to find out Wordsworth's real beliefs about science and poetry, because passages in Wordsworth's poetry seemed ambiguous. Sometimes Wordsworth appeared as an enemy of science. He hated those men who "murder to dissect" and "peep and botanize on their mothers' graves."[12] But he never issued a wholesale condemnation. His criticism was directed against men of small minds, who would sink deeper into the minutiae of matter and fact, rather than rise up to God through contemplation of His universe. In book 4 of the *Excursion* Wordsworth envisioned a new science freed from the "false conclusions of the reasoning power."[13] In an allegory of a child listening to the sound of the sea in a shell, the child's ear is faith, the shell "the universe itself" resounding to the "authentic tidings of invisible things, of ebb and flow, and ever-during power."[14] The scientist who seeks to reveal this harmony is truly worthy of his calling.

> . . . Science then
> Shall be a precious visitant; and then,
> And only then, be worthy of her name:
> For then her heart shall kindle; her dull eye,

> Dull and inanimate, no more shall hang
> Chained to its object in brute slavery;
> But taught with patient interest to watch
> The processes of things, and serve the cause
> Of order and distinctness . . .[15]

Hamilton and Wordsworth held long debates over this portion of the *Excursion*. Hamilton challenged Wordsworth to explain his slight reverence for science. Wordsworth said that he was not opposed to science that "raised the mind to the contemplation of God in works," but that science as a bare collection of facts, or science applied only to material uses of life, was science that "waged war with and wished to extinguish Imagination." That kind of science degraded man. He particularly condemned the popular enthusiasm for "useful knowledge," which he believed had no connection with God and was therefore debasing.[16] Science that flowed from the imagination was the only kind of science that Wordsworth could appreciate.

Hamilton agreed with these criticisms. He hated "useful" knowledge as much as Wordsworth, but he believed also that Wordsworth set too narrow a limit to what was acceptable in science. For Wordsworth the imaginative faculty that was the source of poetic inspiration was the only creative faculty of man. Hamilton insisted that imagination should be expanded to include the *intellect*, a faculty integrally tied to the imagination and one that allowed man to rationalize his vision of harmony into the forms of mathematics. The language of mathematics, according to Hamilton, was as creative and as universal as poetry. He insisted that truth, the goal of the mathematician, and beauty, the goal of the poet, represented two different views of the same nature. Truth, for Hamilton, was the self-consistency or consistency of an object with its place in the universe. Beauty was the fitness of an object to excite tender emotion. The two were quite different, but had an intimate connection in nature.[17] Truth and beauty, the intellect and the imagination, revealed unitary *powers* in nature. Hamilton called these powers "living spiritual energies" or "unific energies"; it was through a discovery of the powers behind the phenomena that he believed an intellectual unification of nature would be possible.

Hamilton almost certainly borrowed his distinction between truth and beauty from Coleridge, who wrote in the *Biographia Literaria* that poetry "seeks pleasure through beauty, science through truth."[18] Poetry, according to Coleridge, flows from the imagination, while science is controlled by reason.[19] Hamilton did not agree with Coleridge, because he did not wish to separate science so completely from the imagination. He believed that the intellect was subordinate to and was included in the imaginative faculty. Thus he believed that science, stemming from the intellect, was also a product of the imagination.[20] In a later introductory astronomy lecture he quoted extensively from Wordsworth and stated that his purpose was "to point out the latent imagination which is involved in the processes of Science":[21] "In all ... physical science, we aim not only to record but to *explain* appearances; that is, we aim to assign links between reason and experience; not merely by comparing some phenomena with others, but by showing an analogy to the laws of those phenomena in our own laws and forms of thought, 'darting our being through earth, sea, and air.' And this appears to me to be essentially an imaginative process."[22] And in a lecture the following year he stated his position even more strongly:

As to the imagination, it results, I think, from the analysis which I have offered of the design and nature of physical science, that into such science generally, and eminently into astronomy, imagination enters as an essential element; if that power be imagination, which 'darts our being through earth, sea, and air'; and if I rightly transferred this profound line of our great dramatist to the faculty which constructs dynamical and other physical theories, by seeking for analogies in the laws of outward phenomena to our own inward laws and forms of thought. Be not startled at this, as if in truth there were not beauty, and in beauty no truth; as if these two great poles of love and contemplation were separated by a diametral space, impassable to the mind of man, and no connecting influences could radiate from their common centre. Be not surprised that there should exist an analogy, and that not faint nor distant, between the workings of the poetical and the scientific imagination; and that those are kindred thrones whereon the spirits of Milton and Newton have been placed by the admiration and gratitude of man. With all the real differences between Poetry and Science, there exists, notwithstanding, a strong resemblance between them; in the power which both possess to lift the mind above the stir of earth, and win it from low-thoughted care.[23]

Hamilton, who was unwilling to sacrifice poetry for science or science for poetry, insisted that they flowed from the same source in the imagination. It was a belief that he held all his life. Helena Pycior has pointed out the fact that the eulogies of Hamilton that appeared shortly after his death all referred to his mathematical imagination, equating it with the imaginative faculty of the poet.[24]

Hamilton insisted that a *science*, as that term is properly used, is a pure creation of the intellect grounded in reason alone. A science thus reflects "our own inward laws and forms of thought" rather than any observed regularities of phenomena. It is deductive rather than inductive. But he also recognized that some sciences, such as logic and mathematics, can obviously be used to describe the physical world. They can be so used because a wonderful accord exists between the laws of thought and the order of the external world. For Hamilton this accord was the work of a wise and benevolent God. It was a theme that he would repeat in his discussions with Whewell about dynamics, and in his system of philosophical triads, which he created ten years later, in 1842. Hamilton was completely serious when he called Lagrange's *Mécanique analytique* a "scientific poem." He believed that poetry and science were akin because they both employ the creative imagination to construct links between the intelligent thinking self and the external world. The goals of truth and beauty are in fact only one goal—there is truth in poetry and beauty in science—poetry and science are complementary expressions of the same imaginative faculty.

These were all arguments that Hamilton had worked out in his discussions with Francis Edgeworth, and Wordsworth quickly recognized the Platonic influence on Hamilton's thought.[25] Wordsworth would have commended Hamilton's attempt to construct a science conformable to a poetic view of the universe, but would not have followed him so far into idealism.[26] Wordsworth's poetry was too concrete, too much concerned with "pedlars and spades," to allow a philosophical interpretation such as the one that Hamilton wished to impose upon it.[27] Hamilton read Kant enthusiastically and absorbed ideas of the *naturphilosophen* through Coleridge; Wordsworth denied any German influence on his own thinking.[28]

Despite their differences, Hamilton persisted in his efforts to incorporate Wordsworth's poetic view of nature into his own philosophy. Wordsworth's search to comprehend the "ever-during power" of the universe became a moral stimulus for pursuing scientific research. In his introductory lecture of 1831 Hamilton referred to Newton as having pursued the same goal by creating a universe of multitudinous but unified energies, reducing the infinite physical world to an intellectual one. "To me," he concluded, "the wonder and sublimity of millions of miles or millions of years is gone; thought has so far outstripped reality that all existing magnitude has dwindled to a point."[29] For Hamilton, then, science was a *mental* activity that succeeded only to the extent that it reduced the external world to forms of thought. Wordsworth would never have strayed so far into the abstract.

Throughout 1829 another Irish nobleman, the Earl of Dunraven, began to make discreet inquiries about Hamilton's willingness to tutor his son the Viscount Adare. Dunraven was much more tactful than the lord lieutenant had been. Anglesey had peremptorily summoned Hamilton to the vice-regal lodge; Dunraven made his wishes known through some of Hamilton's close friends. The young Adare was seventeen years old, a recent graduate of Eton, and a thoroughly likeable young man. He wished to study science, and like many of the Irish nobility he had a passion for astronomical observation.[30] At first Hamilton resisted, but when Provost Kyle of Trinity College put pressure on him he reluctantly agreed. Adare and Hamilton were in correspondence about his studies through the fall of 1829, but because his mother was ill, Adare did not arrive at the observatory until February 10, 1830.[31] He immediately began a rigorous program of observing in the meridian room, as is attested by his frequent entries in the observatory records. He and Hamilton were apparently working with *too* great industry, because Adare was soon having trouble with his eyesight, which forced him to curtail his observing and even his reading, and Hamilton suffered from ill health, the news of which brought a stern warning from T. Romney Robinson at Armagh not to keep such long hours.[32] As a diversion they went together to Armagh to visit Robinson and to look at his instruments. Hamilton was industrious on his pupil's behalf once he got to know him, and obtained his election to the Royal Irish Academy, and then, through Herschel, to the Royal Astronomical Society. Adare was also able to help Hamilton, by bringing him into the homes of prominent nobility both in Ireland and in London when they went there together in 1832.

At Armagh, Adare introduced Hamilton to Lady Campbell, one of his old acquaintances. Lady Campbell was a daughter of the Irish patriot Lord Edward Fitzgerald, and an extremely attractive woman. She and Hamilton talked about the "decline of science" debates and about Coleridge's philosophy and literary criticism. Soon they were on friendly terms. Also living in the neighborhood were Edward Disney and his sister Catherine, now Mrs. William Barlow. It took a special effort for Hamilton to get to see them because the journey had to be made on horseback and was, as he reported to Cousin Arthur, "rather long for an inexperienced horseman."[33] Hamilton returned from his visit visibly agitated and depressed. Catherine had seemed sorrowful. Hamilton imagined that she still loved him, and apparently with some reason, because she soon repaid the visit by coming to the observatory at Armagh. He offered to show her the telescope in the dome and found himself alone with her for the first time since 1824. He was so agitated that when he tried to demonstrate the use of the telescope he broke the wires in the eyepiece.[34] He could say nothing, do nothing. She was married, maybe not happily, but with children about her. His code of honor was such that no intensity of feeling, no matter how great, could bring him to speak about his attachment to her.

Of course he wrote a poem about his feelings. It ends:

> I follow more in patience than in joy;
> Sadly contented, if I may endure
> Life, and in gentle calm await the grave.[35]

But even "gentle calm" evaded him. He could not fight his way out of the depression. Lady Campbell discovered what had happened and managed to lift his spirits enough to keep him from doing anything drastic. Although Hamilton later wrote at great length to many people, both men and women, about his first love, Lady Campbell was the only person outside of the Disney family to whom he revealed Catherine's name. Hamilton thought no one else knew, but considering the volume of his correspondence and the fact that the Disneys and their wives were a numerous clan, it is inconceivable that some of his other correspondents did not figure it out. He decided that Lady Campbell would serve as a tie between himself and Catherine. His plan was to have her cultivate Catherine's friendship without telling Catherine that it was at his initiative. In that way Lady Campbell was to become "a connecting link, a bond of sympathy, a being that we both shall love and that shall have added to the happiness of both." The more he thought about it, the more pleased he was that Lady Campbell had "divined some particulars" of his history that he had sought to conceal.[36] Unfortunately she soon moved to the vicinity of Dublin and could not become the "connecting link" that Hamilton had hoped for. But in Dublin Hamilton saw her more often and gained the advantage of having his sole confidante close by.

In July of that year (1830) Hamilton set out on another journey, this time with his sister Eliza to spend three weeks with the Wordsworths at Rydal Mount. It was a special treat for Eliza, who was allowed to closet herself with Wordsworth in his summer house and receive careful criticism of her poetry. She also met Wordsworth's daughter Dora, with whom she later kept up a long correspondence. The visit consisted of long walks to scenic spots in the lakes and mountains, games of "the Graces" (a nineteenth-century version of charades), and trips to the residence of Lord and Lady Lonsdale, Wordworth's patrons at Lowther Castle, and to Southey at Keswick. They also met Felicia Hemans, who came to Ambleside at the same time. Mrs. Hemans had a substantial reputation as a poetess, and Eliza was eager to meet her, but first impressions were not encouraging. Mrs. Hemans (she was always Mrs. Hemans) was precious and affected in her manner. Men liked her better than women. She had a delicate and ethereal beauty that, when combined with ill health and a tragic and mysterious past (her husband had left her), gave her the character of a suffering saint. Within a year she came to Dublin and was a frequent visitor at the observatory. Robert Graves, who met her at the observatory, was enraptured, but not Eliza. She wrote to Dora Wordsworth that "at present it seems to me that her desperate desire to produce an effect is in danger of injuring the qualities of her heart which must have been, I am sure, originally amiable and good natured."[37]

Hamilton and Eliza were having such a good time at Rydal Mount they extended their visit for an extra week.[38] It was an exciting time for both of them. Hamilton's sagging spirits were lifted and the sense of shared adventures brought brother and sister even closer together. When the steamer reached Dublin in the early morning they walked to the observatory together, a distance of some five miles. Hamilton's letters to Wordsworth after the visit reveal a lightheartedness and enthusiasm that had been missing at Armagh.

Soon after his return to Ireland Hamilton was traveling again, this time to Adare Manor to meet Lord and Lady Dunraven. The visit was in part an obligatory one, but Hamilton had a good time nonetheless. Lord Dunraven was becoming immobilized by the gout and had begun a project for rebuilding the manor house. Lady Dunraven was heiress to a large property in England and had sufficient wealth to support her husband's ambitious program of building. The work began soon after Hamilton's first visit and was continued by Adare after his father's death. The result was one of the grandest houses in all Ireland.[39] The new construction had not yet begun during this first visit of Hamilton's, but the old house and grounds were already grand. Part of the entertainment was a boat trip along the River Maigue, which flows through the grounds past a ruined castle and an old abbey. Lady Dunraven played her harp and one of Adare's friends from Eton played the flute as they drifted along. Pauses between pieces performed in the boat were filled by music from the shore, played by a bugler appointed to follow the boat. Although Hamilton had little appreciation for music, the romantic setting moved him greatly.[40]

A year later Hamilton was back at Adare Manor, this time for a longer stay, and with Cousin Arthur as a fellow guest of the Dunravens. It turned out to be a riotous visit. Adare's youthful friends brought out the most playful side of Hamilton's character. He shot the falls of the weirs in the river with William Smith O'Brien (who was later to lead the uprising of 1848). When O'Brien fell in the river, Hamilton joined him, fully clothed, in a splashing vagary of wild hilarity. He was even more excited by the dancing, which he described in detail to his sister Grace:

This dance illustrates what in astronomy we call the *centrifugal force,* for it required exertion on the part of the gentlemen to prevent their partners, who were outermost, from flying off from the center and from them. Some did so fly off . . . by the rapidity of the motion. . . . We closed the ball by a Coronation Dance, which began with a gentle and solemn music; with motion corresponding; but soon grew fast and furious, till the first couple had held a handkerchief for the ladies to dance under and for the gentlemen to leap over.[41]

There was a serious side to the visit, however. Ever since his trips to Rydal Mount and to Adare the previous year, Hamilton had begun again to contemplate marriage. He made no specific declaration of intent, but he was obviously more aware of the marriageable young women he met. Something he said to Maria Edgeworth made her think he was engaged, and she passed the information around. The rumor reached Wordsworth early in 1831, and Hamilton was put to some trouble to stop it.[42]

In August, 1831, George Biddell Airy visited the observatory and Hamilton was especially eager to introduce him to a Miss Helen Bayly, a cousin of the Rathbornes, who lived at Dunsinea House, across the fields from the observatory. He also thought Helen and her mother might like to see the moon and Jupiter, in case Airy did not prove to be enough of an attraction.[43] Helen was definitely on his mind at the observatory, but at

Adare she was quickly replaced by Ellen De Vere. The De Veres lived at Curragh Chase, near Adare, and were frequent visitors of the Dunravens. Hamilton had met Ellen briefly in 1829 at the house of a neighbor near Dublin, but this time he was able to talk with her at length and to visit her at Curragh Chase.[44] Soon he was writing poems to her, and confessed to his sister that he had been "in another world" for two days since he met her.[45]

Dora Wordsworth described Ellen in a way that confirms the supposition that Hamilton was attracted by fragile and sensitive women:

What an interesting creature she is, but "I think of *her* with many fears"—She is indeed too sensitive, too spiritual a being "to tread the rugged ground, inevitable, of Life's wilderness:" and how she is wrapped up in her brother Aubrey! and tho' it is beautiful to witness, one cannot think without trembling of the sorrow and distress, even this may bring upon her—but it is wicked to anticipate evil—to draw bitter water out of so lovely a fountain—once I would only have looked on the bright side of this fair picture—but years rob one of one's youthful gladness—and now I am too prone to dwell out upon the gloomier picture—so you must forgive me.[46]

Somehow Hamilton persuaded himself that he admired Ellen only for the spirituality of her soul and for her highly refined taste for poetry. He told Wordsworth that he "admired her mind" and that his own feelings were quite platonic, adding that he would not soon again endanger his "philosophic calm."[47] After the pain of losing Catherine Disney, he was not prepared to admit his own feelings. But Wordsworth set him straight with characteristic bluntness: "To speak frankly, you appear to be at least three-fourths gone in love; therefore, think about the last quarter in the journey."[48] Wordworth's advice started Hamilton thinking, and in December he went back to Adare, this time with marriage on his mind. Lady Dunraven invited him to come and gave him encouragement; in fact she was actively engaged in matchmaking, and Adare also favored the marriage. Ellen came to Adare while Hamilton was there and they were able to talk privately. Ellen seemed to be truly fond of him and Hamilton was on the verge of proposing, when she said that she could "not live happily anywhere but at Curragh." Hamilton took these words as a refusal.

When Robert Graves was writing his biography of Hamilton, Ellen was still alive, and he asked her about her relations with Hamilton. Only Graves's side of the correspondence survives, but it is clear that Hamilton should have pressed his suit more vigorously. As in the case of Catherine, Hamilton turned to Ellen's mother to ascertain Ellen's true feelings and received what he believed was a discouraging reply. The reply, as he described it to Eliza, was that although "the friends of Miss De Vere would approve of and desire the union" Lady De Vere knew "every feeling of her daughter's mind" and she could not give "any hope" that Ellen would accept him.[49]

Graves had to admit that these actions did not seem to be "quite the part of a manly man," and at the very least "a mistake in judgment."[50] After receiving the disappointing news Hamilton sent Ellen the following letter from Adare:

In the hints by which you have so kindly sought to reconcile me to your decision, I can see nothing to diminish my desire for our union, or to shake my belief in our fitness for each other; and this continuing and increasing belief and desire, might at times without anything in your manner misunderstood, produce a momentary hope which would uselessly harass and exhaust me. Perhaps, too, I might have the weakness to afflict you by unavailing importunity. Even now I can with difficulty refrain from urging you to tell me, whether the obstacle to our going forth together as companion-spirits on life's way, is one which by any efforts or after any interval, I might hope to remove or surmount. If, as you seem to suggest, those efforts could not be successful it would be a criminal desertion of my duty to neglect any chance of preserving my energy of mind, by removing myself from all outward remembrances of that which I cannot forget. My ambition has long since been satiated, and my income is sufficient for the wants of myself and of those whom I love; when I part from you, I shall feel that I have neither wish nor hope; but I have a disciplined fortitude, a sense of duty, and a habit of study; and of these supports I ought not to risk the weakening by lingering within the sphere of fascination. . . . Be assured, that if I understand myself, my desire for your happiness, is as ardent as that for my own; and that if it were only for your sake, to spare you pain and to preserve your esteem, I shall exert myself to the utmost, that after we have parted, I may not sink under grief.[51]

It appears from this letter that Hamilton had at least talked seriously about marriage with Ellen, but one cannot be sure. He may never have gotten to the point, and may have taken a "hint" from Ellen to mean a refusal. He told Eliza that Ellen had become aware of his feelings only a few days before he spoke to her. It apparently never occurred to him to give her time to consider his proposal. Eliza, who claimed to have "an imagination like a second sight" was convinced that Ellen had another suitor in the background.[52] Hamilton himself thought Ellen would never marry and would never leave Curragh.[53]

The May following his distressful visit to Curragh Chase, Hamilton was hurt in a fall from his horse. Ellen wrote to Eliza in concern, saying, "You must persuade him that it is his positive duty to his country, as well as to his friends, to take the greatest care of a life so valuable. Oh do not let him be so giddy again, dear Miss Hamilton."[54] Such a letter might be taken to indicate a change of mind. Lady Dunraven, who was still hoping to arrange a marriage, wrote to Hamilton that Ellen still showed a *great* interest in him, but she had to admit that there was still family opposition to the marriage, especially from Ellen's youngest brother, Aubrey.[55] Hamilton did not act on any of these hints, however, and Ellen did leave Curragh Chase in 1835, when she married Robert O'Brien, brother of

William Smith O'Brien, with whom Hamilton had romped in the river at Adare.[56]

Considering the fact that Ellen's brother Aubrey strongly opposed her marriage to Hamilton, it is surprising that Hamilton made such a conscious effort to cultivate his friendship. The De Veres were well connected. Aubrey's and Ellen's father, Sir Aubrey De Vere, was a baronet descended from the Earls of Oxford, and their mother was sister of Thomas Spring-Rice, first Lord Monteagle (secretary of war and colonies, 1834, and chancellor of the exchequer, 1835 to 1849). Young Aubrey had two older brothers who succeeded to the title and land, but Aubrey himself was sufficiently well-off to be able to pursue his poetic and metaphysical interests without financial worries. When they met, Hamilton had just turned twenty-six and Aubrey was only sixteen. One might have expected Hamilton to have found one of the older brothers a more congenial companion, but, just as in the case of Francis Edgeworth, Hamilton was attracted by Aubrey's youthful brilliance and his love for metaphysics. Even while he was courting Ellen he was occasionally sidetracked and spent his time "metaphysicizing" with Aubrey instead of talking to Ellen.[57] He and Aubrey became close friends and soon began a long correspondence on poetical and philosophical subjects. At first Hamilton expected Aubrey to act as an emotional link with Ellen, whom Hamilton could not give up entirely. In between talk about poetry and metaphysics Hamilton and Aubrey discussed the practical and ideal aspects of love. After Ellen's refusal Hamilton at first found himself in a strange state, "floating along on the stream of circumstances ... purified from hope and fear, and climbing to a region of unshaken peace [where] the perfect development and harmony of my other faculties to which I have since ascended, and which, I fondly trust, will lead to a deeper tranquility."[58] But soon he descended to the more practical world and began to consider who might be the subject of his next attachment. Hamilton thought it would be someone he had met at a time when he had "mixed more in society," and he contemplated falling in love with one of Catherine's sisters. Unfortunately he remembered them as mere children, and had so far regarded them with only "brotherly affection."[59] Aubrey's theory of love required that Hamilton find a woman with a mind "entirely unworldly and disinterested." Such a mortal would have lived almost entirely in seclusion so as not to be spoiled by mixing with that "universal leveller, society." He suggested several candidates.[60]

Aubrey's theory of love was elaborate. He believed every person sought to complete himself or herself by an affinity to another person possessing those qualities missing in his own character.[61] Thus love was striving for completion of oneself through another. At first Hamilton did not respond. He had sought an antidote from his disappointment in plunging himself into work on his third supplement to the "Theory of Systems of Rays" and in allowing himself to go to seed, staying up all night and letting his beard grow. Aubrey and Adare dragged him out of his rut and persuaded him to go to London in the company of Adare. This was the trip on which he first met Herschel and introduced himself to the British scientific community. It also included a visit to Coleridge. Hamilton obtained an autograph for Ellen; he had not forgotten her yet. And his poetry indicates a revived hope, soon to be dashed by an agonizing dream in which he found himself lying helpless on the ground; Ellen approached him, ignored his entreaties, and walked by. He wrote a poem about it of which he was much ashamed. He told Aubrey that it was "weak and morbid" and "very imperfect."

> Methinks I am grown weaker than of old,
> For weaker griefs prevail to trouble me.
> In dream last night I lay beneath a tree,
> And things around me many a half-tale told,
> Which for a while I could interpret not,
> And knew not where I was, until I heard
> Approaching footsteps, and my heart was stirred
> By power of Voice and Image unforgot.
> Languid and faint I lay, and could not rise;
> She, when she saw me, cared not for my pain,
> But passed on, with unregardful eyes.
> O that I were my former self again!
> Might not the struggle of the Day suffice?
> Must Night add visions false of cold disdain?[62]

Hamilton told Aubrey that he was saved from despair by always being able to work, even while he passed "nearly eight years in a state of mental suffering."[63] In October, 1832, he was visited by an old college classmate, and they spent the evening in riotous rhyming and comic storytelling. The next day he wrote to Aubrey that he could "scarce do anything but laugh" after the "sea and tempest of laughter" in which he had been tossed.[64] He also laughed when Aubrey came to visit him in the same month, but it was hollow laughter. He wrote a poem about this event, too. It ends:

> But of the bitter past we spoke not—no,
> We might have seem'd with mirthful fancies fraught;
> For once we laugh'd, laugh'd! but the rocks around
> Returned that laughter with a ghastly sound.[65]

All of his poems from this period are filled with negatives and dwell constantly on the themes of weakness and helplessness. Hamilton appeared to be in a desperate state, and he may have feared that marriage would evade him forever. In December he began to think again of Helen Bayly, the "girl next door." From the beginning he told Aubrey that she was "not at all brilliant" but that he had "long known and respected and liked" her. There was no mention of love.[66]

Years later Hamilton looked back on the three great attachments of his life, and concluded that he had felt three different kinds of love towards them. Towards Catherine he had felt himself a lover, towards Ellen he had been a "brother," and to Helen he had been a husband. He did not insist that these three emotions were mutually exclusive, but he implied as much.

On April 9, 1833, he married Helen Bayly.

8

Helen Bayly

HELEN BAYLY REMAINS a shadowy figure in Hamilton's life in spite of all the references to her in the manuscripts. Before their marriage Hamilton recognized that she was likely to be chronically ill and that she was almost morbidly shy and timid. Robert Graves, who knew her from frequent visits to the observatory, was harshly critical. She had no "striking beauty of face or force of intellect," could not "manage well the concerns of a household," and was totally unable "to exercise a controlling influence over the habits of [her] husband."[1] As Graves explained to Ellen after Hamilton's death, Hamilton was attracted by her deep piety. "This, I feel certain, is the explanation of his proposal to her; he considered that *piety* to be her all-sufficient qualification. You will see in the end his recognition of her excessive timidity. It was the timidity of one conscious of her utter disparity. . . . After marriage the poverty of her mind and the whole nature must soon have revealed itself to him as not to be ameliorated by all the riches of cultivation which he could bestow upon it."[2]

But Graves's judgment was probably not fair. He idolized Hamilton, and was all too ready to blame his hero's weaknesses on the "lack of controlling influence" from his wife. Certainly Hamilton's problems were exacerbated by Helen's frequent illness, her frequent absences from the observatory, and her general inability to cope with the problems of life, but they were hardly her fault. Nor could she have ever replaced the idealized visions of Catherine Disney and Ellen De Vere that Hamilton always retained in his mind. Just before proposing to Helen, he wrote a revealing sonnet.

> O be it far from me, and from my heart,
> Praising a new love to dispraise the old,
> As if I had before but false tales told,
> In hasty error, or in flattering art!

The ancient images shall not depart
From my soul's temple; the refined gold,
Well proved, shall there remain, though newer mould
Of worth and beauty fill another part.
Sweet Piety, Enthusiasm, Truth,
A several grace for every several brow,
Deck three fair Beings; one in earliest youth
Placed in that fane, one later, and one now:
But sister-like they twine, in love and sooth,
And all in each receive my spirit's vow.[3]

Helen knew that Hamilton married her "on the rebound," so to speak, from two previous romances, and that in beauty, fortune, intellect, and health she was greatly inferior to both Catherine and Ellen. It is not surprising that she married him reluctantly and after much vacillation. There was also the problem of Hamilton's sisters Grace, Eliza, and Sydney, all of whom were still at the observatory. Hamilton seemed to think that they might stay on after his marriage, and was distressed that they all chose to find lodgings elsewhere. Eliza, at least, had grave misgivings about his intended bride. She had felt twinges of jealousy during all of his previous attachments, or so he thought, but her dislike of Helen was more overt. She must have written frankly to Dora Wordsworth, who responded: "I suppose you are now no longer an inmate of the Observatory, and I confess we grieve over this very much more if possible for your brothers's sake than yours. And I will *not* offer my congratulations on an event which has made you think it expedient to leave him as I could not do so with sincerity, but I will say that most sincerely do we trust that his fairest hopes may be realized and your *affectionate* fears prove groundless."[4]

Helen was from Nenagh in County Tipperary, where she lived on a farm with her widowed mother. The family belonged to the minor gentry of Ireland. They lived comfortably but were not wealthy. Helen had many brothers and sisters, which meant that she was left with a meager dowry—a matter of greater concern to her than to Hamilton. There were some problems with the marriage settlement. Hamilton insisted, "I never made fortune in a wife any part of my theory of marriage," and confessed that through inexperience his income seemed "to have flowed out as fast as it flowed in." His intention was to show Helen that he had little concern for money and would love her without it. She was concerned that he took the matter so lightly.[5]

Two of Helen's sisters, Penelope and Jane, had married two brothers, William and Henry Rathborne, who lived at Dunsinea and Scripplestown, both large country houses just across the fields from the observatory. Helen had met Hamilton on visits to her sisters, and the fact that she would be living at the observatory, so close to members of her family, must have been an important factor in her decision to accept his proposal.

When ill, or when Hamilton was away, or when the burden of managing the observatory household became too great for her, she often went to stay with her mother at Nenagh or to Dunsinea or Scripplestown.

Hamilton courted Helen with a forcefulness that was altogether missing from his previous romances. In November, 1832, he made a mighty effort of will to pull himself out of the depression caused by Ellen's rejection. He considered it a "point of duty and conscience" to repel the "mental gloom and languor" that had come over him. On November 12 he wrote to Aubrey that by resolute determination and hard work he had largely succeeded, and was beginning to perceive the "dawn of a new love."[6] A reply from Aubrey four days later was strongly approving, particularly since Hamilton had praised Helen's "truth of character," a quality that Aubrey had placed at the "very foundation" of his theory of women.[7] The next day Hamilton wrote to Helen's mother at Nenagh, asking permission to court her daughter. He made a point of confessing that in the previous year he had "met and become attached to another person," but it had come to nothing. He wanted Mrs. Bayly to know that he had refrained from approaching Helen until his "imagination and affections" were sufficiently free from any previous commitment. Now he wished to propose marriage, but recognized that his success would require Mrs. Bayly's support.[8] She gave her full consent, although she may have asked him to be sure that his former attachments were quite over, because he replied by sending her letters from Aubrey to himself, showing that any possibility of a marriage with Ellen De Vere was quite out of the question.[9] Mrs. Bayly was widowed and elderly, and she had a daughter whose prospects for marriage were slim. She had met Hamilton and had liked him. His proposal must have been welcome, and she did her utmost to persuade Helen to accept.[10]

Having attained Mrs. Bayly's support, he spoke to Helen. Hamilton thought she gave her consent, at least by implication, and was consumed by ardor. The next day he wrote to her:

At last then I may write to you, and call you mine! For are you not *mine* now, in thought, in feeling, by the unuttered vow, by the first kiss of love? . . . Doubtless we shall have sufferings as well as enjoyments, but it is not now that I can realise fear, or think that any bitterness of the future can equal that of the past, or easily persuade myself that any cloud of gloom can soon again overspread me. . . . I began to fear that you wished that we should indeed part forever, and were studying to soften the rejection. The cloud was upon me for but a short time indeed, but in that time I lived over again the two most bitter moments of my past life; the moment when an expression of Miss De Vere's brought home to me first a fear with respect to her; the moment when, alone with the mother of Catharine Disney, I heard the words, "She is going from us, she is going to be married." I remembered all the protracted pain of which those words had been the mournful herald; and I seemed to see a new array of coming hours of pain, a new succession of secret struggles with grief, the breaking up of a new seal which had imprisoned a new fountain of anguish. ... And soon you rewarded me for all ... by your tacit confession the next evening, when we were again alone together. ... Then, for the first time, I touched the lips of any but my nearest relatives, and was filled with an unquiet joy and a trouble sweet, and you too were deeply disturbed, and I kissed away the tear from your cheek, and in the moment that we were together I seemed to enjoy a happiness more exquisite but more untranquil than any I before had known."[11]

At age twenty-seven Hamilton had exchanged his first kiss. To a man of his romantic temperament it was a moment of high passion; it was also a moment of triumph, for he had finally overcome the oppressive fear of another rejection. He frightened Helen, however. She had been preparing to return to Nenagh, but the excitement made her ill and she had to stay in Dublin. She would not permit Hamilton to come to her, and he was full of remorse for having caused her agitation by his own "want of self-control."[12] He could only calm her by saying that he understood her to have given him a profession of love, but not a positive promise of marriage.[13] Helen's illness made her doubt even more her fitness for marriage. She had been ill at Scripplestown during most of the summer of 1832, but illness had only stimulated Hamilton's chivalrous nature and had made him more intent on marriage. He now assured her that he had carefully considered the possibility of her habitual ill-health and that it did not change his feeling for her. If her illness was an affliction on him also, it was an affliction that "our Heavenly Father may have provided ... as a successor to the now finished pain of unreturned affection. That pain has preyed upon me for almost nine years, a third part of my life. It is over now." Helen's love had relieved him of suffering far greater than any that she might cause him in the future. He welcomed the opportunity to share her burden. "Suffering with you who suffer, I may taste more fully than before the consolations by which you are refreshed. And thus chiefly it may be that my hope shall be fulfilled of deriving religious improvement from attachment to a pious wife. ... And thus indirectly but powerfully, may you be conducive to my spiritual progress; and not the less for my having to temper the happiness of our recent and mutual confession of attachment with a mournful sympathy in your sufferings of body and mind."[14]

Helen returned to Nenagh having promised an answer soon after Christmas. Her Rathborne sisters were concerned that she was leaving Hamilton on tenterhooks and wrote to Mrs. Bayly to let her know how matters stood.[15] There was no fear, however, that Hamilton would change his mind, or that he would retire again in the face of minor discouragements. Even his unsuccessful attempts to teach Helen optics and astronomy did not discourage him this time. He still insisted that she had an aptitude for science, and wrote: "Don't fear that I shall change this opinion if I find when we meet that you had forgotten almost all of what we lately went over together. ... I shall remain attached to you even if this little dream should be dissipated of our holding many such conversations hereafter." He also reassured her that he would never force her to go to Court and that they would lead a retired life at the observatory.[16]

True to her word, Helen made her decision shortly after Christmas, and Mrs. Bayly wrote Hamilton telling of her acceptance. He insisted on going to Nenagh at once to see her, even though she urged him not to come. On this and on a subsequent visit in February he appeared very forceful, probably because he had the full support of Helen's mother and sisters. He wrote long letters to Helen to which he receive infrequent and brief replies. He addressed her as "my dearest Helen" and she addressed him as "my dear Mr. Hamilton" and "highly esteemed professor."[17] Helen thawed a bit with time, and as their wedding approached their correspondence took on a teasing and affectionate manner. Lady Dunraven suggested that Hamilton take Helen on a honeymoon trip to Killarney, and stop for some time at Adare Manor on the way or returning. He would probably have accepted, but Helen objected and he immediately agreed with her "that it is foolish in newly married people to go rambling through places which they may never see again, instead of having the first pleasant associations with the home where they are to spend their lives."[18] As it turned out their honeymoon was spent at Bayly Farm, where they stayed for two weeks after a quiet wedding. Helen or her mother had wanted to have the ceremony at home, but Hamilton found that the Church of Ireland frowned on marriages performed outside the church itself. He wrote to Mrs. Bayly that "a marriage in church is considered as a private one, when friends are not invited," and presumably that is how the ceremony was carried out.[19] Hamilton traveled to Nenagh alone; there is no indication that any of his sisters or other close friends or relatives attended the ceremony.

Helen felt insecure in her new position. Frequent references to her timidity are confirmed by her unwillingness to leave Bayly Farm and by her eagerness to extract promises from Hamilton that they would not go into society. She was anxious to make an impression, however, and when Lord Adare came to visit the observatory, she wanted to have one of the servants from Bayly Farm come to supplement the observatory staff. She also hoped that he would bring the Bayly livery; it was something that she would "like the people here to see."[20] This particular servant possessed a dignity that had amused Hamilton. He had gone out of his way at Bayly Farm to impress upon Hamilton the "grandeur" of the Bayly family and its "extensive connexions and possessions."[21] Although Helen usually left little impression behind her, the anecdotes that remain in the family indicate that she was concerned with position and that problems with the servants were not uncommon.[22]

Hamilton's intellectual enthusiasms must also have made her insecure. As a lady she could properly claim ignorance of astronomy and optics, but not of literature. In his enthusiasm Hamilton discoursed in learned fashion on Shakespeare and Chaucer, Greek mythology, and the theory of love.[23] He seemed not to notice that Helen did not respond. When he told her that he was sending her an edition of Shakespeare, she told him "not to think" of sending it because it would be far too expensive a business. "I do not know any place where one is more tempted to spend money in than a booksellers, and I am afraid our small fortune will not [admit] of making indulgences of that kind as well as giving or receiving invitations to dinner parties."[24]

Hamilton must have recognized Helen's feelings of inferiority, because he tried to reassure her that her brief letters were welcome.

> How can you fear that a letter can be uninteresting to me, when it comes from one to whom I have given my heart, and am soon to give my hand? It is not brilliancy that makes one prize a letter from a friend, and what friend can be dearer or closer than a wife? ... But how unwise it would have been if through any mistaken delicacy you had denied me and may I not say yourself the pleasure and profit of correspondence—profit, for it is now a point of prudence that being engaged to each other we should study each other's character, and so increase the likelihood of our future happiness.[25]

With the exception of Eliza, Hamilton's sisters had not shown any extraordinary brilliance, either, and he must have been used to holding one-way conversations with them about his work, but they were capable managers of his household. Grace and Sydney had done a substantial amount of astronomical work, taking transits and reducing them for the records. Usually Grace ran the household. Hamilton once said that if he were told that the observatory was on fire, he probably would respond, "Well, go and tell Grace."[26] In the few months before the wedding Mrs. Bayly gave Helen lessons in housekeeping, but Helen had trouble from the beginning.[27]

When they arrived at the observatory on April 25 Hamilton's three sisters were gone, and so were two of the three servants.[28] The one remaining servant had also given notice, but Hamilton persuaded her to stay until Helen could select a replacement. Housekeeping was made more difficult by the work going on in the observatory. Two huge pillars of granite were being erected right through the house to support the new equatorial in the dome. In June Hamilton left for the British Association meeting in Cambridge. Helen was ill when he left, and it must have been a difficult time for her.

She was often ill, with various complaints that were never precisely described. Before their marriage she wrote about her "awful situation" and complained of complete loss of appetite, restless nights, bad headaches, a painful hip, and severe heart palpitations.[29] Later letters suggest that her illnesses were of a "nervous" sort, but it is difficult to know precisely what that description might mean. She bore three children with no serious physical problems, although each birth seems to have been followed by a period of severe depression.

For Hamilton his marriage was a moment of great triumph. At last his courtship had been successful, and, at the same time, he had crowned his optical researches with the discovery of conical refraction. He was working extremely hard, sometimes laboring through the night. Even at Nenagh he was in constant correspondence with Lloyd about conical refraction. He returned from Cambridge a famous man. He had had to cut his trip short because of Helen's illness, but once he was home she began to improve and there were plans for a visit to Nenagh in July. It was put off, however, because Helen had "not yet been able to put her housekeeping on so satisfactory plan, as to make it quite comfortable to her to leave home," and because Hamilton's trip had been costly.[30]

In September Helen's condition grew worse again, although not severe enough to alarm the doctors.[31] She remained surprisingly cheerful and declared that "my darling William is the sweetest nursetender in the world," and that "his whole happiness seems to be in making others happy; indeed any woman is blessed to be married to such an affectionate kind creature as Hamilton."[32] She was proud of his successes as reported in the newspapers.[33]

Mrs. Bayly suspected that her daughter's illness was caused by pregnancy, and that Helen might be concealing the fact, but Helen reassured her: "I am afraid if I am carrying a child at all it is a dead one, for I would have quickened long ago, and there is no sign of life or size whatever, rather the contrary."[34] Mrs. Bayly's intuition was sound. Helen was already two months pregnant when she wrote the above letter. Mrs. Bayly visited the observatory some time in late October, and it is clear from the correspondence that Hamilton was even more eager to have her come than was Helen, who worried about the chaotic state of the house.[35]

Their first son, William Edwin, was born on May 10, 1834. Letters to Wordsworth and Aubrey De Vere show that the summer was a happy time. Hamilton did not have much idea of how to communicate with a newborn baby. He recited to him Wordsworth's *Ode on Intimations of Immortality* and imagined that the baby liked it, but had to confess that Helen's singing drew a more pronounced response.[36] He told Aubrey that mother and child were "both quite well, and going to her mother in the south."[37] Hamilton was planning to attend the British Association meeting at Edinburgh in September, so it was a good time for Helen to visit. The "visit" extended for nine months, largely because of Mrs. Bayly's poor health. Hamilton went to Nenagh after returning from Edinburgh, and then again for Christmas. College lectures and observatory business forced him to shuttle back and forth between Dublin and Nenagh until the end of the following May (1835). Mrs. Bayly's illness must have been severe, because Hamilton prepared to contact the family lawyers in anticipation of her death.[38]

The only letter from Helen is one written to Hamilton at Edinburgh. It is petulant and cross. She had little to say about the meeting except for the comment: "I am sorry to find you are to be Secretary next year, as I fancy it will be a troublesome office. . . . When next you write let your letter be franked if you can, as my money is all gone. Your nurse costs me 5 or 6 shillings a week for bread and candles, besides I was obliged to buy him a small bedstead."[39] Mrs. Bayly's much more cheerful letter, written on the same sheet, describes William Edwin's antics.

Hamilton found the separation from Helen extremely difficult. He was used to the companionship of his sisters; living in an empty house (even the servants were gone) made him feel lonely. His housekeeping arrangements were primitive. During March Helen and her mother sent him a giant "goose-pye," which he reassured them would "save . . . all trouble about dinner." It was three weeks arriving and it sat in the post office for some time because it was "too heavy and unhandy to be brought out on horseback." It must have been ripe when he began to eat it, but he voiced no complaint, probably because he was able to supplement the pie with dinners at Dunsinea and Scripplestown.[40] He wrote to Aubrey: "It disarranges me much, I find, to live long in so lonely a place as the Observatory still remains awaiting the return of its lady; and even for withdrawing into solitary thought, I want a companion from whom to withdraw. For this, or some such reason, I find my visits to the south, and the weeks that I have spent with my wife and child at the cottage of her mother, much more propitious to study than the intermediate intervals at home."[41] Whatever inconvenience Helen's absence may have caused, it did not hamper Hamilton's work nearly as much as he thought. During these months alone at the observatory he completed much of his work on dynamics and his algebraic theory of number couples.

Hamilton brought his family home to the observatory at the end of May, two months before the British Association was scheduled to meet at Dublin. As one of the secretaries (the other was Humphrey Lloyd), Hamilton was busy making arrangements. To complicate matters, Helen was pregnant again. Their second son, Archibald Henry, was born on August 4, 1835, just six days before the meeting convened.[42] Hamilton could have had little time for either wife or son, and of course Helen attended none of the elaborate banquets and receptions, nor was she present when her husband received his knighthood from the lord lieutenant in the

Long Room of the college library. Hamilton seemed happy in the following months. He extolled the virtues of marriage to Lord Adare, who was planning to marry Miss Goold, the girl that Hamilton had whirled so vigorously during the dance at Adare Manor, and he took obvious delight in William Edwin, who was now old enough to make some kind of communication with his father. Hamilton followed closely his son's struggles with language. William Edwin, known otherwise as "Pinkie," "P.," or "poor Boo," was a constant source of fascination.

In the fall of 1836 the events of 1834 repeated themselves. Hamilton went to the British Association meeting at Bristol, and Helen took the two children to Bayly Farm, where she stayed for ten months. It was another difficult period for Hamilton, but this time he spent more time at Bayly Farm himself, and had a long visit from Aubrey at the observatory, which cheered him greatly. Again the ostensible reason for Helen's absence was her mother's ill health. Hamilton brought his family back to the observatory in June of 1837. This time, when the British Association meeting came around in September, Mrs. Bayly came to stay at the observatory with Helen, although she was still quite ill. Hamilton left for Liverpool soon after Mrs. Bayly arrived. He probably did not realize just how sick she was, but she had some premonition that she would not live much longer. Earlier in the year she had felt the need to make excuses for Helen's frequent absences from the observatory and wrote: "[Helen] is so fearful you should be jealous of her great attachment to me, but that can never interfere with her love for her husband; it will cement it, should such a material be wanting ... Helen says we *three* are united by one chord, when that *chord* is touched it sounds *a unison perfectly in tune.*"[43] And on the eve of her trip to the observatory she wrote: "Helen is so dear to me I get into low spirits when I am not well with the fear of leaving the world without seeing her once more."[44] It was Mrs. Bayly who wrote to Hamilton in Liverpool to tell him about their activities and to send a detailed account of their household expenses.[45] After the Liverpool meeting Hamilton visited Adare at Dunraven Castle in Glamorganshire, apparently unaware that Mrs. Bayly had become gravely ill.[46] When he arrived at the observatory he found her dying.

After the death of Mrs. Bayly, Helen's name appears less and less frequently in Hamilton's correspondence, but there is no indication of any immediate change in their way of life. Hamilton accepted the presidency of the Royal Irish Academy in November, 1837, and with it the burden of administrative chores and social obligations. He made three trips to England in 1838, the first in June to attend the dinner for John Herschel, who had just returned from the Cape of Good Hope, the second in August to attend the British Association meeting at Newcastle and to visit Wordsworth on the way, and the third in October to visit the Marquess of Northampton at Castle Ashby. Lord Northampton was active in affairs of the

Sir William Rowan Hamilton and Lady Hamilton (née Helen Bayly)

British Association and had become friendly with Hamilton from their frequent conversations at the annual meetings. He extended a special invitation to Hamilton and Helen, and, for once, Helen accepted. They were at Castle Ashby for three weeks in November, the only excursion of that nature that Helen ever made. Hamilton reported to Cousin Arthur that they were "pleasantly and fully occupied from morning to night," and that every possible attention had been paid to them, "especially to Helen, to prevent her from feeling shy."[47] Possibly because Helen had just lost her mother, Hamilton felt an especially strong attachment for her at this time. He described to Lord Northampton his return in September from the British Association meeting:

I kissed my sleeping children; then walked down the hill, across two slightly sloping fields, to the home of a married sister [Dunsinea], where it had been arranged that my wife should dine and sleep during my absence, the Observatory at night appearing somewhat large and lonely. She was sleeping with a favourite niece, but woke as I entered the house; dressed in the dark, and soon was with me in the drawing-room. We had, you may imagine, much to talk of; and all that day I thought that she was looking very well. ... My children met me in the fields on our way homeward; they both were wild with delight, and the usually copious (too copious) flow of words of the eldest was changed at first to an inarticulate stammer of happy wonder. ... Dearly did they enjoy the rummaging of my travelling bags, and the searching for things which they might keep as memorials of my last visit to England.[48]

Hamilton kept up his good spirits in spite of the constant demands placed upon him. In addition to time spent on the problems of the Royal Irish Academy, he had the task of trying to settle Mrs. Bayly's affairs. She had asked Hamilton to act as executor of her estate, but he had wisely refused, not wanting to become embroiled in Bayly family matters. One of Helen's brothers contested the will, and Hamilton found that in spite of his resolution to remain aloof he had to look after Helen's interests in the matter. He bent his efforts to obtaining an amicable settlement to prevent "the whole family being exposed by such a paltry squabble."[49] From this time forward Hamilton was constantly called upon to settle estates of deceased family members. He was not particularly good at it and never seems to have gained financially himself, but when Cousin Arthur died in 1840, Hamilton succeeded him as the head of the family and had the responsibility thrust on him whether he liked it or not.[50]

On the whole, the year of 1838 was a happy one at the observatory, but the following year was much more somber, although it is difficult to tell why. On New Years Day, 1840, Hamilton reviewed the events of the previous year.[51] Numerous illnesses—those of Eliza, Helen, Cousin Arthur, and his children—all added to the gloom. A summer of rain and violence in the countryside, particularly the murder of Lord Norbury, had cast a shade on everyone's life. Hamilton had given up all visits, did not go to the British Association meeting, or even to Wicklow for a holiday, and spent more time in religious and metaphysical study. Religion in particular involved his interests. Bible study, religious conversations with Helen, and family prayers became more important in his life. Eliza also wrote an account of the events of 1839 that was exceedingly grim. It was a long list of assassinations, fires, deaths, illnesses, and unfortunate marriages, ending with the statement: "about 17th Decbr, William became a flaming Puseyite!" the first indication that Hamilton's religious interests were connected with the Oxford Movement, which was getting underway at that time.[52]

In the spring of 1840 Helen was again pregnant, but this time she could not go to Nenagh. As the agitation for repeal grew during the late 1830s there was a sharp increase in agrarian crime and terrorism in the countryside. Helen was terrified at the thought of staying at the observatory, even with her husband at home, and Hamilton found lodgings for her in Dublin.[53] Apparently those lodgings had to be given up in August, before the baby was born, for Hamilton tells of visiting "perhaps thirty" places at the last minute, none of which was suitable.[54] Helen returned to the observatory and their only daughter was born there on August 11, 1840. She was named Helen Eliza Amelia after her mother and two aunts. Helen did not make a good recovery. For three weeks she stayed upstairs and soon left the observatory to live with her sister at Scripplestown.[55] Hamilton's letters all mention his concern for her, her sickness, and the fact that he had been able to do little work. The death of Cousin Arthur in December, 1840, pushed him into even deeper gloom. In late January, 1841, Helen went to stay with her sister Charlotte Dana in Shrewsbury. In writing to his old tutor, Charles Boyton, Hamilton apologized for his inactivity, and admitted that he had been in "deep dejection."

I have done very little, for a whole year past, that is, since Lady Hamilton's health obliged her to leave the Observatory—though historical accuracy would require me to state that she came back in August for her last confinement, finding no place in Dublin so quiet for that purpose. But since the christening of our little daughter in September she has not been here at all; and for the last three months has been residing with a married sister in England, which country she, not unnaturally for a timid lady, prefers to Ireland.[56]

Helen stayed in England until January, 1842, a period of slightly less than a year, but, adding the time that she lived in Dublin and at Scripplestown, she was away from the observatory for almost two years. Hamilton's sister Sydney came to help out, and the children stayed with Hamilton, so he was not without companionship, but he remained anxious and depressed. The only surviving letter from Helen during this period is a pathetic one, obviously written under the strain of illness. It contains a minimum of business affairs and ends: "I hope my darling children are well and that Baby is still continuing her lessons in walking. I have got such a palpitation I must lay down the pen."[57]

Hamilton's friends noted a change in him during this period. He became more solitary, more introspective, and much more defensive than he had ever been before. He tended to dwell on past accomplishments rather than show enthusiasm for new pursuits, and he began gradually to drink more than was good for him. Lord Adare began to hear unfavorable gossip in England about Helen's continued absence from the observatory and wrote to find out the true state of affairs. Hamilton replied: "Though we have unhappily (through the state of her health) been much asunder of late, Lady Hamilton and I are in the constant habit of correspondence of the most affectionate kind."[58] If the affectionate correspondence did indeed exist, it has been lost.

The functioning of the household improved under Sydney's management, but declined again when Helen returned in 1842. As usually Graves blames Helen for Hamilton's failings. He had no regular meals, sometimes missed meals altogether, and dispersed the chill of the night with glasses of porter, when he should have had a warm fire and hot coffee. Yet Hamilton's problems could not have been caused entirely by Helen's lack of good management, because he was happier with her than without her. Sydney's controlling hand made him more comfortable, but it did not relieve his depression. Only Helen could do that.[59] In spite of her infirmities she remained the central figure in Hamilton's life, and to some extent she drew him into her own habits of seclusion. After Hamilton's death, Humphrey Lloyd said that he had often been at the observatory, but had *never* seen Lady Hamilton. T. Romney Robinson called her "an abstract idea," because she never appeared in the flesh.[60] And yet Hamilton never wrote a word of complaint about his wife. If he felt her weakness of health and character as a burden, it was a burden that he willingly accepted.

IV
Light and Dynamics

9

Theories of Light

HAMILTON'S COURTSHIP and marriage coincided with his discovery of conical refraction and the beginning of his interest in physical optics. The discovery was well timed, because it occurred right when the debate in England over the undulatory theory of light reached its apex. After his marriage, Hamilton traveled little. He never set foot on the Continent. But he did attend the meetings of the new British Association, where the debate over the physical nature of light had its most public forum.

Hamilton's "Theory of Systems of Rays" had been a theory in geometrical optics. It was potentially of great importance for the design of optical instruments, but it did not address the problem of wave propagation or any of the phenomena associated with it, such as interference. From 1832 onwards, however, Hamilton began to study the physical propagation of light and entered the battle on the side of the undulatory theory.

David Brewster of Edinburgh, who had led in the establishment of the association, was the greatest master of experimental optics in the British Isles and the staunchest advocate of the emission theory of light. By the 1830s he had ranged against him the "Cambridge men," John Herschel, George Biddell Airy, and William Whewell. At Oxford, Baden Powell also supported the wave theory. Brewster and his adversaries differed not only on the physical constitution of light, but also on the proper method to be employed in science and on the purpose of the British Association. And all of these subjects of controversy had a tendency to become confused into a single squabble.

In order to understand the controversy it is necessary to look back to some of its antecedents. The undulatory theory of light was revived in the first years of the nineteenth century by Thomas Young in England

and Augustin Fresnel in France, both of whom demonstrated that the phenomenon of interference could be most easily explained by assuming that light is composed of a wave in the transmitting medium. Both Young and Fresnel ran into determined opposition when they attempted to make their theories public. In a famous series of essays in the *Edinburgh Review* from 1802 through 1805 Lord Henry Brougham attacked Young with such vicious polemic that Young never again attempted openly to convert his fellow scientists away from their Newtonian convictions.[1] In France Fresnel ran up against firmly entrenched opposition. Pierre-Simon Laplace, along with his associates Jean-Baptiste Biot, Siméon-Denis Poisson, and Étienne-Louis Malus, dominated the French Academy. Together they formed a body of scientific talent unequalled anywhere in the world, and they held as a common principle the belief that any adequate explanation in science must necessarily assume the existence of material particles with forces of attraction and repulsion acting between them. Laplace stated his position with regard to light in his "Mémoire sur les mouvements de la lumière dans les milieux diaphanes": "In general, all attractive and repulsive forces in nature are reducible in the last analysis to such forces acting between molecules."[2] Laplace concluded that since Malus had successfully confirmed Huygens's laws using the mechanical Principle of Least Action, it followed that light must be mechanical in nature and that there could now be *no doubt at all* that it was particulate. "Until recently Huygens' law of double refraction had been only the result of observation, approaching the truth within the limits of error of the most precise experiments; but now one may consider it to be a rigorous law, since it fills all the conditions."[3] Hamilton made reference to this article by Laplace at the beginning of part one of his "Theory of Systems of Rays" because Laplace, like himself, had employed the Principle of Least Action, but Hamilton was cautious to state that "the manner in which I have deduced [the Law of Least Time] is independent of any hypothesis about the nature or the velocity of light."[4] Laplace was prepared to use all of his vast skill and authority to suppress any theory conflicting with his, and he effectively blocked Fresnel's approaches to the academy. It was only the willingness of Dominique F.-J. Arago to accept Fresnel's new discoveries that allowed Fresnel to circumvent in part the obstacles that the Laplacian physicists placed in his path.

Fresnel began his work in optics in 1814. Most of the optical phenomena upon which any new theory had to be based had been known since the seventeenth century. The one major exception was Malus's discovery of polarization by reflection in 1808. Prior to that discovery, polarization was known only as double refraction in certain crystals such as Icelandic spar. It was a curious but rather special phenomenon, and was consequently much less important than the more common properties of light such as reflection and refraction. Reflection, however, was probably the most common and most noticeable of all the properties of light, and therefore Malus's discovery that reflected light was polarized attracted new attention to the general problem of polarization. Polarization had always been a problem for the advocates of the wave theory, since it required that the waves be *asymmetric* about the direction of propagation. Malus's discovery gave even further encouragement to the advocates of the emission theory (of which he was one), because there seemed to be no way to account for it following Huygens's theory of waves. Careful tests of Huygens's law of double refraction by William Hyde Wollaston in 1802 and again by Malus in 1810 showed it to be correct, but Huygens gave no mechanism for polarization, and it still seemed more reasonable to explain the law following the emission theory. Young, now writing anonymously, admitted that Malus's experiments on polarization by reflection "present greater difficulties to the advocates of the undulatory theory, than any other facts with which we are acquainted," and in a letter to David Brewster in 1815 he wrote, "With respect to my own fundamental hypotheses respecting the nature of light, I become less and less fond of dwelling on them as I learn more and more facts like those which Mr. Malus discovered."[5]

The inevitable rush to study polarization after Malus's discovery revealed other curious optical phenomena, in particular Arago's discovery of "chromatic polarization" and "optical rotary polarization," both of which were attacked vigorously by Biot. With the wind blowing so favorably for the emission theory, Biot set about creating a light molecule that would answer all the demands of the new phenomena. In his "Theory of Mobile Polarization" he gave to the light "molecules" those ad hoc properties that were needed to explain the observations. The light "molecule" had poles that maintained their directions until the molecule entered a crystal, whereupon it would "begin to oscillate about its center of gravity like the balance-wheel of a watch," the violet molecules turning faster than the red ones. When the molecule emerged from the crystal the oscillations immediately ceased and the poles froze into a fixed position.[6] Biot's light molecules exhibit the failing of all the mechanical models of the emission theorists. They were created to "save the phenomena" in a science where new phenomena kept cropping up with startling frequency. Biot could not "save" them all. He had to revise his theory four times in four years, and when David Brewster discovered biaxial crystals in 1818, the oscillating molecules had to be thrown on the scrap heap already swelling with other similar models.[7]

In 1815 Fresnel sent his first paper on the diffraction of light to the French Academy of Sciences through Arago. Arago told him that it contained nothing new except for the theoretical prediction that the position of any given interference fringe would follow a hyperbolic curve as one moved away from the diffractor.[8] This was a prediction at odds with the emission theory. But the "émissionaires" with Laplace at the head felt confident that diffraction phenomena would eventually succumb to an experimental attack following the lines of the particle theory, and they arranged for the biannual prize competition of 1819 to be on the subject. Arago was the only convinced undulationist on the panel of judges (the others were Gay-Lussac, Laplace, Biot, and Poisson), and it seemed unlikely that Fresnel's essay would receive impartial considera-tion, but as it turned out, there was no serious competitor. Fresnel's work was so superior to any of the other essays that it won easily. His theory explained diffraction with a precision that others could not match. Arago also smoothed ruffled feelings by avoiding any mention of waves in his report, always using the term "ray" in place of "wave."[9] In spite of this victory, Fresnel gained no new converts in the academy. Even after he successfully responded to Poisson's challenge of the "white spot," the émissionaires" refused to budge.[10]

After his attack on diffraction, Fresnel, now aided by Arago, began to study polarization. Arago continued to insist on caution. He delayed two and a half years before publishing Fresnel's discovery that two rays polarized in mutually perpendicular planes will not interfere, and even then he argued that the new rules for polarization had the advantage of being "independent of every hypothesis."[11] There was still the problem of explaining how a light wave could have the asymmetry demanded by polarization. As early as 1816 Ampère had suggested to him that the light wave might partake of transverse motion as well as motion along the direction of its propagation, but Fresnel considered it too far-fetched a notion to count for much at the time. The next year Thomas Young made the same suggestion to Arago. Finally, in 1821, the evidence had grown weighty enough to convince Fresnel that he should commit himself to the notion of transverse vibrations, and he took the bold step of declaring that light waves were *only* transverse and had no longitudinal component at all.[12] Since transverse waves are propagated only through solid substances, the theory was startling, to say the least. Even Thomas Young, who had suggested the notion in the first place, had difficulty accepting Fresnel's decision to make the waves *completely* transverse, and declared that the consequences of such a step would be "perfectly *appalling.*"[13]

It was also in 1821 that the uneasy truce at the French Academy finally blew up. For years Arago had been cautious even to the point of suppressing his and Fresnel's findings in order not to offend the adherents to the emission theory, but in 1821 he clashed with Biot.[14] The exchange was so bitter that it was deleted from the *procès-verbal* of the session, and Biot sought a withdrawal of Arago's report. Although neither side scored a clear-cut victory, Biot had been thrown on the defensive. The year 1821 was the turning point in the French Academy for the wave theory of light. By 1823 even Laplace spoke with praise of Fresnel's work, and although he could not bring himself to adopt the wave theory, his new-found generosity finally brought Fresnel into the academy as a member.[15]

From the beginning of his optical studies Fresnel tackled the major phenomena of optics, one after another, explaining them all by his theory of undulations. Starting with diffraction and interference, he had gone on to study thin plates, reflection and refraction, polarization, and finally, after 1821, double refraction in crystals. He began by calculating the wave surfaces for ordinary double refraction in uniaxial crystals, but soon turned to the more difficult case of biaxial crystals. By 1822 he had found the wave surface that Hamilton later exploited so successfully in his discovery of conical refraction.[16] After 1823 Fresnel's health began to fail and much of his time was spent in academic duties and in directing the installation of new lenses of his design in French lighthouses. When he died in July, 1827, he had just received the Rumford Medal from the Royal Institution in London. Although his colleagues would not admit it, it was clear that any future work in optics, if it stood any chance of success, would have to follow Fresnel's lead. The battle was over in the French Academy; in England it was just beginning.

When Hamilton studied light during his college years between 1824 and 1827, he read the books of Coddington, Wood, and Stack.[17] These were the most recent texts available in English, and they all uniformly explained optical phenomena in terms of the emission theory. Even Humphrey Lloyd in his *Treatise on Light and Vision* stated, "On the whole, the question of the nature of light is still a doubtful one," and this was just one year before he confirmed Hamilton's prediction of conical refraction.[18] British scientists had been particularly active in the study of double refraction and polarization, but before 1827 they had shown no inclination to adopt Fresnel's theory of waves. Wollaston and Brewster never changed their minds, and John Herschel, who studied the polarization of light by crystals in the years 1819 through 1821, took a position close to Biot's theory of "movable polarization."[19] Between 1821 and 1825 Herschel turned his efforts largely to problems of astronomy, but in 1826 he returned to the study of optics in order to write an enormous article on "Light" for the *Encyclopaedia Metropolitana*, in which he showed a distinct change of heart.[20] He was particularly indignant that Fresnel's memoirs had been so effectively suppressed by the French Academy. In a reference to one such memoir, available only in the *Annales de chémie*, Herschel added the footnote: "This memoir was read to the Institute, Oct. 7, 1816. A supplement was received Jan. 19, 1818. Mr. Arago's report on it was read June 4, 1821.

And while every optical philosopher in Europe has been impatiently expecting its appearance for seven years, it lies as yet unpublished, and is known to us only by meagre notices in a periodical journal."[21]

Herschel did not support the undulatory theory whole-heartedly. He called the reader's attention to the fact that *neither* theory "will furnish that complete and satisfactory explanation of *all* the phenomena of light which is desirable," but with respect to polarization and diffraction, he believed the undulatory theory to have a decided advantage.[22] The solid aether required by Fresnel's theory of transverse vibrations could scarcely be declared an established fact, but the undulatory theory was "by far the simplest means yet devised of grouping together, and representing not only all the phenomena explicable by Newton's doctrine, but a vast variety of other classes of facts to which that doctrine can hardly be applied without great violence, and much additional hypothesis of a very gratuitous kind."[23] The wave theory was, in all its applications and details, "one succession of felicities, insomuch that we may almost be induced to say, if it be not true, it deserves to be so."[24]

Herschel saw two classes of phenomena that still were unexplained by the undulatory theory. The first was dispersion—Newton's famous discovery that different colors of light are refracted by different amounts in passing through a prism. According to both theories of light, dispersion is caused by the different colors of rays of light traveling at different speeds in the refracting medium. Unfortunately the undulatory theory had no mechanism to explain this variable speed. The second major problem was that of absorption. It was not clear how a transparent medium could selectively absorb waves of one frequency and transmit others that differed only by a small amount. Nevertheless, because the undulatory theory explained so many other optical phenomena, it was unfair to reject it because of the few things that it could not yet explain.

Although Herschel was leaning towards the wave theory, he still viewed scientific problems in the tradition of the mechanical philosophy. He wished to explain double refraction in crystals by regarding them as an "assemblage of thin, elastic hollow spherical shells in contact."[25] Even if his model of shells was not the real mechanism of crystals, it at least showed that double refraction could be explained on "sound mechanical principles."[26]

Herschel was the best known and best regarded British scientist of his time, and therefore his opinion counted for much. It was a great boon to Hamilton that Herschel closed his article with the statement that he had just received a "powerful and elegant piece of analysis" entitled "Theory of Systems of Rays" from Professor Hamilton of Dublin, "enough of which has reached us, by the kindness of the Author, to make us fully sensible of the benefit we might have derived from its

George Biddell Airy

perusal at an earlier period of our undertaking."[27] The revival of the undulatory theory of light in Britain coincided exactly with Hamilton's first published venture into optics.

The second prominent supporter of the undulatory theory was George Biddell Airy. When he and Hamilton had competed in 1827 for the position of Royal Astronomer of Ireland they were both interested in geometrical optics, Hamilton working out his "Theory of Systems of Rays," and Airy studying the aberrations of telescopic eyepieces.[28] None of these works required the undulatory theory, but in that year Airy began to introduce the theory into his university lectures, and by 1830 he had converted them completely to undulatory explanations. In 1831 he published these lectures in a new edition of his *Mathematical Tracts*, in which he claimed that the undulatory theory was "certainly true" and that "there is no physical theory so firmly established." He carefully distinguished between the mathematical theory of transverse vibrations and the *mechanical* theory of an aether supposed to transmit those vibrations. The former was certain, the latter was extremely questionable.[29] In addition to his lectures Airy also performed two crucial experiments on Newton's rings that strongly confirmed the undulatory theory. Newton's rings are a series of concentric bright and dark circles that appear when a bright light is directed onto a convex lens in contact with a flat glass plate. According to the emission theory, Newton's rings are caused by light alternately reflected and refracted at the curved surface of the lens. According to the wave theory, the rings are caused

John F. W. Herschel

by interference between light reflected from the curved surface and light reflected from the plate on which the lens is resting. The most important difference is that by the emisson theory the phenomenon is caused by light reflected from *one* surface, while by the undulatory theory it is caused by light reflected from *two* surfaces. By resting the lens on a metal plate and by employing polarized light incident at the polarizing angle, Airy was able to insure that the reflected light came only from *one* surface. The rings disappeared, as the undulatory theory predicted.[30] In a second version of the experiment Airy used a lens of low refracting power and a glass plate of high refracting power. Under these circumstances the undulatory theory predicted that if the incident light were polarized perpendicular to the plane of incidence, the *angle* of incidence would determine whether the rings exhibited a dark center or a light center. The observations were exactly as predicted.[31] Only the undulatory theory could explain these experiments satisfactorily.

Hamilton's "Theory of Systems of Rays" did not require him to think a great deal about the physical constitution of light, and it was only his prediction of conical refraction that drew him into the controversy. His correspondence with Airy and Herschel began in 1827, however, and so antedates conical refraction by five years. He met Airy in April, 1827, when they were both competing for the position of Astronomer Royal. Hamilton had to defend his "Theory of Systems of Rays" against Airy's objections at a dinner given by Lloyd.[32] Uncle James was unhappy that his nephew had allowed himself to be drawn into debate and said that it was "very young" of William to respond to Airy's arrogance.[33] Airy was a friendly correspondent, however, probably because he was not that impressed by the position that Hamilton had taken from him. In any case he won the Plumian Professorship of Astronomy at Cambridge the following year without opposition and was more than adequately provided for. With characteristic self-satisfaction he wrote: "[I] made known that I was a candidate and nobody thought it worthwhile to oppose me. ... I had no doubt of success."[34] In temperament Airy was the polar opposite of Hamilton. He was a paragon of order and practicality and had no time for philosophy or other such idle speculations. The Royal Greenwich Observatory under his direction reached the ultimate of efficiency and precision (he became its director in 1835), while the observatory at Dunsink under Hamilton's direction was characterized by moderate chaos and benign neglect. Airy kept careful files of all correspondence, memoranda, and research documents, while Hamilton tended to shove them under the bed. Airy kept his accounts personally by double-entry bookkeeping; working on them was one of his greatest joys. Hamilton never seemed to be quite sure where money came from and where it went. When in 1853 a royal commission investigated Trinity College Dublin, Hamilton was made to answer for what appeared to be irregularities in the operation of the observatory. He did not view with kindness the suggestion of the commission that Dunsink might wish to emulate some of the procedures employed at Greenwich.[35] Hamilton often sent his poems to his regular correspondents, including Airy, but he thought Airy's reaction would probably be to "make three copies and file it."

It was through Airy that Hamilton entered into correspondence with John Herschel. Airy, Hamilton, and Herschel were all astronomers by profession, they were all fine mathematicians, and they all had an intense interest in optics, so it is not surprising that they sought each other's acquaintanceship. Airy wrote in July, 1827, to congratulate Hamilton on his appointment and to ask for his "Theory of Systems of Rays" as soon as it was published. In a later letter he mentioned that Herschel was anxious to see it, too, and Hamilton responded by sending as much of the paper as had been printed.[36] In October, 1827, Robinson wrote from Armagh to tell Hamilton that Herschel and Babbage had been in Ireland and had gone to Dunsink with Robinson as guide, but had arrived while Hamilton was still absent on his excursion with Nimmo. Hamilton then wrote directly to Herschel the first letter in a long correspondence. He met Herschel on his next trip to England, in 1832, when he went to Herschel's house at Slough and saw his instruments.[37] They had become friendly through their correspondence, and

Herschel was responsible for Hamilton's election to the Royal Astronomical Society.

For a *philosophical* companion Hamilton was most closely attached to William Whewell, tutor and later master at Trinity College Cambridge, and professor of mineralogy. Hamilton met Whewell on two visits to England in the spring and summer of 1832. The first visit occurred in March and April—the same trip during which he met Herschel at Slough. Hamilton and his pupil Lord Adare, who accompanied him on the trip, stayed with Airy at Cambridge, and while there Hamilton met Whewell.[38] In June he was back in England for the Oxford meeting of the British Association and met Whewell again. At the Cambridge meeting of 1833 he stayed in Trinity College and saw more of Whewell than before. In particular he and Whewell talked about Immanuel Kant, whose works Hamilton had begun to study. When Hamilton entered the controversy over the theories of light it is not surprising that he was found on the same side as his friends from Cambridge, particularly since his metaphysical idealism led him to deny material explanations of physical phenomena wherever possible.

Optics at the British Association

It was the annual meetings of the British Association that brought Hamilton together with Herschel, Airy, and Whewell on a regular basis, and it was at these same meetings that some of the liveliest debates over the nature of light took place. The British Association for the Advancement of Science, born in controversy and ballyhoo, placed the "master spirits" of science literally on a pedestal before the admiring multitude, and Hamilton found himself elevated to the first rank with astonishing swiftness.[1] The founding of the association coincided with the beginning of the British debate over the nature of light, and therefore optics was a major subject of discussion. Hamilton's discovery of conical refraction accelerated his rise to the rank of "master spirit," and the annual meetings held at different cities around the British Isles drew him out regularly. He was always a prominent participant, delivering papers, chairing sections, and making speeches at the banquets that accompanied each meeting.

The British Association had grown out of a discontent with the state of science in Britain and a minor rebellion within the Royal Society.[2] John Herschel had inadvertently initiated the debate by a statement in his article "Sound" for the *Encyclopaedia Metropolitana* to the effect that in most of the sciences Britain was "fast dropping behind" France and Germany.[3] His friend Charles Babbage, Lucasian Professor of Mathematics at Cambridge and the inventor of a marvelous calculating machine, soon reinforced Herschel's statement with an entire book entitled *Reflections on the Decline of Science in England and On Some of Its Causes*.[4] The book produced controversy, as it was bound to do, especially after David Brewster, who had actually collaborated to some extent in the writing of the book, supported its findings in the *Quarterly Review* with strong language.[5] While Herschel had attributed the decline of British science to lack of contact with foreign research (it should be remembered that Herschel, Babbage, and Peacock had initiated the revolt to introduce continental mathematics into Cambridge in the teens), Brewster attributed it to lack of patronage. Scientists in Italy, France, and Germany were supported by pensions and honors from the government. In England no support of this kind existed. The scientific societies of London, Edinburgh, and Dublin were moribund. Scientific men were "utterly abandoned by the government" and could "find no asylum in our Universities." Worse yet, according to Brewster: "Within the last fifteen years not a single discovery or invention of prominent interest has been made in our colleges, and ... there is not one man in all the eight universities of Great Britain who is at present known to be engaged in any train of original research." These were harsh words, and not very fair. In fact, in a letter to an unknown correspondent written a month after the appearance of Brewster's review, he admitted that he might have done Hamilton an injustice. Brewster had just received a paper from Hamilton (probably the first supplement to the "Theory of Systems of Rays") that indicated that Hamilton had not completely deserted his mathematics for the duties of the observatory. Still, Brewster added that even with this new evidence he was not convinced that his original statement had been incorrect.[6]

While Brewster attacked the universities, Babbage directed his criticism against the Royal Society, where he saw an egregious case of misrule and bad management.[7] Babbage's book appeared just as a strongly contested presidential election at the Royal Society got under way. The traditionalists favored the Royal Duke of Sussex, while a group of reformers pushed for John Herschel. The duke won a narrow victory. Hamilton sent his regrets, but intimated that Herschel was "likely to enjoy more the quiet pursuit of Science at home, than any such situation."[8]

In 1828 Brewster had attended a meeting of the Deutscher Naturforscher Versammlung in Berlin, during which he had been impressed by the kind of persons in Germany prepared to support and pay homage to science. He believed that a British society, organized along the same lines and meeting annually at different cities, might bring the attention and financial support that was so badly needed. He inquired of the officers of the Yorkshire Philosophical Society if they would be willing to hold the first meeting of a newly created British Institution in York, which had the advantage of a central location and an active scientific society that was relatively free from coteries and political infighting. The officers accepted and the meeting was duly held. Beginning with this first meeting, however, Brewster was gradually pushed out of the entire operation. His cohorts in the decline-of-science debates, Herschel and Babbage, refused to attend the first meeting, and the Cambridge men agreed with Whewell that there was no reason "to rally round Dr. Brewster's standard after he has thought it necessary to promulgate so bad an opinion of us, who happen to be Professors in Universities.... It requires all one's respect for Dr. Brewster's merits to tolerate such bigotry and folly."[9] Brewster had an equal antipathy to Whewell, who he felt lacked all judgement regarding the decline-of-science debate and would sabotage any attempt to obtain government support.[10] Hamilton did not concern himself with the debate in 1831, but knew of it through a strong letter from William Wordsworth, who condemned Brewster for believing that a man could not climb the hill of science unless he be "stuck o'er with titles, and hung round with strings." Whewell had defended the universities in the *British Critic* and Wordsworth was very much on Whewell's side.[11]

After Brewster's first initiative, the organization and promotion of the British Association was mostly the work of Vernon Harcourt, who quickly realized that the success of the new enterprise would not be assured by storming the government or railing against the universities. With all the tact that Brewster so obviously lacked, Harcourt smoothed over relations with the Royal Society and with the Cambridge men. As a result he was generally regarded as the founder and primary benefactor of the British Association. The first meeting at York was not especially impressive, but Harcourt's diplomacy began to do its work, and by the second meeting, held at Oxford in June of the following year, Whewell, Herschel, and the Cambridge men had swung around. Whewell shared the vice-presidential chair with Brewster, and five of the reports were written by Cambridge professors.[12] Hamilton was there to represent the Royal Irish Academy. He viewed the association as a kind of fraternal scientific order. At the concluding banquet of the meeting he gave a "hands-across-the-water" speech, mentioning the individuality of Ireland but recognizing at the same time the common culture that the Irish shared with England:

But as the States of Greece, amid their many rivalries, and different and often hostile recollections, had yet their Amphictyonic Council, and their Olympic Games, at which Athenian and Spartan remembered that they were children of one common Mother, speaking one common language, inheritors in common of great historical achievements, descendants of those who had together resisted Persia, and together listened to recited works of Genius, which Time had already stamped immortal; so assuredly must the hearts of Britons and Irishmen be more and more knit together in affection by the fraternal intercourse of their minds in this intellectual and national assembly; this silent sense of sympathy in zeal and love for truth; these mutual expressions of respect, which honour alike the giver and the receiver.[13]

Probably Hamilton's feelings were most accurately expressed when he referred to himself at the end of the speech: "Though an Irishman, and so in part a stranger, I called myself your fellow-countryman." On the whole, Hamilton was enthusiastic about the meeting. He liked to give speeches and listen to them, and the elaborate social functions of the British Association, which repulsed many members, were exciting for him. He could always be counted on for at least one speech, sometimes more, and usually they were well received.

At Oxford, Hamilton met Brewster for the first time. They were guests at the observatory along with Airy, who engaged Brewster in an argument over the management of observatories. Hamilton sat and watched: "As to Brewster, though he and I are as nearly opposite as two persons can well be, whom the world would class together, yet I found it a very tolerable, and even not unpleasant thing, to spend a week in his society, especially as I had the society of so many others at the same time. 'All things are less dreadful than they seem,' and a human interest and kindness can temper usefully the sense of philosophical difference."[14] Hamilton regretted that the "champions of science" he met at the meeting were so indifferent to philosophy, but still it was much to be able to have left Oxford "with recollections of personal and friendly intercourse, of hands clasped in generous trust, and of sitting at table together."[15]

As Hamilton took a more active part in the affairs of the British Association he became increasingly sensitive about its public image. He had been rather proud of his Oxford speech and had sent a copy to Wordsworth. Wordsworth did not comment on it in his reply, but before the next meeting at Cambridge, he wrote that he hoped there would be "less mutual flattery among the men of Science than appeared in that of the last year in Oxford."[16] At Cambridge Hamilton did not resort to flattery, but neither did he restrain his oratory. This time he replaced the "hands-across-the-water" theme with a "climbing-higher-mountains" theme that was more suitable for a member returning for his second meeting of the association.[17] Because of its public character and its invitation to lavish display, each city trying to outdo the previous one, the association was subjected to a constantly increasing flow of ridicule. It began with a series of articles in the *Times* immediately following the Oxford meeting of 1832 and rose to its literary heights in 1837 with the appearance of Charles Dickens's account of the Mudfrog Association for the Advancement of Everything. (The Mudfrog Association held a "supplementary section of umbugology and ditchwateristics" at which members conducted a phrenological analysis of a coconut).[18]

Not all the criticism came from outside the association, however. In 1834 Brewster joined the chorus of critics. In his attack on the government and on the universities Brewster had been in part reflecting upon his own position. He was one of the most important and active scientists in the British Isles, but he held no university position and supported himself by tutoring and by editing a series of literary and scientific journals. In religion he was an evangelical, and he became a member of the Free Church of Scotland after the Disruption in 1843. Oxford, Cambridge, and Dublin were Anglican establishments at a time when religion stood for much; Edinburgh was a center of the Scottish Kirk and orthodox Presbyterianism, so Brewster's antipathy towards universities was understandable. At the Oxford meeting of the British Association the university conferred honorary degrees on Brewster, John Dalton, Michael Faraday, and Robert Brown. Brewster pointed out gleefully that they were all dissenters—there was not a single member of the Church of England in the entire group.[19] Part of Brewster's hostility towards Whewell was caused by Whewell's defense of the Anglican position. In 1834 Whewell actively resisted attempts to liberalize the religious requirements at Cambridge.[20]

Brewster was a foe of "speculation" in science. He wrote: "Speculation engenders doubt; and doubt is frequently parent either of apathy or of impiety."[21] He believed that science rested on the observed phenomena of experience and that idle philosophizing was dangerous not only to science, but to religion, as well. In 1833 Whewell published the first of the famous Bridgewater Treatises, which he entitled *Astronomy and General Physics Considered with Reference to Natural Theology.*[22] In an anonymous review of the book Brewster had full opportunity to air his grievance with Whewell's science and theology. The treatises had been founded and were supported by a large endowment left by the ninth earl of Bridgewater. Brewster began his review by accusing the board of mismanaging the trust and of authorizing the publication of a series of disconnected volumes rather than a single organized work.[23] He then turned his attack on Whewell's understanding of natural theology. According to Brewster, Whewell weakened belief by his assumption that natural theology relied on scientific knowledge of nature. Whewell had said that, because science is incomplete, our natural theology must be "scanty and imperfect." But Brewster believed that one did not need a scientific knowledge of nature in order to admire God's handiwork, and he went off into a rapturous description of the heavens to prove his point. What bothered Brewster most of all was the fact that a substantial part of Whewell's book was an exposition of the undulatory theory of light. By extracting sentences from different parts of the book and arranging them in order, he showed how Whewell gradually moved the status of the undulatory theory from that of a hypothesis, in the beginning of the book, to that of a proven fact, by the end, without adequate evidence. Furthermore, he argued that even the best mathematicians in Europe still regarded the existence of an aether as uncertain. "Professor Hamilton of Dublin, one of the first of our mathematicians, in speaking of the Undulatory Theory, uses the following cautious expressions: 'The verification, therefore, of this Theory of Conical Refraction, by the experiments of Professor Lloyd, must be considered as affording a new and important *probability* in favour of Fresnel's views, that is *a new encouragement to reason from these views in combining and predicting appearances.'"*[24]* Brewster juxtaposed Hamilton's cautious wording against Whewell's statement of fact. Citing Hamilton as an authority, he challenged Whewell's claims for the undulatory theory and closed with these strong words: "When we say, therefore, that we regard such views as supererogatory, we do not sufficiently express our opinion of them. We think them injurious—they lead to idle speculation. They found our Natural Theology on a basis of small considerations; and create a belief in weak minds that its mighty pyramid is in danger."[25]

Hamilton told Whewell that he had seen the review and guessed that it was probably by Brewster. He appreciated the flattering reference to himself, but hastily added that he could not return the compliment. "The whole rhapsody was deficient enough in argument . . . giving up the whole point in dispute."[26] In spite of his caution, Hamilton was definitely on the side of the undulationists. He diverted himself by drawing up an amusing reply to Brewster's review, but was wise enough to realize that any impertinent remark from a young new member of the scientific community was not really advisable. He could not completely resist the temptation, however, and when the British Association met the following year (1834) at Edinburgh, Hamilton found his chance. In his inevitable speech of thanks at the end of the meeting, he praised Brewster, calling him:

A native indeed of Scotland, but one whose name and fame are not confined to Scotland; and who is known over the whole of Europe, as the person who, by his researches and sagacity, has done more perhaps than any other living man for the science of Physical Optics; for that wonderful science which illustrating, each by each, the most beautiful phenomena of light, and the subtlest properties of matter, enables us almost to feel the minute vibrations, the ceaseless heavings and tremblings of that mighty ocean of ether, which bathes the farthest stars, yet winds its way through every labyrinth and pore of every body on this earth of ours.[27]

Since Brewster doubted the existence of the "mighty ocean" and certainly did not believe in the "minute vibrations" and "ceaseless heavings and tremblings," it was a backhanded compliment to grant him credit for having done the work that led to their discovery.

Brewster became increasingly unhappy with the British Association. After the first York meeting he had written in concern to Harcourt about "the Philosophical Frankenstein" that they had called into existence.[28] His original intention had been to found a society that would seek government support of science. Instead the association had taken a different direction from the one he had intended. When the second volume of *Reports* appeared in 1834, he wrote a long review for the *Edinburgh Review,* which seemed to be designed to annoy as many people as possible. Brewster noticed that "the moment . . . the contingent from Cambridge joined the Association, some scheme seems to have been formed to obliterate its origin, to paralyse its objects, and to take it under their own charge as something which their talents and zeal had given a new birth and a new impulse."[29] In particular he resented the continued references by the Cambridge men to Harcourt as the founder of the association.[30] He used the article to document carefully the fact that he, Brewster, had founded the association and not Harcourt. He also hinted darkly at the existence of a "mysterious influence . . . an incubus pressing on the vitals of the Association with its livid weight . . . a congestion somewhere near its heart, impeding its respiration, and disturbing its most vital functions. Is it political, ecclesiastical, or personal, or is it all of them combined? We dare not try to answer the question."[31] Brewster, having lost control of the association, now saw it changed by a conspiracy centered largely in Cambridge.[32]

At the Oxford meeting of 1832 Hamilton had invited the association to Dublin, and after first holding sessions in Cambridge and Edinburgh, the association accepted the Dublin invitation for the year 1835. Hamilton was one of the secretaries for the Dublin meeting, and he was anxious that it be as elaborate and as successful as any of the earlier meetings. (The other secretary was Humphrey Lloyd. The correspondence indicates that Lloyd did most of the work.) He read Brewster's review while in this nervous state of anticipation and took several of Brewster's remarks as personal attacks on himself.[33] He asked Whewell if he had read it, and added: "Hard knocks at all of us discussers and debaters, sly hope that there will be no new influx of eloquence in Dublin!" to which Whewell replied that upon reading the article his reaction had been one of self-congratulation that he had escaped any involvement in Brewster's scheme of "bullying [the government] into giving us pensions. . . . Whoever it was that gave the Association another turn did the State some service."[34]

The barb that struck Hamilton most deeply was Brewster's condemnation of "shallow declaimers" who deal out adulation with "indiscriminate prodigality" to "men of inferior talent with the garnish of a little rank, or wealth, or official importance."[35] This was uncomfortable criticism, coming as it did on the eve of the Dublin meeting. Brewster also criticized a resolution that had been brought forward by the astronomers to request from the government special funds for reducing the observations made at Greenwich.[36] It was the kind of thing that he previously had advocated, but now he seemed to have no good words for any of the association's activities. When the members met in Dublin, Brewster was conspicuous by his absence. His article in the *Edinburgh Review* had been anonymous, but there was no doubt about its authorship, and the anonymity worked both ways. While it protected Brewster, it also meant that critics could be as savage as they liked. T. Romney Robinson publicly praised Harcourt at the Dublin meeting and added that he was *not* a man "who, having reason to complain of our proceedings, instead of making his charges in the face of day, would wield the concealed dagger of a lurking assassin."[37] Hamilton was more cautious in his official report as secretary, but he openly ridiculed the "anonymous censor, who has written in a certain popular review" that astronomical observations are obsolete just because they are "old" and who is willing to see vast sums spent on maintaining an observatory, but nothing on making its observations useful.[38] At the end of his speech Hamilton particularly regretted the absence of Sir David Brewster "who took so active a part in forming this association." He continued: "I am authorized, by a letter from himself, to mention that his absence proceeds entirely from private causes, and that they form the only reason why he is not here."[39]

While the fate of the British Association was being argued by the members and by the popular press, the most exciting scientific issue remained the theory of light. The first series of *Reports,* published in 1832, contained one on light by Brewster; the 1833 meeting at Cambridge was largely taken up with Hamilton's and Lloyd's conical refraction; and at the Edinburgh meeting of 1834 Lloyd presented another "Report on the Progress and Present State of Optics."[40] Brewster's report, of course, defended the particle theory. It devoted a great deal of space to the phenomenon of polarization, and in particular to the work of Malus. Unlike Herschel, Brewster did not see that polarization was strong evidence in favor of the undulatory theory. Airy's experiments with Newton's rings, which had been published shortly before Brewster's report, were a different matter. Brewster agreed that they completely destroyed Newton's theory of "fits of easy reflection and refraction," but he did not believe that they refuted *all* forms of the emission theory.[41] In other words, there was still a chance that another "correction" might make the emission theory sound. Brewster appealed to the statement that Herschel had made in his essay "Light" of 1827 that if the emission theory had been pursued as vigorously as the undulatory theory, it might have produced even more startling results. Brewster also threw up the two greatest obstacles to the wave theory, those of dispersion and absorption. With regard to absorption, he could report a new and startling discovery. He had seen over 1000 dark absorption lines in the spectrum of nitrous oxide gas. Absorption of light by the gas was obviously selective. He believed the emission theory could explain it much better than the undulatory theory.[42] For future research Brewster called attention not only to absorption phenomena, but also to the opportunities provided by double refraction and polarization. And, finally, he mentioned that the French mathematician Augustin Cauchy had developed a theory of dispersion, although he had not seen any detailed account of the theory.[43] Cauchy's optical work was completely unknown in Britain. With Brewster's prompting, Hamilton went to work on it right away.[44]

Lloyd's report on the "Progress and Present State of Optics" came in 1834, just two years after Brewster's, and it provides a revealing contrast. Lloyd was a cautious man and not one to be overzealous, but he had clearly been converted by the undulatory theory from the agnostic position that he had taken in 1831. The wave theory, he concluded, had reached a point almost, if not entirely, as advanced as that reached by the theory of gravitation. The emission theory had led to few predictions by which its validity could be tested. It was loaded with ad hoc hypotheses and exhibited many "symptoms of unsoundness." The wave theory, by contrast, had been confirmed by many phenomena.[45] Furthermore, he denied Herschel's assertion of 1827 that the emission theory, if pursued as vigorously as the wave theory, might have led to comparable results. He pointed out that all the present advocates of the undulatory theory had begun as advocates of the emission theory and had switched over when they found the particle theory inadequate to explain the experimental phenomena. Moreover, the emission theory *had been* vigorously pursued by some of the ablest scientists of all time, including Newton and Laplace (and on a less exalted plane, he could have included Biot, Poisson, Malus, Wollaston, and Brewster).[46] Lloyd believed that the victory of the undulatory theory was complete in 1834.

The emission theorists were not prepared to give up the battle without a struggle, however, and their defense was to seek new optical phenomena that the undulatory theory could not explain. Brewster was the ablest and most indefatigable of these experimenters, but he was not entirely alone. He had several codefenders of the emission theory, most of whom came from outside the universities. These men found the British Association a convenient forum in which to present their views. Richard Potter was a Manchester merchant with an amateur interest in optics and chemistry. For a while he had been tutored by John Dalton, the famous inventor of chemical atoms. At the first meeting of the British Association he read three papers, at the next meeting two papers, and in 1833 he again read three papers. Potter's success convinced him that he should obtain a university education and, after further tutoring, he gained admission to Queen's College Cambridge. Later he became professor of natural philosophy and astronomy at University College, London.[47] He published an important paper in the same number of the *Philosoph-*

ical Magazine as that containing Lloyd's announcement of conical refraction. It was entitled "On the Modification of the Interference of Two Pencils of Homogeneous Light Produced by Causing Them to Pass through a Prism of Glass, and on the Importance of the Phaenomena Which Then Take Place in Determining the Velocity with Which Light Traverses Refracting Substances."[48] The most crucial difference between the undulatory theory and the emission theory was the way in which they predicted the velocity of light in a refracting medium. Both theories accounted for refraction equally well, but according to the wave theory, the light slowed down upon entering a more dense medium (passing from air into glass, for instance), while according to the emission theory the light should speed up in the glass. A direct measurement of the speed of light through different media was not achieved until 1850, but Potter, using a modified version of an apparatus first described by Baden Powell of Oxford, devised what he believed to be an indirect measure of the *relative* velocities of light in air and in glass.

A beam of light was divided into two overlapping beams by a pair of Fresnel mirrors, producing the familiar fringes caused by the interference of the beams of light. He then placed a prism in the paths of the beams quite close to the mirrors, so that one beam passed through the thicker part of the prism near its base and the other beam passed through the thinner part of the prism near its angle. Placing the prism in the beams caused the fringes to shift to the side of the *base* of the prism. It is obvious that the prism should make *some* difference because the light passing through the thicker part of the glass will be affected more than the light passing through the thinner part of the glass, and there will be a relative phase change between the beams emerging from the prism. If the light is *slowed* by passing through the glass, as the undulatory theory predicts, the center of the interference pattern should shift to the side of the angle of the prism; if the light goes *faster* through the glass, as the emission theory predicts, then the fringes should shift to the side of the *base* of the prism. The results of Potter's experiment supported the emission theory. But in fairness it should be mentioned that Potter admitted that the observed shift was *too great* to be explained by Newton's theory, and therefore another correction to the emission theory would be necessary. Since Brewster had already admitted the year before that Airy's experiments on Newton's rings had destroyed Newton's version of the emission theory, Potter was not conceding all that much. The undulatory theory, on the other hand, demanded a fixed ratio between the velocity of light and the index of refraction in a medium. It could not be "patched up" to agree with Potter's findings, and would fail completely if his experiments were correct.

The emission theory was more amenable to such corrections, and Potter created an elaborate system of particles to explain his experiments as well as those of Young and Fresnel. He explained interference by "shells" of particles ejected from the light source at regular intervals. These shells act like wave fronts and pass any fixed point with a determinable frequency. The only difference between Potter's "shells" and waves is that he imagined the particles to actually move from the light source to the eye, while, according to the undulatory theory, the light waves are propagated through a stationary medium. Potter's revision of the emission theory led him to make his particles act more and more like waves. He was even willing to admit that the light particles leaving the source and the particles reaching the eye were not the same, for Dalton had told him that the particles striking a transparent body such as glass were not the same as those coming out the other side.[49]

Potter's experiment brought two replys, one from Hamilton and one from Airy. As might be expected, Hamilton approached the problem mathematically.[50] Potter had assumed that the beams of light from the source retained a spherical front after passing through the prism, but Hamilton, applying his method of the characteristic function, showed that the prism would cause a substantial amount of aberration and that when this aberration was taken into account the undulatory theory predicted not a shift of the fringes to the angle of the prism, but a shift to the base, just as Potter found. As the eye is withdrawn further and further from the prism, the fringes should then begin to shift the other way, towards the angle of the prism, but Hamilton assumed that Potter had not observed from such great distances. He admitted that if Potter's experiments were correct "they would furnish a formidable and, perhaps, fatal objection against the undulatory theory of light."[51]

Unfortunately Hamilton had given Potter the means of rebutting his argument. In his paper he wrote the equation from which the aberration could be calculated directly. Potter had, in fact, made his observations at a rather large distance from the prism (although he had not given that information in his first paper). Under those circumstances, Hamilton's correction for aberration was insignificant. Potter replied that "the difference is so small, that I am sure Professor Hamilton would never have given me credit for being so minutely acute an observer, if he had had recourse to actual quantities."[52]

Before seeing Potter's reply, Hamilton had sent in a second paper to the *Philosophical Magazine* in which he explained in greater detail how he calculated the aberration of the prism using his theory of the characteristic function. When he read Potter's reply, he sent in a third article, agreeing with Potter that prismatic aberration would not account for the phenomenon as Potter *now* described it, but pointing out that there was nothing in Potter's original paper to indicate the magnitude of the effect, and so Hamilton's criticism had been entirely reasonable. He wrote to Baden Powell:

I have just seen Mr. Potter's reply, in which he seems to me to give up the merely mathematical point, on which I ventured to attack him! and therefore I think I shall say nothing more at present lest our amicable discussion should degenerate into an angry controversy. The mere *direction* of deviation I think I account for on the undulatory theory; and the magnitude of the deviation, which Mr. Potter *now* states to be so much greater than my aberrational formula will give, I could not of course know to be so, until my cross-examination, if I may so call it, elicited that (recent) testimony from him. I do not therefore regret that I took the trouble to suggest a small mathematical correction which I still believe to be theoretically just, and which was necessary to be tried, before it could be known to be experimentally insufficient.[53]

While Hamilton was forced to retreat from the debate, Airy came up with objections that were much more formidable. Potter had made a mistake, not in his mathematics, but in his interpretation of the results. He claimed to have performed the experiment with monochromatic light and to have observed an interference pattern consisting of a pronounced central fringe bordered by a symmetrical pattern of smaller fringes. Airy argued that if the light had been truly monochromatic, Potter should have observed only a series of bars, none of which could be identified as the central one. He surmised that the light Potter had used was *not* monochromatic and that the different colors of light dispersed by the prism had created a series of overlapping interference patterns. Since the fringes of different colors should be of differing widths and separated by different amounts, there should be one place where all the colors would coincide, or nearly so, and this would *appear* to be the center of the pattern, although it was certainly *not* the actual central band. The phenomenon could be checked by observing the fringes continuously as one moved away from the prism. The wave theory predicted that the bars would remain more or less stationary, while the white "central" color moved from one bar to the next. According to Potter's account, the bars themselves should move, including the "central" one. Potter blustered in response, but Airy had caught him and there was little that he could say, except to deny that the light he employed had been heterogeneous and to insist that he had not made the mistake ascribed to him by Airy. Airy must have set about checking the experiment right away, because in May, Whewell wrote to Hamilton that he had seen the apparatus: "As you withdraw the eyepiece, you see the bars, *not* move, but *grow* on one side and *dim* on the other so that the centre shifts; just as it ought to do. Indeed one knows very well now, that if any one discovers facts that are at variance with the undulatory theory, they must also be at variance with other facts, and you may get (some) fact which either implies a new property or a contradiction."[54]

Hamilton was anxious to leave the controversy with Potter before it became too acrimonious. Airy had no such concern, and his first paper refuting Potter had been unnecessarily provocative. He praised Potter's openmindedness and urged him to continue his experiments, concluding with the statement that if Potter continued to work in the same spirit, "I can with confidence predict one result:—Mr. Potter will very soon become an undulationist." Potter, of course, replied that he found the Plumian professor's compliment no compliment at all, and that the probability of his becoming an undulationist became daily less rather than greater.[55] Airy apparently realized that it had been a mistake to goad Potter. Hamilton reported to Lord Adare: "Airy has just written to me expressing his confidence on the point, but saying that he does not intend to continue the discussion; so our expectations of a battle-royal vanish into air. He is right, I think, to stop, for it was in danger of becoming too personal a matter."[56] Airy closed the debate with a short paper in the *Philosophical Magazine* entitled "Results of the Repetition of Mr. Potter's Experiment of Interposing a Prism in the Path of Interfering Light."[57] In the note he merely stated that he could now "assert positively, as an experimental fact" that the phenomena were as he had predicted them from the undulatory theory.

In order to give his attack on the undulatory theory all the weight that he could muster, Potter had called attention to statements by Brewster about the absorption of light by gases, a phenomenon that was still unexplained by the undulatory theory.[58] Because he had been called on by Potter, Brewster felt that he had to step forward and express an opinion. However, rather than limit his remarks to the phenomenon of absorption, he took a position on the undulatory theory in general.

I am anxious to state the views which I have taken of this class of phaenomena, in reference to the undulatory theory. I have long been an admirer of the singular power of this theory to explain some of the most perplexing phaenomena of optics; and the recent beautiful discoveries of Professor Airy, Mr. Hamilton, and Mr. Lloyd afford the finest examples of its influence in predicting new phaenomena. The power of a theory, however, to explain and predict facts, is by no means a test of its truth; and in support of this observation we have only to appeal to the Newtonian Theory of Fits, and to Biot's beautiful and profound Theory of the Oscillation of Luminous Molecules. Twenty theories, indeed, may all enjoy the merit of accounting for a certain class of facts, provided they have all contrived to interweave some common principle to which these facts are actually related.

On these grounds I have not yet ventured to kneel at the new shrine, and I must even acknowledge myself subject to that national weakness which urges me to venerate and even to support, the falling temple in which Newton once worshipped.[59]

Brewster was trying to say that no physical theory can ever be entirely "true." Newton's theory of fits and Biot's theory of mobile polarization seemed "true" at the time they were propounded, but both had since been proven wrong. And in spite of its success, the undulatory theory could be expected to meet the same fate. As a *mathematical* theory Brewster was prepared to acknowledge its great predictive power. As a *physical* theory he still found it defective.[60]

But Brewster must not have been thinking clearly when he wrote that "The power of a theory ... to explain and predict facts, is by no means a test of its truth." Airy leapt on this statement and asked what the test of the truth of a theory *is* if it is not its conformity to old facts and its power to predict new ones, and added: "Nothing appears to me more prejudicial to the progress of science than vague statements of such a kind as that to which I allude."[61] He was attempting to turn the tables on Brewster, who frequently railed against the idle speculation and vagueness of much physical theory.

It is also clear from Brewster's remark that the authority of Newton was still haunting the debate. His willingness to continue to worship at the falling temple of Newton was a bit of nationalistic rhetoric that won him little support. Thirty years earlier his patron, Lord Brougham, had successfully called on the authority of Newton in his attack on Thomas Young, but in 1833 it no longer had much force, a fact that Brewster acknowledged when he described Newton's temple as "falling." Nevertheless, Airy accepted the idea that it was good to have Newton on your side in any argument. In his response to Brewster he revived Young's argument that Newton was really a believer in the undulatory theory, and pointed out that Newton himself had warned the reader of his *Opticks* not to connect any *physical* conception with his theory of fits.[62]

As to Brewster's use of absorption phenomena to criticize the undulatory theory, Airy argued that absorption at present fell outside the area described by both the emission and the undulatory theories.[63] It seemed to be different in kind from polarization, reflection, refraction, interference, and so on, and would require a separate or supplemental theory to explain it. Airy believed it was unfair to insist that the undulatory theory account for it. No theory can explain everything at once. If Newton could not have given his theory of gravitation until it also explained capillary action, then he never would have been able to give his theory at all.

Airy was right, of course. No adequate mathematical theory of absorption was possible until the advent of quantum theory. John Herschel made a notable attempt, however, in response to the debate between Brewster and Airy. He began his treatment of the subject by discussing the role of authority in science, and Newton's in particular. Newton's speculations, "however ingenious and elegant, can hardly in their present state of our knowledge, be regarded as more than a premature generalization; and they have had the natural effect of such generalizations, when specious in themselves and supported by a weight of authority admitting for the time of no appeal, in repressing curiosity, by rendering further inquiry apparently superfluous, and turning attention into unproductive channels."[64] The days when one could conjure with the name of Newton were past.

Herschel went on to describe a mechanical theory of absorption by analogy to sound. He compared the propagation of vibrations of light to the transmission of motion from one tuning fork to another and used the phenomena of sympathetic and forced vibrations to explain how the loss of motion could be confined to narrow bands of frequencies. It was the beginning of attempts in England to create a theory of spectroscopy.[65]

Dispersion was the other phenomenon that Brewster had held up against the wave theory in 1832 along with absorption. There was no use arguing that dispersion "fell outside" the undulatory theory, because it was central to the explanation of refraction. In his article on absorption, Brewster reminded the reader that Herschel had written in 1827 that dispersion was the "most formidable objection to the undulatory theory" and that it still remained a formidable obstacle six years after Herschel's remark.[66] His argument was seconded by another amateur optician and advocate of the emission theory, Sir John Barton, controller of the Royal Mint in London.[67]

Barton came forward at the same time as Potter with experiments on dispersion (or "inflexion of light," as he called it) designed to support the emission theory.[68] Baden Powell, who had just begun to study dispersion in some detail, undertook the task of responding to Barton.[69] The task was not a difficult one. It was obvious that Barton did not understand Fresnel's theory, not even the theory of wavelets, "the most unequivocal and elementary part of the theory."[70] He had used Newton's measurements to refute Fresnel, even though Newton had acknowledged that they were rough, while Fresnel's were precise. And he had compared Newton's experiment on diffraction using two *curved* knife edges to Fresnel's experiment done with *straight* knife edges. Powell pointed out that the two experiments were not comparable. The mathematical treatment of diffraction for curved edges was extremely difficult and had never been done. Powell ended his remarks with a scorecard listing all the major phenomena of physical optics and the adequacy of the two competing theories to explain them. The undulatory theory won easily.[71]

Like Potter, Barton had his own version of the emission theory to explain diffraction. He believed that when a stream of light particles passed through a narrow opening, some of the particles struck the edge of the opening and were set into oscillation as they continued on their path.[72] This was a different emission theory from that proposed by

Potter, but it had the same purpose, to make particles act as much as possible like waves. In order to sustain the emission theory, its supporters had to resort more and more to ad hoc hypotheses like those of Potter and Barton. These were the "symptoms of unsoundness" that Lloyd observed in his British Association report of 1834, and they did have a familiar ring. In the eighteenth century the Cartesians at the Paris Academy made similar attempts to patch up Descartes's vortices. By multiplying hypotheses the Cartesians produced an aether that came close to meeting the demands of Newton's *Principia*, but it was so unlike Descartes's original conception that it became easier for philosophers to accept action at a distance than the monstrosity the Cartesian aether had become. The same thing was occurring in the nineteenth century with regard to the emission and undulatory theories of light. In a sense Brewster had been right. Until a medium could be found for the waves, the undulatory theory was incomplete as a physical theory. Just as the light particles began to look more and more like waves, until they might as well *be* waves, so the molecules of the luminiferous aether began to take on strange properties that made them more and more like an electromagnetic field until they might just as easily *be* an electromagnetic field and, eventually, a relativistic electromagnetic field.

By 1834 it looked as if the advocates of the undulatory theory had won the battle in the British Association. David Brewster had not conceded, however, and because he was unquestionably still the greatest contributor to experimental optics, the debate continued. For Brewster mathematical theories were not enough. Until the actual mechanical action of the particles of the luminiferous aether could be discovered by experiment, the question of the nature of light remained open. The major problem facing the supporters of the undulatory theory—experimenters and theorists alike—was to find a model for the aether that would account for all the known phenomena of optics. Beginning in 1833 Hamilton lent his formidable mathematical powers to this effort.

11

The Luminiferous Aether

HAMILTON'S SEARCH for the luminiferous aether began with an attempt to solve the problem of dispersion. Dispersion and absorption of light were the two phenomena that John Herschel had signaled as being the most difficult ones to explain by the undulatory theory. Absorption was not an attractive problem for Hamilton; as yet there was nothing to quantify and calculate. Dispersion, however, was a different matter, and he set about trying to produce an adequate theory. Dispersion is the phenomenon by which the colored rays of light passing through a prism are bent by different amounts. In the undulatory theory dispersion is explained by assuming that the rays of different color (and therefore of different wavelength) travel through the prism at different velocities. But the mathematical theory of wave propagation indicates that waves of all wavelength should travel at the same speed through the same medium. The trick was to create a mechanical model for the medium that would propagate the different wavelengths of light at different speeds and still explain all the other phenomena of optics.

This was basically a *mechanical* problem. Hamilton's characteristic function had been a particularly elegant method for studying the geometry of rays and wave surfaces, but it did not treat the mechanical motion of the medium as the waves passed through it. In studying the problem of dispersion Hamilton was turning from geometrical optics to physical optics—that is, from geometry to mechanics. It was an important new direction, because even though he did not succeed in finding the luminiferous aether, he did find an entirely new abstract theory of mechanics, and as he pursued his investigations he developed important new ideas about how waves propagate in a quiescent medium.

The luminiferous aether that physicists and mathematicians sought in the nineteenth century was a strange material. Because waves of light were known to be transverse and of high frequency and velocity, the aether had to have the characteristics of an elastic solid like steel. It probably had to be incompressible and possess a high resistance to deformation. Obviously the cosmos was not filled with steel or anything like it, but the analogy between the properties of light waves and the properties of waves in an elastic solid made it possible to transfer the mathematics of the latter to the former. In 1828 Augustin-Louis Cauchy had begun the search for an elastic-solid theory that would correctly predict the phenomena of light, and he was soon joined by other mathematicians in what promised to be an exciting new field. The first theories correctly predicted the wave surfaces in crystals. They were then extended to reflection and refraction and to other phenomena, such as diffraction, interference, double refraction, optical activity, metallic reflection, and dispersion.

The elastic-solid theory, however, was deceptively difficult. Cauchy had viewed the problem as an entirely analytical one: "Given the mathematical theory of an elastic solid and the known laws of optics according to the undulatory theory, find the elastic constants and the boundary conditions that would make the waves in the elastic solid follow the known rules of wave optics." There were several problems with this approach. Many of Cauchy's constants and conditions were chosen only to get the right results and did not seem to have any physical meaning. Also, the problem was indeterminate; different models of the elastic solid gave the correct results for some phenomena, but none was adequate to explain *all* the phenomena. Furthermore, all mechanical models predicted the existence of longitudinal waves as well as transverse waves in the aether. Theorists had either to ignore the longitudinal waves (just scratch these terms out of the equations), or to show somehow that they would not be detectable. Another constant source of annoyance was the inability to determine whether the actual vibrations of the light were perpendicular or parallel to the plane of polarization. Theories could be built using either assumption and neither assumption was conclusively more successful than the other. Cauchy tried all variations. Over a ten-year period he produced two theories of crystal optics and three theories of reflection, all giving reasonably good results and all irreconcilable with each other.[1] Much later, in 1858, Peter Guthrie Tait, who had recently entered into correspondence with Hamilton, could still write: "I should like to know your opinion of Cauchy's investigations in the Undulatory Theory—for I have found it possible by apparently legitimate uses of his methods to prove almost *anything*."[2]

Finally it became quite obvious that the luminiferous aether did not behave like an elastic solid after all. The most successful theories either approached the problem in great generality, making few restrictions on the nature of the medium (as Green did in 1837), or they took liberties with the laws of mechanics and gave to their aethers properties that were not permissible in classical mechanics (as MacCullagh did in 1839).

Like most of the aether theorists, Hamilton believed the aether to be composed of fine particles attracting and repelling each other according to fixed laws, and it was to this model that he applied his mathematical methods. A continuous aether might have been easier to treat mathematically (MacCullagh, for instance, assumed a continuous aether), but Hamilton had several reasons for preferring a particulate aether. The first reason was metaphysical. By 1833 he was strongly involved in philosophical study and had come to the conclusion that natural phenomena could best be explained by centers of "power" acting in space and time. These powers he sometimes regarded as point forces after the theory of Roger Boscovich. In October, 1832, Hamilton had written to Coleridge asking if Boscovich's theory was metaphysically sound, because something of the sort seemed to be required by the undulatory theory. He knew that Coleridge detested atomism, but thought that the abstract point centers of force suggested by Boscovich might be a satisfactory alternative: "The ... theory of which I speak is nearly that of Boscovich, and consists in representing all phenomena of motion as produced by the action of localised energies of attraction or repulsion, each energy having a centre in space; and this centre, which is supposed to be a mathematical point, without any figure or dimension, being called an *atom* instead of a point, merely to mark its conceived possession of, or connexion with, physical properties and relations."[3] It was an idea that he repeated conspicuously the following March in his first paper on a "General Method in Dynamics," where he adopted a Boscovichean stance. In describing his new method to the British Association he was even more explicit: "Professor Hamilton is of opinion that the mathematical explanation of all the phenomena of matter distinct from the phenomena of life will ultimately be found to depend on the properties of systems of attracting and repelling points."[4] For metaphysical reasons, therefore, he thought the aether should be composed of mass points with forces between them.

A second reason for Hamilton assuming the aether to be particulate was his interest in the problem of dispersion. Fresnel had suggested that if the aether was composed of particles separated by distances comparable to the wavelength of light, then the waves of different length representing different colors of light would not all travel with the same speed, and this would be an immediate explanation of dispersion. The wave equations that predicted that light of all colors would move through a given medium with the same velocity assumed a continuous medium.

They would have to be corrected if the aether was particulate, and this "graininess" of the aether was the most likely explanation of dispersion in the 1830s.

Professor Baden Powell of Oxford wrote to Hamilton about the problem of dispersion in March, 1834, asking if Hamilton's theory of systems of rays could be used to elucidate Cauchy's theory.[5] The following September, at the Edinburgh meeting of the British Association, Cauchy's theory of dispersion was one of the major topics of discussion.[6] By this time Hamilton was well along in his dynamics and there was some hope that he might be able to use it to unravel the dynamics of light.[7] Humphrey Lloyd had written to Hamilton before the Edinburgh meeting: "I do not despair of seeing a day when the Characteristic Function will be applied to Molecular attraction and when the constants of the integrations will, may-be, tell us something about the precise crystalline form to be assumed."[8] If the characteristic function could unravel the problems of crystalline optics, and especially the problem of dispersion, it would be a powerful tool, indeed.

Powell's first paper on dispersion appeared in the *Philosophical Magazine* two months after his letter to Hamilton.[9] He said he was publishing an account of Cauchy's theory of dispersion partly because Cauchy's papers were "abstruse and scattered," but more especially because they were almost unknown in England four years after they had been published. This first article was merely a note of intention. He began his exposition in January, 1835, reaching the conclusion that the velocity of the light stands in direct proportion to the wavelength and that what was now needed was a quantitative relationship between the intermolecular distances in the crystal and the amount of dispersion.[10] Powell urged anyone with a new theory to bring it forward. The appeal was directed especially to Hamilton, who began to send his ideas to Powell. Cauchy had made certain simplifications in his theory, which Hamilton believed had reduced its precision. In 1836 Powell began to include Hamilton's "exact" formulas along with Cauchy's "approximate" ones.[11] For some reason Hamilton was unwilling to publish his method himself. He may have felt that the subject was "reserved" for Powell, or he may have just decided that his contributions were not important enough to warrant separate publication.[12]

From his own work he published only two papers, both presented to the 1838 meeting of the British Association at Newcastle. "On the Propagation of Light *In Vacuo*" was a study of different possibilities for the law of force between molecules of the aether.[13] He demonstrated to his satisfaction that a force proportional to the inverse square of the distance or to the inverse fourth power, as suggested by Cauchy, would not work. In 1838 he was still studying dispersion along the lines of Boscovich—to find the correct law for a system of attracting and repelling points.[14] His paper "On the Propagation of Light in Crystals" suggested a theory of dispersion based on the supposed interaction between the particles of the aether and the more gross molecules of the crystal lattice, a theory subsequently pursued by Boussinesq and others.[15]

While Hamilton worked with Baden Powell on the theory of dispersion in a particulate aether, a colleague at Trinity College, James MacCullagh, attacked the problem of the aether from an entirely different direction. MacCullagh introduced his ideas at the 1835 and 1836 meetings of the British Association. Airy wrote to Hamilton in 1836 asking what assumptions MacCullagh had made about the structure of quartz, and Hamilton responded to say that MacCullagh had set up differential equations of vibration without assuming any particular crystalline structure—in fact, what he had done was reduce the phenomenon to mathematical, but still empirical, laws without introducing any elastic-solid model at all.[16]

In 1837 MacCullagh made major additions to his theory in an important paper entitled "On the Laws of Crystalline Reflexion and Refraction."[17] In this paper MacCullagh required that the vibrations of the light wave be continuous across the interface between two media. He also eliminated the longitudinal wave by requiring that the aether have a constant density. The theory agreed with experimental values, but it was not derived from any particular molecular order for the crystalline medium. Instead, MacCullagh had worked backwards from the phenomena to create a theory consonant with the known facts of light but at odds with the fundamental laws of mechanics.

Herschel recognized the importance of MacCullagh's work and wrote to Hamilton:

[MacCullagh's paper] has . . . produced a very strong impression on my mind that the theory of light is on the eve of some considerable improvement, and that by abandoning for awhile the *a priori* or deductive path, and searching among phenomena for laws simpler in their geometrical enunciation, and of more or less wide applicability, *without (for a while) much troubling ourselves how those laws may be in apparent accordance with any preconceived notions*, or even with what we are used to consider as general principles in Dynamics—it may be possible to unite scattered fragments of knowledge into such groups and masses as shall afford glimpses of their fitness to combine into a regular edifice. What has most tended to impress me with this is Mr. MacCullagh's temporary abandonment of the *vis viva* principle, and his idea of the equable distribution of the ether in all media. Such assumptions *appear* to those who have prematurely regarded the undulatory theory as complete not a little violent; but in the case of the *vis viva* the abandonment is probably only apparent, and not amounting to a *denial*; while, as to the other, we are far too much in the dark as to the mechanism of the ether to be in a condition to judge *a priori* how far that, or any other hypothesis, may be admissible.[18]

Hamilton recognized the importance of MacCullagh's contribution but could not believe that MacCullagh's approach would ever lead to a complete theory. He had told Airy in 1836 that such a complete theory would have to be based on *dynamic* laws that explained the phenomena by vibrations in some crystalline model.[19] Hamilton called MacCullagh's method *mathematical induction* because MacCullagh had "*not* sought to deduce, from any presupposed attractions or repulsions, and arrangements of the molecules of the ether, any conclusions respecting the vibrations in the interior or at the boundaries of a medium, as necessary consequences of those dynamical principles or assumptions. But he has sought to gather from phenomena a system of mathematical laws by which those phenomena might be expressed and grouped together."[20] Thus, MacCullagh *described* the phenomena of reflection and refraction that others had discovered by experiment. In comparing the progress of the undulatory theory to that of the theory of gravitation, Hamilton called MacCullagh the "Kepler" of optics. Just as Kepler had reduced the observations of Tycho Brahe to law "without seeking yet to deduce these laws, as Newton did the laws of Kepler, from any higher and dynamic principle," so had MacCullagh undertaken the "preparatory but important task of discovering from the phenomena themselves, the mathematical laws which connect and represent those phenomena, and are in a manner intermediate between facts and principles, between appearances and causes."[21] It was not difficult to recognize that Hamilton believed the important advance was not MacCullagh's *mathematical induction*, but some future *dynamic deduction* of the laws of light from real causes. In fact he admitted: "My own habits of thought lead me to feel an even stronger interest in dynamic and deductive researches."[22] Hamilton obviously aspired to be the "optical Newton" to MacCullagh's "optical Kepler."

The distinction that Hamilton made was one that MacCullagh had already recognized in his paper. He admitted that his hypotheses were "nothing more than fortunate conjectures." He believed that they were probably correct because they led to results agreeing with experiment,

but this is all that we can assert respecting them. We cannot attempt to deduce them from first principles; because, in the theory of light, such principles are still to be sought for. It is certain, indeed, that light is produced by undulations, propagated, with transversal vibrations, through a highly elastic ether; but the constitution of this ether, and the laws of its connexion (if it has any connexion) with the particles of bodies, are utterly unknown. The peculiar mechanism of light is a secret that we have not yet been able to penetrate. ... In short, the whole amount of our knowledge, with regard to the propagation of light, is confined to the *laws* of phenomena; scarcely any approach has been made to a mechanical theory of those laws.[23]

Still, MacCullagh's success in reducing the known laws to elegant mathematical form represented a much greater accomplishment than Hamilton's frustrating attempts to discover a dynamic model for dispersion.

In 1839 MacCullagh published an important extension of his theory that brought it much closer to a real dynamic theory of the ether.[24] It was still not a "dynamic deduction" of the kind that Hamilton sought, because it did not assume any particular structure for the aether, or, in MacCullagh's own words: "The reasoning which has been used ... is indirect, and cannot be regarded as sufficient, in a mechanical point of view. It is, however, the only kind of reasoning which we are able to employ, as the constitution of the luminiferous medium is entirely unknown."[25] But it did show how crystalline reflection and refraction could be explained by an aether, if one was willing to make some strange assumptions about the character of that aether. The elastic-solid model would no longer work, because the new theory required that the potential energy depend on the *rotation* of the volume elements of the aether.[26] It was hard to conceive of any mechanical model for such a rotationally elastic substance. A further weakness was an apparent denial of the equality of action and reaction of moments in the rotational terms.[27] Because it differed so radically from the usual mechanical model, MacCullagh's theory had no immediate success. The British aether theorists of the midcentury followed instead the dynamic aether of George Green, whose papers were published in 1837 and 1839, the same two years in which MacCullagh's papers came out.[28]

However, MacCullagh's theory was largely vindicated by the subsequent progress of physics. Those aspects of the theory, such as the rotational elasticity, that made no sense in a theory of the elastic solid were much more suitable for an electromagnetic theory of the aether, and in 1878 G. F. Fitzgerald showed how MacCullagh's theory could be incorporated into James Clerk Maxwell's electromagnetic theory to explain reflection and refraction, a subject that Maxwell had not been able to treat successfully.[29] In the 1890s Joseph Larmor further exploited MacCullagh's aether and tried to correct some of its dynamical deficiencies by introducing electrons into the hypothetical aether.[30]

The successes of MacCullagh, Green, and Cauchy in the years 1836 through 1839 may have persuaded Hamilton that there was little that he could add to the theory of the luminiferous aether. His own approach to the problem by way of assigning some hypothetical structure to the aether and deducing from it optical laws had not been productive. It appeared that for the immediate future the mathematical inductive approach would continue to be more productive than his own dynamic deductive approach. Besides, the field had become uncomfortably crowded, and Hamilton, who was always happier working on some completely new idea, finally found a problem to his liking in January, 1839. All previous wave theories had studied the propagation of light as if the wave were already established in the medium. A moving wave "front" was determined by the successive positions of a wave crest; the assumptions made were that the aether molecules were already in vibration and that one could follow a single wave through the vibrating medium. It was, therefore, a "steady state" theory, describing the motion of the aether some time after the first wave entered the quiescent medium. Hamilton wished to study the action of the light *first entering* a medium and to calculate how the wave acts as the light moves into a darkened region. He called this new science the *dynamics of darkness*, or *skotodynamics*, the second term being suggested by John Herschel. As usual, Hamilton's struggle with this problem produced fruitful new mathematical methods, including a theory of "fluctuating functions" and the asymptotic expansion of what are now called Bessel functions.[31]

But his most important discovery was the difference between the group velocity and the phase velocity in the propagation of waves.[32] Herschel pointed out that this difference can be observed in the expanding circular ripple of waves created by a stone thrown into the water.[33] The ripple consists of waves of different amplitudes, and one can observe that the place of greatest amplitude does not always stay at the same wave crest. The energy of the disturbance is therefore being propagated at a velocity different from that of the waves themselves. Hamilton believed his discovery was important enough to warrant immediate presentation to the Royal Irish Academy as well as a letter to John Herschel inquiring if it was original.[34] He was unwilling to make any speculation about the physical consequences of his new discovery until he had refined it further, but he obviously recognized that it was related to the problem of dispersion, because he concluded the second of his two papers on the subject with an explicit reference to dispersion.[35] These two papers, which were tucked away in the *Proceedings* of the Royal Irish Academy in an incomplete state, were obviously meant as anticipations of a much larger and more extensive memoir for the *Transactions*.[36] But the memoir never appeared and the masses of manuscript calculations that would have made up the body of such a memoir remained unpublished until 1940.[37] Herschel had asked for Hamilton's paper, had received it and lost it, asked for another copy, which he received—but by this time Hamilton had lost interest. A statement in one of his letters to Herschel indicates that he packed up at least part of his manuscript study and sent it off to Herschel to do with as he liked.[38]

The clue to a correct theory of dispersion came with the discovery of anomalous dispersion by F. P. Leroux in 1862, three years before Hamilton's death. Leroux discovered that some substances, such as the vapors of iodine and sodium, will refract the red rays through a larger angle than the violet—just the opposite of the normal case of dispersion.[39] Theories such as Hamilton's and Cauchy's, which had been attempts to explain dispersion by the "graininess" of the aether, required the violet rays to be refracted more than the red rays, so the discovery of anomalous dispersion essentially destroyed these theories. It was soon discovered, however, that anomalous dispersion was always associated with a frequency of strong absorption in the medium. The curve for dispersion (which plots index of refraction against the wavelength of light) approximates the curve predicted by Cauchy until it approaches a wavelength of strong absorption. At that point the index drops rapidly, followed by a leap to a much higher figure, and then it drops back to the Cauchy curve. The phenomenon has a simple mechanical explanation.[40] In a dispersive medium the aether shares the space with the gross molecules of the dispersive substance. As the vibrations of the aether approach the natural frequency of vibration of the molecules of the dispersive medium, the latter will be set into sympathetic vibration, first in phase with the aether, then out of phase, and then back in phase again as the wavelength of the light is increased. Light at the natural frequency of vibration of the molecules will be almost totally absorbed by the molecules of the dispersive medium and then lost to some damping mechanism in the medium. The explanation of anomalous dispersion suggests a new mechanical theory of normal dispersion. Instead of explaining it by the graininess of the aether, Maxwell suggested that it might be caused by the molecules of the crystal absorbing energy from light waves in a continuous aether, and then by their motion generating new waves in the aether, which combine with the original light waves. The result will be a phase shift that, if repeated by all the numerous molecules in the crystal, will appear as a change in the velocity of the light, the amount of change depending on the frequency.

In his skotodynamics Hamilton was on the right track, because his distinction between wave and group velocity was a major step towards understanding the composition of wave motions. But the discovery of anomalous dispersion came twenty years too late, and until mathematicians had this clue to the mechanism of dispersion, they had little to guide them in the right direction.

By the 1840s the debate over the nature of light had been pretty much decided in favor of the wave theorists. There was, however, one last flurry of controversy in 1842 at the Manchester meeting of the British Association, and it involved Hamilton directly. It is striking that the debate could have produced such passion fifteen years after it had first begun, but it is commonly the case with controversies that the old guard refuses to be converted. Potter had reopened the attack in 1840 with a new series of experiments designed to destroy the undulatory theory.[41] Brewster, who had become even more prickly, called Airy's Bakerian Lecture of 1840 "a complete blunder from beginning to end," and in 1841 he charged that the "Cambridge faction" had gained control of the Royal Society, which now rejected all papers not supportive of the undulatory theory.[42] In this intemperate atmosphere the debate opened at Manchester with a description by Brewster of strange screw-shaped fringes that he had seen, with his eye partly covered by a thin plate, in the spectrum of a prism.[43] He declared emphatically that the phenomenon could not be explained by the undulatory theory. Airy tried to find some fault with the experiment, but Brewster insisted that the results were conclusive. Hamilton, in an attempt to please everybody, remarked that "the warmest advocate of the wave theory of light must be gratified with these valuable experiments of Sir David Brewster; even though they should require the wave theory, in its present form, to be abandoned, and yet it was probable they might suggest the very modifications which will adapt it to the enlargement of our knowledge." These words did nothing to placate the hostile feelings. Herschel added that the wave theory was in the position of the theory of gravitation in its early years. It had not been developed far enough to meet every conceivable challenge immediately, but as it continued to be extended by other investigators, it would probably prove to be correct. Therefore it was not proper to put an infant like the wave theory on trial for its life until it had had some time to grow. Brewster replied that he did not wish "to put the undulatory theory on its trial for life or death, but upon one count of the indictment; for he conceived it entirely failed in explaining those facts which he had brought before the section." MacCullagh then joined the debate to say that so little was known about the undulatory theory that "it would be premature to pronounce, that it either could or could not explain every fact." He believed that the undulatory theory would remain incomplete until it was grounded in physical principles, to which Lloyd (also entering the debate at this point) responded that it *did* have a grounding in physical principles. The interchange between MacCullagh and Lloyd was warming up dangerously, and Hamilton stepped in again, this time with a joke, to calm his Irish colleagues. He "hoped it would not be supposed that the wave men were wavering, or that the undulatory theory was at all undulatory in their minds." As to the physical foundation of the various aether theories, he declared that nobody could be a better authority on that subject than MacCullagh.

So far all the major contributors to British optics had spoken— Airy, Brewster, Hamilton, Herschel, Lloyd, and MacCullagh, with only MacCullagh expressing any real support for Brewster. The discussion was reopened on a subsequent day, but when it looked as if another particle-wave debate was developing, Herschel stepped in to change the subject. He announced that "he wished much he could prevail on Sir W. Hamilton to explain to the Section a metaphysical conception, which he had disclosed to him, and which seemed to him, though darkly he owned, to shadow forth a possible explanation of many difficulties." Hamilton then presented his theory, which the unfortunate *Athenaeum* reporter admitted he could not understand: "It appeared to depend on the conception of points, absolutely fixed in space, and endowed with certain properties and powers of transmission, according to determined laws."[44]

The reporter for the *Manchester Guardian* had better luck. At least he reported Hamilton's description, even though he did not understand it:

When he [Hamilton] reflected on the phenomena of light, as explained by the wave-theory, and on the circumstance that no phenomenon had hitherto afforded the least guess as to the *nisus* of the vibration, it occurred to him that there was no *nisus*. Although this might appear to be Irish—a sort of mathematical bull—yet he should take it by the horns. One could easily conceive a *nisus* of motion, without any extent of motion—one of those excessively minute things which was believed to exist, but of the size of which no definite notion was entertained. What he wished to be conceived was, that each vibration had some determinate direction; that at each moment the *displacement* was towards *one* part of space; the displacement . . . had a latitude and a longitude; in fact, two angular polar co-ordinates might represent the direction in which the point was supposed to be displaced. In short, there would be a polarity of the *nisus*—an intensity of it, and that polarity would be expressed by two angles and *nil*. For each particle of the ether (if they chose to retain that word), there would be three functions of four quantities—namely, three things considered as functions of the three co-ordinates, and the points of space which marked the time. He thought the conception of those three functions as clear as any thing he knew in the mathematics. He did not wish to be considered as offering more than a guess, which he thought might be thus clothed in mathematical language.[45]

These two accounts do not agree in their entirety, because the *Athenaeum* mentions "points absolutely fixed in space" while the *Guardian* talks about "displacements" of the particles. Hamilton probably meant that there was no actual motion of the particles, but that the intensity or activity representing light was passed from one stationary point to the next. Although the particles were mathematical points, they had polar properties, because their action determined both a direction of propagation and an orientation at right angles to the direction of propagation, which gave the plane of polarization.

In support of Hamilton's theory, Herschel said that "a power exerted in a point of space, or a quality temporarily in a point of space, was a perfectly distinct and consistent conception. He (Sir John Herschel) could also conceive such powers moving into another point of space, so as to cause the propagation of a wave of power, without any movement of a wave of substance."[46] Herschel described the theory in greater detail in one of his *Popular Lectures on Scientific Subjects*. The aether is "conceived as an indefinite number of regularly arranged equidistant points (mathematical localities) *absolutely fixed and immovable in space*, upon which, as on central pivots, the molecules of the ether, supposed *polar* in their constitution, like little magnets (but each with *three pairs of poles*, at the extremities of three axes at right angles to each other), should be capable of oscillating freely, as a compass-needle on its center, but in all directions."[47] The theory, as Herschel described it, is more of a mechanical theory than a metaphysical one, but it retains the Boscovichean notion of point centers of force that Hamilton had described to Coleridge in 1832. These points were absolutely fixed in space, and transmitted "powers" by rotation. Because the points are fixed, the aether would be incompressible and would not transmit longitudinal waves. Also, the energy would be transmitted by rotation, not displacement of the unextended points. This was just what MacCullagh's "mathematical induction" required. It was now expressed by Hamilton, however, as a "dynamic" theory, albeit a peculiarly metaphysical one.

After hearing Hamilton's presentation Brewster and MacCullagh objected to the fanciful nature of Hamilton's theory. MacCullagh said that he "had indulged in speculations . . . involving this very conception of Sir W. Hamilton, and had even followed out some of its consequences . . . but he had abandoned it as mere speculation."[48] "He thought . . . that the *safest* way was not to start with metaphysical speculations, and *a priori* views, which were generally barren; but to reason upwards from the facts."[49] Brewster, upon seeing an opportunity to get in a few additional licks against the undulatory theory, rose and seconded MacCullagh, saying that "these speculations tended to repress experimental research, and to turn men's minds from what was solid to what was fanciful. He conceived also that indulgence in them, and mere abstract mathematical research, by rendering men averse from the more humble and laborious pursuits of experiment absolutely produced a distaste for these subjects; and to this he attributed the fact that, while learned societies frequently overlooked, and even refused to publish in the *Transactions* experimental papers, the transcendental flights were always sure to find a welcome place."[50]

Since Herschel had urged Hamilton to present his views, he felt obliged to defend them, and replied to Brewster and MacCullagh that "there could be no true philosophy, without a certain degree of boldness in guessing; and such guessing, or hypothesis, was always necessary in the early stages of philosophy, before a theory has become an established certainty; and these bold guesses, in their proper places, he conceived, should be encouraged, and not repressed. Sir W. Hamilton's conception, he thought, perfectly clear in its metaphysics, and should not be thrown overboard merely because it was metaphysical."[51]

Much of the quarrel at the British Association meeting came from Hamilton's use of the word *metaphysical*. What he had created was a model halfway between MacCullagh's mathematical induction and his own hoped-for dynamic deduction. His theory was not entirely mathematical, because it assigned a structure to the aether and gave laws for the activity of the particles. But neither was it a dynamic deduction, because it did not obey the laws of motion and because it had not been deduced from a priori principles. Hamilton created his "metaphysical" theory to provide just those properties that MacCullagh's mathematical theory required, and MacCullagh's theory had in turn been drawn from experiment. Such concepts as force fields and point centers of force in space were becoming acceptable in physics, and Hamilton could have used the concepts without calling them metaphysical. MacCullagh and Brewster were probably right in their objections. The model had too much of the flavor of an ad hoc hypothesis. On the other hand, it did suggest that there could be a physical model for the aether other than that of material atoms acting according to the accepted laws of motion. The elastic solid model had obviously failed, and it would eventually be necessary to create some other model that would give physical meaning to MacCullagh's equations. So to this extent Herschel's reply to MacCullagh's objection was justified.

Hamilton expected the reaction he received from Brewster, but he was annoyed at MacCullagh for claiming once again to have anticipated him in a discovery or important idea. He wrote to Robert Graves:

The peculiarity of memory of the friend to whom you allude is one with which I do not well know whether to be amused or annoyed. It reminds me of a remark once made by Wordsworth respecting another great poet of modern times, [Coleridge] that, with reference to compositions which he had in his mind, he was apt to confound the future with the past, and to imagine that he had already written them. In the person you refer to [MacCullagh] there seems to be a sort of Platonic reminiscence, by which any discovery or suggestion that strikes him in other people is recognised as having been known in some former state of existence.[52]

Hamilton was obviously annoyed by MacCullagh's frequent claims of priority. MacCullagh thought that he had "suggested" to Hamilton not only conical refraction, but also the system of polar molecules and quaternions. Ever since 1837, when he became president of the Royal

Irish Academy, Hamilton had been trying to improve his relations with MacCullagh, but with no marked success.[53] Then, in 1847, without any warning, MacCullagh committed suicide. Hamilton was deeply disturbed by his death and realized that MacCullagh's exaggerated claims were the product of a deranged mind. The whole matter made him touchy about questions of priority. Much of the preface to his *Lectures on Quaternions* was devoted to a historical sketch, carefully apportioning credit for advances in algebra previous to his quaternions. Hamilton explained his reasons for the long preface to Augustus De Morgan, to whom he applied for much of the historical information:

This is a delicate point to touch on—I had been made cautious, perhaps sensitive, by my intercourse with poor MacCullagh, who was constantly fancying that people were plundering his stores, which certainly were worth the robbing. This was, no doubt, a sort of premonitory symptom of that insanity which produced his awful end. He could inspire love, and yet it was difficult to live with him; and I am thankful that I escaped so well as I did, from a quarrel, partly perhaps because I do not *live* in College, nor in Dublin. I fear that all this must seem a little unkind; but you will understand me.[54]

Hamilton had nothing to feel guilty about, but he felt guilty just the same.

Obituaries appearing in the *Encyclopaedia Britannica* and in the *Abstracts of the Papers Communicated to the Royal Society of London* give sharply conflicting evaluations of MacCullagh's work. The writer for the *Encyclopaedia Britannica* dramatized with little tact the circumstances of MacCullagh's death:

His methods, which, in less known subjects, were almost entirely tentative, were altogether inadequate to the solution of the more profound physical problems to which his attention was mainly devoted, such as the theories of double refraction, of crystalline reflexion, etc. Here not only are the utmost powers of analysis required, but also the highest physical knowledge, and in consequence MacCullagh's work was entirely overshadowed by that of contemporaries such as Cauchy and Green. ... The story of his later days painfully suggests the comparison of a high-bred but slight racer tearing itself to pieces in the vain endeavour to move a huge load which a traction-engine could draw with ease and promptitude. He wasted, on problems altogether beyond his strength, powers of no common order, which, had they only been suitably directed, might have immensely extended our knowledge. Such at least is the estimate which we cannot avoid forming from a perusal of his published works.[55]

Although anonymous, this article employs the style of P. G. Tait, a frequent contributor of articles to the *Britannica*.[56]

By contrast, the obituary in the *Abstracts*, which was written by Hamilton, claimed that MacCullagh had given "to posterity a perfect and complete mechanical theory; that is to say, analytically complete:

so that any one who in future may attempt to discover in this region of science, can only do so by treading in his steps, and adopting his principles, but can never supersede them. In fact he has discovered and handed down the general principles which must hold in all cases."[57]

Both of these evaluations of MacCullagh's work are obviously exaggerated, and the debate still goes on, although in decidedly more measured tones. Sir Edmund Whittaker says that "there can be no doubt that MacCullagh really solved the problem of devising a medium whose vibrations, calculated in accordance with the correct laws of dynamics, should have the same properties as the vibrations of light, ... thereby affecting that reconciliation of the theories of light and dynamics which had been the dream of every physicist since the days of Descartes," while Kenneth F. Schaffner wrote his recent book, *Nineteenth-Century Aether Theories*, at least in part to correct "some severe defects in Whittaker's monograph, particularly in connection with his unwarranted idolization of MacCullagh's aether."[58] Measuring MacCullagh's technical achievement should not be all that controversial, but measuring its historical "influence" will probably continue to generate debate.

Considering the importance of the debate over the nature of light, Hamilton's contributions to the subject are surprisingly few. Certainly he did not attack the problem with anything like the energy that he had expended on his systems of rays or that he would expend on his later studies of quaternions. The few papers that he did publish on the subject were extremely tentative, and he often allowed his ideas to be presented by others (Baden Powell and John Herschel), or left them in manuscript. One would have thought that his great mathematical power, his previous interest in optics, and his sympathy for Faraday's views would have commended the problem to his attention. But Hamilton was not as much interested in Faraday's physics as in his metaphysics—Faraday's denial of material atoms and his concept of forces acting in space and time.[59] Hamilton did not show any inclination to study Faraday's experimental researches in detail. The problem of the luminiferous aether required a mathematical physicist, not a mathematical metaphysician. Hamilton was not close enough to the experimental work to see his way through to a solution.

When James Clerk Maxwell finally solved the problem of the aether in the 1860s by identifying light with waves in the electromagnetic field, Hamilton was near the end of his career and had moved a long way from optics. The electromagnetic theory of light was a solution that he would have applauded, because it unified disparate parts of physics by a single theory of forces acting in space and time. But it was not the "dynamical deduction" that he had expected, because it depended substantially on experiment. In 1841, one year before the

Manchester meeting, Hamilton had described to John Herschel what he believed to be the most fruitful approach to the problem of the luminiferous aether:

I have been struck by a remark of Faraday's which was somewhat to this effect, that the mechanical motion of an electrically excited body must be considered to produce, or be equivalent to, a current of electricity. Light, heat, chemistry, electricity, crystallography (with galvanism, magnetism, etc., as more obviously related therewith), have been this good while suspected not to say known, to be all branches of one science, as yet imperfectly discerned, but to be probably established in this century—may it be in this generation, and in our own time! ... But to make any approach to such a consummation, it will (I think) be absolutely necessary to give fair play to metaphysical as well as to mathematical thought— experiment and observation may be trusted to take care of themselves.[60]

If Hamilton had been willing to give more room to experiment in his own attempts to construct an aether theory, his work might have been more fruitful. It is important to note, however, that strictly mechanical theories of the aether did not succeed, either. Maxwell depended on

Michael Faraday

Faraday, and Faraday's success lay in his ability to conceive of highly abstract models of force fields, models that differed drastically from the usual mechanical explanations. Few mathematicians or physicists before Maxwell had any real appreciation for Faraday's field theory, and Hamilton was one of those few. His preference for forces acting in space and time was in the right direction, but because he had no familiarity with, or real appreciation for, the experimental side of physics, he was never in a position to make the synthesis that was finally accomplished by Maxwell.

12

The Metaphysical Foundations
of Mechanics

HAMILTON'S "General Method in Dynamics" is the most important of all his major mathematical discoveries; it is also the one on which he spent the least time. It grew directly out of his "Theory of Systems of Rays," and it exhibited to an extraordinary degree his ability to create rapidly an enormous body of theory almost unparalleled in generality and abstraction. Hamilton was not skillful in applying his own theories to physical problems. It is not that he totally disregarded the application of his ideas. But for one reason or another they never seemed to find their way into print in a usable form. He applied his "Theory of Systems of Rays" to the analysis of the symmetrical optical system with notable success, but left it in manuscript. He made little use of the important analysis of wave motion that appeared in his skotodynamics, and as we have seen, he was largely unsuccessful in his attempt to construct a satisfactory model for the luminiferous aether. But Hamilton's success was not in what he contributed directly to physics, but in the opportunities that he left for others. James Clerk Maxwell, who finally solved the puzzle of the aether, understood this very clearly. He wrote to Peter Guthrie Tait: "The good of H.[amilton] is not in what he has done but in the work (not nearly half done) which he makes other people do. But to understand him you should look him up, and go through all kinds of sciences, then you go back to him, and he tells you a wrinkle."[1] The "General Method in Dynamics" contained the most important "wrinkle" of them all.

Hamilton began to apply his optical theory to dynamics early in 1833, just after the successful confirmation of conical refraction by Humphrey

Lloyd. He turned from optics to dynamics because the theory of the luminiferous aether was basically a *dynamic* problem—to find the forces and motions of a medium that would explain the phenomena of physical optics. The "Theory of Systems of Rays," on the other hand, had been a theory of *geometrical* optics. It studied the geometrical paths of light without regard for the motion of the waves or particles transmitting the light, and, as a geometrical theory, it had no dynamic content.

In their application, geometrical ray optics and dynamics are quite different sciences, but Hamilton found a beautiful theory that would connect the two at an extremely abstract level. The theory was largely mathematical, but it had practical applications, particularly in celestial mechanics, which Hamilton explored at the same time he was working out his theory.

Hamilton's first mention of his new theory was in a letter to Lloyd written on February 9, 1833. He told Lloyd that he had found the form of a new dynamical function for elliptical motion and that he wished to try it out on the difficult problem of planetary perturbations. He also said that by using his new system, he had worked out the equations of motion for any number of mutually attracting points.[2] Throughout 1833 Hamilton was busy with his study of the foundations of algebra, his controversy with Potter, marriage, and the British Association meeting in Cambridge, but he kept coming back to mechanics. In September, 1833, he wrote again to Lloyd: "You may remember my telling you in February of an extension of my optical method and function to physical astronomy which I had just then been thinking of. I have since found that nearly the same idea had occurred to me in 1826, a time when I made some of my chief general steps in optics, which the cycle of thought has lately caused me to retread, after quite forgetting some of them amid the details of calculation. I had even determined the explicit form of the dynamical function for the case of ordinary projectiles in a void."[3] Hamilton had found his journal of 1826, where he had worked out his Principle of Constant Action for the cases of projectile motion and atmospheric refraction while studying for the college examinations. It must have been an exciting moment when he recovered these old ideas, especially since he was now prepared to recast them in an entirely new light.

In the same letter he also informed Lloyd that he was writing a popular article for the *Dublin University Review* with which he said he was taking "some pains." The Board of Fellows, at their annual visitation to the observatory, had asked Hamilton to write an "easy" book on his optical methods.[4] He regarded his article in the *Dublin University Review* as "practice and preparation" for the book on light, which, in fact, was never written. True to character, Hamilton was unable to write about anything that was not foremost in his mind, and the article appeared as a paper "On a General Method of Expressing the Paths of Light, and of the Planets, by the Coefficients of a Characteristic Function."[5] It begins with a long historical and philosophical introduction that stresses the prime importance of the Principle of Least Action for optics and mechanics, then outlines Hamilton's application of that principle to optical rays, and concludes with a brief application of the theory to planetary orbits. It is an uneven paper. The historical introduction could entertain the average reader, but the exposition of the characteristic function would stop all but the most hardy, and his final application of it to planetary orbits was too brief to carry the reader very far. By December he was churning out pages of manuscript on the famous three-body problem.[6] Although he found no solution superior to those already known, he did sharpen his new theory using the three-body problem as a whetstone. Through the summer and fall of 1834 he wrote the two famous papers "On a General Method in Dynamics."

William Whewell was another catalyst for Hamilton's work in dynamics. Whewell had begun his extraordinarily varied career by writing a series of books on mechanics and dynamics. Beginning in 1819 (with an *Elementary Treatise on Mechanics*), these books went through many editions, each with changes of title and organization.[7] On the whole they can be described as straightforward textbooks appropriate for an undergraduate course. They relied heavily on Newtonian mechanics and contained little on the philosophical or logical foundations of mechanics. However, when Hamilton and Whewell first met in April, 1832, Whewell was already collecting material for his two most famous works, the *History of the Inductive Sciences* (1837) and the *Philosophy of the Inductive Sciences* (1840), and he engaged Hamilton in a discussion of the foundations of dynamics.[8] At that time Hamilton was still completely absorbed in optics. Within nine months, however, he had begun to apply the characteristic function to dynamics, and it may well have been Whewell who got him to think about the subject. Hamilton found Whewell to be a congenial philosopher. Most British scientists did not share Hamilton's idealist views, so it was a real pleasure for him to be able to write to his friend Aubrey De Vere: "In Whewell at Cambridge, I thought with delight that I perceived a philosophical spirit more deep and true than I had dared to hope for"; and to Coleridge he wrote: "Professor Whewell, a man of great variety of mind, appeared to me to have more of the philosophical spirit than any other whom I met there [at Cambridge]."[9]

Hamilton and Whewell were together again at the British Association meeting at Oxford in June, 1832, and at the Cambridge meeting the following year. In anticipation of another discussion at the Cambridge meeting Whewell sent Hamilton his recently published *First Principles of Mechanics*, a book that Whewell said he much wanted Hamilton to see. Hamilton responded by giving his own position on the foundation of mechanics—a position that differed considerably from that of Whewell.[10]

The opinion which I think I tried to express when I had the pleasure of talking the thing over with you, and which I have not since reflected on enough to alter it, was in substance this: that there are, or may be imagined, two dynamical sciences: one subjective, *a priori*, metaphysical, deducible from meditation on our ideas of Power, Space, Time; the other objective, *a posteriori*, physical, discoverable by observation and generalization of facts or phenomena: that these two sciences are distinct in kind, but intimately and wonderfully connected, in consequence of the ultimate union of the subjective and objective in God, or, to speak less technically and more religiously, by virtue of the manifestation which he has been pleased to make of himself in the universe to the intellect of man; so that the two sciences are never wholly separate, but may and ought to advance together, and use many common expressions, and each possess an analogon to many if not to all of the results and theorems of the other. For example, it is, I think, a subjective, *a priori*, metaphysical theorem, that when we think of a point as moving in a curve, we must think of it as changing direction, and must consider this change as an effect, and must attribute the effect to a cause; which may fitly be called a deflecting force or power. And it is, I think, another theorem, objective, *a posteriori*, physical, discovered by observation and generalisation of facts, that when a body is found to move in a curve, we may expect to find some other body, resisting or attracting, or somehow physically influencing the former. This physical result, and not the analogous, but distinct metaphysical theorem, I take to be the law of natural rectilinear motion which Newton meant to propose (agreeing, I believe, with you in this interpretation of the *Principia*); and I account it a great physical and inductive discovery, the merit of the establishment of which seems mainly due to Newton, whatever may have been guessed by others. The metaphysical theorem I account indeed higher in dignity; but do not consider it as including the other, or as an adequate ground to us for the expectation of any one appearance, though in the Divine mind, indeed, there may be some mysterious union between the causes or first springs of our thoughts and our sensations; of the ideas which seem necessary and eternal, and of the phenomena which seem casual and changing.[11]

Hamilton's argument is perplexing as it stands. He certainly cannot mean that a physical cause must be given for the shape of a mathematical curve. His "subjective, *a priori*, metaphysical theorem" is a theorem about the limits of possible experience. It is a statement that all sequential appearances must be linked in time by a relationship of cause and effect. Therefore without any recourse at all to experiment or observation, we must conclude that any change in the state of uniform motion of a point requires a cause. Hamilton recognizes that Newton's First Law of Motion, on the other hand, is founded on experience. Whenever a physical object in uniform motion is observed to change its speed or direction, it has always been possible to find another physical body *causing* the change of motion of the former body, either by friction, collision, attraction, and so on, and we conclude by induction from past experience that we will always be able to find a cause in all future cases. Hamilton concludes that the agreement between our *mental* rules and the observed laws of motion must result from the fact thay they are both the creations of one God and are unified in His single will.

In his paper "On a General Method of Expressing the Paths of Light, and of the Planets, by the Coefficients of a Characteristic Function," Hamilton gave a brief description of his dynamical method for the first time. The long introduction to this paper is largely historical rather than philosophical, but Hamilton does say something about method. He quotes directly from Newton to describe the method of induction and deduction, analysis and synthesis, whereby "we . . . gather and group appearances, until the scientific imagination discerns their hidden law, and unity arises from variety: and then from unity must re-deduce variety, and force the discovered law to utter its revelations of the future."[12] The highest induction so far attained for optics, according to Hamilton, was the Principle of Least Action, from which he believed the prediction of phenomena had to be deduced.

Hamilton was unwilling to adopt Maupertuis's metaphysical proof of the existence of God through the Principle of Least Action, but his own metaphysical studies were advancing further into idealism. Every year during Michaelmas Term he gave a series of lectures on astronomy. These lectures formed his major duty as Andrews Professor at Trinity College. The lectures tended to be technical, but he always began each series with a popular "philosophical" lecture, which he delivered in a high rhetorical style and which often attracted outside auditors. (Ladies began attending this initial lecture, which caused some problems at an all-male university.) In December, 1833, he began his lectures by expanding on his earlier comments to Whewell about the two theories of dynamics.

And so say I with respect to the observation of phenomena, even when combined with mathematical calculation: that the visible world supposes an invisible world as its interpreter, and that in the application of the mathematics themselves there must (if I may venture on the word) be something meta-mathematical. Though the senses may make known the phenomena, and mathematical methods may arrange them, yet the craving of our nature is not satisfied till we trace in them the projection of ourselves, or that which is divine within us; till we perceive an analogy between the laws of outward appearances and our inward laws and forms of thought; till the Will, which transcends the sphere of sense, and even the sphere of mathematical science, but which constitutes (in conjunction with the conscience) our own proper being and identity, is reflected back to us from the mirror of the universe by an image mentally discerned. This it is, and not [merely] the beauty of the mathematical reasoning, nor the practical accordance with phenomena, great and important as they are, which gives the highest value and the deepest truth to the dynamical theory of gravitation. Do you think that we *see* the attraction of the Planets? We scarcely see their orbits. . . . We observe, or rather we *make*, the configurations and arrangements of these visibles by mathematical moulds of our own minds.[13]

Theory is both metaphysical and metamathematical. It is created by "our inward laws and forms of thought" and reflected back to us by the phenomena of the universe. As he continued his lecture with a brief history of modern astronomy (still fresh in his mind from the paper "On the Paths of Light and the Planets"), Hamilton came to Newton's concept of universal gravitation, which he termed an "external image of the will," implying that the dynamic *power* of gravity was a creation of the human mind. The mind attributes "causes" to perceived events that are really only extensions of our own subjective feelings of power and volition.

In concluding his lecture, Hamilton claimed that it was this mental or metaphysical construction of theory that men really sought. Even if scientists discovered all the forces of nature and were able to predict all nonmiraculous phenomena, still "we should desire to fuse them all and interpenetrate them all with mind, and throw over them all the poetry of Science; and from the seemingly finished work there would rise up a new and growing enterprise, an unexplored and unimagined world of genius."[14]

All of these ideas are loosely expressed, and it is impossible to know precisely what Hamilton meant by the "inward laws and forms of thought" and by an "external image of the will," but it is easy enough to see the tendency of his thought towards idealism. By March, 1834, Hamilton had completed the first part of his "General Method in Dynamics." It was communicated to the Royal Society on April 1 by Francis Beaufort and read April 10. Hamilton sent it to the Royal Society with Uncle James's blessing. Hamilton and his uncle had long since learned that anything sent to the Royal Irish Academy would be entombed for months, perhaps years, and when finally released would enjoy only a modest circulation in England and none on the Continent.[15] James thought it perhaps inaccurate to call this paper a "General Method in Dynamics," since Hamilton had applied it only to problems of celestial mechanics, but Hamilton responded that the title was correct because it expressed his "hope and purpose to remodel the whole of Dynamics, in the most extensive sense of the word, by the idea of my Characteristic Function or central law of relation."[16]

Hamilton was fully aware of the importance of his new dynamics. Not only was it important for the value that it might have in solving dynamical problems in optics and astronomy, it also extended that "invisible world" described in his lecture, which served as interpreter of the visible world. To Whewell he simply stated that "whenever it shall be caught by others, it will make, perhaps, a revolution."[17]

Hamilton said that he had not ventured into metaphysics in the paper. "It is merely mathematical and deductive. I ventured, indeed, to call the *Mécanique Analytique* of Lagrange 'a scientific poem'; and spoke of Dynamics, or the science of force, as treating of 'Power acting by Law in Space and Time.' In other respects it is as unpoetical and unmetaphysical as my gravest friends could desire. Yet it is unpractical enough to excite, perhaps, the contempt or pity of many worthy people. After so much Algebra, I intend to refresh myself awhile with other things—the stars and Kant."[18]

Whewell congratulated Hamilton on his achievement and sent a paper that he had recently written, entitled "On the Nature of the Truth of the Laws of Motion."[19] Hamilton liked the paper, because it indicated that he and Whewell were now in much closer agreement than they had been. To Whewell he said that he did not know whether this was due to a change in Whewell's attitude or in his own, but in writing to Adare he stated no such doubts: "Whewell has come round almost entirely to my views about the Laws of motion."[20]

As a logical study of the laws of motion Whewell's paper leaves much to be desired, but it is easy to see why Hamilton liked it.[21] Whewell argued that the laws of motion have both an a priori content and a content from experience. The a priori content of mechanics he states in the form of three "axioms of causation," the first of which approximates Hamilton's "subjective, a priori, metaphysical theorem" that the motion of a body along a curve must necessarily have a cause.[22] It was, apparently, Whewell's a priori theorems that represented the change that Hamilton noticed. There did remain some differences, however, between his position and that of Whewell. Where Hamilton believed that there are two theories of dynamics, one a priori and one a posteriori, which agree because they are both creations of God's will, Whewell believed that there is one theory of dynamics, constructed in part from a priori and in part from a posteriori elements. Hamilton's arguments probably did persuade Whewell to change his mind. In 1837 Whewell stated that since his *History of the Inductive Sciences* covered much more fully the material contained in the *First Principles of Mechanics,* he now considered the book superseded.[23] The *First Principles* was the book that Whewell had sent to Hamilton for his opinion. Whewell's ideas had changed enough so that he now wished to repudiate the work.

It is possible that the greater coincidence between the views of Hamilton and Whewell in 1834 came from their common study of Kant. Both men found the a priori content of dynamics in the principle of causality. This is a bit surprising, because it was more customary, at least among the French philosophers, to find an a priori basis for the laws in geometry and to deny any knowledge of "causes" of motion.[24] But Kant had argued in the "Second Analogy of Experience" of the *Critique of Pure Reason* that experience is only *possible* through the representation of a necessary connection of perceptions, thereby providing a unity of all perceptions in time. The Law of Cause and Effect is the representation that allows us to link events in time. Without it no dynamics would be possible, because without it there could be no rules linking the observed motions of bodies. Moreover, the need for a relationship of cause and effect is known a priori without any recourse to experience. It is, as Hamilton claimed, a "subjective, a priori, metaphysical theorem." Both Whewell's and Hamilton's writings reflect a familiarity with Kant's "Second Analogy of Experience." Whewell defended Kant's "Second Analogy" in his *Philosophy of the Inductive Sciences*, and Hamilton used the same doctrine in arriving at his view of algebra.[25] In their theories of dynamics Whewell's assertion that dynamics has both an a priori and an a posteriori content was closer to Kant's theory than was Hamilton's belief in two separate sciences joined only by the benevolent act of God.

There is one last bit of evidence indicating that Hamilton may have been instrumental in getting Whewell to pursue his study of Kant. In 1842 Robert P. Graves wrote a short sketch of Hamilton's career for the *Dublin University Magazine*.[26] Graves mentioned that Hamilton was one of the first British philosophers to study the philosophy of Kant, "now made familiar to the English reader by 'Whewell's Philosophy of the Inductive Sciences.'"[27] The notice by itself proves nothing, but among Hamilton's manuscripts is a set of notes, in his hand, that he had prepared for Graves to use in writing his article. It was *Hamilton* who provided the information on Kant and Whewell, and Graves merely included it verbatim.[28] Therefore it was *Hamilton's* opinion that he had begun the study of Kant and that Whewell had followed him or at least had not preceded him.

Hamilton plunged deeper into the study of Kant after sending his first paper "On a General Method in Dynamics" to the Royal Society in March, 1834. His correspondence was full of metaphysics, particularly those letters to friends who shared his enthusiasm for metaphysical idealism. H.F.C. Logan, a tutor to the son of one of Maria Edgeworth's friends, seemed to have been the best informed, but Aubrey De Vere and Lord Adare also shared Hamilton's excitement as he went through the *Critique of Pure Reason*. The meeting of the British Association in Edinburgh lifted Hamilton out of his almost total immersion in Kant, and after returning to Ireland he joined his wife in Nenagh, where she was caring for her mother. Hamilton always found Bayly Farm a good place to study. He wrote to Lloyd that he "enjoyed a most luxurious quiet, far greater than any I have ever had at home; and to be ungrateful, I have been scribbling or rather writing in a copy-book with a decently fair hand, some twenty pages of a *Second Essay on Dynamics*, which is likely to go to press after some months without much farther alteration."[29] Hamilton was better than his word. The "Second Essay on a General Method in Dynamics" was received at the Royal Society before the month was out.[30]

13

The General Method in Dynamics

HAMILTON'S TWO PAPERS on dynamics contain little philosophy. The introduction to the first paper hints at a new, dynamic, Boscovichean view of the universe, but as he told Whewell, apart from this introduction, "It is as unpoetical and unmetaphysical as my bravest friends could desire." And Whewell caught the sense of Hamilton's theory correctly when he remarked that Hamilton would drill right through the science of mechanics and out the other side with his "long analytical borer."[1] The two papers are masterpieces of analysis; there is almost no physics as such in either one.

Hamilton begins the first paper by showing how the characteristic function in optics can be used analogously in dynamics. He derives the mechanical equivalent of the Principle of Constant Action and shows how the characteristic function can be found from the simultaneous solution of two partial differential equations of the first order and the second degree—the same method that he had employed in the "Theory of Systems of Rays." He then shows how different forms of the equations of motion can be deduced from his characteristic function and also shows that the equations are invariant under coordinate transformation. Finally, he applies his method to problems in celestial mechanics, the subject on which he first tested his ideas.

The second essay contains material more familiar to the modern student of Hamiltonian dynamics. At the end of the first essay Hamilton had derived an "auxiliary" function, which, in the second essay, he named the "principal function." He then derives the "canonical equations" of motion, "Hamilton's principle" and Hamilton's version of the

Hamilton-Jacobi equation. The rest of the paper is devoted to a study of perturbations of planetary orbits following the method of his dynamics.

Although both papers are long, the actual description of the method is brief and highly condensed.[2] He develops the entire theory from an equation that he calls the *"equation of the characteristic function* or the LAW OF VARYING ACTION." Although modern books on Hamilton's dynamics seldom mention the Law of Varying Action, he regarded it as the foundation of his new theory, as did his contemporaries. The Law of Varying Action does not appear anywhere in Hamilton's optical papers, at least not under that name. The first mention of it is in "On the Paths of Light and the Planets," where he states: "From this known law of least . . . action, I deduced (long since) another connected and coextensive principle, which may be called, by analogy, the LAW OF VARYING ACTION, and which seems to offer naturally a method such as we are seeking; the one law being as it were the last step in the ascending scale of induction, respecting linear paths of light, while the other law may usefully be made the first in the descending and deductive way."[3] Hamilton argues here that the Principle of Least Action is the "highest and most general axiom" that we can obtain by induction from the phenomena of optics and dynamics. His own Law of Varying Action is a *deductive* extension of the Principle of Least Action—an entirely mathematical consequence of a known mechanical principle.[4]

Moving backwards through Hamilton's papers on dynamics and optics we can see where the Law of Varying Action came from. In the first paper "On a General Method in Dynamics" (1834) he also calls it the "equation of the characteristic function." In the third supplement to his "Theory of Systems of Rays" (1832) there is no mention of the Law of Varying Action, but Hamilton does use the name "equation of the characteristic function" to designate his fundamental equation of optics.[5] In the first supplement (1830) he identifies the equation of the characteristic function with the Principle of Constant Action, which tells us that the Law of Varying Action in dynamics is the same as the Principle of Constant Action in optics.[6]

Moving from the supplements back to the three-part "Theory of Systems of Rays" (1827) we find no mention of the Principle of Constant Action, the Law of Varying Action, or the equation of the characteristic function, as Hamilton later named them, except in the last part of the table of contents, which describes the unpublished third part of the "Theory of Systems of Rays." There he says that in the text he gives "reasons for calling this principle the PRINCIPLE OF CONSTANT ACTION" and an "analogous principle respecting the motion of a system of bodies."[7] In the introduction to his first paper "On a General Method of Dynamics" Hamilton calls this brief mention of an analogous principle an "announcement" of his intention to apply his principle to dynamics, but because he also told Lloyd that he had forgotten about it during the intervening years, it cannot have been an announcement that he took seriously.

The Principle of Constant Action, the equation of the characteristic function, and the Law of Varying Action are all the same thing mathematically—the variation of the action integral written as a function of the initial and final coordinates. In the published first part of the "Theory of Systems of Rays," Hamilton treated only *straight* rays, and the full generality of his method was not obvious to him. But in the summer of 1826, as we have seen, he first began to study curved rays passing through a nonhomogeneous medium. At this point he explicitly defined the characteristic function (*V*) as the action integral, and it was at this point that he first began to talk about his Principle of Constant Action.[8]

He called it the *Principle of Constant Action* because it showed immediately the existence of surfaces orthogonal to the rays and the fact that the rays originating from a point source reach each of these surfaces together, having expended the same amount of action. They are therefore "surfaces of constant action." When he came to apply it in dynamics, however, he wanted to emphasize a different aspect of the principle. The existence of the orthogonal surfaces of constant action was not so important for dynamics as it had been for optics. Instead he wanted to indicate that his principle was an extension of the familiar Principle of Least Action and that it differed from that principle by allowing a variation of the end points of the particle's path. Thus the apparent contradiction involved in calling the same principle both a Principle of *Constant* Action and a Law of *Varying* Action was in fact only a difference in emphasis.

Because the Principle of Least Action applies to both mechanics and optics (understood as Fermat's Principle of Least Time in the optical case), the paths of a bundle of light rays and the paths of a system of particles can be described by completely analogous mathematical expressions. Hamilton derives his Law of Varying Action in the following way: By the "celebrated law of living force" (we would now call it the law of conservation of energy), Hamilton writes, $T = U + H$, where T = kinetic energy (whole living force), U = the negative of the potential energy, and H = total energy (constant in time, since he is here dealing with conservative systems). If we differentiate this expression with respect to time we get:

$$\frac{dT}{dt} = \frac{dU}{dt}.$$

The H drops out because it is constant in time. But, if instead of taking the time derivative along the path, we vary the path itself, including the initial point of the motion, the total energy will also vary, and we must write

$$\delta T = \delta U + \delta H. \tag{13.1}$$

The next step is to find values of δT and δU. From the definition of kinetic energy,

$$T = \frac{1}{2} \Sigma\, m\,(\dot{x}^2 + \dot{y}^2 + \dot{z}^2),$$

where we are summing over all the particles in the system. The variation of the kinetic energy is therefore

$$\delta T = \Sigma\, m\,(\dot{x}\,\delta\dot{x} + \dot{y}\,\delta\dot{y} + \dot{z}\,\delta\dot{z}).$$

We find δU from the fact that the work done by the particles in the system is equal to the change in potential energy.

$$\delta U = \Sigma\, m\,(\ddot{x}\,\delta x + \ddot{y}\,\delta y + \ddot{z}\,\delta z).$$

Therefore, substituting in 13.1, we get

$$\Sigma\, m\,(\dot{x}\,\delta\dot{x} + \dot{y}\,\delta\dot{y} + \dot{z}\,\delta\dot{z}) = \Sigma\, m\,(\ddot{x}\,\delta x + \ddot{y}\,\delta y + \ddot{z}\,\delta z) + \delta H.$$

Hamilton next multiplies each term by dt:

$$\Sigma\, m\left(\frac{dx}{dt}\,\delta\dot{x} + \frac{dy}{dt}\,\delta\dot{y} + \frac{dz}{dt}\,\delta\dot{z}\right) dt$$

$$= \Sigma\, m\left(\frac{d\dot{x}}{dt}\,\delta x + \frac{d\dot{y}}{dt}\,\delta y + \frac{d\dot{z}}{dt}\,\delta z\right) dt + \delta H\,dt,$$

and, integrating, gets

$$\int \Sigma\, m\,(dx\,\delta\dot{x} + dy\,\delta\dot{y} + dz\,\delta\dot{z})$$

$$= \int \Sigma\, m\,(d\dot{x}\,\delta x + d\dot{y}\,\delta y + d\dot{z}\,\delta z) + \int \delta H\,dt. \tag{13.2}$$

So far he has only performed analytical manipulations on the well-known relationship between kinetic and potential energy. There is nothing new here.

He now introduces the characteristic function defined as the action integral. It is the same characteristic function as in the optical case except that here the expression v in the integral stands for the velocity of the particle while in the optical case it stood for the index of refraction, and, of course, Hamilton has to include the mass in the mechanical case. Since he regards it as constant throughout, the mass is unimportant in the development of his theory and does not affect the analogy between mechanics and optics.

$$V = \int mv\,d\rho = \int \Sigma\, m(\dot{x}\,dx + \dot{y}\,dy + \dot{z}\,dz)$$

$$= \int \Sigma\, m\left(\dot{x}\,\frac{dx}{dt} + \dot{y}\,\frac{dy}{dt} + \dot{z}\,\frac{dz}{dt}\right)dt = \int_0^t 2T\,dt,$$

and taking its variation gives

$$\delta V = \int \Sigma\, m[(\delta\dot{x}\,dx + \delta\dot{y}\,dy + \delta\dot{z}\,dz) + (\dot{x}\,\delta dx + \dot{y}\,\delta dy + \dot{z}\,\delta dz)],$$

or, by separating the integrals,

$$\delta V = \int \Sigma\, m(\delta\dot{x}\,dx + \delta\dot{y}\,dy + \delta\dot{z}\,dz) + \int \Sigma\, m(\dot{x}\,\delta dx + \dot{y}\,\delta dy + \dot{z}\,\delta dz). \tag{13.3}$$

Hamilton next exchanges the signs of variation and differentiation in the second integral (permissible in this case) and integrates it by parts:[9]

$$\int \Sigma\, m(\dot{x}\,d\delta x + \dot{y}\,d\delta y + \dot{z}\,d\delta z) = \Sigma\, m(\dot{x}\,\delta x + \dot{y}\,\delta y + \dot{z}\,\delta z)\Big|_{(a,b,c)}^{(x,y,z)}$$

$$- \int \Sigma\, m(\ddot{x}\,\delta x + \ddot{y}\,\delta y + \ddot{z}\,\delta z)\,dt$$

or

$$\delta V = \int \Sigma\, m(\delta\dot{x}\,dx + \delta\dot{y}\,dy + \delta\dot{z}\,dz) + \Sigma\, m(\dot{x}\,\delta x + \dot{y}\,\delta y + \dot{z}\,\delta z)\Big|_{(a,b,c)}^{(x,y,z)}$$

$$- \int \Sigma\, m(\ddot{x}\,\delta x + \ddot{y}\,\delta y + \ddot{z}\,\delta z)\,dt. \tag{13.4}$$

We notice that the first integral of equation 13.2 is identical to the first integral of equation 13.4; therefore, by substituting from 13.2 into 13.4, we get

$$\delta V = \int \Sigma \, m \, (d\dot{x}\, \delta x + d\dot{y}\, \delta y + d\dot{z}\, \delta z)$$

$$+ \int \delta H \, dt + \Sigma \, m \, (\dot{x}\, \delta x + \dot{y}\, \delta y + \dot{z}\, \delta z) \Big|_{(a,b,c)}^{(x,y,z)}$$

$$- \int \Sigma \, m \, (\ddot{x}\, \delta x + \ddot{y}\, \delta y + \ddot{z}\, \delta z) \, dt.$$

In this expression the first and last integrals are the same (because $d\dot{x}\, \delta x \equiv \ddot{x}\, \delta x \, dt$), but of opposite sign. Also, because H is independent of the time, $\int \delta H \, dt = t \, \delta H$, and

$$\delta V = \Sigma \, m \, (\dot{x}\, \delta x + \dot{y}\, \delta y + \dot{z}\, \delta z) \Big|_{(a,b,c)}^{(x,y,z)} + t \, \delta H. \tag{13.5}$$

This is the *Law of Varying Action*. If we can solve it we get V as a function of the $3n$ coordinates of the n particles of the system and the energy H. The function V completely determines the mechanical system and gives us its state at any future time once the initial conditions are specified—exactly what we want to know. It is a direct consequence of the Law of Varying Action that if the initial and final points of the path are fixed so that δx_1, δy_1, δz_1, δx_2, $\ldots$, δz_n are all zero, *and* if H is constant so that δH is also zero, then $\delta V = 0$, and we have proved the Principle of *Least* Action. Hamilton's important discovery was to notice that if the end points and H are allowed to *vary*, the function V expresses an important new dynamic relationship. The analogy to his optical Principle of Constant Action is obvious. In the optical case we do not have to consider the varying energy (H). Omitting this term the Law of Varying Action is completely analogous to the Principle of Constant Action for an isotropic, but nonhomogeneous, medium, the momenta $m\dot{x}$, $\ldots$ in the mechanical case replacing the product of refractive index and the direction cosine $v\alpha$, $v\beta$, $\ldots$ in the optical case.

Comparing the coefficients of the variables in equation 13.5 to the coefficients in the variation of V we get $6n + 1$ separate equations:

$$\left.\begin{array}{l} \dfrac{\partial V}{\partial x_1} = m_1 \dot{x}_1, \ \ldots, \ \dfrac{\partial V}{\partial x_n} = m_n \dot{x}_n \\[2.5ex] \dfrac{\partial V}{\partial y_1} = m_1 \dot{y}_1, \ \ldots, \ \dfrac{\partial V}{\partial y_n} = m_n \dot{y}_n \\[2.5ex] \dfrac{\partial V}{\partial z_1} = m_1 \dot{z}_1, \ \ldots, \ \dfrac{\partial V}{\partial z_n} = m_n \dot{z}_n \end{array}\right\} \text{(for final coordinates)}$$

$$\frac{\partial V}{\partial a_1} = -m_1 \dot{a}_1, \ldots, \frac{\partial V}{\partial a_n} = -m_n \dot{a}_n$$

$$\frac{\partial V}{\partial b_1} = -m_1 \dot{b}_1, \ldots, \frac{\partial V}{\partial b_n} = -m_n \dot{b}_n \quad \text{(for initial coordinates)}$$

$$\frac{\partial V}{\partial c_1} = -m_1 \dot{c}_1, \ldots, \frac{\partial V}{\partial c_n} = -m_n \dot{c}_n$$

$$\frac{\partial V}{\partial H} = t \quad \text{(for total energy).}$$

If we substitute these values into our expression for the conservation of energy,

$$T = \frac{1}{2} \sum m (\dot{x}^2 + \dot{y}^2 + \dot{z}^2) = U + H,$$

we get

$$\frac{1}{2} \sum \frac{1}{m} \left\{ \left(\frac{\partial V}{\partial x} \right)^2 + \left(\frac{\partial V}{\partial y} \right)^2 + \left(\frac{\partial V}{\partial z} \right)^2 \right\} = U + H$$

for the final coordinates and

$$\frac{1}{2} \sum \frac{1}{m} \left\{ \left(\frac{\partial V}{\partial a} \right)^2 + \left(\frac{\partial V}{\partial b} \right)^2 + \left(\frac{\partial V}{\partial c} \right)^2 \right\} = U_0 + H$$

for the initial coordinates. Thus the problem of finding the characteristic function V (which is itself the solution of the dynamic problem) reduces to one of solving simultaneously these two partial differential equations of the first order and second degree.

At the end of the first essay "On a General Method in Dynamics," Hamilton transformed his function to create an "auxiliary" function that he later called his *principal function*.[10] The advantage of this new principal function (S) over the old characteristic function (V) is that in place of the variable energy (H), the principal function takes a new variable time (t). Defining the new function $S = V - Ht$, the old Law of Varying Action transforms as follows:

$$\delta V = \sum m (\dot{x} \delta x - \dot{a} \delta a + \dot{y} \delta y - \dot{b} \delta b + \dot{z} \delta z - \dot{c} \delta c) + t \delta H \quad (13.5)$$

$$\delta S = \delta(V - Ht) = \delta V - t \delta H - H \delta t$$

$$\delta S = \Sigma \, m (\dot{x} \, \delta x - \dot{a} \, \delta a + \dot{y} \, \delta y - \dot{b} \, \delta b + \dot{z} \, \delta z - \dot{c} \, \delta c) - H \delta t. \quad (13.6)$$

Taking the partial differential coefficients from the variation of this new function (S)

$$\delta S = \frac{\partial S}{\partial x_1} \, \delta x_1 + \cdots + \frac{\partial S}{\partial x_n} \, \delta x_n + \frac{\partial S}{\partial y_1} \, \delta y_1 + \cdots + \frac{\partial S}{\partial y_n} \, \delta y_n$$

$$+ \frac{\partial S}{\partial z_1} \, \delta z_1 + \cdots + \frac{\partial S}{\partial z_n} \, \delta z_n + \frac{\partial S}{\partial t} \, \delta t$$

and equating them to the corresponding coefficients in the Law of Varying Action (13.6) gives another set of equations that is comparable to that for V.

$$\left. \begin{array}{l} \dfrac{\partial S}{\partial x_1} = m_1 \dot{x}_1, \ldots, \dfrac{\partial S}{\partial x_n} = m_n \dot{x}_n \\[2ex] \dfrac{\partial S}{\partial y_1} = m_1 \dot{y}_1, \ldots, \dfrac{\partial S}{\partial y_n} = m_n \dot{y}_n \\[2ex] \dfrac{\partial S}{\partial z_1} = m_1 \dot{z}_1, \ldots, \dfrac{\partial S}{\partial z_n} = m_n \dot{z}_n \end{array} \right\} \text{(for final coordinates)}$$

$$\left. \begin{array}{l} \dfrac{\partial S}{\partial a_1} = -m_1 \dot{a}_1, \ldots, \dfrac{\partial S}{\partial a_n} = -m_n \dot{a}_n \\[2ex] \dfrac{\partial S}{\partial b_1} = -m_1 \dot{b}_1, \ldots, \dfrac{\partial S}{\partial b_n} = -m_n \dot{b}_n \\[2ex] \dfrac{\partial S}{\partial c_1} = -m_1 \dot{c}_1, \ldots, \dfrac{\partial S}{\partial c_n} = -m_n \dot{c}_n \end{array} \right\} \text{(for initial coordinates)}$$

$$\left. \frac{\partial S}{\partial t} = -H \right] \text{(for time } t\text{).}$$

Again substituting into the expression for the conservation of energy,

$$T = \frac{1}{2} \Sigma \frac{1}{m} (\dot{x}^2 + \cdots) = U + H \quad \text{(final values)}$$

$$T = \frac{1}{2} \Sigma \frac{1}{m} (\dot{a}^2 + \cdots) = U_0 + H \quad \text{(initial values)}$$

gives

$$\frac{1}{2} \sum \frac{1}{m} \left[\left(\frac{\partial S}{\partial x} \right)^2 + \left(\frac{\partial S}{\partial y} \right)^2 + \left(\frac{\partial S}{\partial z} \right)^2 \right] = U + \frac{\partial S}{\partial t}$$

$$\frac{1}{2} \sum \frac{1}{m} \left[\left(\frac{\partial S}{\partial a} \right)^2 + \left(\frac{\partial S}{\partial b} \right)^2 + \left(\frac{\partial S}{\partial c} \right)^2 \right] = U_0 + \frac{\partial S}{\partial t},$$

and we again get two partial differential equations of the first order and second degree, which, when solved simultaneously (if it can be done) give the principal function (S), relating the initial and final coordinates.

Hamilton wrote the "Second Essay on a General Method in Dynamics" to exploit this new principal function. He found that it greatly simplified the problems of perturbation that he had attacked in his first essay, and so in his second essay he added this new "wrinkle" to his theory. The first few pages, in which he briefly develops the theory of the principal function, contain the most important (and most familiar) parts of his theory. His presentation is extremely condensed, so condensed, in fact, that the reader must have a great deal of determination to get through these first five pages. Every word counts, and important results pop up as if by magic. Hamilton sent an enthusiastic account of his second essay to Herschel, who was busy observing the southern heavens from the Cape of Good Hope.[11] His letter was twenty pages long, but apparently it was still not detailed enough, because Herschel replied: "Alas! I grieve to say that it is only the general scope of the method which stands dimly shadowed out to my mind amid the gleaming and dazzling lustre of the symbolic expressions in which it is conveyed. ... I could almost regret that you had taken so much trouble for one who can now only look on as a bystander, and mix his plaudits with the smoking of your chariot wheels, and the dust of your triumph."[12]

Hamilton begins his second essay by giving a new derivation of Lagrange's equations for generalized coordinates.[13] If a problem can be stated and solved in generalized coordinates, then the solution is valid in any coordinate system that one might choose. Often the choice of the coordinate system will make a great deal of difference in the ease with which the problem can be solved. Hamilton writes Lagrange's equations as

$$\frac{d}{dt} \left(\frac{\partial T}{\partial \dot{\eta}_i} \right) - \frac{\partial T}{\partial \eta_i} = \frac{\partial U}{\partial \eta_i},$$

where T = kinetic energy, U = negative of the potential energy, and η_i = generalized coordinates of position. (Hamilton expresses the velocity as η'_i. I have written it $\dot{\eta}_i$ to conform to modern custom, but have not changed any other notation.) The kinetic energy may be a function of both position and velocity coordinates, $T(\eta_1, \ldots, \eta_{3n}, \dot{\eta}_1, \ldots, \dot{\eta}_{3n})$, but the potential energy U is only a function of the position coordinates $U(\eta_1, \ldots, \eta_{3n})$, which allows us to write Lagrange's equations in their more familiar form,

$$\frac{d}{dt}\left(\frac{\partial L}{\partial \dot{\eta}_i}\right) - \frac{\partial L}{\partial \eta_i} = 0,$$

where $L = T + U$, although Hamilton does not use the "Lagrangian" L. Solving for the motion of n particles requires finding the relationship between $3n$ coordinates, 3 for each particle, at any time t. Each coordinate has its own Lagrange's equation, so the solution requires one to integrate $3n$ equations, each one of the second order and first degree. Since each generalized coordinate has its own Lagrange's equation and since these equations all have the same form, we can think of the motion of the system not as n points moving through 3-dimensional space, but as *one* point moving through $3n$-dimensional space. Mathematically it comes to the same thing, and this so-called "configuration space" seems to be a better representation of the system of generalized coordinates. After deriving Lagrange's equations, Hamilton immediately introduces a new system of coordinates that allows him to write the equations of motion in an entirely new way. Using Euler's Theorem for a homogeneous function f of order n, of a set of q variables

$$nf = \sum_i q_i \frac{\partial f}{\partial q_i},$$

he applies it to the kinetic energy T, which is a homogeneous function of order 2, in the variables $\dot{\eta}_1, \ldots, \dot{\eta}_{3n}$:

$$2T = \sum_i \dot{\eta}_i \frac{\partial T}{\partial \dot{\eta}_i}. \tag{13.7}$$

Now taking the variation of $T(\eta_1, \ldots, \eta_{3n}, \dot{\eta}_1, \ldots, \dot{\eta}_{3n})$,

$$\delta T = \sum_i \left(\frac{\partial T}{\partial \eta_i}\delta\eta_i + \frac{\partial T}{\partial \dot{\eta}_i}\delta\dot{\eta}_i\right), \tag{13.8}$$

and the variation of 13.7,

$$\delta(2T) = \sum_i \left[\delta\dot{\eta}_i \frac{\partial T}{\partial \dot{\eta}_i} + \dot{\eta}_i\delta\left(\frac{\partial T}{\partial \dot{\eta}_i}\right)\right], \tag{13.9}$$

and subtracting 13.8 from 13.9, one gets for the variation of the kinetic energy:

$$\delta(2T) - \delta T = \delta T = \sum_i \left[\dot{\eta}_i \delta \left(\frac{\partial T}{\partial \dot{\eta}_i} \right) - \frac{\partial T}{\partial \eta_i} \delta \eta_i \right].$$ (13.10)

At this point Hamilton gives a new symbol ω for $\partial T/\partial \dot{\eta}$ and introduces it as a new independent variable. This expression $\omega = \partial T/\partial \dot{\eta}$ later became known as the "conjugate momentum" because in Cartesian coordinates it would represent the linear momentum of the particle. Hamilton creates a new symbol F for the kinetic energy in the new system of coordinates:

$$T(\eta_1, \ldots, \eta_{3n}, \dot{\eta}_1, \ldots, \dot{\eta}_{3n}) = F(\eta_1, \ldots, \eta_{3n}, \omega_1, \ldots, \omega_{3n}).$$

Equation 13.10 becomes

$$\delta F = \sum_i \left[\dot{\eta}_i \delta \omega_i - \frac{\partial T}{\partial \eta_i} \delta \eta_i \right].$$ (13.11)

By introducing the momentum as a new independent variable, Hamilton has doubled the number of coordinates, but it is not as great a complication as it might at first seem, because in Lagrange's equations the velocities were already treated as independent variables. This new "phase space," as it is called, gives important transformation properties for Hamilton's equations of motion.

He derives new equations of motion in the following manner. Taking the variation of $F(\eta_1, \ldots, \eta_{3n}, \omega_1, \ldots, \omega_{3n})$ gives

$$\delta F = \sum_i \left(\frac{\partial F}{\partial \eta_i} \delta \eta_i + \frac{\partial F}{\partial \omega_i} \delta \omega_i \right).$$ (13.12)

Comparing this to 13.11 we see that $\dot{\eta}_i = \partial F/\partial \omega_i$, or, since the potential energy is not a function of the momentum coordinates, we may write it as

$$\dot{\eta}_i = \frac{\partial(F - U)}{\partial \omega_i} = \frac{\partial H}{\partial \omega_i}.$$

Again comparing 13.11 and 13.12 we get

$$\frac{\partial F}{\partial \eta_i} = -\frac{\partial T}{\partial \eta_i}.$$

Lagrange's equations then become

$$\frac{d}{dt}\omega_i + \frac{\partial F}{\partial \eta_i} = \frac{\partial U}{\partial \eta_i}$$

or

$$\frac{d\omega_i}{dt} = \frac{\partial(U - F)}{\partial \eta_i} = -\frac{\partial H}{\partial \eta_i},$$

giving us two sets of equations:

$$\dot{\eta}_i = \frac{\partial H}{\partial \omega_i}, \qquad \dot{\omega}_i = -\frac{\partial H}{\partial \eta_i}. \tag{13.13}$$

These equations became known as Hamilton's *Canonical Equations of Motion*. There are $6n$ of them compared to the $3n$ equations of Lagrange, but they are equations of the first order, while Lagrange's equations are of the second order. This is not a great advantage in actual practice, but it leads to important consequences in the theory of dynamics.

Hamilton then reintroduces his principal function into the theory. In the first essay he had already derived the integral expression for the principal function from the definition of the characteristic function,

$$V = \int mv\,ds = \int_0^t 2T\,dt,$$

the second form of the integral being that used by Lagrange. The principal function S represents a similar integral

$$S = V - Ht = \int_0^t 2T\,dt - Ht = \int_0^t (2T - H)\,dt$$

$$= \int_0^t (T + U)\,dt = \int_0^t L\,dt,$$

where L is now called the "Lagrangian" of the motion and H is now called the "Hamiltonian." In the second essay Hamilton uses this last expression as a definition of the principal function, but writing it now as

$$S = \int_0^t \left(\Sigma\,\omega\,\frac{\partial H}{\partial \omega} - H\right)dt. \text{[14]}$$

There then follows a bit of mathematical skulduggery that works, but looks as if it should not. He writes the variation of the principal function as follows:

$$\delta S = \int_0^t \delta \left(\frac{dS}{dt} \right) dt.$$

This assumes that the end times are fixed so the time is not affected by the variation; in other words, a particle covers any of the varied paths in the same time, but the *positions* of the end points are assumed to vary. Then

$$\frac{dS}{dt} = \Sigma \, \omega \, \frac{\partial H}{\partial \omega} - H$$

$$\delta \left(\frac{dS}{dt} \right) = \Sigma \, \left[\omega \delta \left(\frac{\partial H}{\partial \omega} \right) + \delta \omega \left(\frac{\partial H}{\partial \omega} \right) - \frac{\partial H}{\partial \omega} \, \delta \omega - \frac{\partial H}{\partial \eta} \, \delta \eta \right]. \quad (13.14)$$

Cancelling the like terms of opposite sign and substituting the canonical equations 13.13, one gets

$$\delta \left(\frac{dS}{dt} \right) = \Sigma \, \left[\omega \delta \dot{\eta} + \dot{\omega} \delta \eta \right] = \frac{d}{dt} \, \Sigma \, \omega \delta \eta,$$

and since

$$\delta \left(\frac{dS}{dt} \right) = \frac{d}{dt} \, (\delta S),$$

$$\delta S = \Sigma \, (\omega \delta \eta - p \delta e), \quad (13.15)$$

where p and e are the momenta and position coordinates at time $t = 0$. Comparing these to the differential coefficients for the variation of δS, one gets another set of $6n$ equations:

$$\omega_1 = \frac{\partial S}{\partial \eta_1}, \, \omega_2 = \frac{\partial S}{\partial \eta_2}, \, \ldots, \omega_{3n} = \frac{\partial S}{\partial \eta_{3n}} \text{ (for final coordinates)}$$

$$(13.16)$$

$$p_1 = -\frac{\partial S}{\partial e_1}, p_2 = -\frac{\partial S}{\partial e_2}, \, \ldots, p_{3n} = -\frac{\partial S}{\partial e_{3n}} \text{ (for initial coordinates).}$$

Hamilton draws the following conclusions:

1. The canonical equations give the *differential equations* of the motion, because they give the time derivatives of the position and momentum coordinates. Equations 13.16 are the *integrals* of those equations, because they give the *actual* initial and final momenta as functions of the position coordinates. Equations 13.16 can be found easily by

differentiating the principal function S, so everything rests on the ability to find S. Hamilton tells the British Association (and also Herschel in a separate letter) that his principal function S differs from Lagrange's function L in that "Lagrange's function *states* [while his] function would *solve* the problem. The one serves to form the *differential* equations of motion, the other would give their *integrals*."[15]

2. The integral defining the principal function S now becomes an important dynamical principle in its own right. The defining integral is:

$$S = \int_0^t (T + U)dt = \int_0^t L\,dt.$$

Its variation is, by 13.15,

$$\delta S = \delta \int_0^t L\,dt = \Sigma\,(\omega\,\delta\eta - p\,\delta e).$$

The end times are fixed, and the quantity under the summation sign represents variation of the initial and final positions. If the end points are fixed in space and time so that particles covering the different paths all leave from one initial point at the same time and arrive at one final point at the same time, then the variations of the end positions described by 13.15 are zero and

$$\delta S = \delta \int_0^t L\,dt = 0. \tag{13.17}$$

The Germans soon named this "Hamilton's Principle." It is similar to the Principle of Least Action, but has one advantage over that principle in that it does not require all the varied paths to have the same energy. It *does* require that the particles cover all the paths in the same time, a condition that does not apply to the Principle of Least Action.

3. Hamilton finally observes that the integral $S = \int_0^t L\,dt$ serves a double purpose in his theory. If it is varied as in 13.17, with fixed end positions and times, it leads directly to Lagrange's equations, which are the *differential* equations of motion. (Lagrange's equations *are* the Euler conditions for the integral $\int_0^t L\,dt$ to be stationary.) If, on the other hand, the end positions are allowed to vary, it leads to equations 13.16, which are *integrals* of those differential equations.[16]

The principal function S as he has used it throughout the second essay has been understood to be a function of the generalized coordinates of position and the time

$$S(\eta_1, \ldots, \eta_{3n}, t),$$

just as he defined it in the first essay. In deriving equation 13.15, Hamilton did not vary the time, since he was only interested in finding the integrated equations 13.16, which give the momenta as functions of the coordinates of position. In order to find the complete variation of S it is necessary to add the term $(\partial S/\partial t)\delta t$ to equation 13.15. He finds this term in the following manner. The differential of S gives

$$\frac{dS}{dt} = \frac{\partial S}{\partial t} + \Sigma \frac{\partial S}{\partial \eta} \frac{d\eta}{dt},$$

and

$$\frac{dS}{dt} = \Sigma \omega \frac{\partial H}{\partial \omega} - H$$

from 13.14. Therefore

$$\frac{\partial S}{\partial t} = \Sigma \omega \frac{\partial H}{\partial \omega} - H - \Sigma \frac{\partial S}{\partial \eta} \frac{d\eta}{dt},$$

but from 13.16 we have that $\partial S/\partial \eta = \omega$, and from 13.13, $d\eta/dt = \partial H/\partial \omega$, so

$$\frac{\partial S}{\partial t} = \Sigma \omega \frac{\partial H}{\partial \omega} - H - \Sigma \omega \frac{\partial H}{\partial \omega} = -H.$$

The canonical equations show immediately that H is constant, because

$$\frac{dH}{dt} = \Sigma \left(\frac{\partial H}{\partial \omega} \frac{d\omega}{dt} + \frac{\partial H}{\partial \eta} \frac{d\eta}{dt} \right) = \Sigma \left(\frac{d\eta}{dt} \frac{d\omega}{dt} - \frac{d\omega}{dt} \frac{d\eta}{dt} \right) = 0.$$

Now if we again write the expression for the conservation of energy, but this time write the kinetic energy as a function of the momentum ω and the position coordinates η, we get

$$F(\omega_1, \ldots, \omega_{3n}, \eta_1, \ldots, \eta_{3n})$$

$$= U(\eta_1, \ldots, \eta_{3n}) + H(\omega_1, \ldots, \omega_{3n}, \eta_1, \ldots, \eta_{3n})$$

and by substituting from 13.16 for the momentum coordinates in the expression for the kinetic energy, and by using the expression just proven that $\partial S/\partial t = -H$, we get

$$F\left(\frac{\partial S}{\partial \eta_1}, \ldots, \frac{\partial S}{\partial \eta_{3n}}, \eta_1, \ldots, \eta_{3n}\right) = U(\eta_1, \ldots, \eta_{3n}) - \frac{\partial S}{\partial t}$$

$$F\left(-\frac{\partial S}{\partial e_1}, \ldots, -\frac{\partial S}{\partial e_{3n}}, e_1, \ldots, e_{3n}\right) = U(e_1, \ldots, e_{3n}) - \frac{\partial S}{\partial t_0},$$

or, since $F - U = H$ and the momentum coordinates do not enter into the expression for the potential energy, we can write

$$H\left(\frac{\partial S}{\partial \eta_1}, \ldots, \frac{\partial S}{\partial \eta_{3n}}, \eta_1, \ldots, \eta_{3n}\right)$$

$$= -\frac{\partial S}{\partial t} \text{ (for final coordinates)} \quad (13.18a)$$

$$H\left(-\frac{\partial S}{\partial e_1}, \ldots, -\frac{\partial S}{\partial e_{3n}}, e_1, \ldots, e_{3n}\right)$$

$$= -\frac{\partial S}{\partial t} \text{ (for initial coordinates).} \quad (13.18b)$$

Again we have two partial differential equations of the first order, which, if we can solve them, will give the principal function S.

As a practical problem, the solution of these two differential equations can be extremely difficult. In 1837 C.G.J. Jacobi discovered an easier way to find S using only the first of the two equations 13.18,[17] and because of his contribution to the theory, equation 13.18a is usually referred to as the "Hamilton-Jacobi equation."[18] The integration of Hamilton's two partial differential equations creates $6n$ constants that, by Hamilton's theory, are the $3n$ initial momenta and the $3n$ initial position coordinates. Jacobi discovered that through the theory of contact transformations it was possible to show that *any* integral of equation 13.18a containing as many arbitrary constants as there are independent variables can be used and that there is no need to integrate *two* partial differential equations—one will suffice. Hamilton used contact transformations in the second essay when applying his method to the particular problem of perturbation, but it was Jacobi who developed the theory in its generality.[19]

In his first important paper on Hamilton's dynamics, Jacobi was somewhat critical; he stated that Hamilton had presented his theory in a "false light" and had "unnecessarily complicated and limited" his theory by requiring the solution of two partial differential equations when one would have sufficed.[20] The criticism is valid if integrating the differential equations is the main object, but that was not Hamilton's purpose. Cayley defended Hamilton in his 1857 report to the British Association: "It is not a *method of integration*, but a theory of the representation of the integral equations assumed to be known. I venture to dissent from what appears to have been Jacobi's opinion, that the author missed the true application of his discovery; it seems to me, that Jacobi's investigations were rather a theory collateral to, and historically arising out of the Hamiltonian theory, than the course of development which was of necessity to be given to such a theory."[21] Hamilton first heard of Jacobi's paper from H.F.C. Logan, who wrote saying that in his opinion neither Poisson nor Jacobi had given Hamilton's theory the justice it deserved.[22] In a later letter Logan changed his evaluation to say

You are of course aware that Jacobi has fully appreciated your Dynamical memoirs and made your important discovery a stepping stone to a new method for the Integration of Partial Differential Equations of the first order and degree. I have not yet seen his memoir, but it is printed in the last number of *Crelle*. Have you printed the work you were preparing upon your Calculus of Principal Functions? I hope you will not delay much longer so valuable a present to the mathematical world, and one which those who have read your former writings so eagerly expect.[23]

Hamilton had not seen Jacobi's paper when he received Logan's letter, and did not see it until six months later, at which time he replied to Logan that he had read a translation of some of Jacobi's remarks in Liouville's *Journal de mathématiques*.[24] He was flattered by Jacobi's praise, but unhappy that Jacobi had found his theory unnecessarily restricted. "When I can find leisure to take up the subject again," he wrote, "I do not despair of showing that in the way of generalisation the tables may be turned." Hamilton had in mind his "Calculus of Principal Relations," which he had announced in the introduction to his "Second Essay on a General Method in Dynamics" and which Logan was urging him to publish soon.[25]

Beginning in January, 1836, he filled a "huge blank notebook" with his researches and began a treatise on his calculus that he planned to publish at the Dublin University Press.[26] His decision to publish in Ireland was in part a patriotic one. By sending his dynamics papers to the Royal Society of London he had missed a chance to bring luster to Dublin and to the Royal Irish Academy. "A *book* is commonly thought more of a publication than the printing of Papers in *Transactions*, and I should like to contribute my mite, or shall I say, my *stone* to throw upon the pile which hides the buried slander against the 'Silent Sister'."[27] Hamilton never finished his book. He gave a brief account of his method to the British Association at Bristol in 1836, but that was all that ever appeared in print.[28] The Calculus of Principal Relations was a method of integrating differential equations that grew out of his dynamics. In addition to this purely mathematical theory, he was also working to apply his dynamics to the theory of the moon. He began this work in 1836, too, partly at the instigation of the British Association, and he elaborated it in several long letters to J. W. Lubbock between July and November, 1837.[29]

Hamilton met Jacobi in 1842 at the Manchester meeting of the British Association. In reporting on the meeting he said that he had seen "a great deal of Bessel and some of Jacobi."[30] Of the two German visitors, Bessel apparently held a greater attraction for Hamilton than did Jacobi. The German visitors almost certainly heard the debate over the theory of light and Hamilton's theory of polar points fixed in space, because both were present at the meetings of the mathematical and physical sections. Jacobi was certainly flattering, at least one gets that impression from the proceedings of the section that were reported in the newspapers. In response to Hamilton's paper on fluctuating functions, Jacobi said that "Lagrange stated it as his opinion, that it was not possible to express these functions by any mathematical formulae. It appeared, however to him (Professor Jacobi) that Sir William Hamilton had shown that it was possible," and in a paper of his own on the "New General Principles of Analytic Mechanics," he referred to Hamilton as "the illustrious Astronomer Royal of Dublin" and, later, as "the Lagrange of your country."[31] It would be difficult to think of an association that would have pleased Hamilton more.

14

The Fate of the
Optical-Mechanical Analogy

HAMILTON'S METHOD was mentioned with respect in treatises on mechanics throughout the nineteenth century, but in actual practice it was seldom employed, because other simpler methods would do just as well in most cases. As a formal structure it possessed admirable elegance and was something that a mathematician or theoretical physicist would know about even if he did not find himself using it constantly. Hamilton's optical papers, however, gradually sank from view and were little known in Germany, where much of the advanced work on mechanics and theoretical physics was taking place.

With the advent of the quantum theory in the twentieth century, Hamilton's theory found new value, because it was the one form of classical mechanics that could be transferred almost directly to quantum mechanics. In the second part of his famous paper of 1926 on wave mechanics, Erwin Schrödinger wrote at length about the optical theory that Hamilton had published almost a century earlier. What Schrödinger found of greatest importance was Hamilton's discovery that the laws of geometrical optics and the laws governing the motion of particles could be expressed in the same mathematical form. In the nineteenth century it was nothing more than an analogy, and Hamilton had no experimental evidence that would indicate a more fundamental unity between the properties of light and the properties of material particles. But by 1926 the situation was quite different. Classical mechanics had failed in the realm of the very small and had been replaced by quantum mechanics. The work of Louis de Broglie suggested that, in the world of the atom, material particles might have wave properties like those of light. The optical-mechanical analogy was becoming not only an analogy of mathematical formalism, but also an analogy relating the essential properties of matter and light.

In his paper of 1926 Schrödinger exploited the Hamiltonian analogy to show how his famous wave equation was related to the principles of classical mechanics and optics. Hamilton's classical mechanics as it appeared in his essay "On a General Method in Dynamics" and his classical ray optics of the "Theory of Systems of Rays" had both failed at the atomic level. The diffraction and interference of light passing through small apertures could not be explained by the geometry of rays; only the wave theory could account for the observations satisfactorily.[1]

In the transition from the world of our common experience into the world of the atom, classical mechanics had to give way to quantum mechanics just as ray optics had to give way to wave optics. Schrödinger speculated that these failures might be related; that there was some fundamental connection between the physics of optics and the physics of particles that caused their classical formulations to fail at approximately the same place in the scale of magnitude. If such a fundamental connection existed, it would explain in part why Hamilton was able to find a common formulation for optics and mechanics in his characteristic function. These are not two different sciences at all, but different aspects of the same science. Schrödinger wrote:

We know today, in fact, that our classical mechanics fails for very small dimensions of the path and for very great curvatures. Perhaps this failure is in strict analogy with the failure of geometrical optics ... that becomes evident as soon as the obstacles or apertures are no longer great compared with the real, finite, wave length. Perhaps our classical mechanics is the *complete* analogy of geometrical optics and as such is wrong and not in agreement with reality; it fails wherever the radii of curvature and dimensions of the path are no longer great compared with a certain wavelength, to which in q-space a real meaning is attached. Then it becomes a question of searching for an undulatory mechanics, and the most obvious way is by an elaboration of the Hamiltonian analogy on the lines of undulatory optics.[2]

If the Hamiltonian analogy were a *complete* analogy, then the wave properties that allow one to pass satisfactorily from the macroscopic to the microscopic world in the case of light could be applied equally well to particles and provide a similar satisfactory transition for mechanics. A "wave mechanics" might be expected to apply equally well to the macroscopic and microscopic worlds. This is exactly what Schrödinger found (see figure 14.1).

A closer look at the optical-mechanical analogy reveals that there are really two analogies involved—or perhaps it would be better to say that the analogy has two different aspects. These two aspects are more pronounced in Hamilton's writing than in Schrödinger's and therefore it is worthwhile keeping them distinct. The first part of the analogy concerned the debate

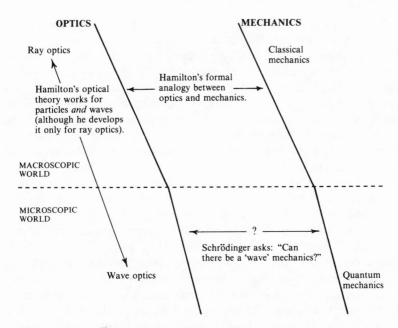

Fig. 14.1. The optical-mechanical analogy.

over the physical theory of light that was fought, as we have seen, first at the Paris Academy in the 1820s and then in Britain during the 1830s. The fact that Hamilton's theory applied equally well to particles and waves might have been regarded as a shortcoming of the theory in the nineteenth century, but for Schrödinger in the twentieth century it took on an entirely new significance. It is not a matter of "either-or," but a matter of "both-and." Light quanta and subatomic particles exhibit both particle and wave properties simultaneously. At the close of his Nobel Prize Address of 1933, Schrödinger asks us to understand this anomaly by appealing to a model similar to Hamilton's rays and surfaces of constant action. A subatomic "object," light quantum or particle, has "longitudinal continuity" along a ray and "transverse continuity" on a surface orthogonal to the ray.[3] It is these two forms of continuity that best conform to the dual nature of elementary particles.

The second part of the optical-mechanical analogy as derived by Hamilton did not necessarily have anything to do with waves. He showed that he could describe the geometry of light rays and the motion of particles by the same mathematical formula. The analogy is a mathematical one, and is revealed by comparing Hamilton's optical papers of the years 1827 through 1833 with his papers on mechanics dated 1834 and 1835. In this part of the analogy we remain entirely in the classical world of geometrical light rays and billiard-ball particles.

The analogy is one of mathematical formalism. Two completely different sciences—geometrical optics and classical mechanics—are shown to have a mathematical, but not a physical, equivalence. It is this second aspect, the formal mathematical equivalence of optics and mechanics, that is usually given the name of the optical-mechanical analogy. For an appreciation of Hamilton, it is necessary to keep in mind the difference between the mathematical theories and the physical theories of optics. In one sense it could be said that Newton had an optical-mechanical analogy, because he explained light by the mechanical motion and interaction of particles. He sought for each optical phenomenon an explanation in terms of the mechanical action of particles. But Hamilton's optical-mechanical analogy was quite different from this. By the application of mathematics, he was able to show that the *entire science* of geometrical optics and the *entire science* of rational mechanics could be given the same formulation.

In 1926 Schrödinger complained that Hamilton's theory had been misunderstood and that his optics had been unjustly ignored. He lamented the fact that other contemporary descriptions of Hamilton's theory had failed to emphasize its connection with wave propagation: "Unfortunately this powerful and momentous conception of Hamilton is deprived, in most modern reproductions, of its beautiful intuitive raiment, as if this were a superfluous accessory, in favor of a more colourless representation of the analytical connections."[4] This criticism immediately raises the questions of what the actual fate of Hamilton's ideas was, as well as how Schrödinger came to revive them in 1926. If the optical-mechanical analogy was "lost" in the years between Hamilton's paper and those of Schrödinger, it was because the optical papers slipped from notice, while the mechanical theory, with the additions of C.G.J. Jacobi, received a great deal of attention. H. Bruns's "discovery" of the Eikonal in 1895 would suggest that Hamilton's optical papers had not reached a wide audience.

Yet the analogy did survive in British books on mechanics. The famous *Treatise on Natural Philosophy* (1867) by William Thomson (Lord Kelvin) and Peter Guthrie Tait contained a lengthy discussion of Hamilton's Law of Varying Action and the differential equations derived from it. The authors explained that "irrespectively of methods for finding the 'characteristic function' in kinetic problems, the fact that any case of motion whatever can be represented by means of a single function . . . is most remarkable, and, when geometrically interpreted, leads to highly important and interesting properties of motion, which have valuable applications in various branches of Natural Philosophy. *One of the many applications of the general principle made by Hamilton led to a general theory of optical instruments,* comprehending the whole in one expression" (italics mine).[5] And there follows a reference to Hamilton's optical papers and an example of his principles applied to "common optics." The analogy is there, but it is treated as a curiosity—not as a matter of significance. The authors continue: "The now abandoned, but still interesting, corpuscular theory of light furnishes a good and exceedingly simple illustration [of Hamilton's theory]." Obviously Thomson and Tait believed that the victory of the wave theory of light deprived the optical-mechanical analogy of any real physical significance.

An equally famous book on mechanics, E. T. Whittaker's *Treatise on the Analytical Dynamics of Particles and Rigid Bodies* (1904), contains a lengthy development of Hamilton's principle from the theory of contact transformations. In the preface Whittaker states: "I may mention that the new explanation of the transformation theory of Dynamics . . . sprang from a desire to do justice to the earliest great work of Hamilton's genius. . . . The origin of the method is to be found in a celebrated memoir on optics, which was presented to the Royal Irish Academy by Hamilton in 1824: the principles there introduced were afterwards transferred by their discoverer to the field of dynamics."[6] Whittaker goes on to say: "In order to follow Hamilton's thought, we must refer to the connexion between dynamics and optics—a connexion which is perhaps less obvious in our day than in his when the corpuscular theory of light was widely held."[7] There is great irony in these lines by Whittaker. He recognized the connection between Hamilton's dynamics and optics, but, like Thomson and Tait, he could not grant to it any physical meaning. Light was composed of waves in an electromagnetic field, matter was composed of particles, and the two were always distinct. He published a year too early. In 1905 Einstein showed that the photoelectric effect could only be explained by granting a particle nature to light.

While the British writers emphasized the dynamic aspect of Hamilton's theory, they never lost sight of the optical-mechanical analogy, or of the fact that Hamilton had first derived his theory from a study of optics. In Germany, however, mathematicians emphasized Hamilton's mechanics and neglected his optics. Jacobi's influence was largely reponsible, because his method of integrating the "Hamilton-Jacobi equation" tended to obscure the analogy to optics; and yet the analogy was scarcely "lost," even in Germany, in spite of Jacobi's supposedly nefarious influence and in spite of the general ignorance of Hamilton's optical papers. We can more properly say that although it was ignored, it remained available to anyone who might wish to use it. At the suggestion of Helmholtz, Thomson and Tait's *Treatise of Natural Philosophy* was translated into German in 1871, and it became an important source for German physicists. Whittaker was known, too; Schrödinger cites him in his 1926 papers.[8]

A major effort to "recover" the optical-mechanical analogy for physics was made during the 1890s by Felix Klein. In his *Vorlesungen über die Entwicklung der Mathematik im 19. Jahrhundert,* he recounts the history of his attempts. Klein tells us that Hamilton's optical foundations were ignored and his results "snatched away" by Jacobi, whose name replaced that of Hamilton, leaving Hamilton as only an insignificant precursor of Jacobi.[9] Klein is quite indignant about the whole matter. He claims that he learned of the true state of affairs by his travels, but was unsuccessful in his attempt to make Hamilton's optical and mechanical works known in Germany. Since Schrödinger acknowledges these efforts in 1926, we know that Klein's labor was not all in vain. In the summer of 1891 Klein worked out the Hamilton method, making all of the mechanics an art of optics in *n*-dimensional space, including Jacobi's later developments. And in the same year he lectured on the subject to the Naturforscherversammlung in Halle. He complained that his notes lay around for twenty years in the reading room at Göttingen but attracted no enthusiasts.[10] Only two mathematicians joined in the effort to exhume Hamilton's optics. The first was Eduard Study, who in 1905 wrote a biographical sketch of Hamilton in honor of the one hundredth anniversary of the date of his birth, followed by an article in which he developed Hamilton's geometrical optics using the theory of infinitesimal contact transformations.[11] The second attempt to revive Hamilton was made by Georg Prange, who entitled his *Habilitationsrede* "W. R. Hamilton's Significance for Geometrical Optics."[12] Prange echoed the criticism of Cayley that the integration of the Hamilton-Jacobi equation had not been Hamilton's major interest. He claimed that the great oversight of those who followed Jacobi was missing Hamilton's work in geometrical optics, both for its theoretical and for its practical importance.

Neither of these articles by Klein's protégés had any noticeable effect. Klein attributed this neglect in part to the debate in Germany over the teleological implications of the Principle of Least Action. For the confusion over this subject he blamed Helmholtz and Planck.[13] Helmholtz spent a great part of his later years studying action principles because he believed it probable that least action was the universal law pertaining to all processes in nature.[14] His pupil, Heinrich Hertz, took an intermediate position and argued that Hamilton's method is "not based on any physical foundation of mechanics, but that it is fundamentally a purely geometrical method, which can be established and developed quite independently of mechanics, and which has no closer connection with mechanics than any other of the geometrical methods employed in it."[15] And yet Hertz also worried about the teleological nature of variational principles and recognized that the problem would not go away just by claiming that teleology has no place in physics.[16]

Planck, on the other hand, was willing to go much further into metaphysics, which elicited a few predictable snorts from Klein. In his most extreme statement, Planck claimed that the Principle of Least Action "possesses an explicitly teleological character. . . . In fact the least action principle introduces a completely new idea into the concept of causality: the *causa efficiens* . . . is accompanied by the *causa finalis* for which . . . the future—namely a definite goal, serves as the premise from which there can be deduced the development of the processes which lead to this goal.[17] Although Planck was not consistent in his position (at other times he rejected all teleological principles in mechanics),[18] his occasionally misguided enthusiasm brought action principles under suspicion, or so Klein argued.

The great enthusiasm for action principles that reached its height in the 1880s can also be observed in British physics. Sir Joseph Larmor published an article in 1884 entitled "On Least Action as the Fundamental Formulation in Dynamics and Physics," in which he argued that all phenomena can be unified to the extent that they can be described by a common action principle.[19] He then proceeded to show how disparate parts of physics, including optics, can be brought under this single umbrella. (The Second Law of Thermodynamics was the one troublesome exception.) And yet Larmor, as had Thomson and Tait before him and as did Whittaker after him, failed to see any particular significance in the optical-mechanical analogy. Larmor was an Irishman with an interest in the history of physics and a large reservoir of pride in his homeland. He heaped praise on Hamilton, but apparently never read his optical papers. In 1927, after Schrödinger had revived the optical-mechanical analogy with his wave mechanics, Larmor dug out the optical papers and described them in a "Historical Note on Hamiltonian Action." He found the papers "very dishevelled in form, doubtless from the distraction of [Hamilton's] wide range of philosophical and poetic interests." The later optical papers "enforc[ed] the fundamental ideas, but perhaps also still further confus[ed] their application by excess of detail only partially relevant." These criticisms were all perfectly valid, but they sound suspiciously like an excuse for Larmor having overlooked the papers in the 1880s.[20]

Actually Larmor needed no excuse. In the 1880s quantum mechanics did not exist and there was no experimental evidence to indicate that the wave theory of light might not be the complete explanation of optical phenomena. Moreover, Hamilton's analogy was limited to geometrical optics and was not adequate to deal with the phenomena that had confirmed the wave theory. It is Klein's position that needs explaining, and the only explanation that I can give is that the optical-mechanical analogy seemed more important to a mathematician like Klein, who recognized its beauty and unifying power, than to a physicist who could not yet think of photons and matter waves.

The question that now presents itself is: "How did Schrödinger become aware of the optical-mechanical analogy, and how did it lead him to wave mechanics, if, indeed, it had any influence at all?" Schrödinger derived his famous wave equation in the first paper from the Hamilton-Jacobi equation without any reference at all to Hamilton's optical-mechanical analogy. It appears to depend far more on the work of Louis de Broglie than on that of Hamilton. The analogy appears only in the second paper, where it is elaborated on at length. In a footnote Schrödinger thanks Arnold Sommerfeld for calling his attention to Felix Klein's campaign to foster Hamilton's ideas.[21] It looks as if Schrödinger may have first derived his wave equation on his own, then showed it to Sommerfeld, who mentioned Klein's work to him, and, finally, developed the connection between wave mechanics and classical mechanics through the optical-mechanical analogy. In other words, the second paper may have been merely an attempt to justify a new theory that had in fact been obtained in a different way.

Sommerfeld was the obvious person to put Schrödinger onto Klein's ideas. He had been Klein's assistant and "clerk" of the famous Göttingen reading room, where Klein claimed that his papers on the optical-mechanical analogy collected dust for twenty years.[22] It is hardly possible that Sommerfeld did not read them. Moreover, Sommerfeld's own work in 1916 had shown the close link between the action integral and the quantum conditions for the hydrogen atom. The importance of Hamilton's method for the older quantum theory had indicated that it was likely to be the most successful approach for any future theory.

Schrödinger also cites another important paper, one by A. Sommerfeld and J. Runge on geometrical optics, that follows closely Hamilton's arguments. The most interesting section (and the section Schrödinger cites) is entitled "The Eikonal and the Limits of Geometrical Optics."[23] Sommerfeld recognizes that the Eikonal and the Hamiltonian theory of the characteristic function are equivalent in geometrical optics, and, following a suggestion by Pieter Debye, he works out the limits at which wave optics may be correctly approximated by ray optics. Schrödinger goes out of his way to emphasize Debye's role in this paper, because he was working closely with Debye at Zurich at the time. Sommerfeld's suggestions were obviously important for Schrödinger, but probably not crucial, and we still do not know if Schrödinger's pursuit of Hamilton's optical-mechanical analogy came before or after his discovery of the wave equation.

As is usually the case with historical investigations of this kind, the truth does not seem to lie at either extreme, but somewhere in the middle. The immediate search for the wave equation was probably a response to the work of de Broglie, but this does not mean that Hamilton's analogy was not also a contributing factor. Schrödinger was thoroughly acquainted with the optical-mechanical analogy long before Sommerfeld suggested that he look at Klein's papers. More than that, his training and previous research in large part determined the direction his thoughts took in 1926.

Several authors have given careful attention to the events that led up to Schrödinger's discovery.[24] In 1925 Schrödinger was working on the quantum statistics of gases. While reading a paper by Einstein on the same subject, he was "suddenly confronted with the importance of de Broglie's ideas," which were mentioned in the paper. In his dissertation for the Sorbonne, de Broglie had created the concept of "matter waves" ascribing wave properties to electrons. Schrödinger first followed Einstein's lead in trying to apply de Broglie's new concept to the quantum statistics of gases. By early November, 1925, he wrote that he now regarded the particles of a gas as merely wave crests on a background of waves. Soon afterwards he was working to apply the same theory to the electron orbits of the hydrogen atom, and discussed his efforts with Pieter Debye, who was also in Zurich, and Alfred Landé and Wilhelm Wien. But finding the correct wave equation for his theory was a big problem. His first relativistic equation was a failure (because it did not take into account the then-unknown phenomenon of electron spin), and he confessed to Wien: "I must learn more mathematics in order to fully master the vibration problem—a linear differential equation, similar to Bessel's, but less well known, and with remarkable boundary conditions that the equation 'carries within itself' and that are not externally determined."[25] He then turned to search for a nonrelativistic wave equation, and was successful in January, 1926.[26]

Schrödinger readily confessed that his inspiration had come from de Broglie and that his wave mechanics should be regarded as an extension of de Broglie's theory.[27] In his thesis de Broglie had returned again and again to emphasize the importance of the optical-mechanical analogy, although he attributed it to Fermat and Maupertuis rather than to Hamilton.[28] Anyone reading the thesis could not miss a point made so emphatically. There were several reasons why Schrödinger was much more receptive to de Broglie's ideas than were most German physicists. De Broglie had been brashly critical of Bohr and Sommerfeld, and had earned the reputation in German circles of being an ill-tempered crank. Also, Schrödinger's interest in quantum statistics and his objections to the probabilistic interpretation of quantum theory put him in the same camp with Einstein against the Copenhagen school. But surely de Broglie's use of the optical-mechanical analogy must have made his theory congenial to Schrödinger.[29]

These were the immediate events leading up to Schrödinger's discovery. It was a straightforward attempt to find a wave equation for the electron orbits of the hydrogen atom that satisfied the quantum conditions. Because of the great importance given to the optical-mechanical analogy by de Broglie, we can scarcely regard it as an afterthought that entered Schrödinger's mind only after the first paper was finished. Less than a month separates the first and the second papers. It is difficult to believe that the ideas in the second paper were worked out entirely in such a short interval. Moreover, Schrödinger mentions that he originally intended to give a more physical or "intuitive" representation of the wave equation, but decided instead to put it in a more "neutral" mathematical form, indicating that he probably did have a vibrational model in mind when he wrote the first paper.[30] It is likely that the "intuitive" model was something close to the presentation that Schrödinger actually made in his second paper.

In the short autobiographical sketch attached to his Nobel Prize address of 1933, Schrödinger paid special tribute to his first teacher at Vienna, Fritz Hasenöhrl, who had been killed at a relatively young age in the First World War. Schrödinger believed that if he had lived, Hasenöhrl might well have been receiving the prize for the discovery of wave mechanics in his place.[31] Cornelius Lanczos has recently judged the importance of Hasenöhrl in these words: "It is no accident that Schrödinger repeatedly and emphatically referred to his outstanding teacher at the University of Vienna, Fritz Hasenöhrl. Hasenöhrl was one of the few theoretical physicists who fully recognized the importance of Hamilton's work and gave full account of it in his lectures. Thus it was completely in line with Schrödinger's theoretical background to take de Broglie's geometrical optics and change it into a physical form of optics, with the result that he arrived with necessity at his famous equation."[32]

Beginning in 1906 Schrödinger sat through a four-year cycle of daily lectures by Hasenöhrl on theoretical physics. The emphasis was on Hamiltonian mechanics and the theory of eigenvalue problems in the physics of continuous media.[33] The direction of Hasenöhrl's lectures is indicated by one of his articles in the Boltzmann *Festschrift* of 1904 entitled "On the Application of Hamilton's Partial Differential Equation to the Dynamics of Continuously Distributed Masses."[34] In this paper Hasenöhrl claimed that he was making the first attempt to employ Hamilton's partial differential equation in the solution of problems of motion in a continuous medium. He selected the simplest example—that of the vibrating string—and obtained a solution using Hamiltonian methods. While he did not derive a wave equation (in fact the beauty of the method is that it permits a solution *without* deriving a wave equation), Hasenöhrl was obviously headed in the right direction for Schrödinger.

Among Schrödinger's manuscripts are three university notebooks entitled "Tensor analytische Mechanik." It is possible that some of the notes were taken at Hasenöhrl's lectures, although they are more likely from the years 1918 to 1920, when he returned to Vienna after The First World War.[35] Schrödinger had the ambition to continue the tradition of Hasenöhrl's lectures in a new teaching career at Czernowitz, but when that opportunity collapsed with the Russian occupation of the city after the war, he returned to teach at Vienna.[36] These notes were probably prepared for lectures there. The third book contains a section entitled

"Analogy to Optics, Huygens' Principle and Hamilton's Partial Differential Equation." It is a beautiful development of the optical-mechanical analogy, with the mechanical version of the equations on the left-hand side of the page and the optical version on the right-hand side. He develops the optical theory in detail, derives Hamilton's "Principle of Constant Action," derives from it the optical equivalent of the Hamilton-Jacobi equation, and proves the existence of "wave" surfaces orthogonal to the optical rays. There is no attempt to derive a wave equation, of course, but that would have been premature in 1918. The passage concludes with a section on the "Direct Transfer [of the Optical Theory] to Mechanics."

When Schrödinger came to search for the wave equation in 1925 his familiarity with problems of wave propagation and his exposure through Hasenöhrl to the methods of Hamiltonian mechanics applied to motion in a continous medium must have stood him in good stead. With de Broglie pointing the way, exploiting Hamilton's analogy was the obvious path to follow.

We can conclude that Hamilton's optical-mechanical analogy was not as "lost" or "misunderstood" as Schrödinger (and Klein before him) had claimed. Schrödinger may have borrowed Klein's complaint (in almost the same strong words) to dramatize the novelty of his discovery. But it is unlikely that he learned anything from Klein directly. Schrödinger was well versed in the analogy long before he began his search for the wave equation. There is no evidence that he ever read Hamilton's optical papers, but he did not have to. Hamilton's optical-mechanical analogy remained "alive" through the century following his formulation of it in the 1820s, first in the writings of British physicists, and then in the campaign by Klein. Hasenöhrl elaborated the theory in lectures that Schrödinger followed closely. As experimental evidence mounted for a dual wave-particle description of light and electrons, the optical-mechanical analogy was the obvious theoretical construct to exploit, and it is no coincidence that both de Broglie and Schrödinger gave it such a prominent place in their work.

Could Hamilton have discovered wave mechanics for himself in the 1830s and 1840s? His metaphysical view of power acting in space and time would certainly have permitted "matter waves," and he had the necessary mathematical equipment, including the concept of group velocity that appears prominently in Schrödinger's derivation. But the answer to this hypothetical question has to be "no." Without any of the experimental evidence that later demanded a new mechanics, Hamilton had no reason to construct a wave theory of matter; and even if he had done so, it could only have been another "idle speculation" of the kind that annoyed Brewster so in 1842.

V
Politics and Religion

15

Reform and Religious Turmoil

DURING THE 1830s Hamilton's life revolved around his mathematical and metaphysical research, his work for the Royal Irish Academy and the British Association, and his own personal problems of marriage and family. He paid little attention to politics even though this was a period of intense political activity in Ireland. Before and during college he had followed political events carefully, but once at the observatory that interest rapidly dwindled. One reason was his relative seclusion. He was far enough from Dublin to escape the turmoil of the city, and his position and salary were secure enough for him not to be threatened by the vagaries of patronage. The famine scarcely touched him at all. As for his political inclinations, to the extent that he thought about it, he could best be described as a liberal Tory.

Hamilton always had a profound reverence for royalty. When George IV visited Dublin in 1824, Hamilton told his aunt Mary Hutton that he believed the king "to be by prescriptive right the Lion of England."[1] It was not easy to overlook George's selfishness and bad judgment (Graves described him for what he was—"a self-indulgent and hollow-hearted monarch"), but on the subject of monarchy Hamilton had a closed mind.[2] The scandalous affair of George's wife Queen Caroline was the subject of much debate at the observatory. George had been trying to get rid of her since 1806, charging her with various derelictions, and when he became king in 1819 he brought divorce proceedings against her. The queen, supported by the great Whig politician Lord Brougham (codefender, with David Brewster, of the particle theory of light), refused to be bought off and returned to London, even attempting to force her way into the coronation at Westminister Abbey. Hamilton's sisters felt strongly on the subject. Grace in particular was violently "anti-queenite" and thought that Caroline's timely death in 1820 had saved the country from civil war. Hamilton had milder views on the subject, but found it expedient not to argue the point with his sisters.[3]

When Queen Victoria came to the throne in 1837 Hamilton wrote a sonnet for the occasion, and twelve years later he added another when she visited Ireland. He attended a levée in her honor at the Viceregal Lodge, and sent both sonnets to the queen through Lady Clarendon, wife of the lord lieutenant, whom Hamilton had come to know quite well.[4] Victoria and Albert returned to Dublin in 1853, and this time Hamilton had a private interview with the Prince of Wales, who had supposedly expressed an interest in Hamilton's quaternions, although it is more likely that Hamilton had applied for the honor of presenting the prince with a copy of his recently published *Lectures on Quaternions*.[5] Hamilton had additional reason to feel gratitude to Victoria and to Robert Peel's Tory ministry. In 1843 Peel had placed him on the civil list of pensioners, granting him £200 per annum for life, which eased some of his growing financial problems.[6]

Hamilton's devotion to the English monarchy was a measure of his commitment to the union of Ireland with England, and that union was the chief determining factor of Irish politics. Nineteenth-century Ireland was not a time or place where one could be oblivious to politics, no matter how much one might try to avoid it. Hamilton believed it his duty to vote in all elections for which he was eligible; and elections in Ireland were often violent affairs. On one of his early visits to Dublin from Trim, Hamilton had witnessed an election at which Henry Grattan was "every day attended by great multitudes at the Hustings—and on the last day, on which he was finally defeated, he was drawn in triumph by an immense mob, who broke all the windows of the principal friends of Ellis [the rival candidate]."[7] With the limited franchise that existed in Ireland, it was possible for electioneers to follow every vote and perform the acts of bribery and extortion necessary to gain a majority. Hamilton was not a likely subject for bribery, nor was he especially vulnerable to political pressure. Nevertheless, the mere act of voting took a certain amount of courage. The polling was conducted openly. Each voter had to declare his identity, the right by which he obtained his franchise, and then his choice of candidate.[8]

There were usually large mobs present to intimidate the voters as much as possible. In the countryside, being enfranchised was not always an advantageous right. The Catholic farmer inevitably found himself caught between his landlord, who would insist on controlling his vote, and his priest, who was prepared to send him to hell for voting with his landlord. Maria Edgeworth described the election of 1831 at Trim:

There never was such a terrible tyranny as that under which the poor wretched people are here. It is literally a reign of terror: at the very end of every table where the votes were to be given stood a priest, threatening every poor Catholic that came with perdition if he didn't vote for Mr. Grattan and any one who did was shivering with fright as he gave his name to Mr. Bligh. The Protestant farmers and shopkeepers they threatened with—"Remember the harvest's not out" and "No man will deal any more with you." They seized hold of any doubtful freeholder and dragged him off to Mr. Grattan's Committee and locked others up safe out of the way, who were delighted to be secure from the dread alternative of losing their souls and bodies to the [Catholic] Association if they voted against them, or their farms if they voted against the landlord.[9]

Maria was one of the most enlightened landlords in all of Ireland, but she reluctantly gave in to her brother-in-law's insistence that the tenants at Edgeworthstown be punished for voting the "wrong way."[10] Maria was distressed that her tenants might *want* to vote against her. One or two she could excuse—Gaffery, for instance, because he was "almost beat to a jelly last election for voting with his landlord."[11] Others she accused of arrant cowardice, or of telling lies—even *superfluous* lies. And yet the Protestants were not eager to introduce the secret ballot, which would have protected the voters from the mob and threatening priests, but would also have removed them from the control of their landlords. Moreover, the ballot was an "un-English" way of doing things. On one occasion, when Hamilton attended a dinner given by the Whig Lord Lieutenant Lord Normanby (against whose party he had just voted), he told Under Secretary Thomas Drummond: "With what satisfaction or comfort could I, as a gentleman, sit now at Lord Normanby's table if I had voted *secretly*, instead of openly, against his candidates in Dublin a week ago."[12] He regarded the ballot as a dishonorable procedure, whatever relief it might have given to the electorate.

Hamilton and most of his friends had considerable sympathy for the cause of Catholic emancipation, even though they were shocked by the violent and abusive language that characterized Daniel O'Connell's campaign. William Plunkett, the member for Dublin University (Trinity College) had fought for Catholic emancipation before O'Connell founded his Catholic Association in 1823, and some Protestants, such as Maria Edgeworth, had held the naive hope that emancipation would bring an end to sectarian conflict in Ireland. A more pessimistic, but more accurate, judgment was that presented to the House of Lords in 1829 by Lord George Beresford, the Anglican archbishop of Armagh. His address was delivered the year before Hamilton and Adare talked with him. "Are you prepared, my lords, to go the length to which you will be urged, after you have conceded all that is now demanded? Are you prepared to sacrifice the Irish church establishment and the protestant character of the Irish portion of the empire—to transfer from Protestants to Roman Catholics the ascendancy of Ireland?"[13] Emancipation was forced on parliament by O'Connell and his Catholic followers. If it had been granted gracefully at the time of the Act of Union it might have relieved some of the hostility between Catholics and Protestants, but coming as it did in 1829, after years of bitter struggle, it appeared to Catholics not as a compromise or as a concession from the Protestants, but as a victory in battle to be succeeded by more such victories.

Catholic emancipation was soon followed by agitation for reform of Parliament to make the representation in the House of Commons more nearly reflect the population distribution of the country. "Rotten boroughs" that had once been populous but now contained only a handful of voters were bought and sold like commodities, while the growing industrial cities of Manchester, Birmingham, Leeds, and Sheffield had no representation at all. Defenders of the system argued that the opportunity to buy seats in Parliament meant that political power was held by those with the greatest economic power, just as it should be. Wordsworth, who violently opposed the reform bills, could get heated about the subject. In October, 1831, Dora Wordsworth reported that her father was quite weighed down by the evil that he foresaw coming from that "dreadful Reform Bill."[14] The previous January Wordsworth had asked Hamilton: "How came you not to say a word about the disturbances of your unhappy country? O'Connell and his brother agitators I see are apprehended; I fear nothing will be made of it towards strengthening the government"; the following June he expressed pleasure that "the educated classes in England and Ireland" were so strongly against "this rash and unprincipled measure." He went on to write: "You, I trust, will be glad also to hear that a large majority of the *youth* both of Cambridge and Oxford disapprove the measure; and this proof of sound judgment in them I think the most hopeful sign of the times."[15]

Hamilton tried tactfully to avoid the subject in his correspondence with Wordsworth, arguing that politics required a practical mind and experience that he lacked, but finally his "honor" forced him to reveal himself: "The confession is that I am a Reformer, though not from any confidence in the present ministry of England [the Whig cabinet of Earl Grey], and though I have not by any public act expressed my leaning, opinion I can hardly call it, formed, as it has been, after so slight attention to politics, and avowed, as it is now to you who have made politics so much your study."[16] The expected response soon arrived. Wordsworth assured Hamilton that the reformers were set on destroying the Constitution of England, that "Fount of Destiny, which if once poisoned away goes all hope of quiet progress in well-doing." "The Constitution of England," he wrote, "offers to my mind the sublimest contemplation which the history of society and government have ever presented to it."[17]

Hamilton continued to try to avoid the subject. The following March, during his trip to England with Lord Adare, he had an opportunity to follow part of the debate in the House of Commons. Aubrey De Vere's uncle Thomas Spring-Rice (later Lord Monteagle) got him a place on the speaker's list so that he could sit under the gallery close by the members. The powerful oratory of the Tory opposition to the Reform Bill had a pronounced effect. Hamilton reported back to Eliza one particularly violent jeremiad delivered by Colonel Perceval that nearly swept him away. The colonel announced that "the storm which was even now whistling about their walls would descend and desolate the land. The pestilence, which they had despised, would rage, and the sword would be let loose. The Church would be swept away along with that State with which it had formed an adulterous and unholy alliance," to which Hamilton added: "When also you remember that I am a reformer chiefly because I prefer a gradual to a sudden revolution, you will not wonder that I was strongly and awfully reminded of him who ran for years about the devoted city of the Jews crying 'Woe, Woe, to Jerusalem!' "[18]

As the agitation for reform increased Hamilton began to waver. He confessed to Wordsworth that while he had thought it wise to concede reform he had no desire to see Ireland broken off from England and reconstituted as a republic, which seemed to be the direction that events were taking.[19] The first Reform Bill had been brought forward in March, 1831, and had passed the House owing to the decisive vote of the Irish members, who stood behind O'Connell as a separate party. The king dissolved Parliament to prevent passage of the bill, but the following election gave the reformers an even larger majority, and the king was finally forced to accept the third Reform Bill in June of 1832. It gave little advantage to the Irish, however, because it added only five Irish members to the one hundred already sitting, while at the same time it maintained the existing restrictions on the franchise.[20]

Once the Reform Bill became law, O'Connell again took up the cry for repeal of the Union, a move that Hamilton could never support, and as O'Connell moved into alliance with the Whigs during the years from 1834 through 1840, Hamilton fell into sympathy with the Conservative Party, headed in Dublin by his old tutor, Charles Boyton. The *Dublin Evening Mail* and the *Dublin University Magazine*, both Tory publications, began to mention him favorably. On August 19, 1834, he joined the Conservative Society, his first and only public political action, and made a speech to the effect that the agitators in Ireland wished to establish the Roman Catholic Church on the ruins of the Establishment. It would be a mistake, he said, for anyone to believe that "the seclusion of our libraries shall afford us effectual protection for they would continually be afraid of hearing 'the step of the inquisitor on the stairs.'"[21]

Although he joined the Conservatives, Hamilton never became an "ultra-Tory" or "pigtail Tory." He was more in sympathy with the aims of Robert Peel's "Tamworth Manifesto," which was published in December, a little less than a month after Hamilton joined his party. In the Tamworth Manifesto, Peel accepted the Reform Bill and said that he favored reform to the extent that it meant "a careful review of institutions, both civil and ecclesiastical" and "the correction of proved abuses and the redress of real grievance," but he was not prepared to accept a violent overthrow of established religious and political institutions merely at the whim of the mob.[22] Aubrey De Vere was unhappy with Peel's attempt at compromise. He wrote to Hamilton: "I hope the Whigs are not going to turn Radicals; though they have just as good a right to do so as the Tories have to turn Whigs. [Peel's party now called themselves "Conservative," rejecting the old party name of Tory.] I am no admirer of the late ministry: but nothing can be a greater proof of the utter ruin with which they overthrew the Tory party than the manner in which the latter, since their return to office have been obliged to abandon their old principles and even their name."[23] Peel's ministry only survived for a few days, and was succeeded by Lord Melbourne's second Whig cabinet.

The political events that distressed Hamilton and Aubrey the most were attacks on the established Church of Ireland, particularly the Tithe Wars, which were ruining Uncle James. O'Connell endorsed the refusal to pay tithes and was given strong support by the Roman Catholic clergy. Hamilton first mentioned the problem in January, 1832; by 1835 James was in real distress.[24] Hamilton appealed to Bishop Whately and to the Archbishop Lord Beresford, arguing that he had done everything in his power to help enforce the payment of tithes, and to help with the relief of the clergy, and added: "To my Uncle, in particular, by whom I was educated, I am bound by every tie to give what assistance I can."[25] This plea suffered the same fate as all his former appeals in support of Uncle James—neither bishop nor archbishop gave any help.

Protestants such as Hamilton reserved their greatest hostility for the Roman Catholic clergy who led the political agitation against the tithe and the established Church in general. They saw much of the trouble coming from Maynooth, the Roman Catholic seminary established in 1795 for the training of the Catholic clergy.[26] Most of the Roman Catholic hierarchy had been trained on the Continent, as had been all the Catholic clergy before the foundation of Maynooth, but Maynooth drew its seminarians from the peasantry, and its graduates retained peasant prejudices and peasant aspirations. Maria Edgeworth found them "so vulgar no gentleman can, let him wish it ever so much, keep company with them. This puts them in a class by themselves, hence they feel looked down on by the gentry and so long to pull them down to their own level and teach the people nonsense about destroying the aristocracy."[27] Hamilton was more moderate, and tended to favor the continuation of the endowment for Maynooth, although with some reluctance. In 1852 he wrote to Augustus De Morgan: "I am old enough to remember foreign-bred ecclesiastics of the Roman faith. They were extremely agreeable people, and much welcomed in Protestant society. The case is very different now. The education at Maynooth is, I believe, *anti-*English. But I repeat that I do not set up for being an unprejudiced man."[28] The success of O'Connell's campaign for Catholic emancipation had shown for the first time what an organized Catholic opposition could accomplish in Ireland. Unfortunately, it also greatly increased the sectarian conflict. Neither O'Connell nor his followers were clear in their goals. To many peasants O'Connell seemed to be promising the elimination of the landlord class altogether and a complete separation from England, although O'Connell himself was not prepared to take any such step.

The sectarian conflict would have been even worse in the 1830s if it had not been for the able administration of Thomas Drummond, who served as undersecretary between 1835 and 1840. Drummond served under the Whig ministry of Melbourne and therefore did not share Hamilton's particular political complexion, but he managed to suppress much of the disorder in the countryside through the establishment of a new Irish Constabulary. During the tithe wars the Catholic "Ribbonmen" and Protestant "Orangemen" held frequent bloody encounters in the countryside, which produced constant unrest and fear of civil war. Nothing pleased the Catholic peasantry more than Drummond's efforts to break the power of the Orange Order, and for five years the most extreme Protestants felt that the administration at Dublin Castle had gone over to the enemy.[29] Hamilton had met Drummond in 1828, when he visited the Ordnance Survey where Drummond was working. They corresponded on occasion during Drummond's administration on matters regarding the Royal Irish Academy, of which Hamilton was president, and sometimes on mathematical questions.[30]

Of all the distressful events of the 1830s nothing shook Hamilton as violently as the murder of Lord Norbury in January, 1839. Lord Norbury was an elderly nobleman with no political ambitions whatsoever. He was murdered in broad daylight on his own grounds.[31] The effect in Ireland was profound, and it echoed all the way to Westminster, where a committee of enquiry was established in the House of Lords. Wordsworth blamed the murder directly on O'Connell's inflammatory rhetoric and asked Hamilton, "How long is the reign of this monster over the British Islands to endure?"[32] Hamilton, who appears to have been in a hanging mood, blamed the murder on the conciliatory policy of the government. He wrote to Lord Northampton:

The murder of Lord Norbury grieved many, but surprised few in Ireland. All Protestants, all friends of England, have been too familiar, for some years, with the probability of a violent death, to experience any such surprise. But I do mourn deeply over the accumulation of national guilt, which cries unto Heaven against Britain. . . . But England, probably, will still reply, "Am I my brother's keeper?" The easy pardoner of felons, the discourager of the discouragers of murderers, the friend of the enemies of England, is still entrusted with the privilege of pardon, may still discourage in the royal name, wields still the power of England [Lord Normanby, second earl of Mulgrave, lord lieutenant of Ireland, and the man who had bestowed knighthood on Hamilton in 1835]. Attempts will still be made to conciliate, instead of punishing, assassins; to swell a parliamentary majority, the jail and gibbet will be cheated still; all will be done which God allows, to carry to extremities a state of things which now, through a large part of Ireland, makes glad the bad, alarms the timid, puts on their guard the brave.[33]

Northampton was quick to defend the lord lieutenant, pointing out that violence in Ireland long preceded the administration of Lord Normanby. Northampton attributed it to "ignorance and poverty—especially the *last*—and want of employment."[34]

Another issue that stirred Ireland in the 1830s was that of education. Since 1815 the government had supported attempts by the Kildare Place Society, a nondenominational foundation, to set up nonsectarian elementary schools in Ireland. These schools had only limited success in attracting Catholics, and they became more and more under the direction of Protestants. An attempt in 1831 to restore a nondenominational character to elementary education through a series of "national schools" had only a little more success in bringing together the warring factions of Presbyterians, Anglicans, and Catholics. Its only real accomplishment was to greatly discourage the Irish language, the use of which rapidly dwindled over the next fifty years.[35]

Eliza felt strongly about the new schools. While Hamilton was listening in 1832 to the debate on the Reform Bill in the House of Commons, Eliza was attending a tumultuous meeting at the Rotunda in Dublin, which she described to him in a letter.

High on the platform so as to be seen by all sat about 30 poor Irish peasants all Roman Catholics, they were dressed in frieze coats and entered in a body with one solitary woman among them. These men had come as a deputation from one of the wildest parts of Ireland with a declaration against the Proposed System of Education signed by 3000 (or some large numbers over) Roman Catholics. . . . But *how* shall I describe the speech of one of those poor men themselves a teacher in one of the Irish schools, it was eloquent, it was beautiful, it was affecting beyond anything I ever heard. Not a Clergyman on the Platform could contain or well conceal his tears. *There* was the [print] of their silent and patient labours (blessed by God) for years back. But now at the decree of an Infidel Ministry it was all to be trampled under foot! When our eloquent Irish peasant turned to his Irish companions and after describing the persecutions they were prepared to expect such as what he had himself known of, when the tongue was cut out of one Irish teacher and his barbarous mutilator said "Now you'll no more read your Irish to the people" When I say he turned to his companions and said in Irish which he afterwards translated "We fear not those who kill the body etc." and when they joyfully expressed assent, the *power* of the Gospel was indeed felt.[36]

Hamilton tended to avoid educational controversies that had strong political overtones, but he was concerned about them. Education had been the subject of much of Maria Edgeworth's lifework, and he was certainly familiar with her writings. He directed his interests to the new College of St. Columba at Stackallan, which was founded in 1843 by a group of Anglicans led by William Sewell from Oxford and J. H. Todd, professor of divinity at Trinity College Dublin. The purpose of the college was to provide a school comparable to the great public schools of England for the sons of Irish Protestants. Aubrey identified Sewell to Hamilton as "one of the High Church Oxford Divines."[37] When they met, Hamilton and Sewell found a common interest in metaphysics; Hamilton was probably instrumental in introducing Sewell to Professor Todd.[38] Lord Adare also took an active part in the founding of the college. In 1843, soon after the college opened, Hamilton delivered an address to the students on "The Changing Aspects and Unchanging Laws" of the heavenly bodies. He had originally intended to send his eldest son, William Edwin, to Stackallan in 1846. By that time, however, the college had acquired the reputation of being dangerously Anglo-Catholic, in part because Sewell, its founder, was a major figure in the Tractarian movement.[39]

With the coming of the Great Famine political questions in Ireland were all connected with the struggle for survival. In 1846 Aubrey De Vere returned from almost three years of travel abroad to find his brother deeply involved in famine relief. To everyone's surprise, Aubrey moved out of his poetry and metaphysics and into the soup kitchens. When he heard about Aubrey's efforts, Hamilton felt obliged to justify his own activities (or lack of them):

Though I have been giving, and shall continue to give, through various channels whatever I can spare in the way of money to the relief of those wants, yet I am almost ashamed of being so much interested as I am in things celestial, while there is so much of human suffering on this earth of ours. But it is the opinion of some judicious friends, themselves eminently active in charitable works, that my peculiar path and best hope of being useful to Ireland, are to be found in the pursuit of those abstract and seemingly unpractical contemplations to which my nature has so strong a bent. If the fame of our country shall be in any degree raised thereby, and if the industry of a particular kind thus shown shall tend to remove the prejudice which supposes Irishmen to be incapable of perseverance, some step, however slight, may be thereby made towards the establishment of an intellectual confidence which cannot be, in the long run, unproductive of temporal and material benefits also to this unhappy but deeply interesting island and its inhabitants.[40]

Aubrey confirmed Hamilton's judgment, thereby easing his guilty conscience, and speculated in his characteristically metaphysical fashion on the general doctrines of the abstract and the practical.[41] The famine for Hamilton was also a bit of an abstract idea. As he wrote to Herschel: "I try to abstract my thoughts from useless brooding over the state of the country, that state is enough to sadden anyone in Ireland. However this neighborhood is not the worst part of it; the subscriptions and other funds have hitherto seemed to meet the distress of the parishes immediately adjacent."[42] While Aubrey faced the stark reality of the famine in the fever sheds of Limerick, Hamilton remained in relative isolation at the observatory in an area that did not face serious deprivation. His only recorded reaction to the famine, besides giving money to relief committees, was keeping an eye on the potato supply at the observatory.

Although Hamilton did not suffer physically from the famine, the political and religious controversies of the 1840s seemed to take their toll on his emotional strength. The unrest of the times, combined with the strain of Lady Hamilton's prolonged illness, put him in an unhealthy nervous state. In November, 1844, John Graves wrote to Robert about the visible signs of strain that he had observed in Hamilton's behavior. There was a "certain nervous irritability in his temperament," a tendency to run on almost desperately with abstruse mental calculations in a way that indicated a morbid overactivity of work.[43] Robert Graves saw Hamilton the following June at the Cambridge meeting of the British Association, and he also noticed that Hamilton was excitable. Hamilton was taking religious issues more and more seriously. He was terribly worked up by the early controversies of the Oxford Movement and had begun a strict observance of all religious fasts and festivals specified in the Prayer Book.

He was also beginning to drink more. A high consumption of alcohol was customary in mid-nineteenth-century Ireland. Any dinner or banquet was an occasion for drinking large quantities of wine in numerous toasts, and Hamilton was never one to hold back. There is no indication that he recognized any problem in his drinking habits before 1845, but at the British Association meeting of that year he discovered after one of the banquets that he had had too much to drink. He sought out Robert Graves in great agitation and asked if he could kneel at

Robert's feet, confess his sin and receive absolution. Robert decided that a little excess alcohol was not sin enough to require a formal confession and absolution, particularly since it might aggravate rather than relieve Hamilton's distressed state of mind. Besides, private confession to a priest was a Roman and high-church Anglican practice, and he did not want to do anything that would encourage Hamilton to move in a Romish direction.[44] Later Robert had reason to regret the fact that he had not taken Hamilton's appeals more seriously.

On the eleventh of February, 1846, Hamilton attended a dinner at which members of the Geological Society of Dublin prepared plans for their anniversary meeting. During the dinner he talked excitedly about a plan to use the instruments at the observatory to detect earth movements. After leaving the dining room he fainted at the top of a flight of stairs. As he described it: "I was seized with a giddiness and rush of blood to the head, which totally incapacitated me from ... keeping my ideas under control." In writing his apology to the president of the society, he did not admit that he had been drunk, but did say that he was adopting "a regimen so severe, as to make it unlikely if not impossible, that such a state of things should ever occur again," which would indicate that alcohol had been the cause of his collapse.[45] Although Hamilton probably could not remember it, he had done more than just faint and tumble down the stairs. Graves, who was close to the events, reports that "his reason was disturbed for a time. The result was that he became violent, and had to be restrained: I forbear going into further details."[46] It must have been an excruciatingly humiliating experience for Hamilton. The story quickly made the rounds of Dublin gossip, and Hamilton knew the kind of whispering that was going on behind his back. Robert's brother Charles, who was the mathematics professor at Trinity College and also in holy orders, came out to the observatory to talk to Hamilton about his "problem." Hamilton decided to stop drinking entirely, but did not take any vow of abstinence. It put him in a difficult position, because refusing to drink in public further advertised his weakness. He described how he handled the problem at a dinner given by the lord lieutenant:

I dined with Lord Heytesbury on Saturday, not thinking it right to decline another invitation. His Excellency was very kind in manner and had the air of entering with the interest of a friend into my concerns, as respected my health, the Academy, and the Observatory. He asked me at dinner to drink champagne with him, and took it very good-humouredly when, on his then inquiring whether I preferred any other wine, I said that with His Excellency's permission I should prefer to pledge him in water. This was the sixth or seventh time of my dining in company since I adopted the water system.[47]

To measure his progress under the new regimen he resolved to give a shilling to the Society for the Propagation of the Gospel in Foreign

Parts for each week that he avoided all alcohol.[48] He was successful for nearly two years. It was not easy for him because, as he wrote to George Peacock, "I had previously taken my full share, when stimulated by society, of what are called the pleasures of the table." He thought he noticed an improvement in his health: "I feel by this time a more firmly settled health, though possibly somewhat less strength of body, and drink with real relish a mug of new milk every morning. If health alone had been my object, perhaps I should have taken less tea and coffee than I do, and have gone through a less amount of study and intellectual exertion."[49]

Hamilton remained on the edge of alcoholism for the rest of his life, but his condition was never as bad as the Dublin gossips would have it. There were no more incidents like the one at the Geological Society, and there is no indication that he ever lost his capacity for sustained mathematical work. The gigantic volumes of the *Lectures on Quaternions* and the *Elements of Quaternions* could never have been written by a man in an alcoholic stupor.

At the time when Hamilton's drinking problem first became known, he had already requested that he be allowed to resign his presidency of the Royal Irish Academy. The timing was fortunate. Although the presidency of the academy was decided by an annual election, it had been the custom for presidents to serve extended terms. Hamilton was president from 1837 through 1846. In 1841 he had already proposed that the president of the academy be limited to a four-year term and had indicated his willingness to resign. He reached this decision during the time that Helen was living in England; it appears that the times when he felt depressed and isolated at the observatory were also the times when the duties of the academy were particularly burdensome.[50]

His suggestion in 1841 was not entirely his own idea. He had got wind of a plan by an unfriendly cabal within the academy to limit the term of the presidency. Since Hamilton was already contemplating resigning, the best thing seemed to be to bring forth a similar plan himself.[51] The council of the academy turned down his proposal and he remained in office. He wrote to Adare, however, that in the future he did "not intend to give up so much time . . . to the routine business of the Body."[52] Throughout 1844 and 1845 Hamilton talked about resigning his position, and on November 17, 1845, he sent a formal letter to the council expressing his intent to resign the following March.[53] Thus he was able to leave the presidency without generating the suspicion that he had been forced out because of the unfortunate event at the Geological Society the previous month.

The academy went out of its way to recognize Hamilton's contributions. He was toasted twice at the Academy Club with the members standing, which was an extraordinary honor, and in its formal report the council emphasized that Hamilton's letter of resignation had been received long before recent events could have influenced his or the academy's actions.[54] Hamilton was happy to see his scientific colleague Humphrey Lloyd voted in as his successor, and letters came in from all directions praising Hamilton's contributions, regretting his resignation, and expressing concern for his health. Augustus De Morgan congratulated him on finally ridding himself of the burden: "I was glad to hear that you are to resign the Presidentship. You have no business there at all; there are plenty of people who can do all that a President, as such, has to do; and I maintain that any man who is fit for original research has no business to be a president, or secretary, or treasurer, at the *expense* of his researches."[55] Hamilton must have had mixed feelings about resigning the presidency of the academy. For eight years he had been constantly active in its affairs. Just the volume of correspondence written while he was president indicates that he took his duties seriously. He had a more detailed knowledge of the academy's affairs than any previous president and could frequently resolve disputes by his superior knowledge of the regulations and precedents of the academy.[56]

Under Hamilton's direction the academy had made its greatest gains in the areas of Irish archaeology and antiquities. The Cross of Cong, the most spectacular of all the antiquities collected by the academy,

Sir William Rowan Hamilton with one of his sons (circa 1845)

was donated by James McCullagh in 1839. Hamilton also led in the collection of old Irish manuscripts, and during his administration the academy acquired several important collections, including those of Hodges and Smith and Sir William Betham.[57] He was fortunate in having George Petrie, the real founder of Irish archaeology, as an active member of the academy. Lord Adare was also an avid enthusiast of Irish architecture, and, with Petrie, founded the Irish Archaeological Society.

Of course Hamilton was not able to escape entirely the bickering that Maria Edgeworth had warned him about when he first took office, and there were some famous examples of it during his administration. Probably the most dramatic was an exchange between Sir William Betham and George Petrie. The Committee of Publication of the academy refused to publish a series of papers by Betham claiming to show a connection between ancient Etruscan inscriptions and the Irish language. Hamilton's diplomacy was not up to the task of calming the disturbed Betham, who resigned from the council in a huff.[58] When the Medal for Polite Literature was awarded to George Petrie for his work "On the History and Antiquities of Tara Hill," Betham attacked it vigorously and published two even more violent letters that charged Hamilton with speaking on a subject about which he knew nothing.[59] Hamilton kept silent, as he usually did in such situations, and allowed Petrie to respond.

Other controversies of a more political nature were not so easily resolved. In 1836 the government attempted to obtain changes in the constitution of the Royal Dublin Society. Hamilton was only a nonvoting honorary member of the society and not yet president of the Royal Irish Academy, but several members of the society called on him to help seek a compromise between the government and the more recalcitrant members, who resented the government interference. William Smith O'Brien was on the parliamentary committee looking into the affairs of the society, and he in particular asked for Hamilton's opinion.[60] But the Royal Dublin Society continued its resistance, and in 1840 and 1841 its annual grant of £5300 was in serious jeopardy. The lord lieutenant, Lord Ebrington, appointed Hamilton to a committee that would recommend to the government how best to use the £5300.[61] As president of the Royal Irish Academy Hamilton's opinion now carried greater weight, and he was able to effect a compromise. He later claimed to have saved the Royal Dublin Society from destruction.[62]

William Smith O'Brien put Hamilton in a more difficult situation in 1844. An ardent nationalist and reformer, O'Brien favored repeal of the Union. When Hamilton in 1844 requested gifts to the academy for the purchase of antiquities, he was surprised to receive a gift from O'Brien in the name of the Repeal Association. Hamilton went through Adare, now a member of Parliament, to find out what the government's attitude would be if the academy accepted the money.[63] Then he turned the problem over to the council of the academy, where he would not have to vote. As he wrote to Adare, "The trap was certainly well laid, and the Academy can scarcely come out of it in either way without being charged with a political bias."[64] The academy refused the gift.

O'Brien's attempt to embarrass the academy coincided with a political shift in Ireland that saw the radical leadership pass from Daniel O'Connell to the Young Irelanders, a group that included O'Brien among its leaders. In a series of mass meetings of ever increasing size and by the use of increasingly inflammatory speech, O'Connell had brought his campaign for repeal of the Union to a climax in 1843. He had planned the greatest meeting of all for October 8, 1843, at Clontarf. A huge crowd of emigrants and old Irish patriots was expected to return for the event, as were thousands of supporters from all parts of Ireland. But the government stood firm and declared the meeting illegal. When O'Connell saw that further defiance would certainly lead to bloodshed, perhaps a great deal of bloodshed, he cancelled the meeting, and the government, pressing its advantage, arrested him and put him in jail for a year. The momentum was lost and O'Connell was never again an important force in politics. The leadership of the Irish nationalist cause fell into the hands of the Young Irelanders, whose romantic idealism was quite unlike O'Connell's brand of popular oratory and practical politics. The Young Irelanders Hamilton knew, like O'Brien and the poetess "Speranza" (later Lady Wilde, mother of Oscar Wilde), were Protestants, as were the better-known leaders of the movement, Thomas Davis and John Mitchel. They broke with O'Connell over the establishment of the nonsectarian Queen's Colleges in 1845. The Catholic hierarchy and O'Connell were violently opposed to the new colleges, while the Young Irelanders, who placed Irish nationalism above all sectarian differences, supported them.

The Young Irelanders were not as close to the peasantry as O'Connell, and they lacked his rough vigor. Their influence was through their revolutionary newspaper the *Nation*, which Thomas Davis had started in 1842 and which enjoyed enormous success. They were more in tune with the nationalistic idealism of Mazzini and the other "liberators" who fomented revolutions throughout Europe in 1848. The Great Famine, coming as it did between 1845 and 1848, both weakened the revolutionary movement in Ireland and pressed its leaders into desperation. John Mitchel was the most radical of the Young Irelanders in 1848 as well as the one most firmly committed to revolution. When the French led the way with their successful revolution in February, 1848, the Young Irelanders began to prepare an uprising. But Mitchel was arrested on a charge of sedition in May, 1848, and their original plans were thwarted. The tension in the countryside was extreme. Uncle James wrote from

Trim that "this country has been proclaimed; and is in a very excited state, though as yet there has been no outbreak."[65] William Smith O'Brien was left as leader of the organization, and in August he led the Young Irelanders out in Kilkenny. He must have known that the rebellion had no chance of success. O'Brien was captured, charged with high treason, and condemned to death, but the government commuted his sentence to transportation in order to avoid alienating public opinion by an execution. Long after the event Hamilton wrote: "I distinctly remember, that William Smith O'Brien had *publicly* announced before that outbreak in 1848, which if otherwise conducted might really have been a dangerous one, that he *knew* himself to be deficient in the element of personal *courage*. A curious *leader* to select! but he would have made and indeed has made an admirable *martyr*: and I think, but since he became a rebel I have had no communication with himself (whom I once knew), that he was rather disappointed at not being beheaded."[66] As a military operation the revolt was a farce, but as a stimulus to national consciousness it was a success. The Young Irelanders kept alive the "spirit of '98" and added a new series of names to the martyrology of Irish republicanism.

Hamilton, as might be expected, stood firmly on the government side, and even enrolled in the spring of 1848 among those who were ready to carry arms in support of the government cause.[67] But he also worked to mitigate the sentence of O'Brien, and he specifically asked Aubrey not to write to him about Irish politics so that he could truthfully say that he had not discussed the matter with any of the De Veres.[68] Smith O'Brien was Ellen De Vere's brother-in-law and a friend from Hamilton's youth. However much he despised his politics, Hamilton wanted to help him escape the gallows.

16

In and Out of the
Oxford Movement

MORE IMPORTANT for Hamilton than political events in Ireland were
the religious controversies that disturbed the Anglican church thoughout
the century. These controversies were often closely associated with
political events, because the Anglican church and its Irish branch, the
Church of Ireland, were the religious establishment of the United
Kingdom and were subject, at least in part, to its control. Many Irish
Protestants held the extreme Erastian view that the church *was* the
nation, and that only true believers, subscribers to the Thirty-nine
Articles and followers of the Book of Common Prayer, could be full
members of the body politic. By the time Hamilton was born, Dissenters
and Catholics had won limited rights, but these had been granted
grudgingly, and the Anglican church still held a monopoly of higher
education, a control over Parliament, and a huge fund of patronage to
dispense to right-minded clerics.

The great religious controversies of the century originated in England,
but often their political significance was most keenly felt in Ireland.
In England the Established church was supported by most of the inhabitants; in Ireland, according to the census of 1834, the population
was 80.9% Roman Catholic, 10.7% members of the Established church,
8.4% Presbyterian (most of whom lived in Ulster), and 0.27% Dissenters
of other varieties.[1] It was in Ireland that the Establishment was under
the greatest pressure for reform, and it was there that reform came most
rapidly. The Catholic Emancipation Bill was passed in 1829; a nonsectarian system of national education was founded in 1831; ten Irish
bishoprics were suppressed by Parliament against the resistance of the
Irish church in 1833; and in 1845 three new nonsectarian colleges were
founded to make university education more accessible and to break the

monopoly held by Trinity College Dublin. (Trinity College admitted Dissenters and Roman Catholics but reserved fellowships and scholarships for members of the Established church.) Finally, in 1869, four years after Hamilton's death, the Irish church was disestablished entirely. The political implications of these changes are obvious; they bear out the words of the primate, Lord Beresford, that the decline of the Protestant political and social ascendancy could be measured exactly by the decline of the religious establishment.

Political assaults on the Anglican church brought about movements for reform not only of the church's structure, but also of its internal life. The great questions became what constituted the Christian church and what the position of the Established church of England and Ireland was, relative to the universal church of all believers. The various groups within the Anglican communion were defined by the way in which they answered these questions. The Broad Churchmen wished to grant the widest possible latitude of belief, while the Tractarians wished to narrow that latitude to the confines of the Medieval church; the Evangelicals saw the church as an invisible and mystical body drawing its authority entirely from Scripture, while the Erastians thought it was a creation of Parliament at the time of the Reformation. Outside of the Establishment were the Roman Catholics, who rejected all Christians not in communion with the Pope; and the nonconformists, who saw the church as an aggregate of separate congregations.

Before 1830 the Church of Ireland existed in a comfortable but somnolent state. From 1830 until Hamilton's death in 1865, there seemed to be nothing but controversy. Through it all Hamilton managed to remain a staunch defender of the Establishment, swaying slightly in the blasts of controversy between high and low, but keeping pretty much to the same character of churchmanship. He had to witness some violent changes about him, however. His sister Sydney (and probably also Eliza) became a Calvinist. His two closest friends, Aubrey De Vere and Lord Adare, went to the other extreme in the 1850s and converted to Roman Catholicism. Just standing still was not easy during those years for a person truly concerned about religious questions.

During Hamilton's youth there were two major currents of religious opinion within the Church of Ireland. One current was the traditional Anglicanism of the "high and dry" variety. Its followers adhered to the Prayer Book carefully, shunned ostentation, and abhorred emotionalism of any kind. Their sermons were learned and orthodox. Many were devoted and sincere pastors of their congregations, but many others used patronage to pry out for themselves positions of comfort and sometimes of real wealth. Often they were country gentlemen who had obtained a benefice through family connections; there were, for instance, eight beneficed clergy who were members of the Kilkenny Hunt Club.[2]

While riding to the hounds was not an entirely improper recreation for a clergyman, it became so when the clergymen was more interested in the hounds than in his own parishioners. Many clergy did not reside in their parishes. In 1824 Bishop John Jebb told the House of Lords that of the fifty-one benefices in Limerick Diocese, twenty-six had incumbents resident, and he considered that to be a commendable record.[3] Uncle James had a small benefice from the parish of Almorita at some distance from Trim. During the Tithe Wars he lost most of his income from this benefice, and what tithes were collected were sequestered for the repair of the glebe house, which James had not occupied and had allowed to decay into ruin. Patronage, privilege, and a certain "quiet worldliness" were the chief failings of the traditional clergy in Ireland.[4]

The second tradition of Irish churchmanship was Evangelical. The teachings of John Wesley had found their way into Ireland and had caught the enthusiasm of many clergy. Their Methodism tended to be more orthodox than that of the Evangelical clergy in England, but it had the same emphasis on ministry and mission. The Evangelicals were "low church." They avoided elaborate liturgy, wore simple vestments, and emphasized personal devotion and service. They were responsible for the founding of many missionary societies, such as the Hibernian Bible Society, and had their own religious journal, the *Christian Examiner*.[5] Hamilton himself took an active part in another Bible society, the Society for the Propagation of the Gospel in Foreign Parts, and served as president of the local chapter for some time.[6]

The greatest weakness of the Evangelicals was their lack of a systematic theology. They were unwilling to adopt the teachings of Luther or Calvin, but at the same time were not comfortable with the dusty conventionality of Anglican theology. As a result they tended not to stress theology at all, but to concentrate on ministry and service. Hamilton tended towards the Evangelical side of the church and adopted the doctrines associated with it (salvation through faith, with free forgiveness, and with the sole mediation of Christ); even towards the end of his life, in 1858, he still referred to himself as an Evangelical Anglican.[7] But for a man so interested in theology, the Evangelical belief was incomplete, and he was attracted by other movements within the church that emphasized doctrine and systematic theology.

As the Church of Ireland came under pressure from Roman Catholicism and from political reformers in England, groups from within the Anglican church came forward in an attempt to breathe new life into its moribund body. The Oxford Movement began in 1833 and was led by Keble, Pusey, Froude, and John Henry Newman. This was the most dramatic of all the attempts to revive the Anglican church. The "Tractarians," as they were called after their most important publication, the "Tracts for the Times," saw the Reform Movement by the Whigs as a direct attack on the free and independent institution of the Anglican church. The suppression of ten Irish bishoprics in 1833 was the occasion for their first foray against the policies of the government. The Tractarians placed great emphasis on the authority and tradition of the Anglican church as continued by the apostolic succession. To the question "What is the church?" they answered that it was the early church, one, catholic and apostolic, a divine religious society, the true uninterrupted Church of the apostles. To the horror of many Protestants, the Tractarians found much good in the Roman church, and some bad in the English church as it had emerged from the Reformation. To the Evangelicals they appeared Popish, not only because of their sympathetic treatment of Roman Catholicism, but also because of their emphasis on authority and their high liturgical practices.

But the Tractarians owed much to the Evangelicals. It was the Evangelical emphasis on personal piety and commitment to a personal religion that made the Oxford Movement possible. Without this sense of urgency, Newman, who began as an Evangelical, would never have begun writing the Tracts in 1833. The Tractarians wrote directly and earnestly and with great learning. Even those who distrusted them felt the force of their arguments. As a result, the Oxford Movement drove a wedge between high and low churchmen, forcing members of the Anglican church to take a stand on one side or the other.

Alongside the Oxford Movement stood another, less well-defined group called the "Broad Churchmen," who argued for tolerance of diverse religious opinions. They emphasized the need for freedom of religious inquiry, and insisted that no individual or sect could claim to have a corner on the truth. Some of the best-known Broad Churchmen, such as Richard Whately and Thomas Arnold, were also at Oxford. Newman had been a protege of Whately at Oriel College, but from the publication of the first of the Tracts Whately stood in strong opposition to the Oxford Movement. The Broad Churchmen at Oxford also included such figures as Edward Copleston, Blanco White, Baden Powell (Hamilton's optical correspondent), and R. D. Hampden, whose appointment as Regius Professor of Divinity in 1835 caused great controversy. Whately stood out as their leader. A vigorous, assertive man, he shocked convention by his slight reverence for college custom and by his unusual enthusiasms, such as his tree-climbing dog and his favorite sport of throwing the boomerang. Whately wrote an important textbook on logic and another on rhetoric, and for a short time he held the professorship of political economy at Oxford. There was not a spark of transcendentalism or mysticism in him anywhere. He was a practical man who insisted that the Anglican church as a divine appointment must be open to changes dictated by "common sense." Christianity, for him, was a matter of fact to be decided by examination of the available evidence, and he delighted in disputation with the young fellows at Oxford and later, when he became archbishop, with his clergy. He called them his "anvils" on which he would hammer out his ideas. In 1835 Aubrey gave a good picture of Whately to his sister Ellen:

So you have been reading Whately's works—they are certainly very clever. I was at a party at his Grace's last night; it consisted entirely of clergymen, and was amusing beyond measure. The Archbishop lectured them, generally about practical matters, sometimes about doctrinal, in a style so characterised by originality, rapidity, and fearlessness, that the rest used now and then to stare at each other a little, and always to halt behind in a most lamentable manner. Now and then some one would oppose him; upon which the Archbishop would come down upon him in all the pride and power of Logic and Rhetoric, and roll him over as a greyhound rolls over the little dogs in play. He certainly possesses all the *inferior* faculties in a combination and perfection that is almost miraculous: whether he possesses the higher in any degree is a question which I should be afraid of deciding.[8]

As a high churchman, a supporter of the Tractarians, and a transcendental poet, Aubrey could credit Whately with little appreciation of the higher faculties of mind.

In 1831 Whately was appointed archbishop of Dublin by Lord Grey. The appointment was unpopular, and Whately did little to endear himself to the Irish clergy. He was a libertarian in politics and religion, but was extremely strict in his administration within the church. He had been in favor of Catholic emancipation, and worked hard to promote nonsectarian schools, which were opposed by most of his own clergy. The Evangelicals regarded him as a dangerous latitudinarian, and he cordially despised them in turn; the traditional Anglican clergy found him dangerously liberal, and the Tractarians could not help but regard him as their most dangerous critic.

Hamilton could not have been happy with Whately's religious and political liberalism, nor could he share Whately's utilitarian and business-like approach to philosophical and religious matters. They met frequently at the Royal Irish Academy in apparent amicability until 1837, when they both competed for the presidency of the academy. Hamilton was ready to concede to Humphrey Lloyd his claims to the position, but he stated bluntly that he would not be willing to step aside for the archbishop. In fact he and Lloyd were anxious to prevent any division of the votes between the two of them that might give Whately a chance to gain the election.[9] Hamilton won, and although Whately graciously accepted a vice-presidency when Hamilton offered it, there remained some friction between the two men. In 1843 their relationship was still uncomfortable, as Hamilton reported to Eliza:

The Archbishop has certainly taken some offence either against me, or against the Academy. For so able a man, I suspect that he is rather touchy in little things, and that he has either never forgiven me for not dining at the Palace about two years ago, or else is displeased with the Council for a change of plan which will probably remove him from the office of Vice-President. At all events, though he shook hands with me, when I came into the room, he was freezingly cold, and, as if suspecting that he might not be welcome to the highest place at dinner, he was about to pass on to my left hand when I requested him to occupy his usual seat.[10]

Hamilton also saw Whately at his parish church at Castleknock, where the archbishop preached occasionally, especially after 1842, when Whately's close friend Dr. Samuel Hinds became rector.[11] One gets the impression that during the first twenty years of Whately's service as archbishop he and Hamilton were on terms of chilly cordiality. After 1853 they became much better friends.

There was another branch of the Broad Church Movement that had greater appeal for Hamilton than Whately's rational liberalism. This second group had its center at Cambridge and its members were all followers of Samual Taylor Coleridge and Immanuel Kant. They had greater sympathy for the Tractarians than did the Whately branch, but never joined them. Besides Coleridge himself, this second group included J. C. Hare, John Sterling, F. D. Maurice, and Charles Kingsley.[12] Hamilton did not meet or correspond with members of this group except for Coleridge, but his philosophical studies and his commitment to transcendentalism put him into their camp. Coleridge's *Aids to Reflection, The Friend,* and *Biographia Literaria* were the works in which this particular version of Anglican theology was worked out.

When Hamilton met Coleridge in 1832 and 1833 most of their discussion dealt with religious questions, especially the doctrine of the Trinity. A sample from Coleridge's letter following their first meeting shows that his theology was not easy to comprehend. A conversation with him was a monologue, and he rambled wildly through metaphysics, religion, politics, and aesthetics. He wrote to Hamilton:

He is God, the I AM, the God that heareth prayer—the Finite in the form of the Infinite = the Absolute Will, the Good; the Self-affirmant, the Father, the I AM, the Personeity;—the Supreme Mind, Reason, Being, the *Pleroma*, the Infinite in the form of the Finite, the Unity in the form of the Distinctity; or lastly in the synthesis of these, in the *Life*, and *Love*, the Community, the Perichoresis, the Inter[cir]culation. ... And I believe in the descension and condescension of the Divine Spirit, Word, Father, and Incomprehensible Ground of all—and that he is a God who *seeketh* that which was lost, and that the whole world of Phaenomena is a revelation of the Redemptive Process, of the Deus *Patiens*, or Deitas *Objectiva* beginning in the separation of Life from Hades, which under the control of the Law = Logos = Unity—becomes *Nature*, i.e., that which never *is* but *natura* est, is to be, from the brute Multeity, and Indistinction, and is to end with the union with God in the Pleroma.

It is difficult to sort out the ideas in these wild ejaculations, but Hamilton tried mightily, and must have been pleased by Coleridge's closing compliment to the effect that Hamilton was a "man of profound science" who realized that his chosen discipline needed a *"Baptism*, a Regeneration in Philosophy [and] . . . Theosophy."[13]

Coleridge believed that truth was spiritual in nature, and that it was available equally to all men. Unlike the Understanding, which depended on the senses and was cultivated to different degrees in different men, the truths of Reason, which were all basically moral truths, were grasped directly and intuitively by the human mind. The Broad Church, in Coleridge's view, was the church universal, the body of Christian truth that transcended all sectarian differences. He hated sectarian religion, and claimed that Christianity was "neither Anglican, Gallican, nor Roman, neither Latin nor Greek."[14] Yet if forced to choose between the sects, he believed that the Anglican church came closer to fulfilling his idea of the church universal than any other. The Anglican church, he admitted, had spots, but they were "spots on the sun." It was the most Apostolic of all the churches and the most truly tolerant.[15] Roman Catholicism could never unite the sects because it was too exclusive.

The *Aids to Reflection* contained words that, had he lived longer, Coleridge would certainly have directed against Newman and the Tractarians. He wrote: "He, who begins by loving Christianity better than truth, will proceed by loving himself better than all."[16] The High Anglicans were vain and as dangerously exclusive as the Romans. On the other side, Coleridge had an equal suspicion of the Aristotelian logic-choppers (such as Whately) who wished to substitute rationalism and formal logic in the place of real spiritual truth. Formal logic was, for Coleridge, "the rustling, dry leaves of the reflex faculty."[17] His own *Logic*, which he worked out during the last years of his life, borrowed heavily from Kant and attempted to show how the correct employment of the reasoning faculty could lead to meatier truths than could the syllogistic games of Whately.[18]

Both Newman and Whately criticized Coleridge in turn. Newman appreciated Coleridge's efforts to bring theology back to the center of philosophy, but he could not accept Coleridge's view of the Christian church. Coleridge had "indulged a liberty of speculation, which no Christian can tolerate, and advocated conclusions which were often heathen rather than Christian."[19] Whately was even more critical. He believed that Coleridge's philosophy was doing "incalculable mischief" because it mixed truth and error in a way that made them both tantalizingly palatable. Coleridge's transcendentalist philosophy taught that the evidence for Christianity is internal to the mind and not founded on any historical or traditional evidence. The danger, according to Whately, lay in Coleridge's claim that subjective internal "feelings" should replace church doctrine. Internal evidence was part, but not all, of the evidence for Christianity, and Whately feared that if one followed Coleridge's theology, subjective "truths" would become little more than matters of opinion. Christian truth, like any other truth, had to be demonstrated by argument and by real evidence. He wrote:

I have ... fought against the tendency in the present day to discard all moral reasoning and to encourage the practice of making one's opinions of all moral and religious questions a matter of taste. ... Men did indeed formerly reason on little and ill; but they [at least] professed and attempted to reason; they sought ... some rational ground for their conclusions; and though no doubt often biassed by their feelings, they did not, as now avow and glory in this [their biased feelings]. The evidences of Christianity ... were contemned; but it was by avowed unbelievers; not, as now, by persons professing a veneration for Christianity.[20]

Between transcendental and common-sense Christianity there was a great gulf that was extremely difficult to straddle.

Hamilton took the side of transcendentalism. Even before he began to study Coleridge and before the Oxford Movement got underway in 1833, he had been exposed to similar teachings by a lay theologian named Alexander Knox. Knox had been a special friend of John Wesley and a former secretary to Lord Castlereagh. He withdrew from public life after the rebellion of 1798 and lived at his lodgings in Dublin almost as a recluse, devoting himself to theological study. After a trip to England he returned to Ireland in 1803 and paid a visit to the La Touche family in Wicklow. It was meant to be a brief visit, but he stayed for nearly twenty-five years. Hamilton's Aunt Elizabeth (Uncle James's wife) was a niece of Mrs. Peter La Touche, and so Hamilton was welcomed at Bellevue (the La Touche estate) almost as a kinsman. He was eager to absorb whatever learning he could, and during his first visit in 1824 he wrote to Uncle James that he was spending every spare moment with "several interesting and valuable books. ... One of these was Knox: a book that wore spectacles. With him I had a great deal of conversation, and was a good deal together."[21] In a much later letter to Lord Northampton he recalled frequent visits to Bellevue and remembered Knox as a man "who possessed one of the most thoughtful, cultivated, and pious minds that have ever appeared among men."[22]

Newman credited Knox with having anticipated the ideas of the Oxford Movement and with having predicted its eventual emergence in the Anglican church long before it actually occurred.[23] In spite of his Evangelical background and in spite of his close association with Wesley, Knox argued that the Church of England was not Protestant, but a reformed branch of the Church Catholic. In fact he believed it was the most strictly Catholic church of all. It represented the spirit of the Greek fathers and should seek union with the *Greek* church rather than with the Romans or with other Protestant denominations. Some of these early teachings may be reflected in Hamilton's later interest in the liturgy of the Greek church, particularly the liturgy of Saint Chrysostom. Knox also despised Calvinism in every form. His Anglo-Catholic stance was close to that taken later by the Tractarians, but, unlike them, he never showed any inclination to adopt Roman Catholicism.

Hamilton's exposure to Knox probably prepared him for the Oxford Movement and tended to draw him away from his early Evangelical beliefs. Certainly he shared Knox's hostility towards Calvinism. He later told the Reverend Mortimer O'Sullivan that if Calvinism had been imposed upon him as a child it would have probably made him an infidel. He urged O'Sullivan to "denounce and batter" Calvinism with "a sledge-hammer power of reasoning." He added that he had studied the infidelity of Burns, Byron and Shelley, and had concluded that it was not Christianity, but Calvinism that had revolted them, "May God have mercy on their souls and on those who misrepresent Christianity."[24]

The Oxford Movement began in 1833 with a sermon by Keble on "National Apostasy" and the publication of the first of the "Tracts for the Times" by Newman. Hamilton paid little attention to the Tracts at first, and did not take a serious interest in them until the late 1830s. He was afraid of religious controversy, and in his letters he often put off correspondents who asked him theological questions.[25] But in February, 1840, Lord Adare wrote to ask him directly what he thought of the most recent Tracts. Particularly he was concerned about Newman's call for a return to the Nicene church and his increasing tendency towards Romanism.[26] Hamilton responded by saying: "With respect to the theological points, I should much more willingly talk than write; so many things are liable to misinterpretation, if not accompanied with the full and ready comment which a free conversation supplies." He then went on to confess that he had read the collected volumes of the earlier Tracts, but had not seen the most recent ones.[27] It is difficult from his reply to Adare to judge whether he had just ignored the Tractarian controversy, or whether he considered it too sensitive a subject for correspondence. The latter is probably the case, because Hamilton had begun to show enough sympathy for the Tractarians the previous year for Eliza to brand him as a "flaming Puseyite," and his correspondence with Aubrey De Vere had contained references to high church practices, usually discussed in a bantering tone.[28] In August of the following year (1841) Pusey himself came to visit Hamilton at the observatory and made a favorable impression.[29] By this time the controversy was warming up considerably. Pusey's enormously learned Tracts represented the heavy artillery of the Oxford Movement. He was one of the few British theologians who could handle the works of the great German theologians Schleiermacher, Tholuck, and Ewald with comparative ease, and he brought the weight of all their learning to support his arguments.

In February, 1841, Newman published his famous Tract 90, which tried to show that the Thirty-nine Articles of Anglican faith did not conflict directly with any Roman Catholic doctrine. The Heads of Houses at Oxford censured him in March. Adare's concerns about his orthodoxy appeared to have been well founded. Although Hamilton was probably cautious in his discussions with Pusey when he met him five months later, his pleasure at their meeting indicates that he probably still sympathized with the Oxford Movement. Then, in 1843, Pusey was delated for heresy and forbidden to preach at the university for two years. W. G. Ward, the editor of the party journal *British Critic*, led a campaign in support of Tract 90 and was ousted from Balliol College in 1845. In October, 1845, Newman converted to Roman Catholicism and was joined by Ward and many of his followers.

Closer to home Hamilton was shaken by the conversion of his local curate, the Reverend George Montgomery, who had become a close friend and confidant in matters of religion. Montgomery's conversion came in March, six months before Newman's and Hamilton regarded it as close to treasonous. When he heard of Montgomery's conversion he wrote to him that the event had caused him "the most exquisite pain." He then cut off the relationship.

Our intimacy ... has been one of the greatest pleasures of my life, and I believe that you have been good enough to regard it as not disagreeable to you; but neither of us could desire, under the circumstances you mention, that it should continue such as it has been. Indeed, a decided diversity of religious sentiments can scarcely co-exist with real and cordial intimacy; and, while I acknowledge and feel my great inferiority to you in all theological learning, and am sure that you might entirely vanquish me in argument, I must own that we have been of late receding from each other—a Romeward tendency in some, producing always a Protestant reaction in others.[30]

This was the year that Hamilton's friends noted his excitability over religious matters. He became much more strict in all religious obser-vances. Robert and John Graves were particularly worried about him, enough so that after John Graves married the following year he was cautious in telling Hamilton that his wife was a Puseyite.[31] But Hamilton had cooled off by that time and he told Graves not to worry, that he would probably get along perfectly well with Graves's new wife, and that his excitement the previous year had been caused by Montgomery's conversion, which, as he said, "alarmed me as to the tendency of a movement, in which I still see much to sympathise with and profit by, but also much to fear and guard against."[32] Hamilton had been carried along by the Tractarian bandwagon; now he had to jump off or be carried over the brink.

It is not difficult to see why the Tractarians had attracted him. The year before Pusey's visit he had met the Reverend William Sewell, the founder of the College of Saint Columba at Stackallan. Sewell was another of the Tractarians at Oxford who shared Hamilton's enthusiasm for German transcendental metaphysics. He was introduced to Hamilton by Adare and Aubrey.[33] Hamilton found Sewell's metaphysics extremely seductive and said that at last he was discussing important subjects with someone who shared his own viewpoint. After reading Sewell's writing, Hamilton wrote to him that "an intellectual friendship is inseparably established between us—."[34] These were what Hamilton's son William Edwin later called his "high church days," when Hamilton felt strongly about proper religious observance. Even their greyhound "Smoke" was subject to special religious discipline and received a beating for chewing on an old Book of Common Prayer in the library, the implication being that chewing a sacred book deserved a more severe punishment than chewing a nonsacred one.[35]

Looking back on these years Hamilton claimed that he never was in danger of converting to Catholicism. In describing them to Robert Graves he wrote: "You may possibly have heard that some people were pleased to call me a Puseyite, some years ago. However, I never pleaded guilty to the charge, though I had certainly leanings to high-churchism. But I have never allowed my views and feelings of religion to harden into any system; nor have I ever joined any party in the Church. The creeds and collects seem to me to contain a sufficient summary of doctrine; though I felt no scruple whatever in subscribing the Articles also when I received a Doctor's Degree in Cambridge in 1845."[36] The fact that Hamilton subscribed to the Thirty-nine Articles in 1845, the year of the first great wave of conversions by the Tractarians, means that he was ready to stand firmly with the Establishment in spite of his high-church sympathies. Hamilton had begun to sort out his religious opinions in 1843, and wrote to Eliza that his attitude towards the Oxford Movement was that held by William Palmer in his *A Narrative of Events Connected with the Publication of the Tracts for the Times*, a book that he sent to Eliza along with his letter. Manuscripts from 1843 contain notes from his careful reading of Newman's *Lectures on Justification*.[37] The more extreme views of the Tractarians he opposed, however, in particular the *British Critic*, which had taken an increasingly strong position under the editorship of Ward. Hamilton was perplexed by Newman, who had seemed to be edging closer to Roman Catholicism in 1843. But until he actually called himself a Roman Catholic, Hamilton had refused to criticize him.[38]

During these years Hamilton also brought his mathematical skill to bear on several religious problems that he regarded as being important. He made an elaborate calculation of the date of the equinox in the year of the Council of Nicea, which he printed in the *Proceedings* of the Royal Irish Academy for May 9, 1842. Soon afterwards he published an article in the *Irish Ecclesiastical Journal* computing the time of Christ's ascension into heaven. Because Christ ascended in body as well as in soul (as indicated by the Gospels and as required by the fourth Article), the ascension must have taken place in time. The soul, being a non-material thing, might move instantaneously through space, but no material object could do so. Hamilton concluded that Christ's body probably reached heaven on the day of Pentecost, and that the Holy Spirit then descended instantaneously upon the Apostles.[39] It is a strangely medieval argument, but just the kind of narrow doctrinal point that interested Hamilton during these years.

The Great Famine during the years 1846, 1847, and 1848 put religious controversy in the background, at least in Ireland. Aubrey De Vere and his brother Stephen developed a greater sympathy for the Catholic peasantry through their work in famine relief. In 1848, as the famine finally lifted, Stephen converted to Roman Catholicism in Canada, where he had gone to aid Irish immigrants. Aubrey wrote to Hamilton that he did not approve of his brother's decision, but understood that Stephen acted on what he believed to be an imperative duty.[40] Aubrey himself responded to the famine in a more political manner, by publishing a book entitled *English Misrule and Irish Misdeeds*, in which he vigorously attacked English policy, or lack of it, towards Ireland during the famine.[41]

Then, in 1850, another church quarrel on an abstruse doctrinal point produced a second wave of conversions to Roman Catholicism. In the wake of the "Gorham decision," by which the Judicial Committee of the Privy Council, a lay court, pronounced on a matter that the Tractarians said was a matter of doctrine, Henry Edward Manning (later Cardinal Manning) went over to Catholicism and carried Aubrey with him. Hamilton must have known what was coming. In a letter of February 27, 1851, after a long break in their correspondence, Aubrey wrote to ask Hamilton's opinion of Newman's *Essay on Development*, a book that Aubrey found compelling. He believed that the Gorham decision had cut the ground out from under him and all others holding high church principles. Aubrey, on the verge of conversion, wanted some direction from Hamilton. They had not discussed religion for a long time, but Aubrey placed great weight on Hamilton's opinions. It was an urgent matter. He hoped to hear from Hamilton immediately. "Few are at once qualified by intellect, philosophic habits, and the absence of prejudice, to consider so grave a question."[42] Hamilton responded that he was happy to hear from Aubrey and would respond more fully as soon as possible, but could only say in the meantime that his own convictions were as strongly in support of the Anglican church and as strongly opposed to the Roman Catholic church as they ever had been, although he despaired of being able to state the logic of his belief in any form that Aubrey would find compelling.[43] There is no indication that Hamilton ever followed up on his promise to write more fully. Aubrey and Manning traveled to Rome together both physically and theologically in that summer, and on November 15, 1851, Aubrey took the step that he had already taken in thought, if not in deed, when he wrote to ask Hamilton's opinions.

Just before Aubrey's conversion Hamilton had been caught up in a religious controversy closer to home that had caused him a great deal of trouble. Beginning in 1850 with the Gorham decision, questions of ritual and liturgy became increasingly important in the Anglican church. In that year Hamilton reluctantly agreed to serve as one of the two wardens of the church at Castleknock. Soon after he accepted the position the widow of a former parishoner gave a stained glass window to the church as a memorial to her deceased husband, and it was duly installed. But Castleknock was in the diocese of Archbishop Whately, and Whately had grown increasingly hostile to the Tractarians and to any high-church practices. The window at Castleknock contained the emblems of the Lamb, the Pelican, and the Dove and the inscription "To the Glory of God," all of which he regarded as dangerously popish. He demanded that they be removed, and Hamilton, as churchwarden, was caught in the midst of a nasty quarrel. When he discovered that he could not escape by resigning his position, he entered into a long series of negotiations that could only make everyone involved—the widow, his rector, and the archbishop—angry at him.

The churchwardens discovered that they had a legal liability and could not touch the window until they were served with a formal monition from the archbishop. Whately furnished the necessary document, written with all his customary forcefulness.[44] When the affair was over, Hamilton was able to describe it humorously to Augustus De Morgan. A frightened messenger arrived at the observatory with the monition. Hamilton suspected that he had heard of other unecclesiastical uses of windows when such monitions had been served in times past. (This is a reference to the Defenestration of Prague, 1618, which commenced the Thirty Years War. The governors were thrown out the window by the rebel faction.) Although Hamilton believed that the entire affair was a misunderstanding, Whately probably had some reason for concern. Hamilton admitted that the other churchwarden was an even higher churchman than himself, that their rector had given his consent to the installation of the window rather hastily, and that the congregation might be prepared to defy the archbishop. The whole business must have appeared to Whately as a dangerous drift in the direction of popery. De Morgan agreed. The pelican was a notoriously Roman symbol, and symbols meant what they were understood to mean, no niceties of logic to the contrary.[45]

As for the inscription "To the Glory of God and to the Memory of . . ." Hamilton found it highly amusing that Whately wished to have the name of God stricken from the inscription, which would leave only the name of the deceased. A compromise was finally reached. The emblems were removed and the inscription boarded over, leaving the rest of the window in place.

Actually Hamilton's greatest worry was not the window, but the possibility that Whately might object to an expensive organ that he had just accepted from one of the Rathbornes as a gift for the church. Fortunately Whately did not notice it or else decided that organs, unlike "painted windows" did not classify as popish adornments. Removal of the organ would have caused Hamilton much greater embarrassment than the window had.[46]

From the time of Aubrey's conversion and the controversy over the window, Hamilton began to move more decidedly away from the Roman Catholic church. He had made a careful study of the liturgy of Saint Chrysostom, translating it directly from the Greek, and he consulted with Bishop Longley to determine if Chrysostom's references to the Virgin Mary represented a deification. Longley responded that the passage quoted was in full accord with the second Article of the Anglican faith and that it afforded one more proof of the wrong done by the Church of Rome when it abused the language of the Fathers in such a way as to prevent the Anglican church from using the same expressions.[47] In 1853 Hamilton sent his translation of Chrysostom to Mrs. Whately, with whom he was exchanging poems.[48] He seemed to have developed a personal friendship with the Whatelys by this time. His work for the Society for the Propagation of the Gospel in Foreign Parts was especially commendable in the archbishop's eyes.[49] Political events also strengthened his Protestantism and increased his hostility to the Roman Catholic clergy, especially to those priests who led the agitation of the Tenants' Rights League. Hamilton complained to De Morgan of their ignorance, their insolence, their willingness to support violence, and especially their breaking of oaths solemnly sworn.[50] He was also indignant that so few of the "Irish Papists in Parliament . . . appeared to be even *hampered* in their conduct, by their equally formal and solemn oath to respect, or not to injure, the temporalities of the Protestant Establishment."[51] The letter reveals that the "temporalities" of the Establishment were just as important to Hamilton as its "spiritualities." His move back from high churchmanship was a reconfirmation of his conservative political views.

In 1855 Hamilton and Aubrey renewed their correspondence after another long silence. They discussed religion openly, but there was a constant sparring in their letters. Aubrey spoke with all the zeal of a recent convert, and Hamilton defended himself with occasional jibes and attempts at humor. Once they had agreed completely in their views on Catholicism, but since the shock of Montgomery's conversion, they had moved steadily apart.[52] Starting as an Evangelical, Hamilton had been drawn to Coleridge's vision of the church universal. The Oxford Movement then drew him towards high-church practice and belief, until the shock of the first conversions to Roman Catholicism turned his path back towards the Broad Church.

Hamilton's wife, Helen, witnessed all these changes with concern. Aubrey, in particular, she regarded as a threat, and she opposed his visiting the observatory. Hamilton told Aubrey that he did not show their letters to Helen, "because I knew that she felt somewhat afraid of your opening a vast vulpine throat, and swallowing me down, body and soul."[53] In 1855 Lord Adare, who had first warned Hamilton of the direction the Tractarians were taking, also converted to Roman Catholicism, and Hamilton lost his two closest friends to the "other side."

In 1856 Aubrey tried hard to convert Hamilton, or at least to explain in greater detail why he had gone over to Rome. In a series of long letters, obviously carefully written, he presented an eloquent defense of the Roman position. But Hamilton remained unmoved. Only one aspect of Roman practice attracted him, and that was private confession. He quizzed Aubrey on the point, and Aubrey explained the doctrine in a way that must have seemed attractive to Hamilton. Since 1848 Hamilton had carried a burden of guilt caused by his continued affection for Catherine Disney. When he finally unburdened himself of that guilt it was Aubrey who received the most detailed "confession."[54]

Other aspects of Roman Catholicism repulsed Hamilton. The Roman church was intolerant. Was not Aubrey, as a Roman Catholic, committing an error by even discussing religion with Hamilton, who, according to the Romans, must be a heretic? The Roman church was idolatrous. Were not Aubrey's own poems beginning to dwell on the "glories of Mary" in an idolatrous fashion? Early in 1856 Aubrey thought he had a chance to persuade Hamilton of the superiority of Roman doctrine. He pointed out that most of Hamilton's objections to Romanism were "impressions" or prejudices caused by social differences and were not objections to theological doctrine. When Hamilton admitted that "early *impressions* must in my case, as in most others, leave a very lasting trace ... I do not pretend to be *unbiassed* ... indeed who can be?" Aubrey pressed him hard, arguing that "*Duty, Honour,* and *Safety*" required every Christian to throw off prejudice and seek the true faith.[55] If Hamilton had been even close to choosing in favor of Rome he might have yielded to Aubrey's eloquence, but in fact he did not *want* to plunge himself into theological controversy, and he attempted to turn Aubrey away from the subject.

In July, 1856, he wrote out a series of *Theses on Romanism,* which he thought of enclosing in a letter to Aubrey, but decided not to when he realized that they would only provoke further controversy. The theses were:

1. The Roman Pontiff is not the Vicar of Christ; nor is there any human Head of Christendom.
2. The Roman Communion is not the Holy Catholic Church; nor is it even an incorrupt branch thereof.
3. The Roman System is unscriptural, and is expressly denounced by anticipation in the Scriptures.
4. The Roman Worship is idolatrous in tendency; and its followers are in peril of the sin of actual idolatry.[56]

The first and the fourth points—the supremacy of the Pope and the veneration of Mary—were the real sticking points. If his objections to them were mere prejudices, then they were prejudices that he dearly cherished. Aubrey's constant importunity tired him. In 1858 Aubrey visited him at the observatory and they took their favorite walk along the River Tolka to Abbotstown. It was not a satisfactory excursion. The old spell of camaraderie was broken. Hamilton reported: "Aubrey talked with or to me, for about two hours ... continuously; no beauty of Nature seemed able to win him, for even a moment, from his intense contemplation of what he regards as the 'Glories of Mary'; and I confess that I parted from him with a feeling of fatigue."[57]

John Henry Newman, the greatest light of the Oxford Movement, came to Dublin in 1854 as rector of the new Catholic University. He stayed until 1858, walking the same streets as his old friend and now implacable foe Whately, but never making any contact. Hamilton never met him either, although these were the years when he and Aubrey were discussing religious questions most actively. Aubrey regarded it as a tragedy that the two men whose minds he most greatly admired could never meet. He wrote to Hamilton: "That two such persons should have been so long near each other, without even meeting, is a piece of *bizarre* irony on the part of that Social Fate which holds us in her iron meshes."[58] But Hamilton was happy with the religious position that he had finally found. In the last years of his life he moved more firmly into Whately's camp and became close to the Reverend John O'Regan, vicar of nearby Finglass, chaplain to Whately, and later archdeacon of Kildare. In 1869 Hamilton's only daughter, Helen, married O'Regan.[59] The only remaining vestige of Hamilton's high-church days was a portrait of the Virgin that hung over the mantlepiece in his drawing room. It had been designed and given to him by the daughter of the marquis of Northampton, and Hamilton would not part with it.[60] He remarked on several occasions that visiting clergy took no notice of the picture.[61]

VI

"Algebra as the Science of Pure Time"

17

The Foundations of Algebra and the Coleridgean View of Science

ONE WOULD expect that Hamilton's theology and metaphysics would not be carried over into his mathematics. A physical theory always reflects some world view that often has a basis in metaphysics, but mathematics, at least "pure" mathematics, does not seem to depend on any physical or metaphysical notions. It is surprising, therefore, to find that Hamilton's metaphysics came to his aid more directly in algebra than in physics. Hamilton's first major work on algebra carries the title "Theory of Conjugate Functions, or Algebraic Couples; With a Preliminary and Elementary Essay on Algebra as the Science of Pure Time" (1835). And in the preface to his *Lectures on Quaternions* (1853) Hamilton stated that he was stimulated to write this "Essay" by reading passages in Kant's *Critique of Pure Reason*, which indicated that such a science of pure time was possible.[1] As for the quaternions, his second major discovery in algebra, Hamilton wrote about them as follows:

The quaternion [was] born, as a curious offspring of a quaternion of parents, say of geometry, algebra, metaphysics, and poetry. ... I have never been able to give a clearer statement of their nature and their aim than I have done in two lines of a sonnet addressed to Sir John Herschel:
"And how the one of Time, of Space the Three,
Might in the Chain of Symbols girdled be."
It is not so much to be wondered at, that they should have led me to strike out some new lines of research, which former methods had failed to suggest.[2]

In these and other places Hamilton claimed a metaphysical basis for his algebra, but it is not clear how such a logical discipline as algebra could benefit from the poetic reveries of Coleridge, or even from the more rigorous metaphysics of Kant.

Hamilton *could* make use of metaphysics because British mathematicians during the 1820s and 1830s took an intense and renewed interest in the foundations of algebra. They asked where the concepts of algebra came from and how they could be used. It was in the investigation of these concepts, particularly the concept of number, that Hamilton found a bearing for his metaphysical ideas.

Throughout the seventeenth and eighteenth centuries algebra had been regarded as an extension of arithmetic. The symbols of algebra stood for real numbers and for operations on numbers (such as addition, multiplication, equality and so on). The symbols allowed the use of unknown quantities, and they also allowed a generalization of arithmetical equations by revealing their mathematical form (such as linear, quadratic, cubic, and so on). Algebra was, then, the science of quantity, or the "science of magnitudes in general," as it was usually called in contrast to traditional geometry, which was the science of extension. Problems arose, however, when mathematicians attempted to derive general theorems about algebraic equations. If, for instance, one were allowed to substitute any magnitude for the letters representing constant quantities in an equation, it is quite possible that one might end up subtracting a larger magnitude from a smaller one. In the "science of magnitudes" this would be an absurdity. A "negative magnitude" just does not make sense. An even greater absurdity is the square root of a negative number, because the square of any number is always positive. Like negative numbers, these "imaginary numbers" have no meaning in a "science of magnitudes in general." There were two alternatives open to the mathematician. One alternative was to say that subtracting a larger quantity from a smaller one and taking the square root of a negative number were impossible operations, comparable to that of dividing by zero, and therefore not allowed in algebra. The other alternative was to deny that algebra was the science of magnitudes and search for a new meaning for its symbols that included the mysterious negative and imaginary quantities.

A few mathematicians took the former course and boldly proscribed any operations that were not consistent with an algebra described as a science of magnitudes. But from within algebra itself these restrictions seemed arbitrary and awkward. If negative and imaginary numbers were allowed, it could be proved that every polynomial equation must have at least one root, that in general an equation has as many roots as its order, and other theorems of comparable value. Without negative and imaginary numbers all this theory would have to be given up.[3]

By the nineteenth century the theory of equations with its negatives and imaginaries had obviously come to stay. It was too valuable and too powerful a theory to do without. But there was not yet an adequate explanation of what the "impossible" quantities of algebra might *mean*.

Hamilton showed his concern over this state of affairs in a letter to John Graves in 1828:

I have often persuaded myself that the whole . . . logic of analysis (I mean algebraic analysis) would be worthy of [ra]dical revision. But it would be for a person who should attempt this to go to the root of the matter, and either to discard negative and imaginary quantities, or at least (if this should be impossible or unadvisable, as indeed I think it would be) to explain by strict definition, and illustrate by abundant example, the true sense and spirit of the reasonings in which they are used. An algebraist who should thus clear away the metaphysical stumbling-blocks that beset the entrance of analysis, without sacrificing those concise and powerful methods which constitute its essence and its value, would perform a useful work and deserve well of Science.[4]

The logical or "metaphysical" foundations of algebra were in urgent need of revision, and it was to this task that British algebrists such as George Peacock, Augustus De Morgan, Duncan Gregory, John Graves, and Hamilton bent their efforts. From the 1820s through the 1840s the study of the foundations of algebra became almost a British monopoly. Hamilton also worked on the traditional problems of algebra contained in the theory of equations. In particular he became the greatest British expert on attempts to find a general solution to the equation of the fifth degree, the most important problem in the theory of equations during Hamilton's lifetime. But it was in the foundations of algebra that he made his most important contributions.

Because the British algebrists were all interested in the same problem it is tempting to include them all in the same school and to assume that Hamilton shared the views of his colleagues at London and Cambridge, but this would do violence to Hamilton's own view of the situation. Although he worked on the same problems as they did, he found their formal logical approach to be the complete antithesis of his own metaphysical approach. In 1831, when his enthusiasm for algebra was being rekindled, he wrote to Aubrey De Vere: "I differ from my great contemporaries, my 'brother-band,' not in transient or accidental, but in essential and permanent things: in the whole spirit and view with which I study Science."[5] Hamilton hoped to give meaning to the "impossibles" of algebra by creating numbers out of the intuition of pure time described by Immanuel Kant. For him algebra was a pure science of number, where those numbers were grounded in a mental intuition. The British algebrists, following George Peacock, argued that algebra was the manipulation of symbols according to prescribed rules, and that the symbols could be given any interpretation consistent with the rules. In his famous *Treatise on Algebra* (1830) and in his "Report on the Recent Progress and Present State of Certain Branches of Analysis" before the British Association in 1833, Peacock distinguished between "arithmetical algebra" and "symbolical algebra." In arithmetical algebra the symbols used must stand for arithmetical quantities, but in symbolical algebra they can have any meaning that one might wish to assign to them. Symbolical algebra does not deal with numbers or magnitudes at all. It is true that the rules of symbolical algebra are *suggested* by arithmetical algebra, but they are not logically dependent upon it. Symbolical algebra is a separate science in which "impossible quantities" may be used freely without concern.

Peacock's approach to algebra revolted Hamilton when he read about it in the early 1830s. He later told Peacock: "When I first encountered [your *Treatise on Algebra*] now many years ago, and indeed for a long time afterwards, it seemed to me ... that the author designed to reduce algebra to a mere system of symbols, and *nothing more*; an affair of pothooks and hangers, of black strokes upon white paper, to be made according to a fixed but arbitrary set of rules: and I refused, in my own mind, to give the high name of *Science* to the results of such a system."[6] For Hamilton the symbols of algebra had to stand for something *real*. They did not have to be real in the sense of material objects, but they had at least to be mental constructs. To say that they were nothing more than scratches on a piece of paper was to remove all significance from the science of algebra. The harmony of number was too much a part of his world view for him to allow algebra to degenerate into a game of logic.

Their different opinions led Peacock and Hamilton to emphasize different aspects of algebra. Because he believed the letters of algebra were only arbitrary signs, Peacock emphasized the *operations* of algebra. If both the elements *and* the operations of algebra were completely arbitrary, then algebra would be whatever one chose to scribble on a piece of paper. Peacock was not willing to go that far. In order to insure that algebra at least remained part of mathematics he felt it necessary to restrict the *operations* of algebra to the operations of arithmetic. They were not *identical* with the operations of arithmetic, because they were open to a much wider interpretation, but as rules for the manipulation of symbols, they were the same as those of arithmetical algebra. It seemed that without this restriction, algebra would have no logical structure at all.[7] Peacock called it the "Principle of the Permanence of Equivalent Forms" and stated it as follows: "Whatever form is algebraically equivalent to another when expressed in general symbols, must continue to be equivalent whatever those symbols denote. Whatever equivalent form is discoverable in arithmetical algebra considered as the science of suggestion, when the symbols are general in their form, though specific in their value, will continue to be an equivalent form when the symbols are general in their nature as well as in their form."[8] This fuzzy principle boils down to the statement that only those operations permitted in arithmetical algebra will be permitted in symbolical algebra.

In contrast to Peacock, Hamilton emphasized the *elements* of algebra.

He wanted to find a real meaning for them in some mental intuition. From this intuition he wished to *construct* the concept of number. Kant's "form of inner sense," which was the name he gave to the pure intuition of time, was an obvious candidate, particularly because Kant specifically said in the *Critique of Pure Reason* that time was the intuitive ground for the concepts of arithmetic and algebra. This emphasis on the real meaning of the elements—not on the signs, but upon the things signified—meant that the operations of algebra were of secondary importance. Once the correct meaning could be assigned to the symbols, then the operations would be determined by the meaning of the symbols.

While Peacock's formal logical approach to algebra seemed to give the greatest freedom to the algebrist, Hamilton's metaphysical approach had more advantages at the time he was writing. From the concept of pure time he would construct numbers, not just the real numbers, but also couples, triplets, quaternions, and "polyplets" with any number of elements. He was not limited to the real numbers of arithmetic, but was free to create all sorts of strange and wondrous beasts as long as they could be built from his one primitive intuition of time. More importantly, these numbers did not have to follow the rules of arithmetical algebra. They determined their own operations. Hamilton was not bound down by the Principle of the Permanence of Form. One of the main reasons why he of all the British mathematicians discovered quaternions was because he could give up the commutative law when his quaternions demanded it. The algebrists who were wedded to the Principle of the Permanence of Form never even considered the possibility.

Most historians of mathematics have not been willing to recognize any real difference between Hamilton's approach to algebra and that of the other British algebrists. They regret that Hamilton's "philosophical and even metaphysical speculations [make] the reading of his publications sometimes excessively tiresome."[9] These authors believe the concept of algebra as the science of pure time was a complete waste of time. Eric Temple Bell, whose *Men of Mathematics* contains the best-known account of Hamilton and his work, states that "this queer crotchet has attracted many philosophers, and quite recently it has been exhumed and solemnly dissected by owlish metaphysicians seeking the philosopher's stone in the gall bladder of mathematics. Just because 'algebra as the science of pure time' is of no earthly mathematical significance, it will continue to be discussed with animation till time itself ends."[10]

G. J. Whitrow is less polemical than Bell, but he remains equally skeptical of the mathematical value of Hamilton's metaphysics. He writes: "Thus, the ultimate conclusion of his train of thought was the revelation that algebra is not unique, a view very difficult to reconcile with his Kantian conception of its nature and a powerful argument in favour of the formalist philosophy of mathematics to which he was so resolutely opposed."[11] The paradox as Whitrow sees it is that because Hamilton sought the source of algebra in a primitive intuition, algebra for him would have to be unique; while the formalists, who did not insist on any particular interpretation of their symbols, were much more free to create new algebras. The conclusion of this argument is that Hamilton was saying one thing and doing another. He opposed Peacock, De Morgan, Gregory, and Graves on metaphysical grounds, but actually followed their approach when it came to doing the real mathematics. This argument overlooks the fact that Hamilton used the primitive intuition of *time* and not of *number*. He constructed numbers *from* time and was, therefore, able to create different kinds of numbers and to accept different kinds of algebraical operations. He actually had greater flexibility than the formalists, who held to the Principle of Permanence of Form.[12]

Once Hamilton had shown the way, the formalists abandoned the Principle of the Permanence of Form and created many new algebras. In the long run their emphasis on mathematics as a form of logic was more valuable than Hamilton's intuitive approach, a fact that Hamilton began to realize soon after his discovery of quaternions. But as a method of discovery, Hamilton's approach was more valuable. From the very beginning of his work in algebra, Hamilton had wished to regard it as "a branch and almost a type of the Philosophy of the mind in general."[13] He thought he might succeed in his efforts, because time as "the form of inner sense" is the most primitive of all intuitions, and a science of this pure intuition must then be the most basic science of human thought. The answer that Hamilton sought came out of Ireland during his lifetime, but it came from Cork and not from Dublin, and it came from one of the formalists, George Boole, whose *Mathematical Analysis of Logic* (1847) and *Investigation of the Laws of Thought* (1854) showed that algebra could best be understood as a system of formal logic.[14] Hamilton did not win over the formalists to his view, but he did give them a way to see "around the corner," so to speak, into a new avenue of mathematics. Modern algebra begins with the number couples and the quaternions, because they were the first algebras to break away from the real number line. They were, therefore, the first meaningful algebras that were not based solely on arithmetical numbers and operations.

As we have seen, Hamilton's metaphysics, including his metaphysics of mathematics, borrowed heavily from the romantic philosophy of Samuel Taylor Coleridge. Romanticism is a notoriously difficult "-ism" to define, and romanticism in science presents even greater difficulties than romanticism in literature. Hamilton never became one of the *Naturphilosophen*, the ultimate romantics of science, who, in the wake of Goethe, Kant, and Schelling, wished to make the whole world out of a clash of primitive forms and opposing egos. His philosophy remained more strictly Kantian than theirs, probably because mathematics gave him fewer opportunities to indulge in speculations about opposing forces than did physics, chemistry, and biology.

Still, Hamilton had the makings of a *Naturphilosoph* in him. L. Pearce Williams finds that the *Naturphilosophen* were all much alike: "They were all highly sensitive men, seeking both beauty and truth in their philosophy. Most wrote poetry and expressed their emotions in verse. All had a deep sense of form and thought architectonically. The whole was always more important and more than the sum of the parts. All recognized the importance of and all felt the near ecstasy of creativity springing from the active mind. Spirit was as real to them as body. All underwent youthful crises and discovered Kant as the answer to their personal *Angst.*"[15] In this passage Williams is writing about Hans Christian Oersted, Johann Ritter, F.W.J. Schelling, Christian Samuel Weiss, and Samuel Taylor Coleridge, but he has unintentionally outlined a biography of Hamilton with uncanny accuracy. The similarity of Hamilton's character to that of Coleridge was remarked upon at the time, both by William Wordsworth and by his widow.[16] Wordsworth said that Hamilton reminded him more of Coleridge than any other man he ever met. Hamilton considered himself a disciple of Coleridge and tended to express himself in Coleridgean language when on the subject of metaphysics, which probably made the similarity seem greater than it actually was. His similarity to the *Naturphilosophen* was a function of his basic personality and the pervasive influence of Coleridge.

Hamilton never revelled in obscurity as Coleridge did, nor did he ever reach Coleridge's heights of poetic ecstasy, but his metaphysics was drawn more from Coleridge than from anyone else. Of course he admired Coleridge's poetry, but he was even more strongly attracted by the prose works, where Coleridge expounded his philosophy in greater detail. These works were not always very readable or very comprehensible. His metaphysics tended to have a deadening effect, which he himself acknowledged. "When I wished to write a poem, [I] beat up Game of far other kind—instead of a Covey of poetic Partridges with whirring wings of music ... up came a metaphysical Bustard, urging its slow, heavy, laborious, earth-skimming Flight, over dreary and level Wastes."[17] And yet Hamilton found these metaphysical works enlightening, and he studied Coleridge's work carefully, particularly the prose works, *The Friend*, the *Aids to Reflection*, and the *Biographia Literaria*.[18]

There are few direct references to mathematics in Coleridge's writings, and on the occasions when he did use algebraic equations to illustrate his metaphysics, Hamilton was sharply critical.[19] Coleridge was almost totally ignorant of mathematics and Hamilton did not like to see his mentor making a fool of himself. But when Coleridge talked about the construction of science and the origin of knowledge, Hamilton paid close attention. In the preface to his "Preliminary and Elementary Essay on Algebra as the

Science of Pure Time" Hamilton distinguishes among three different schools of algebra. These are the practical, the philological, and the theoretical schools.[20] The practical algebrist seeks a rule that he can apply to a practical problem, one that will give him a correct result with the greatest ease of calculation. The philological algebrist regards algebra as a language and concerns himself with the simplicity of its notation and the symmetrical structure of its syntax. Philological algebrists are the formalists who manipulate algebraic signs without concern about what the signs might mean. Peacock and the other British algebrists were in this category.

The theoretical algebrist, on the other hand, constantly "looks beyond the signs to the things signified." He does not ask merely whether a theorem gives correct results, or whether it is a permissible form in the language of algebra, he also asks if it is *true*. To know if it is true he must know its *meaning*. It is not enough to manipulate negative and imaginary numbers as symbols. The theoretical algebrist wants to know what *real thing* is signified by them. Hamilton obviously placed himself among the theoretical algebrists—in fact, he stood there in splendid isolation. He argued that only by discovering the real signification of algebraic symbols can one make algebra into a *science,* "properly so called; strict, pure, and independent; deduced by valid reasonings from its own intuitive principles; and thus not less an object of priori contemplation than Geometry, nor less distinct, in its own essence, from the Rules which it may teach or use, and from the Signs by which it may express its meaning."[21]

In this "Preliminary Essay" the word *science* always appears with an initial capital and has a special meaning taken directly from Coleridge. Coleridge defines *Science* as "any chain of truths which are either absolutely certain, or necessarily true for the human mind, from the laws and constitution of the mind itself. In neither case is our conviction derived, or capable of receiving any addition, from outward experience, or empirical *data*—that is, matters of fact given to us through the *medium* of the senses—though these *data* may have been the occasion, or may even be an indispensable condition, of our reflecting on the former, and thereby becoming conscious of the same."[22]

A *Science* (or *Pure Science,* as Hamilton would prefer to call it) is a creation of the mind working on its own intuitive powers without reference to the external world. Sense data cannot be *part* of a science, although they may bring into play the faculties that do form a science. For example, in the *Transcendental Aesthetic* Kant claims that time and space are the necessary *forms* of sensible intuition. All of our experience of the external world is ordered in space and time. Space and time are not *in* external objects, but only in our way of perceiving them. They are mental molds into which we fit our sensible intuitions. The forms are called into being by the act of seeing and feeling and so on, but they are not part of what is seen or felt. Space in particular is the "pure form of all outer intuition."[23] It is the mold by which the mind shapes and orders sensations, but it is also the ground from which one constructs the science of geometry. Geometry is a science whose theorems are shown to be necessarily true from "the constitution of the mind itself." It does not depend on experience for its validity, although it is only through experience that the forms of intuition from which geometry can be constructed are called into being.

The one passage in Kant's *Critique of Pure Reason* that most attracted Hamilton's attention was the place where Kant states: "Time and Space are, therefore, two sources of knowledge, from which bodies of *a priori* synthetic knowledge can be derived. (Pure mathematics is a brilliant example of such knowledge, especially as regards space and its relations.) Time and space, taken together, are the pure forms of all sensible intuition, and so are what makes *a priori* synthetic propositions possible."[24] Picking up Kant's hint, Hamilton wished to ground algebra on the form of inner sense called *time*, just as geometry was grounded on the form of outer sense called *space*. Only in this way could algebra become a *Science* in the sense that Coleridge had given to the word.

Coleridge insisted that science was the province of reason. And here he introduced another distinction that also came originally from Kant—the distinction between reason and understanding. Coleridge claimed: "All the labors of my life [will have] answered but one end if I shall have only succeeded in establishing the diversity of Reason and Understanding."[25] He labored hard and long to make the distinction clear. Hamilton realized that this goal was the major purpose of Coleridge's *Aids to Reflection* when he first read it in 1831.[26] It was not just a philosopher's distinction. Coleridge was quite prepared to prove that "every heresy which has disquieted the Christian Church, from Tritheism to Socinianism, has originated and supported itself by arguments rendered plausible by the confusion of these faculties."[27]

Reason, according to Coleridge, is "the power of universal and necessary convictions, the source and substance of truths above sense, and having their evidence in themselves."[28] Reason is identical with its object. God, soul, and eternal truths are objects of reason. In fact they *are* reason. Reason is absolute and unchanging. It does not admit of degree, and therefore is equally knowable to all persons. It is the "scientific faculty" because it is the "intellection of the *possibility* or *essential* properties of things by means of the Laws that constitute them."[29] In geometry, for instance, every man has the intuitive power of perceiving the truth of the fundamental properties of the circle and the triangle. Men may differ in the power of attention required to link these truths into theorems, but the intuitive perception of primary truths is the same for all.[30]

Understanding, on the other hand, is "the faculty judging according to sense."[31] It works with particulars and can never ascend to the unity of the whole.[32] Ants building a nest exhibit understanding but not reason. They form judgements that allow them to construct a suitable home for themselves, but they can never ascend to principles that are universal and abiding.[33] Men partake of understanding in various degrees and are therefore able to manipulate the things of this world with varying degrees of success, depending on their possession of this faculty. But understanding will never save their souls. The faculty of reason, on the other hand, gives us spiritual truths. All truth for Coleridge is essentially spiritual, and only abiding spiritual truth can be the ground for a *Science*.[34]

Coleridge believed that the Oxford Movement, the Reform Movement, and all the other religious and political difficulties of England during the 1830s were brought about by the unfortunate subjection of reason to understanding: "My object is to draw the attention of my countrymen, as far as in me lies, from expedients and short sighted tho' quick sighted Experience [gained from the Understanding], to that grand algebra of our moral nature, Principle and Principles [gained from the Reason]—in public as in private life, in criticism, ethics, and religion ... [for] no great principle was ever invaded or trampled on, that did not sooner or later avenge itself on the country, and even on the governing classes themselves."[35]

England was tottering, according to Coleridge, because its citizens had transferred to the understanding the primacy that was rightly due to reason: "In no age since the first dawnings of science and philosophy in this island have the truths, interests, and studies which especially belong to the reason, contemplative or practical, sunk into such utter neglect, not to say contempt, as during the last century."[36] Hamilton read this passage from the *Aids to Reflection* in 1831. He remembered it the following year in writing to Lord Adare about his distaste for the science of his practically minded British colleagues. A visit from Airy had left him dejected. Airy had claimed that the Liverpool and Manchester Railway was the highest achievement of man, which caused Hamilton to remark that Airy's mind was an unfortunate instance of "the usurpation of the Understanding over the Reason, too general in modern English Science," and then, echoing Coleridge's lament, "When shall we see an incarnation of metaphysical in the physical! When shall the imagination descend, to fill with its glory the shrine prepared for it in the Universe, and the Understanding minister there in lowly subjection to Reason!"[37] Both Hamilton and Coleridge believed that what passed for *science* in Britain would become *Science* in truth only when the physical was injected with the metaphysical.

The construction of an algebra from the intuition of pure time was the work of reason. Not only did it make algebra a *Science*, that is, a structure of abiding truth, but is also gave *meaning* to the signs and symbols used.[38] "Algebra as the Science of Pure Time" is built into the very foundation of the way we look at the world. If all sensible intuition is ordered in space and time, then it is ordered by our minds according to the relations of geometry and algebra. Is there any certainty that an algebra of arbitrary signs will serve the same function? As a disciple of Coleridge, Hamilton was bound to disagree with his "brother band," and insist on an intuitive basis for algebra.

18

"Algebra as the Science of Pure Time"

HAMILTON FIRST suggested that algebra might be based on the intuition of time in a memorandum dated November 15, 1827. He entitled his memorandum "Consideration on Some Points in the Metaphysics of Pure Mathematics," and he stated in it that the concept of time might serve "to give greater precision and simplicity to our notion of ratio."[1] His first use of time steps to define the operations of algebra was in a memorandum dated June 9, 1830, and the first lengthy statement on the subject appears in a notebook entry dated February, 1831, and entitled "Metaphysical Remarks on Algebra."[2] The fact that Hamilton's first mention of algebra as the science of time was in 1827, long before he had read the *Critique of Pure Reason*, confirms his later statement that what he learned from Kant was more a process of recognizing than discovering.

When he wrote the following in his "Metaphysical Remarks" he still had no direct knowledge of Kant's work:

In all Mathematical Science we consider and compare relations. In algebra the relations which we first consider and compare, are relations between successive states of some changing thing or thought. And numbers are the names or nouns of algebra; marks or signs, by which one of these successive states may be remembered and distinguished from another. ... *Relations between successive thoughts thus viewed as successive states of one more general and changing thought, are the primary relations of algebra.* ... For with Time and Space we connect all continuous change, and by symbols of Time and Space we reason on and realise progression. Our marks of temporal and local site, our *then* and *there*, are at once signs and instruments of that transformation by which thoughts become things, and spirit puts on body, and the act and passion of mind seem clothed with an outward existence, and we behold ourselves from afar. And such a transformation there is, when in Algebra, we contemplate the change of our own thoughts as if it were the progression of some foreign thing, and introduce Numbers as the marks or signs to denote place in that progression.[3]

The similarity between Kant's arguments in the *Critique of Pure Reason* and these ideas expressed by Hamilton makes it easy to understand why Hamilton had become dissatisfied with the secondhand accounts of Kant that he read in Madame de Staël's *Allemagne* and in Dugald Stewart's *Philosophical Essays.*[4]

Kant's works were hard to come by in Ireland, or in England, for that matter, until the late 1830s. But Hamilton was tempted by what he had read in Coleridge and in the British reviewing journals. In 1830 he copied out long sections from a review of Novalis's *Schriften* that appeared in the *Foreign Review and Continental Miscellany.*[5] The author was Thomas Carlyle, although the review was published anonymously. It was less a review of Novalis than a review of all idealist philosophy during the previous one hundred years, and it marked Kant as a crucial figure. Berkeley, Boscovich, Coleridge, and Kant all received strong if occasionally misdirected praise from Carlyle, who concluded that idealism, immaterialism, and reason were the strongest defenses against atheism. Reason, wrote Carlyle, was "the pure, ultimate light of our nature, wherein ... lies the foundation of all Poetry, Virtue, Religion; things which are properly beyond the province of the Understanding to *contradict* Reason."[6] With this kind of inspiration Hamilton concluded that he must go to the source itself and read the *Critique of Pure Reason* as Kant had written it and not as it was presented through the eyes of his interpreters.

Hamilton's godson, William Wordsworth, the son of the poet, was studying in Germany, and Hamilton enlisted his help in trying to find a copy of the *Critique.*[7] On October 29, 1831, Hamilton wrote to the senior Wordsworth: "I have got Kant's *Kritik der Reinen Vernunft,*" but he did not indicate in the letter whether he had received it from Wordsworth's son or from some other source.[8] He immediately set about trying to read it, but in spite of his great linguistic skill, he found the German intractable. (His manuscripts contain six attempted translations between October, 1831, and sometime in 1845. Four are translations of the first few pages of the introduction, one is a translation of the preface, and the sixth is a translation of the table of contents.)[9] Hamilton was still working at it the following summer, and took the book along on a visit to Oxford. He always carried a sizable library with him wherever he went; this time he stored it in a pillowcase. While he was riding atop an omnibus in Birmingham, the pillowcase fell open. He caught Laplace's *Calculus of Probabilities* just as it was disappearing over the side of the coach, but Kant's *Critique* had already slipped away. He was not able to replace it until 1834.[10] In August of that year he reported that he was approximately one third of the way through the book.[11]

In the meantime he met Coleridge, first in 1832 and then again in 1833. Hamilton had hoped to obtain a letter of introduction to him from Wordsworth, but Wordsworth was strangely reticent to write one, either because he knew that Coleridge was ill, and did not want to impose on him by sending unknown visitors to his door, or because he was somewhat annoyed by Hamilton's great enthusiasm for Coleridge's philosophy.[12] Coleridge, in the later years of his life, rambled on to visitors in an almost incoherent fashion, usually about his *Opus Maximum*, which he had outlined in 1828. Visitors, including Hamilton, were submitted to long harangues on the logic, which was the most nearly completed part of the work and was taken largely from Kant.[13]

In the letters he wrote to his friends, Hamilton said little about his visits to Coleridge. It is difficult to discover what transpired between them. Certainly there was a discussion of Kant, because Coleridge offered to loan Hamilton his copies of Kant's *Miscellaneous Essays* and wrote later that he was searching for the volumes that he had not been able to find in his library.[14] After the 1833 visit Coleridge also sent Hamilton his copy of Kant's *Critique of Judgment*, and gave his copy of the *Critique of Practical Reason* to Hamilton's pupil, Lord Adare.[15] Hamilton's previous reading of Kant had been superficial and secondhand, but after his visits with Coleridge he began to read the *Critiques* more seriously.

When Hamilton first met Coleridge in 1832 he was completing his work on the characteristic function as applied to optics. Much of the conversation must have turned on Hamilton's theory of matter, but it is also likely that he talked with Coleridge about the nature of science, particularly Coleridge's distinction between reason and understanding as it had appeared in his *Aids to Reflection*. Hamilton wrote a memorandum on the *Aids* in June, 1831, in which he concerned himself with precisely this question.[16] Whether he talked with Coleridge specifically about his idea of "Algebra as the Science of Pure Time" is not certain. There is some evidence, however, that he did, and that Coleridge might have been a sympathetic listener. One thing that Hamilton did mention in his letters describing the meeting was Coleridge's enthusiasm for one of his own poems entitled "On Time, Real and Imaginary."[17] The poem is an allegory on the difference between time as real and absolute, and time as felt or experienced.[18] It does not contain any reference to mathematics, but it may well have come up as a result of a discussion over Hamilton's meaning of "pure" time. In 1819 and 1820 Coleridge had held a position close to Hamilton's. In the marginal notations to several of his presentation copies of *The Friend*, Coleridge anticipated Hamilton by claiming that arithmetic, at least, was derived from our intuition of time. In this passage Coleridge claimed that the science that transcends the evidence of physical phenomena should properly be divided into two parts, metaphysics and mathematics. Mathematics "has for its department the acts and constructions of the *necessary* Imagination and is subdivided into Geometry as the correspondent to Space or the Outer Sense, and Arithmetic, correspondent to Time, or the Inner Sense: while Algebra may [be] considered as the conversion of the one into the other by principles of equation and compensation."[19] Coleridge was completely ignorant of mathematics and apparently mistook algebra for analytical geometry, but the reference to geometry as the correspondent to the outer sense and arithmetic as the correspondent to the inner sense is significant, since it indicates that Coleridge used the notions of time and space as understood by Kant to give an intuitive basis for geometry and algebra. With Coleridge and Kant as authoritative support for his position, Hamilton was able to summon sufficient courage to present his metaphysical ideas in spite of the opposition they were certain to draw from most mathematicians.

At the same time that Hamilton was learning Coleridgean metaphysics he was also occupied with other, more purely mathematical, questions. Optics and mechanics, of course, occupied the lion's share of his attention, but his college friend John Thomas Graves never let him forget about algebra completely. Almost all of Hamilton's work in algebra was inspired and maintained by the constant needling of Graves. In 1827 Graves moved to England, where he was called to the English Bar in 1831, but this did not interfere with his mathematical conversations with Hamilton, which continued to be carried on by correspondence. The first extant letter of this correspondence is one from Hamilton to Graves dated April 5, 1825, on the subject of logarithms.[20] Sometime in 1826 Graves concluded that the general logarithm must necessarily contain two arbitrary integers—the discovery that Hamilton would later defend in his "Essay on Conjugate Functions, or Algebraic Couples." Ever since the seventeenth century there had been controversy between mathematicians over the possible existence of logarithms of negative and complex numbers.[21] The difficulty lay in the fact that the logarithm of a complex number has many values. John Graves had shown that in the most general case, that of the logarithm of a complex number taken to a complex base, two arbitrary integer numbers must be specified to determine the solution, since the complex number of the logarithm and the complex base each introduce an infinite number of possible solutions. He and Hamilton continued to discuss the subject through 1827, and in 1828 Graves submitted a paper to the Royal Society entitled "An Attempt to Rectify the Inaccuracy of Two Logarithmic Formulae." Graves feared that his discovery might have been anticipated; Hamilton assured him that John Herschel would certainly know if it was new, because he read so widely in current research in algebra.[22] While Herschel did not know of any precursor to Graves's paper, both he and George Peacock were unconvinced by Graves's arguments.[23] Herschel had hoped to avoid commenting on the paper, because, as he candidly admitted to Graves, he could not determine if it was correct.[24] But the council of the society responsible for the *Transactions* appointed him judge, and he concluded that the only honorable thing to do was to ask Graves to withdraw his paper.[25]

Hamilton intervened at this delicate and sensitive point in the negotiations, without obtaining Graves's prior consent, and tried to convince Herschel of the paper's worth. It was published the following year in the *Philosophical Transactions.*[26] Peacock continued to oppose the results in his report to the British Association of 1833, but De Morgan was able to show that Graves's results were correct if one accepted Graves's definition of the logarithm.[27]

The controversy over the logarithms of complex numbers was stimulated in part by the possibility that a new kind of imaginary number might be discovered, that is, that the complex numbers might not have the property of closure under some of the ordinary operations of algebra, a possibility that John Graves had urged on Hamilton in 1829.[28] De Morgan admitted later that he had half-expected the same thing.[29] Hamilton doubted that such a new imaginary number was likely to arise from the study of logarithms (a surmise that proved to be correct), but he did not hesitate to consider new hypercomplex numbers *defined* as such, and it was the search for such hypercomplex numbers that eventually led to his discovery of quaternions in 1843.

Another reason for the great interest in complex numbers during the 1820s and 1830s was the realization that such numbers could be represented geometrically by a pair of axes in what is now called the complex plane. This discovery was first published by Casper Wessel in 1797 and was rediscovered independently by several others, but Hamilton knew nothing about it until 1829, when he read it in a book by John Warren entitled *A Treatise on the Geometrical Representation of the Square Roots of Negative Quantities.*[30] Typically, it was John Graves who called Hamilton's attention to Warren's book and stimulated him to search for an algebra of hypercomplex numbers that could be represented geometrically in three-dimensional space.[31]

While Hamilton and Graves were debating the question of how to write the logarithm of a complex number, in France Augustin-Louis Cauchy was working out the whole theory of complex functions. He wrote his first important paper on the subject in 1814, but it was not published until 1827. The book that Hamilton saw was Cauchy's *Cours d'analyse,* published in 1821. It contained material bound to catch his eye, such as the statement that "every imaginary equation is only the symbolic representation of two equations between real quantities."[32]

In 1853 Hamilton referred to this particular quotation with the remark that "that valuable work of M. Cauchy was early known to me: but it will have been perceived that I was induced to look at the whole subject of algebra from a somewhat different point of view, at least on the metaphysical side."[33]

According to Cauchy, the passage from complex quantities to real quantities was to be made through a pair of real equations that have come to be known as the "Cauchy-Riemann equations." Hamilton discovered them before he ever saw Cauchy's work and called them the *equations of conjugation.* They had first been used by d'Alembert in 1752 to solve for the motion of a body through an ideal fluid, but Cauchy was the first to recognize their importance for the theory of complex functions.[34] In a letter of June 11, 1829, to John Graves, written between the fellowship examinations at Trinity College, Hamilton closed with the remark, "I may observe that I have lately resolved some functional equations between real quantities, which seem to contain the statement of your exponential problems, when divested of imaginary symbols."[35] Later in the summer, in a letter discussing the geometrical interpretation of complex numbers, Hamilton was more explicit:

I have sometimes thought that this theory might be illustrated by the consideration of *couples of curve surfaces.* Every real function of a single variable, $y = \phi(x)$, can be constructed by a plane curve; and when by changing x to $\alpha + \beta \sqrt{-1}$ we deduce two real functions of two real variables

$$\phi(\alpha + \beta \sqrt{-1}) = \phi_1(\alpha, \beta) + \sqrt{-1}\ \phi_2(\alpha, \beta)$$

connected by those partial Differential Eqns which I believe I first perceived and which I would propose to call the *Equations of Conjugation,* namely

$$\frac{\partial \phi_1}{\partial \alpha} = \frac{\partial \phi_2}{\partial \beta}, \qquad \frac{\partial \phi_1}{\partial \beta} = -\frac{\partial \phi_2}{\partial \alpha}.$$

We can construct these two new functions by two curve surfaces

$$\gamma_1 = \phi_1(\alpha, \beta), \qquad \gamma_2 = \phi_2(\alpha, \beta)$$

which form what I call a *couple,* being connected by geometrical relations which correspond to the Eqns of Conjugation.[36]

Because the equations of conjugation are *real* functions, while at the same time they are necessary conditions for the differentiability of a *complex* function, they seem to hold a possible clue for a method of writing complex numbers without imaginary symbols. In particular they indicate that it might be possible to substitute couples of real numbers for the traditional complex numbers.[37] This would be a real advance, because, according to Hamilton, a complex number written as the sum of a real part and an imaginary part can have no meaning. It expresses the sum of unlike things, and unlike things cannot be added. A number couple would be a new *kind* of number, but it would be meaningful if operations on the number couples could be defined consistently and unambiguously. For some historians of mathematics Hamilton's representation of complex numbers by number couples is his greatest achievement in algebra, even more significant than his later discovery of quaternions.[38] It meant breaking away from the real number line as the basis for algebra, and constructing new elements defined in a different way.

Hamilton began by using his number couples to explore Graves's problem of the logarithm of a complex number. As he worked on the problem he found that he had to define the operations of addition, subtraction, and raising to a power for number couples.[39] The equations of conjugation entered in only peripherally, and Hamilton began to place greater emphasis on his metaphysical notion that all numbers, including his number couples, and possibly numbers of higher orders, could be constructed from the intuition of pure time.

Hamilton broached his theory of "Algebra as Pure Time" with caution. First he tried it out in November, 1832, in the first lecture of his astronomy course at Trinity College. The lecture was published with Hamilton's permission, and possibly at his instigation, in the *Dublin University Review* for January, 1833.[40] Then, in a bizarre hoax, he wrote a review of his own public lecture, which he sent to Viscount Adare and to Aubrey De Vere to test their reactions. The review was amusingly critical, but attacked severely his notions about algebra and his taste for "the ravings of the German school, and the unintelligible mysticism of Coleridge."[41] In July of the following year he told Adare that the passage that he had submitted to the greatest ridicule "might almost be taken as a summary of Kant's view of Space and Time."[42]

The following year he again cautiously tested the reactions of other mathematicians by presenting a sketch of his theory to George Peacock at the Edinburgh meeting of the British Association. Hamilton was able to write to Graves that Peacock spoke "handsomely" of the sketch, and it was duly published in the *Reports* of the meeting.[43] It contained no metaphysics and no reference to time as the fundamental intuition of algebra. Peacock would have been certain to object to any such antics, so Hamilton presented his mathematical findings without any philosophical support.

In 1835, just before he completed his "Essay on Algebra as the Science of Pure Time," he wrote a series of letters on the subject in order to clarify his own ideas and to instruct Adare. The letters contain an exposition of Kant's philosophy and an explanation of his own ideas about algebra. The seventh and last letter in the series was never sent. Hamilton expanded it with the inclusion of some material from the first letter to form the "General Introductory Remarks" to his essay.[44]

In August the British Association met at Dublin, and in honor of the occasion Hamilton presented his ideas in their totality—metaphysics and all—in a paper entitled "Theory of Conjugate Functions, or Algebraic Couples; With a Preliminary and Elementary Essay on Algebra as the Science of Pure Time." In writing to his friend Aubrey De Vere he called it "the first installment of my long-aspired-to work on the union of Mathematics and Metaphysics."[45]

The paper that Hamilton completed for the meeting was in three parts. The last part, the "Theory of Conjugate Functions, or Algebraic Couples," was the first to be finished and had been read to the Royal Irish Academy on November 4, 1833. The middle and longest section was completed in the spring of 1835 and read to the academy on June 1 with the title "Preliminary and Elementary Essay on Algebra as the Science of Pure Time." The first part, entitled "General Introductory Remarks," was finished sometime during the summer. Although portions of the work had been read as early as November, 1833, the entire "Essay" was first published in the volume of the *Transactions* for 1837.[46]

Hamilton was pursuing three related directions of mathematical research in his paper. First, he wished to give an algebraic definition of complex numbers that avoided the concept of imaginary numbers; second, he wished to confirm John Graves's general expression for the logarithm of a complex number taken to a complex base, and third, he wished to place the algebra of real numbers on a more secure logical foundation. In the middle section of the "Essay" Hamilton attempted to state the properties of the real number system and to define negatives by the use of steps in the ordered relations of time. Thus he attempted to base algebra on the *ordinal* character of the real numbers. Evaluated by the standards of rigor demanded in modern algebra, Hamilton's attempt was highly intuitive and badly flawed. Yet his attempt to define the natural numbers by appealing to our intuitive sense of progression in time has been revived in this century by the Dutch "intuitionists" and still provides subject for debate among philosophers of mathematics.

The entire "Essay" is an effort to get away from the intuitive concept of number and to base algebra on another intuitive concept, that of pure time. Therefore, Hamilton resists the temptation to introduce the ordinal and cardinal integers until he has developed the operations on time steps. As he says, he wants to "treat these spoken and written names of the integer ordinals and cardinals, together with the elementary laws of their combinations, as already known and familiar."[47]

In developing his notion of algebra as the science of pure time, Hamilton presents us with one of the earliest attempts to list systematically the properties of the real number system.[48] He comes surprisingly close to defining an algebraic field. There are certain points of confusion where he passes from steps to cardinal numbers expressed as multipliers of time steps, but the paper as a whole is a remarkably successful beginning. The commutative and distributive properties are described and definitions of zero, the additive and multiplicative inverse, and the law of closure are all there. What he misses is the associative rule, probably because it is more subtle than the others, and because it did not occur to him that there might be an algebra that did not follow it. Later, when he studied systems of hypercomplex numbers, he found that the associative law did not always hold, and he stated it for the first time in 1843 as an important property of real numbers, complex numbers, and quaternions.[49]

After showing how the real numbers may be built from the intuition of steps in time, Hamilton then does the same thing for number couples, and shows that by proper definition all the operations of real numbers also hold for number couples. The third and concluding part of the "Essay" repeats the arguments of the previous year regarding the logarithm of a complex number, although this time Hamilton presents the arguments in the metaphysical context of algebra as the science of pure time.

Hamilton's paper showed great mathematical imagination. By using time steps and number couples he demonstrated that algebra could be more than the ordinary algebra of real numbers, and that real meaning could be given to negative and imaginary numbers. At the end of his paper Hamilton stated that he hoped "to publish hereafter many other applications of this view [of algebra as the science of pure time]; especially to Equations, Integrals, and to a Theory of Triplets and Sets of Moments, Steps and Numbers, which includes this Theory of Couples"— a hope that he finally fulfilled in 1843 with his discovery of quaternions after a long and fruitless search for triplets.[50] With the quaternions Hamilton created an algebra that did not obey the commutative law. It was a bold step that opened the way for a rash of new algebras.[51] Although the quaternions appeared as a startling discovery, Hamilton claimed that they were merely "a continuation of those speculations concerning algebraic couples, and respecting algebra itself, regarded as the science of pure time, which were first communicated to the Royal Academy in November 1833. ... The author has thus endeavored to fulfill, at least in part, the intention which he expressed in the concluding sentence of his former Essay."[52] As Hamilton saw it, the couples, the ill-fated triplets, the quaternions, and all other new hypercomplex numbers were part of the same speculation, which he grounded in his metaphysical notion of pure time, a notion that had been growing in his mind for many years. Once we understand the metaphysical and mathematical background to the "Essay" it is easier to see why Hamilton brought together such diverse subjects into a single paper.

19

The Kantian Content of the "Essay on Algebra as the Science of Pure Time"

HAMILTON FOLLOWS the Kantian notion of time closely in his "Essay on Algebra as the Science of Pure Time." Since the inner sense of time is more general than the outer sense of space, Hamilton concludes that algebra is a more general and fundamental branch of mathematics than geometry.[1] Moreover, time is not merely one of many ways to illustrate the rules of algebra. The intuition of pure time (corresponding to Kant's use of the term *pure intuition*)[2] "will ultimately be found to be co-extensive and identical with Algebra, so far as Algebra itself is a Science."[3] Hamilton takes care to explain why he uses the expression *pure* time. Pure time is to be carefully distinguished from apparent time. It is the pure form of sensible intuition described by Kant, not the objects of the intuition or the order of events. When it is thus purified from all sensations, leaving only the a priori form of sensibility, time becomes the ground for a mathematical science. It can be used as the rudiment for algebra when it is "sufficiently unfolded, and distinguished on the one hand from all actual Outward Chronology (or collections of recorded events and phenomenal marks and measures), and on the other hand from all Dynamical Science (or reasonings and results from the notion of cause and effect)."[4]

In spite of this debt to Kant, Hamilton believed that he had gone beyond the *Critique* in asserting that algebra is the one science of pure time. He believed that Kant had recognized that there *might*, or *ought* to be, a science of pure time analogous to geometry but never suspected that such a science lay ready at hand in algebra.[5] His reading of Kant is essentially correct. Only in the *Prolegomena* did Kant openly assert that arithmetic was the science of time.[6] Elsewhere, and especially in the *Critique of Pure Reason*, he was much more reticent. While time as the form of inner sense was of greater general importance than space, which provided material for the science of geometry, Kant did not take the additional step of making time the ground for a pure science.

In the "Transcendental Analytic," where Kant presents the schematism of the pure concepts of understanding, the intuition of time appears in a crucial role. The transcendental schemata link the categories of understanding with sensory intuitions, that is, the schemata mediate between the pure concepts of understanding and sensations. According to Kant the schema of a pure concept is "a product which concerns the determination of inner sense in general according to conditions of its form (time)."[7] Thus the schema of each of the twelve categories "contains and makes capable of representation only a determination of time."[8] Of particular importance is the category of magnitude: "The pure schema of magnitude, as a concept of the understanding, is *number,* a representation which comprises the successive addition of homogeneous units. Number is therefore simply the unity of the synthesis of the manifold of a homogeneous intuition in general, a unity due to my generating time itself in the apprehension of the intuition."[9]

Thus for Kant, number is constructed by the generation of time in intuition, much as Hamilton later argued. Not all of Hamilton's contemporaries agreed that magnitude must necessarily be generated in time. In his paper "On the Foundations of Algebra," De Morgan asserted that it is possible to conceive of a line segment as generated instantaneously, "no portion of it coming into thought before or after another."[10] Hamilton objected. Quoting Kant as an authority, he insisted that we cannot even think of a line without drawing it in thought.[11] This conviction that every extensive magnitude is apprehended in space and time by a succession or continuous sequence of acts of perception persuaded Hamilton that without the pure intuition of time we should have no concept of magnitude or number, and it confirmed his belief that algebra had to be based on the ordinal property of number. At the urging of "mathematical friends" he had considered basing algebra on continuous progression alone, without reference to time, in order to avoid the criticism of De Morgan and the other formalists. But as he realized that progression itself had its intuitive basis in the notion of time, he felt that he could not exclude this primary intuition from his presentation.

Two notions that are crucial for Kant's mathematics and important for Hamilton as well are those of "intuition" and "construction." Both of these notions are less ambiguous when applied to geometry than to algebra, and this is probably one reason why Kant preferred geometrical illustrations to arithmetical ones.[12] Kant argued that the

Augustus De Morgan

great advantage of mathematical reason over philosophical reason is the fact that in mathematics we *construct* the concepts we need from pure intuitions, while in philosophy we must confine ourselves to reasoning on universal concepts.[13] The intuition required for geometry is that pure intuition of space that appears in the "Transcendental Aesthetic." From this pure intuition we can construct "mental images" and actually display our constructions mentally or on paper to reach conclusions regarding these figures.[14] But what is the analogous case for arithmetic and algebra? The reference of arithmetic is not necessarily to objects in space and time. For example, mathematical objects, which we can count, do not appear to have an existence in space and time.[15] What are the pure intuitions that we need for arithmetic and algebra, and how do we construct numbers or other mathematical objects of algebra from these intuitions? There is no doubt in Kant's mind that time is the one essential intuition for the construction of number, but it is not clear what kinds of intuition we need—that is, instants in time, portions of time, succession in time—or how the construction is to be accomplished.[16] Until Kant could remove these ambiguities he could suggest, but not create, a science of time.[17] Hamilton's "Essay" attempts to make precise the construction of number where Kant left it ambiguous, and it is this emphasis on the *construction* of numbers that brings him closest to the arguments of the modern intuitionists.[18]

Hamilton begins his construction by saying: "If we have formed the *thought* of any one moment of time, we may afterwards either *repeat* that thought, or else think of a different moment."[19] By comparing two such moments we can create the notion of a "step" either forwards or backwards in time. A time step may be greater, equal to, or less than another given step. It is part of the original intuition that time steps can be equal by analogy even if they are not actually coincident. Number is obtained from a sequence of equal time steps, either positive or contrapositive, taken from an arbitrarily chosen zero moment. A "contrapositive" step is a step backwards in time. It allows Hamilton to define negative numbers by the concept of opposite direction in time, rather than by the meaningless notion of negative magnitude. The concept of positive number is originally ordinal, that is, one denotes the first, second, third (and so on), time step in the progression. It is also proper to call these numbers (or "multipliers" of the time steps) cardinal numbers, since one may ask "how many" steps there are between two given moments in time. The answer is one, two, three (and so on), but these cardinal numbers are obtained by counting in progression the steps from the zero moment, and therefore they are dependent on the ordinal relationship of the steps.[20]

Any multiple step, or number of steps in progression, may be treated as a new base or new unit step for a different system. A new series of multipliers specifies the order of these steps in time according to the new base, thereby defining multiplication. If, for example, three steps of base (a) are counted off and specified as the unit base (b) of a new system, and five of these new unit steps (b) are counted off to reach a specific moment in time (T), then the number of time steps according to base (a) from the zero moment to moment (T) will give 15, or the "product" of 3 and 5 (see figure 19.1).

In his earlier manuscripts and letters Hamilton had frequently argued that number should be defined as the *ratio* of steps in a progression of time.[21] For the integers this amounts to the ratio of a time step to a unit step by which it is counted, but to define the real numbers and to provide for division, Hamilton has also to consider ratios of incommensurable time steps. This he does in a lengthy discussion of ratio.

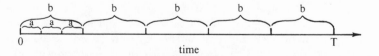

Fig. 19.1. Multiple time steps.

He is particularly concerned to show that the intuitive idea of the continuity of the progression from moment to moment in time also provides the idea of a *continuous progression* in ratio, thereby assuring the continuity of the real number system. [22]

When Hamilton comes to consider complex numbers, he constructs his number couples in a manner similar to that of his construction of the integers. He begins by comparing couples of moments in time (A_1, A_2) rather than single moments, as was the case for ordinary algebra. This "comparison" of two moment couples $(B_1, B_2) - (A_1, A_2)$ is a complex relation that Hamilton chooses to define as

$$(B_1, B_2) - (A_1, A_2) = (B_1 - A_1, B_2 - A_2),$$

thus generating a couple of time steps, much as he did in the case of ordinary algebra. [23] Numbers (or in this case number couples) are again multipliers of time steps, and Hamilton derives a "reasonable" and not wholly arbitrary definition of multiplication of number couples that is equivalent to the multiplication of complex numbers. [24]

This construction of number couples from steps in time assumes that progression in time is in every way comparable to the progression of real numbers. Hamilton worried about the fact that time appeared to have an objective reality that did not exist in algebra. In a manuscript from 1832 he argued that our knowledge of time seems to contain much more than pure order or progression because it refers to events happening, while algebra is grounded in a purely mental idea of order *that can be changed at will*. Hamilton noted that the entire theory of functions was based on this notion of changing relation or continuous progression altered at will. In a short memorandum he wrote:

The idea of *time* is that of an *order objective. Subjectively* viewing it, that is endeavoring to attend to thoughts rather than to things, we form the *nearest approach* to the idea of time when we think of one order as the mental basis of another, and consider the latter arrangement, which in this view resembles the course of events, as reducible to a mental dependence on the former arrangement which corresponds to the course of time. But though we may thus approach ... the idea of time, we do not quite attain nor adequately express it thus, nor by any other method which excludes the ascription of objectivity, or makes the order of time a voluntary arrangement of the mind. Do I then reject the doctrine of Kant that time is subjective, and in the mind? No, if I rightly believe the doctrine to be that time is a form of human thought, a result of the mechanism of our Understanding. But surely, even according to this doctrine, thus interpreted, our thought of time is involuntary, and the arrangement or order of time is so far objective. I am aware that most persons would call all this a play upon words, but I am of opinion with Me de Stael and with others, that when men dispute about words there is always some difference of ideas, too, and the question whether time be not in part objective, serves at least to make more clear the meaning attached by the inquirer to the word objective, if not [to] their idea of time.[25]

The relations of algebra depend on the notion of continuous order in succession, which can be found in time. But is the order of time something that we can vary at will? In 1832 Hamilton thought that it was not. Even if we remove from our intuition of time all actual events and all notions of cause and effect, there would still remain an *objective* content not subject to will. In time we recognize a past, present, and future—the notion of an inexorable progress beyond human control. By contrast, in arithmetic and more especially in algebra, the mind creates its progressions at will, and to this extent mathematical time is a more subjective creation than time as it is known through the act of perception.

Hamilton also worried about the *existence* of mathematical objects, a problem closely related to the objective character of time. A notebook entry dated April 24, 1835, contains the following passage: "When we consider any moment in the indefinite succession of time and regard it as an *object* of thought, we must think of it as having some certain place of its own in that succession by which it is distinguished *as an object* from all other moments of time ... the possibility of thus treating moments in algebra, and points in Geometry as *objects,* seems to be an essential postulate of these sciences and a condition of their possibility as such."[26] The existence of moments is guaranteed if they are objects of thought, that is, if they are intuited directly by a mental act and held in the memory. They must have this kind of objective existence and permanence (albeit an existence in the mind) if we hope to construct numbers from them.

Hamilton's arguments on the objectivity of time show an interesting similarity to some of Kant's arguments in the "Analogies of Experience." The purpose of Kant's analogies is to prove that experience is possible only through the representation of a necessary connection of perceptions, thereby providing a *unity* of all perceptions in time. Of the three analogies, the second is the most important for algebra, because it deals with the *succession* of appearances in time. Kant argues that "experience itself ... is ... possible only in so far as we subject the succession of appearances, and therefore all alteration, to the law of causality."[27] This law or formal rule is an a priori condition for ordering appearances in time. Kant gives the famous examples of a man viewing the facade of a large house and a man viewing a boat as it moves down stream.[28] In both cases the manifold of appearances must be generated in the mind successively. The appearance of the house is obtained by synthesizing a succession of apprehensions. The eye wanders over the facade of the house at will as the mind synthesizes the apprehensions received into the appearance of a house. Kant asks if the order of apprehensions is in the manifold of the house. Clearly not. This order of apprehensions must be a subjective succession, for the observer might well view the parts of the house in a different order and still obtain the same appearance of a house.

The case of the boat moving downstream is quite different. In this case the sequence of apprehensions is "bound down," as Kant says, to a determined order, because the observer is perceiving a series of *events.* He recognizes that this order of events is a *necessary* order, determined by the formal a priori rule of cause and effect that unifies experience. Without this rule the "succession of perceptions would be ... merely subjective, and would never enable us to determine objectively which perceptions are those that really precede and which are those that follow. We should then have only a play of representations, relating to no object; that is to say, it would not be possible through our perception to distinguish one appearance from another as regards relations of time."[29] In the case of the boat moving downstream, it is important to make a careful distinction between the subjective succession of apprehensions (the kind of subjective succession that took place in viewing the house) and the objective succession of appearances. Kant says that in the case of the boat, we must derive the subjective succession of apprehensions from the objective succession of appearances.[30]

Kant's distinction between subjective and objective succession in time is similar to Hamilton's distinction between subjective and objective order. The subjective order is arbitrary and subject to the will, while the objective order must conform to a determined rule that the observer does not create. Algebra requires the subjective order and cannot be created from a notion of time bound down to the law of causality. In his memorandum of 1832 Hamilton expressed the concern that if time could not be separated from the objective order of appearances, it would not be an adequate basis from which to construct mathematical objects.

Of course the objective order of appearances in time that Kant discusses in his second analogy is an order of *events,* while the doctrine that he presents in the "Transcendental Aesthetic" identifies time as the form of inner sense. Time as the form of inner sense is an a priori condition of any appearance. He says: "Appearances may, one and all vanish, but time (as the universal possibility) cannot be removed."[31] Whether his statements about time in these two parts of the *Critique* are consistent is a matter that need not be argued here.[32] It is enough that, for Hamilton, Kant's argument in the "Transcendental Aesthetic" brought conviction. In the "General Introductory Remarks" he does not raise the objections that appeared in his memorandum of 1832, and he concludes that pure time can be distinguished from all physical events and from all reference to cause and effect. Pure time becomes "coextensive and identical" with algebra. He retains the notion that moments of time must be "objects of thought." To that extent time remains objective, but he relinquishes the notion that it is in any way determined by an objective order of events.

Hamilton did not necessarily believe that Kant had resolved all the metaphysical problems regarding time, but he did believe that his doctrine was adequate for the creation of algebra. As a mathematician this was his real concern. "There is something mysterious and transcendent in the idea of Time," he wrote, "but there is also something definite and clear: and while the Metaphysicians meditate on the one, Mathematicians may reason from the other."[33] From the "definite and clear," he proceeded to create what he called the *Mathematical Science of Time.*

Hamilton's metaphysics was not popular with mathematicians at the time, nor was the importance of his number couples fully appreciated. Number couples provided another representation of complex numbers, and thereby helped clarify the foundations of algebra, but did they give mathematicians anything new? The algebra of number couples was valid for all the operations on real numbers, but the same thing was true of complex numbers. It was not clear that Hamilton had created anything more than a new representation of what was already known. If, however, his "Algebra as the Science of Pure Time" could be extended to triplets, quaternions, and other numbers of higher orders, then he would be creating truly new algebras. Hamilton was convinced that it could be done. In fact, his metaphysics convinced him that it *had* to be possible.

20

The Equation of the Fifth Degree

WHILE HAMILTON WAS developing his theory of algebraic couples and his metaphysical foundations of algebra he was also involved in a much more traditional field of algebra, the solution of algebraic equations. By the end of the sixteenth century, mathematicians had found general methods for solving polynomial equations of every degree through degree four. Attempts to find general solutions for equations of higher degrees met with no success, and by the eighteenth century the most important problem in algebra was recognized to be the polynomial equations of the fifth and higher degrees. A series of illustrious mathematicians had tried their hand at the problem, and the question naturally arose as to whether a general solution was possible. The Italian mathematician Paolo Ruffini attempted in 1799 to prove the impossibility of a general solution for the fifth degree, but his proof was defective.[1] Gauss succeeded in proving the "Fundamental Theorem" of algebra, which states that every polynomial equation with real coefficients has at least one real or complex root, but the proof of the *existence* of a root did not give a method for *finding* the root.[2]

The first successful proof of impossibility was by Niels Henrik Abel. Abel had thought at first that he had actually found a general solution, but on further reflection saw that his proof contained an error and he went on to demonstrate that *no* general solution was possible. He completed his proof in 1824 and published it in 1826.[3]

The most revealing work on the solution of algebraic equations was by Évariste Galois, who illuminated the whole problem by his creation of the theory of groups. But Galois was singularly unsuccessful in getting an audience for his new ideas, and there is no indication that Hamilton ever heard of them.

At the 1835 meeting of the British Association in Dublin, where Hamilton first presented his "Essay on Algebra as the Science of Pure Time," the British mathematician G. B. Jerrard presented a paper in which he claimed to have found a general solution to the equation of the fifth degree.[4] Jerrard had just finished the publication of a three-volume collection, *Mathematical Researches,* which was known to Hamilton and which also contained Jerrard's supposed solution.[5] At the meeting Hamilton was asked to report on Jerrard's paper, which he did on the day following the request. He concluded that Jerrard had exhibited a great deal of mathematical ingenuity, but had not found a solution.[6] Preparing a report in one evening on such a complicated subject was recognized as a great tour de force, even by the disappointed Jerrard and his brother, who was also at the meeting. Hamilton had probably known Jerrard for a long time, because they were the same age and had been at Trinity College Dublin together. From the correspondence it is obvious that Hamilton went out of his way not to offend Jerrard.

The following month, however, John Graves reported that the Jerrard brothers were still claiming to have solved equations of all degrees, and he asked Hamilton to clarify their results.[7] By the end of the year Hamilton was in correspondence with J. W. Lubbock, who also urged Hamilton to publish his findings.[8] Hamilton sent a brief response to the *Philosophical Magazine* in May, 1836, showing how methods like Jerrard's fail to solve the equation of the fifth degree, but without mentioning Jerrard's name.[9] Thus far Hamilton had been writing only in response to Jerrard's work, but at the end of May, 1836, he was obviously prepared to consider the entire problem in greater detail in preparation for the upcoming British Association meeting at Bristol. On May 30 he asked Lubbock to send him a copy of Abel's proof of the impossibility of a solution to the equation of the fifth degree, and the following day he wrote an enormous letter of 124 pages to Jerrard, describing his objections in detail. He said that he felt called upon to make his views and differences from Jerrard known, but was not clear how best to do it. An oral presentation at the British Association would be unintelligible, since the argument was too complicated to follow orally.[10]

As the summer wore on Hamilton prepared for the meeting. He received Abel's proof from Lubbock and a request from the secretary of the association to report on Jerrard's work.[11] Hamilton agreed reluctantly to report, but insisted that he would not be dragged into any argument. He told Jerrard that he would not debate him publicly and would discuss the matter orally only if Jerrard wished it.[12] The Jerrards appreciated Hamilton's tact and repaid his kindness by arranging grand accommodations for him at Bristol.[13] Hamilton's caution was justified, because Jerrard had obviously become touchy. Just before the meeting in August he wrote to Hamilton complaining about a knot of unscrupulous mathematicians who had great influence in the journals and who were attacking his work with insulting remarks.[14] Hamilton replied: "The resolving of equations of all degrees would certainly have been a brilliant achievement; but it is really a less interesting problem than that which you have proposed, and (in my judgment) resolved. For I hope, that acquitting me of all moral obliquity, you acquit me also of the intellectual error, of conceiving that your object was not rather to establish a general method for the transformation of the mth degree than to make the very easy and obvious application of it."[15]

What Jerrard had actually accomplished was not a solution to the general equation of the fifth degree, but a method of transformation by which the equation could be reduced to a "trinomial" or "normal" form:

$$x^5 - x - a = 0.$$

As it turned out, Jerrard had been anticipated in his method. E. S. Bring had made the same observation in 1786, but this fact had remained unknown until 1861, just at the end of Jerrard's life.[16] Hamilton honestly believed that Jerrard's results were important even though they did not solve the equation of the fifth degree, but it was difficult to persuade Jerrard of that fact. In the meantime it was a matter of getting through the British Association meeting at Bristol without a row. Apparently he succeeded, because after the meeting he wrote to his sister Sydney: "With Mr. Jerrard I got off very well; indeed, he made no reply to my arguments, and Peacock, one of the best judges on such subjects, expressed himself as entirely on my side. Nevertheless, I fear that Mr. J. is not yet convinced; but we all spoke of him in such high terms that he appears to be personally quite satisfied, and not to think that there was any design to run him down, which certainly there was not."[17] Indeed, Jerrard remained unconvinced throughout his life and continued to defend his solution in the *Philosophical Magazine* and in his *Essay on the Resolution of Equations* of 1858.[18]

Meanwhile Hamilton published his *Report* to the British Association and went on to study Abel's proof on the impossibility of solving the equation of the fifth degree.[19] He found two errors in Abel's proof, neither of them crucial to the argument, which he was able to repair. The result was a long paper entitled "On the Argument of Abel, Respecting the Impossibility of Expressing a Root of Any General Equation above the Fourth Degree by Any Finite Combination of Radicals and Radical Functions."[20] This paper was extremely difficult (L. E. Dickson calls it "a very complicated reconstruction of Abel's proof"), and it was replaced in 1879 by a simpler proof of Leopold Kronecker and the further development of Galois's theory of solvability.[21] Hamilton's analysis of Abel's proof essentially settled the issue in the minds of most mathematicians, although Hamilton was called on at least twice more to respond to the claims that a solution had been found.[22]

The fact that able mathematicians continued to go astray when they tackled this particular problem says something about the difficulty of Abel's proof as amended by Hamilton. It is also revealing that Hamilton, who was known in 1835 for his great powers of analysis but not for any previous work in algebra, was called on by the British Association to settle such a momentous question. It indicates the speed with which his star was rising during the 1830s.

VII

Quaternions

21

From Algebraic Couples
to Quaternions

HAMILTON'S CONTROVERSY with Jerrard over the solution of the equation of the fifth degree had diverted him somewhat from his study of the foundations of algebra, but not entirely and not for long. The "Essay on Algebra as the Science of Pure Time" ended with Hamilton's expressed intention "to publish ... many other applications of this view; especially to Equations and Integrals, and to a Theory of Triplets and Sets of Moments, Steps and Numbers, which includes this Theory of Couples."[1] Hamilton never forgot his intention to generalize his theory of couples to numbers of higher order. The theory of triplets was the most obvious extension of the theory of couples, partly because triplets came after couples in the hierarchy of "complex" numbers, and partly because an algebra of triplets would give a natural mathematics of three-dimensional space comparable to the two-dimensional geometrical representation of number couples. Such an algebra might have profound implications for applied mathematics since it would provide a new way of describing our three-dimensional world.

Over a period of thirteen years Hamilton searched off and on for the elusive triplets. We know now that they do not exist, at least not with the algebraic properties that he was seeking, but Hamilton did not know that, and time after time his mathematical and metaphysical instincts brought him back to the quest.[2] During the 1830s and early 1840s his search for the algebraic triplets became interwoven with a more general triadic scheme for all categories of knowledge. This combined search finally came to fruition, first on April 19, 1842, when Hamilton composed a complete system of philosophical triads, and again on October 16, 1843, when he discovered or invented quaternions, the final fruit of his search for triplets. Both discoveries came as flashes of inspiration, and both were written down and elaborated in a single sitting.

The mathematical triplets first make their appearance in manuscripts of February and May, 1830, entitled "Triads" and "Triads as the Fundamental Idea of Algebra."[3] In this setting Hamilton writes that the idea of algebraic numbers is that which exists between one triad and another; that is "between the mutual relation of three states of a progression and the corresponding mutual relation of three other states of the same or another progression." The triads in this context are basically expressions of ratio from which Hamilton wished to construct number. In his final description of number couples, Hamilton found time steps easier to use than triads, and the triads do not appear in his essay of 1835. But the manuscripts indicate that he continued to search for algebraic triplets while developing his theory of couples. In 1831 he made the first of several unsuccessful attempts to find a way to multiply lines in three-dimensional space.[4] He was stopped in this early effort by the failure of his triads to follow the distributive law. He also attempted to extend his conjugate functions to include triplets, but discovered that it was not possible to establish a theory of triplets from partial differential equations.[5]

In February, 1835, as he was finishing his "Essay on Algebra as the Science of Pure Time," Hamilton returned again to the triplets and attempted an algebraic solution.[6] This time he was stopped by the discovery that in his system the product of two finite triplets (or of two finite lines representing the two triplets) could give a zero result. And even more serious was his discovery that division of these triplets (or of lines representing them) could give an ambiguous quotient.[7]

John Graves did not get a copy of Hamilton's "Essay on Algebra as the Science of Pure Time" until the spring of 1836.[8] When he read it he was immediately interested in the triplets and proposed several systems of his own. Graves wrote, "I hope that you have not abandoned the subject, and shall be glad to see how you define the multiplication of triplets in the continuation of your theory. In the mean time I send you some of my speculations, which may interest you notwithstanding their crudeness, though I dare say they contain nothing of the slightest importance that you have not already fully investigated."[9] None of the systems was successful, but each came close enough to give cause for hope, and Graves urged Hamilton to pursue the theory. Then, in a strange diversion, Graves plunged into the realm of metaphysics: "It seems to me that Time possesses this quality of continuity only in common with Space and colours and sounds, and even Passions and affections of our minds. We may symbolize *Love* as a single, passing on from Indifference into ardour or veering round through jealousy into hatred, or we may treat algebraically of the continuous properties of

Love considered as a couple, varying with these separate continuities." [10] After another excursion into the algebra of triplets Graves turned back into metaphysics: "As connected with the doctrine of triplets, I must communicate to you what you will perhaps deem one of the wildest of my theories about colours and vision ... all colours may be considered as triplets of which the constituents may be denoted algebraically by numbers proportional to the quantities or intensities of the component primitive colours, or all colours may be represented geometrically by lines whose coordinates are referred to three positive chromatic axes." [11] He goes on to suggest negative or "counter-colors" and the like.

John Graves's appeal was lost on Hamilton, at least for the time being—the two men did not return to the triplets until 1840. But Graves's letter reveals a metaphysical interest that had long been growing in his mind and in Hamilton's as well. From his study of Coleridge Hamilton could not help but be impressed by the constant emphasis that Coleridge placed on the polar and triadic arrangement of nature, and as Hamilton sought his own philosophical position, the dynamic triad of metaphysical idealism took an increasingly important position in his thought. A good example of Coleridge's triad appears in this passage from *The Friend:* "EVERY POWER IN NATURE AND IN SPIRIT *must evolve an opposite, as the sole means and condition of its manifestation:* AND ALL OPPOSITION IS A TENDENCY TO RE-UNION. This is the universal Law of Polarity or essential Dualism. ... The Principle may be thus expressed. The *Identity* of Thesis and Antithesis is the substance of all *Being*; their *Opposition* the condition of all *Existence,* or Being manifested; and every *Thing* or Phaenomenon is the Exponent of a Synthesis as long as the opposite energies are retained in that Synthesis." [12] Here is a clear statement of the dynamic triad of thesis, antithesis, and synthesis. The phenomena that we observe must be the product or synthesis of opposite and conflicting powers. These powers have real *being* through their identity and unity behind the phenomena, but we cannot know them except through the phenomena that their conflict creates. Only the phenomena can be said properly to *exist*, but our intellect cannot be content with the phenomena alone. It seeks a unity and a constancy in the phenomena, a unity that can only be discovered in the dynamic forces producing the phenomena.

Hamilton copied out this entire passage from *The Friend* in the fall of 1831 and added his own commentary. It is helpful to read Hamilton's words, because his prose is considerably less murky than Coleridge's. He wrote:

So far as I understand this principle, I would perhaps express it thus:—Power can be manifested only by its effects, that is, by overcoming Resistance, which is Contrary Power. Existence is manifested by the struggle between two opposite tendencies . . . . Each particular Phenomenon, or individual Manifestation of Existence, is determined to be such as it is, and no other, by the kind and degree of its producing Power, that is, by its own particular combination or synthesis of two opposite tendencies. The thought of *Being* or *Existence general* (a new name, the propriety of which may demand a special inquiry), as distinguished from phenomena, that is, from individual manifestations of existence, arises in us along with, and as a realization or externalization of, our belief in a common ground, a hidden principle of unity, belonging to the two opposite tendencies. [13]

Hamilton grasps Coleridge's meaning, but he is unclear as to the nature of the opposing powers creating the triad.

Coleridge's works contain other frequent references to the need to trichotomize all the categories of understanding, and he criticized Kant's arrangement of the twelve categories under four rather than three major divisions. [14] Following Schelling, Coleridge wished to make the triad a basic key to philosophy. In the *Aids to Reflection* he further refined the triad of thesis, antithesis, and synthesis into a "noetic pentad" of the following arrangement:

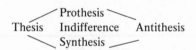

and related it both to trichotomous logic and to the mystical Tetractys of the Pythagoreans. [15]

It is difficult to comprehend just what Coleridge meant to signify by his Pentad. Muirhead calls it an "eccentricity" that can be ignored; [16] Orsini says that it was "too cumbersome for its author himself to make much use of;" [17] but Hamilton and his friend Aubrey De Vere strained to catch the meaning of that "terrible note of the Pentad and the Idea" in the *Aids to Reflection.* [18] Hamilton was still struggling with it in the spring of 1835 when he was working on his "Essay on Algebra as the Science of Pure Time." In May he reported to Aubrey De Vere:

In the course of a long silent train of thought some months ago, on my own view of Algebra, I suddenly arrived at a stage and state of speculation, and as it seemed of insight, which reminded me so strongly of that terrific note about the *Pentad* that I started from my seat and ran to search my library for the *Aids,* to try whether, now at last, I comprehended what I had so often thrown aside in despair. The book was with my sisters, but the delay, while it excited, served me, by forcing me to make my thoughts more clear and more my own; and when at last I met the page again, I had the satisfaction of feeling that they had attained in their own process of development to a point which seemed to differ little from that at which Coleridge had been. [19]

Unfortunately, he does not go on to explain this sudden illumination, and we are left only with the knowledge that Hamilton sought in the tripartite division of the categories some significance for his own mathematical work.

In October, 1840, Graves returned to the algebraic triplets and showered Hamilton with letters proposing different geometrical approaches.[20] He also resumed his speculations about visible space as a "natural algebra" and related it to the phenomenon of colors.[21] But with these last letters, Graves's enthusiasm began to flag. He confessed that on the subject of triplets he had reached "the extent of his tether." He stated, "From want of time, knowledge and grasp of intellect, I can do no more than sting the rhinoceros hide of the representation of algebraic quantities by points in space"—and he urged Hamilton to turn his attention more specifically to algebraic triplets.[22]

Hamilton's reaction, however, was to take the triplets more deeply into the realm of metaphysics. During the years from 1839 to 1842 he became seriously depressed. This was the time that his wife left Ireland to live with a sister in England, leaving their three children at the observatory. Hamilton's cousin Arthur, his closest family friend, died in 1840, and Hamilton became more retired, seldom leaving the observatory, and suffered from his first difficulties with alcoholism. He told Graves, "The illness of my wife has been much upon my spirits, and I have done little lately in the intellectual way, except think of the metaphysic of physics," and to his old tutor Charles Boyton he wrote the following spring about his continued deep dejection:

Still when a man is driven from mathematics, he may take up, or rather cannot part with, metaphysics. And to one of a metaphysical bent the line of our old friend Horace applies:—

Purae sunt plateae, nihil ut meditantibus obstet;
[The streets are clear, so that nothing stands in the way of thinkers.]

not in the ironical sense, but in a sober seriousness. Accordingly, my metaphysical propensities developed themselves largely for many months, while my scientific tendencies were dormant.[23]

An astronomy lecture from October, 1840, shows that Hamilton had come to place greater emphasis on the dynamic opposition of forces in his philosophy. He had come to regard this opposition as "a strife of which the importance and intellectual dignity are such, that hardly can one speak of it without emotion."[24] At the end of his draft for this lecture Hamilton made his own first attempt to set up triadic categories. He used as his primary triad the categories of will, mind, and life. A manuscript scrap of jottings is the only record of Hamilton's meetings with Coleridge in 1832 and 1833, but this scrap contains the triad of will, mind, and life. It proves that Hamilton had discussed the trichotomous logic with Coleridge and had remembered the discussion eight years later.[25]

Metaphysical questions were obviously turning in Hamilton's mind. In January, 1842, he traveled to England to find his wife in much better health, which brought about a marked improvement in his own spirits. They returned to Dublin together, and in April Hamilton produced his most spectacular writing on metaphysical subjects. His ideas fell into place and he wrote out a complete system of philosophical triads in a long letter to his former pupil Viscount Adare.[26] He considered the triads to be important enough to warrant writing out four additional copies, one for John Graves, a second for John's brother Robert (which he sent with a message asking that he pass the letter on to Wordsworth), and two more copies for another close friend, Captain Larcom.[27]

Hamilton's letter shows the strong impress of Kant's *Critique of Judgment*. He reminds Adare that in the *Critique of Judgment* Kant confirmed the triadic arrangement of all the mental faculties much more emphatically than in the *Critique of Pure Reason*.[28] The *Critique of Pure Reason* is primarily a study of the *understanding*. The *Critique of Practical Reason* is a study of the *reason* as the basis for morals, and the *Critique of Judgment* is a study of the faculty of *judgment*. Judgment is the "middle term" between the understanding and the reason. Thus the three great critiques of Kant explore the triad of cognitive faculties, one for each.

Hamilton concluded that Coleridge's primary triad of will, mind, and life should be associated directly with Kant's three cognitive faculties. Mind should be identified with the faculty of understanding, which Kant establishes as the a priori basis for the study of nature. Will should be identified with reason, which he applies to the moral problems arising from the freedom of will. And finally, life should be identified with judgment, which Kant associates with the feeling of pleasure and displeasure. The faculty of judgment is, therefore, the basis for aesthetics and also for our inclination to seek final ends in the operations of nature.[29]

Hamilton goes beyond Kant in his desire to construct a completely trichotomous arrangement of all mental powers. The three moments of each triad "subsist in intimate union, and are in some sense *one* though *three*," an obvious echo of the Trinity. Furthermore, each moment of the primary triad may be subdivided into a secondary triad analogous to the first. Each moment of the secondary triad may then be again subdivided into a tertiary triad, and "this process may be performed again, perhaps without limit, though the distinctions thus obtained may soon become minute, and may appear to be vague."[30] Hamilton limits his own system to three stages of division, providing a final system of twenty-seven categories. As might be expected, Hamilton takes greatest interest in the categories under the heading of mind, because it is these categories that are the a priori basis for our understanding of nature.

An investigation of all of Hamilton's categories would carry us into too great detail. It is more profitable to explore how Hamilton borrowed ideas from Kant's *Critique of Judgment*, for this work more than any other opened the way for teleological explanation in science and related science to aesthetics.

Kant explains that the faculty of judgment is necessary to bridge the gap between the reason and the understanding. The understanding provides for a philosophy of nature, while reason provides a philosophy of morals. These two faculties are separated by a "great gulf," because the philosophy of nature is a philosophy of the sensible, while the philosophy of morals, based as it is on the concept of freedom, is a philosophy of the supersensible.[31] But there *must* be a common ground of unity of the supersensible that lies at the basis of nature, because by our free will we can act on nature. The problem facing Kant is, then, to determine how the world of freedom can influence the world of sensible reality, and he finds the answer in the judgment, which is, as it were, a bridge across the gap.

Any inquiry into nature presupposes the possibility of unifying phenomena to create natural law. Without this presupposition we cannot even begin an investigation of nature. "For, were it not for this presupposition, we should have no order of nature in accordance with empirical laws, and consequently, no guiding-thread for an experience that has to be brought to bear upon these in all their variety, or for an investigation of them."[32] Furthermore, since we recognize the need for a unity of natural law, while at the same time we know that natural phenomena are contingent, we are forced to accept in nature a principle of finality or ultimate ends. Kant cautions us to remember that this principle of teleological judgment is entirely *subjective,* that is, it gives us no right to legislate for nature, nor does it state a priori conditions for the existence of objects. It is merely a heuristic principle, but still, it is a principle without which we cannot pursue an inquiry into the natural world. And while the accordance of phenomena with our principle of uniformity can only be, as far as we can know, a purely contingent fact, it is extraordinary, nevertheless, that nature conforms so conveniently to our needs.[33] For Hamilton it was more than extraordinary; the agreement of the laws of nature with the laws of thought could only be the result of Divine will.

Turning to Hamilton's exposition of his triads we see Kant's notions repeated with Hamilton's own peculiar philosophical twist.

Science, at least Natural Science, supposes a *Nature* to be known. It involves not only a System of Observations and a System of Reasonings, but also a System of INTERPRETATIONS. It refers not merely to Appearances and to Thoughts but to (*believed*) REALITIES; to Existences outside ourselves. Such is, in *Science,* an inevitable *attitude* of mind, however clearly it may be shown by metaphysicians that the evidence for the existences of an external world is not of the same kind with logical or mathematical proof, ... such a reference is *dynamic*. It supposes an antagonist *power*, and something like a foreign *will*; although we conceive, and cannot avoid conceiving, this power to act by *laws*, and in no arbitrary manner. Besides, this reference of Appearance to Existence is a passage from the seen to the unseen, and partakes of the character of *Faith*: but every form of faith is *ethical*, and derives itself at least partly from the WILL. [34]

Although Hamilton does not directly identify "faith" in science with Kant's principle of teleological judgment, he means much the same thing. No science is possible without some belief in the unity of natural law. The exact statement of natural law can be derived only from experience, but for any progress at all we must assume that nature is organized and acts toward some ends. For Hamilton this assumption is an act of faith, and since the faculty of judgment connects the realm of morals and free will to the realm of nature, it is a faith that must necessarily have an ethical content. Hamilton's reference to "believed realities" or "existences outside ourselves" would seem at first to contradict Kant's assertion that the things-in-themselves are forever unknowable. But Hamilton insists that these "existences" are dynamic. They are powers external to our own will, but necessary for the unification of phenomena, and in that sense he is following Kant's assertion that for science we must regard nature as if it were organized by a supersensible being. Hamilton's "unific energies" are directly analogous to Kant's "highest intelligences." The difference comes in their conclusions regarding theology. Kant insists that we must not assume that our need for a teleological principle in nature proves the existence of God, while Hamilton believes that his "unific energies" could only be explained by a divine creator. [35]

Hamilton's triads were an important step in his own attempt to develop a speculative physics. By identifying the powers or "existences" or "existing realities," as Hamilton had called them in his letter on the triads, the unity of science will be accomplished. Light, heat, chemistry, and electricity may be distinct physical phenomena, but they are merely syntheses of more fundamental conflicting powers. [36] The phenomenon is, so to speak, the apex of the triangle of triadic elements, generated from the other two angles. The object of the science of powers is to identify the conflicting powers generating the phenomena. Hamilton insisted that the route to this synthesis lay through metaphysics and mathematics. Experiment and observation could play only a minor role. In pursuing the metaphysical triads and mathematical triplets he was engaged in a truly "scientific" enterprise. For him only this kind of activity would lead to a true science of nature.

22

The Creation of Quaternions

HAMILTON'S "TRIADIC FANCIES," as he called them, were in part a reflection of his increased interest in metaphysics and in part a reflection of his frustration at being unable to extend his theory of number couples to three dimensions. We can assume that the triplets were in the back of his mind even when he was not consciously searching for them. But in the autumn of 1843 he began again to think earnestly about the problem. We do not know what brought him back to the triplets at this particular time, although there are several obvious possibilities. In a letter to his son written shortly before Hamilton's death, he recalled the circumstances of the discovery.[1] He said that when he returned from Cork, where he had attended the meeting of the British Association, "the desire to discover the laws of multiplication [of triplets] regained with me a certain strength and earnestness, which had for years been dormant." Every morning during the month of October, he tells us, he would come downstairs to breakfast and would be asked by his elder son, "Well, Papa, can you multiply triplets?" and every morning he had to confess that he could only add and subtract them. This episode has the characteristics of a charming, but unreliable, anecdote. Even if it is false in detail, however, it indicates that Hamilton was intensely engaged again in thinking about the triplets.

There was no subject discussed at the Cork meeting that would have attracted Hamilton specifically to the triplets; certainly none of the papers that he delivered at the meeting had anything to do with the subject.[2] One possible inspiration was occasioned by the visit of young Gotthold Eisenstein to the observatory during the summer of 1843.[3] Eisenstein was only nineteen at the time. He wrote to Hamilton asking about possible teaching positions in Ireland, and followed the letter by a visit. The next year Eisenstein made a spectacular mathematical debut by the appearance of *twenty-five* of his papers in Crelle's *Journal*. It is likely that the mathematical ideas contained in these papers were subjects of discussion at the observatory, in particular a paper entitled "Allgemeine Untersuchungen über die Formen dritten Grades," in which Eisenstein introduced an early matrix notation. Since the invention of matrices is usually credited to Arthur Cayley, and supposedly did not occur until 1858, Eisenstein's paper may be regarded as an early anticipation of some of the matrix ideas.[4]

Among these ideas is the observation that matrices form the elements of an algebra that is much like ordinary arithmetic except that multiplication is not commutative. Eisenstein wrote: "Incidentally, an algorithm for calculation can be based on this; it consists in applying the usual rules for the operations of multiplication, division and exponentiation to symbolical equations between linear systems; correct symbolical equations are always obtained, the sole consideration being that the order of factors, i.e., the order of the composing systems, may not be altered."[5] It is possible that mathematical discussions with Eisenstein brought Hamilton back to the study of triplets or at least alerted him to the possibility of a noncommutative algebra.[6]

Another likely inspiration was Hamilton's resumed study of the work of the German mathematician Martin Ohm. In 1834, when he had been writing his "Essay on Algebra as the Science of Pure Time," Hamilton had encountered Ohm's *Versuch eines vollkommen consequenten Systems der Mathematik*.[7] He read it carefully; in fact he said that it was the desire to read Ohm even more than the desire to read Kant that caused him to refurbish his knowledge of German. A year later he stated that probably he was the only one in Great Britain to have read it and to have recognized its importance.[8] Ohm was one of the algebraic formalists, and therefore his work had a different metaphysical and logical foundation than that of Hamilton, but Ohm had made a thorough study of the foundations of algebra, and therefore his book raised questions about the nature of number that may have turned Hamilton's attention to the triplets.

In May, 1843, eight years after his first encounter with Ohm's *Versuch*, Hamilton received a letter from Alexander John Ellis enclosing an English translation of Ohm's most recent work, *Der Geist der mathematischen Analysis*. Hamilton was asked to give his opinion of the book with the purpose of persuading the publisher to bring out other works of Ohm in translation. Hamilton read the work and discussed it with both John Graves and Lord Adare. He could not publicly support the book, because he differed with Ohm over the foundation of number, but he recognized it as a work of great value nevertheless.[9]

At the same time that he was profiting from the inspirations of Ohm and Eisenstein, Hamilton had a chance to take part in a change of professorships in the late summer of 1843, and this may have inspired him to renew his mathematical efforts. Humphrey Lloyd resigned as professor of natural philosophy to become senior fellow. It was suggested that Mac-Cullagh take Lloyd's place and that Hamilton might replace MacCullagh as professor of mathematics, thereby escaping from the observatory, which was becoming increasingly burdensome to him. His unhappiness had been caused by new demands from the Board of Fellows for a more strict accounting of the operations of the observatory. Hamilton thought for awhile that he might have to give up the study of mathematics altogether.[10] The chance to move entirely into mathematics must have been attractive to him, but the hostility of the board, which had caused him so much trouble at the observatory, also prevented him from obtaining the professorship. It was given instead to John Graves's younger brother Charles.

Whatever the cause of Hamilton's renewed interest in the triplets, his mind was full of them in the autumn of 1843. Since 1830 he had attempted many different systems of triplets, both algebraic and geometric, and in no case had he been able to obtain a triplet algebra that would allow an unambiguous method of multiplication and division, but he was still convinced that such an algebra *must* be possible. On the sixteenth of October, 1843, he and his wife were walking into Dublin along the Royal Canal to attend a council meeting of the Royal Irish Academy. Suddenly a possible resolution of the problem of triplets leapt into his mind. He described the event as follows:

An *under-current* of thought was going on in my mind, which gave at last a *result*, whereof it is not too much to say that I felt *at once* the importance. An *electric* circuit seemed to *close*; and a spark flashed forth, the herald (as I *foresaw, immediately*) of many long years to come of definitely directed thought and work, by *myself* if spared, and at all events on the part of *others*, if I should even be allowed to live long enough distinctly to communicate the discovery. Nor could I resist the impulse—unphilosophical as it may have been—to cut with a knife on a stone of Brougham Bridge, as we passed it, the fundamental formula.[11]

Hamilton wrote the above description in 1865, the year of his death. One might expect him to have embellished the story in the years since the event, but a letter to P. G. Tait written seven years earlier gives essentially the same account.

I then and there felt the galvanic circuit of thought *close*; and the sparks which fell from it were the *fundamental equations between i, j, k; exactly such* as I have used them ever since. I pulled out on the spot a pocket-book, which still exists, and made an entry, on which, *at the very moment*, I felt that it might be worth my while to expend the labour of at least ten (or it might be fifteen) years to come. But then it is fair to say that this was because I felt a *problem* to have been at that moment *solved*—an intellectual want relieved—which had *haunted* me for at least *fifteen years before*.[12]

The pocket book remains to verify this story, which casts doubt on his later claim to have scratched the quaternion formulas on the bridge. Hamilton would scribble on any handy surface if he did not have paper along, but on this occasion he did have a notebook. Maybe he felt that the solemnity of the moment required an inscription in stone.

Hamilton's commitment to quaternions became even greater than he had foretold. They consumed the rest of his mathematical career. At the time of his death in 1865 he was completing the *Elements of Quaternions*, his second major work on the quaternion theory.

After the discovery flashed across Hamilton's mind he quickly elaborated it while riding in a car the rest of the way to the council meeting. Once at the meeting he showed his discovery to MacCullagh and to William Sadlier, and obtained permission to make a brief presentation of

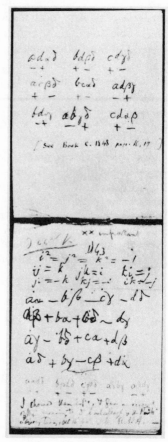

The pocket book in which Hamilton first entered the fundamental formulas of quaternions.

it at the next general meeting of the academy, scheduled for November 13, 1843.[13] The day after the discovery he sent a full account of it to John Graves.[14] This letter to Graves, a memorandum made the day of the discovery, and a later preface to his *Lectures on Quaternions* (1853) all contain attempts by Hamilton to reconstruct the train of thought that led him to his discovery. These documents make the moment of truth on the bridge in Dublin one of the best-documented discoveries in the history of mathematics.[15]

Considering the fact that Hamilton had tried so many different forms of triplets and had found them all wanting, it is surprising to find him in 1843 working with the most obvious form of triplets, and approaching the problem in the most obvious way. It was a problem that he had been turning over in his head and struggling with for some time prior to his discovery by the bank of the canal. What occurred to him there was the sudden realization that the "obvious" system of triplets that he had under consideration at the time would succeed if extended to a quaternion of four numbers.

The course of his thought was as follows. He tried a triplet containing one real and two complex parts $(x + \mathbf{i}y + \mathbf{j}z)$. Since the complex number $(x + \mathbf{i}y)$ can be represented by two rectangular axes in a plane, it was natural to present the $\mathbf{j}$ term by a third axis that was perpendicular to the other two axes. In complex numbers $\mathbf{i}^2 = -1$; therefore the new term $\mathbf{j}$ should by symmetry have the same property, and $\mathbf{j}^2 = -1$. The square of such a triplet would then be

$$(x + \mathbf{i}y + \mathbf{j}z)(x + \mathbf{i}y + \mathbf{j}z) = (x^2 - y^2 - z^2) + \mathbf{i}(2xy) + \mathbf{j}(2xz) + \mathbf{ij}(2yz).$$

$$(22.1)$$

This result was not entirely satisfactory, because the product of two triplets should be another triplet, while the product that Hamilton obtained had four terms. He had to do something with the $\mathbf{ij}$ term—either set it equal to zero, or add it somehow to one of the other three terms. Hamilton thought he might find an answer in a geometrical analogy between triplets and complex numbers.

In the geometry of complex numbers, the product of two lines each representing a complex number is a third line (Hamilton calls it a *third proportional*) in the same plane, whose length is the *product* of the lengths of the multiplier lines and whose angle with the real axis is the *sum* of the angles of inclination of the multiplier lines (figure 22.1). The *length* of such a line was called the *modulus*. The modulus of the complex number $(a + \mathbf{i}b)$ is $\sqrt{a^2 + b^2}$, as can be seen immediately by the Pythagorean Theorem. The law of the modulus for complex numbers states that the product of the moduli of two complex numbers equals the modulus of the product of the two numbers. Thus if

$$(a + ib)(c + id) = (e + if),$$

the law of the modulus states

$$\sqrt{a^2 + b^2}\,\sqrt{c^2 + d^2} = \sqrt{e^2 + f^2},$$

or written instead as a law of the *norms*,

$$(a^2 + b^2)(c^2 + d^2) = (e^2 + f^2). \tag{22.2}$$

Hamilton was particularly anxious that the law of the norms hold for triplets as well as for number couples, because it would determine the length of the product line in the geometrical interpretation of triplets and would guarantee that division of one triplet by another would always give a unique quotient. In order to have a law of norms for triplets that enjoys the same symmetry as the law of norms for couples (equation 22.2), one would have to have, using equation 22.1,

$$(x^2 + y^2 + z^2)(x^2 + y^2 + z^2)$$

$$\overset{?}{=} (x^2 - y^2 - z^2)^2 + (2xy)^2 + (2xz)^2 + (2yz)^2, \tag{22.3}$$

but in fact,

$$(x^2 + y^2 + z^2)(x^2 + y^2 + z^2)$$

$$= (x^2 - y^2 - z^2)^2 + (2xy)^2 + (2xz)^2. \tag{22.4}$$

The law of the norms does not hold unless the **ij** term is removed from equation 22.1, and so here is another reason for setting **ij** $= 0$, although such a choice seemed "odd and uncomfortable" to Hamilton.[16] It would also be possible to get rid of the **ij** term by setting **ij** $= -$**ji**, because the

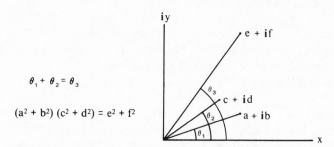

Fig. 22.1. Geometrical representation of complex numbers.

term **ij**($2yz$) should be written yz(**ij** + **ji**) if the actual order of multiplication is to be maintained. In fact Hamilton had already shown a willingness the previous year to dispense with commutativity in one of his unsuccessful triplet systems.[17] Denial of the commutative law was a drastic step, but it seemed more natural to him than to have the product of two unit vectors at right angles to each other vanish entirely. If one considers the products **ij** and **ji** as lines of equal magnitude, but oppositely directed, then it is not surprising that they should add to zero.

So far Hamilton had been trying to apply the law of the modulus to the square of a triplet. He then asked the question, "Will the law for the multiplication of vectors in the complex plane still hold if the plane is in three-dimensional space?" He imagined the plane rotated about the real axis so that the lines being multiplied, though still in a plane, were represented by triplets (figure 22.2). (To simplify the calculation he took two lines with the same complex coordinates, **i**y and **j**z.) Taking the two triplet lines as $(a + $ **i**$y + $ **j**$z)$ and $(x + $ **i**$y + $ **j**$z)$, he wished to see if the product of these two lines would remain in the same plane and if it would be in the same place on the plane as that given by the geometry of complex numbers. Or, as he described it, he wanted to know if the product line was a "fourth proportional to the three lines of which the ends are 1, 0, 0; a, y, z; x, y, z" on Warren's principles.[18] Setting **ij** $= -$**ji** as before, the law of the norms is quickly confirmed, but the calculation of the angle that the product line makes with the real axis is tedious, and we can assume that in spite of his vaunted ability as a mental calculator, Hamilton did not do it in his head. The result, however, was favorable. Triplet lines lying in a plane containing the real axis have the same product line as they would have if the plane were in two-dimensional complex space.

So far the assumption of **ij** $= -$**ji** had carried Hamilton a long way. He had shown that the square of a triplet and the multiplication of two triplets in a plane behave in a reasonable manner. The next step was to "try boldly the *general* product of two different triplets" not lying in a plane containing the real axis.[19] But this time the result was not so favorable. Letting **ij** $= -$**ji** as before,

$$(a + \mathbf{i}b + \mathbf{j}c)(x + \mathbf{i}y + \mathbf{j}z)$$

$$= (ax - by - cz) + \mathbf{i}(ay + bx) + \mathbf{j}(az + cx) + \mathbf{ij}(bz - cy). \qquad (22.5)$$

Now multiplying the moduli,

$$(a^2 + b^2 + c^2)(x^2 + y^2 + z^2)$$

$$= (ax - by - cz)^2 + (ay + bx)^2 + (az + cx)^2 + (bz - cy)^2. \qquad (22.6)$$

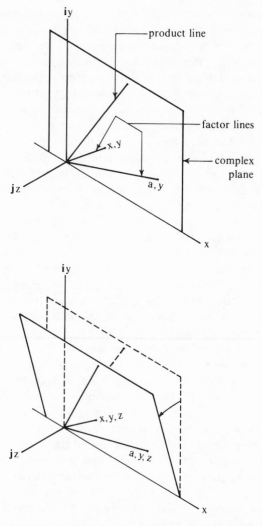

Fig. 22.2. Hamilton asks: "Does the product line remain in the same place on the complex plane when the plane is rotated into triplet space?"

These results are encouraging. The norms seem to have the correct form, but there are still several awkward problems. Setting **ij** = −**ji** does not remove the **ij** term from the general product of two triplets. And since the coefficient of **ij** appears squared in the product of the norms, one cannot simply remove the **ij** term from equation 22.5 by setting **ij** = 0. This is because the product of the norms is the sum of *four* squares, not of three squares. But the sum of four squares cannot be the modulus of a triplet. The only thing to do was to set **ij** equal to some unknown **k** and see if it might somehow be removed.

Hamilton had probably reached this far in his calculations before he set forth on his walk along the Royal Canal. He was musing over his triplets and their unfortunate inclination to make a quaternion of terms in the product, talking to his wife, but with the triplets in the back of his mind. Suddenly it occurred to him that if the product of two triplets had four terms and if the product of their norms also produced a norm with four terms, maybe he would have better luck multiplying expressions with four terms than he had had multiplying triplets. If **k** were taken to be a third imaginary in addition to **i** and **j**, his elements would become quaternions and maybe they would work better. He thought that possibly $k^2 = 1$, but could see immediately that this would not work, because some of the cross terms in the calculation of the norm would not cancel out. They would cancel, however, if $k^2 = -1$.

Although Hamilton does not tell us what in particular caused him suddenly to think of quaternions in place of triplets, I think it was probably the realization that the right members of equations 22.5 and 22.6 contained four related terms. If the right member of 22.6 was the norm of the right member of 22.5, then the right member of 22.5 would have a conjugate quaternion analogous to the conjugate of a complex number. It is an easy matter to check, and Hamilton, who had great powers of mental calculation and who had been mulling over the calculation in his mind for years, probably did it in his head. Having already assumed $i^2 = j^2 = -1$ and $ij = -ji = k$, he got

$$(a + ib + jc + kd)(a - ib - jc - kd)$$

$$= a^2 + b^2 + c^2 + d^2(-k^2) - bd(ik + ki) - cd(jk + kj).$$

The expected norm is $a^2 + b^2 + c^2 + d^2$, and it is obviously obtained by setting $k^2 = -1$ and $ik = -ki$, $jk = -kj$. These substitutions are entirely analogous to the values that Hamilton had already obtained for i^2, j^2, and ij. The symmetry of the equations for multiplication of imaginaries must have been a powerful impetus in favor of quaternions.

The next step was to check to see if the law of the norms held for quaternions. In order to do this he needed values for **jk** and **ik**. (He had already assumed $ij = k$.) He decided that $ik = -j$, because $ik = i(ij) = (ii)j = (-1)j$. And in like fashion probably $kj = -i$, because $kj = (ij)j = i(jj) = i(-1)$. Then summarizing his "multiplication assumptions," as he called them:

$$i^2 = j^2 = k^2 = -1$$

$$ij = -ji = k, \quad jk = -kj = i, \quad ki = -ik = j.$$

These are the equations that he scratched on the bridge in great excitement. They would preserve his priority at least until he could announce his discovery that afternoon at the academy.[20]

On the way into the council meeting he checked the law of the modulus, writing out the terms in his pocket notebook. He discovered that the law holds for quaternions. With the law of the modulus confirmed, Hamilton was convinced that his discovery had vast significance for mathematics: "At this stage, then, I felt assured already that quaternions must furnish an interesting and probably important field of mathematical research: I felt also that they contained the solution of a difficulty, which at intervals had for many years pressed on my mind, respecting the particularisation or useful application of some great principles long since perceived by me, respecting *polyplets* or sets of numbers."[21] Hamilton quickly found that his quaternions preserve all the operations of arithmetic except for the commutative law. His willingness to give up this law is often regarded as a stroke of real genius, because it recognized that Peacock's Principle of the Permanence of Forms was not an absolute requirement for an algebra to be meaningful. For Hamilton the elements of algebra were to be constructed from the intuition of pure time, and the operations on those elements would be determined by the nature of the elements themselves. Still it is a surprise to see him write $\mathbf{ij} = -\mathbf{ji}$ in such a cavalier fashion, insisting all the while that it is a more natural expression than $\mathbf{ij} = \mathbf{ji} = 0$. Possibly Hamilton had been prepared for the rejection of the commutative law by discussions with Eisenstein, or possibly his attempts to find a geometrical representation of triplets had shown him that rotations in three-dimensional space do not commute either, or, what is even more likely, sacrificing commutativity seemed to him to be the only way to get any results at all.

The quaternions came as a shock to those with whom he had corresponded about the triplets. The reactions of John Graves and Augustus De Morgan, both formalists of the "philological school," betray their surprise. Neither of them had realized that one might *construct* new elements of algebra rather than find them from the elements of existing algebra. John Graves wrote: "There is still something in the system [of quaternions] which gravels me. I have not yet any clear view as to the extent to which we are at liberty arbitrarily to create imaginaries, and to endow them with supernatural properties. You are certainly justified by the event ... [but] what right have you to such luck, getting at your system by such an *inventive* mode as yours?"[22] Graves obviously had not thought it permissible for Hamilton to *create* quaternions, but he could raise no logical objection against what Hamilton did. Yet it "graveled" him. De Morgan was also caught by surprise, but he immediately recognized the importance of Hamilton's approach. "[John] Graves gave me some extracts from your letter now published. ... He never dropped a hint about *im-*

agining imaginaries. On such little things do our thoughts depend. I do believe that, had he said no more than 'Hamilton *makes* his imaginary quantities,' I should have got what I wanted."[23] These two spontaneous remarks by Hamilton's friends show how quaternions were the gateway to modern algebra. The Principle of the Permanence of Form was shattered by Hamilton's discovery, and the road was open to a wide variety of algebras that did not follow the rules of ordinary arithmetic.

The same day that Hamilton discovered quaternions, he tried to find a geometrical interpretation for them similar to that he had sought for the triplets. He found immediately what we now call the scalar or dot product and the vector or cross product of two directed lines in space.[24] He also returned to earlier speculations about a fourth dimension and concluded that the **i**, **j**, **k**, terms of the quaternion probably represent the three dimensions of space while the real term represents time.[25] He also considered a "semi-metaphysical" interpretation of quaternions, which he described to John Graves: "There seems to me to be something analogous to *polarized intensity* in the pure imaginary part; and to *unpolarized energy* (indifferent to direction) in the real part of a quaternion: and thus we have some slight glimpse of a future Calculus of Polarities."[26] These speculations are similar to some of Hamilton's earlier ideas regarding the philosophical triads. Those also had a polar character as he had described them to Graves. Needless to say Hamilton would have liked to have joined his new algebra to the philosophical triads if that could have been possible, but all his metaphysical interpretations of quaternions, their polar character, and their connections to a fourth dimension have the flavor of explanations after the fact. The actual path that he followed towards his discovery on that October day was algebraical. And yet his metaphysical speculations had given him insights that were not so obvious to De Morgan and Graves. It was algebra as the science of pure time that had put him on the right track, and had convinced him that triplets and even polyplets of higher orders were possible. It was also algebra as the science of pure time that allowed him to dispense with the commutative law and construct elements that had different properties from those of ordinary arithmetic.

The philosophical triads of 1842 serve to indicate the extremely metaphysical character of Hamilton's quest for the triplets. Both he and Graves combined metaphysical and mathematical concepts in an extraordinary way in their correspondence. Hamilton was convinced that the operations of the mind, including mathematical operations, are in perfect harmony with the operations of the physical world as God created it. To understand the mind is to understand the world, and vice versa. The triplets and triads were two aspects of the same harmony between thought and nature, and it was his faith in this harmony that had compelled Hamilton to continue his search for triplets over so many years.

23

The Fate of Quaternions

THE QUATERNIONS HAD not been quite the revelation that Hamilton had expected, and therefore it was not immediately apparent what their significance would be. They followed most, but not all, of the rules of ordinary algebra. Therefore there remained some doubt as to whether quaternion algebra was consistent. Since quaternions contained four elements rather than three, it was not obvious how they could be applied to the analysis of three-dimensional space. The geometrical properties of quaternions remained to be discovered. Also, the quaternions suggested the possibility of other algebras of even higher order. Could such algebras be created, and were the triplets for which Hamilton had searched so long still a possibility?

All of these questions led to a flurry of activity among British mathematicians, who were now prepared to forsake the Principle of Permanence of Form and follow Hamilton in the search for new algebras.[1] Hamilton, however, stayed with the quaternions. His original intuition included a conviction that the quaternions were likely to be the most fruitful of all possible algebras beyond the number pairs, because he suspected that other algebras would require the sacrifice of more rules of algebra than just the commutative law. This intuition proved to be correct. In fact, the failures of other algebras helped to define more precisely just what the fundamental rules of ordinary algebra were.

Over the next ten years Hamilton produced an enormous number of long quaternion papers, usually in series, such as "On Quaternions: Or on a New System of Imaginaries in Algebra," which appeared in the *Philosophical Magazine* in eighteen installments from 1844 to 1850, and "On Symbolical Geometry," which appeared in the *Cambridge and Dublin Mathematical Journal* in ten installments from 1846 to 1849.[2]

Both of these series ended with the note "To be continued," but there were in fact no continuations. In 1848 Hamilton had given a series of lectures at Trinity College on his new algebra, and during the next five years he published less in journals, but worked to expand his lectures into an enormous book, the *Lectures on Quaternions,* which appeared in 1853. The *Lectures* proved to be too cumbersome for the ordinary reader, and at the urging of other mathematicians he began in 1858 a simple introductory manual of quaternions. But, like the *Lectures* and the series of papers that preceded it, the manual refused to keep within reasonable bounds. Hamilton's early resolve to write a brief treatise collapsed, and the manual expanded to become the gigantic *Elements of Quaternions,* which was even longer than the *Lectures* and was unfinished at the time of his death in 1865.

In one sense Hamilton's vision was wrong. The quaternions never gained the importance that he expected, nor did they become the mathematical key to the universe that his metaphysics had anticipated. His long devotion to their study is seen by some historians as a tragic waste of time and talent. But while the quaternions themselves did not remain long at the forefront of mathematics, they did open the way to several fruitful areas of study. Most immediately and most obviously, they led to a study of other systems of hypercomplex numbers. Secondly, geometrical properties of quaternions led to modern vector analysis, and thirdly (and somewhat less directly), they led to the more general concept of a linear vector space.

Within two months of Hamilton's discovery of quaternions John Graves found that hypercomplex numbers composed of eight elements, which he called *octaves,* also satisfied the law of the modulus. He described the law for their multiplication in a letter to Hamilton dated December 26, 1843, which was followed by other letters describing the algebra in greater detail.[3] Hamilton offered to make public Graves's new algebra, an offer that Graves accepted in a letter of January 18, 1844. Unfortunately, Hamilton did not act right away. Involved in his own researches on quaternions, he put Graves's letter aside and did not study its contents until the following July. At the time he found a serious weakness in the octaves. They were not associative: "In general, in my system of quaternions (containing only three imaginaries) it is indifferent where we place the points, in any successive multiplication: $A \cdot BC = AB \cdot C = ABC$, if A, B, C be quaternions; but not so, generally, with your octaves. Perhaps you may alter your binary products so as to get over this difficulty; but I suspect that then you will have to give up the law of the moduli."[4] Since this is the first clear statement of the associative law and apparently the first realization that an algebra might *not* enjoy this important property, it is in its own right an important moment in the history of mathematics.[5]

Hamilton took a holiday in late July to the Lake Country, where he visited Wordsworth and stayed with John Graves's brother Robert. On returning to the observátory he wrote to Robert that he had "not yet found a good open for mentioning John."[6] Then he became ill and had to miss the meeting of the British Association at York—another opportunity for advertising the octaves had passed by.[7] Hamilton plunged back into quaternions, and the octaves were again set aside. In the March, 1845, issue of the *Philosophical Magazine* Arthur Cayley, who had been reading Hamilton's papers, ended a paper of his own with a postscript describing an algebra essentially identical to Graves's octaves.[8] Graves reacted immediately and wrote to Hamilton, "I find that Cayley is near my octaves. I must publish my extension of Euler and Lagrange's theorems of products of squares."[9] In the following number of the same journal he added a postscript to one of *his* papers on number couples claiming that he had known of the octaves since Christmas, 1843—but it was too late.[10] Graves's octaves became known instead as "Cayley Numbers," and Hamilton could only proffer his apologies for not having acted more quickly.

The law of the modulus held for octaves, because, as Graves had shown, any product of two sums of eight squares is itself the sum of eight squares. He also attempted in January, 1844, to prove a similar theorem for the sum of sixteen squares, but met with an "unexpected hitch."[11] In the summer of 1847 Hamilton began a correspondence with J. R. Young, professor of mathematics at the Belfast Institution. Young had arrived at the eight-square theorem independently from Graves, and also believed at first that he had extended the theorem to the sums of any 2^n squares. In July, however, he found that there was no way around Graves's "hitch," and went on to prove the impossibility of the sixteen-square theorem.[12] Hamilton arranged for the publication of Young's paper in the *Transactions* of the Royal Irish Academy and took the occasion to make a belated claim for John Graves's priority in the discovery of the eight-square theorem.[13]

While these attempts to create algebras of eight and sixteen elements were being made, the search for triplets continued, not by Hamilton, who believed that in quaternions he had discovered what he had been looking for, but by Augustus De Morgan and by John and Charles Graves. Of course Hamilton had already expended an enormous amount of time and energy searching for triplets, and it is not surprising that he believed his quaternions to be a more fruitful line of inquiry. De Morgan and the Graves brothers, on the other hand, saw a new opportunity as a result of Hamilton's discovery of quaternions. They realized for the first time that a mathematician is free to *invent* algebraic symbols and to *define* permissible operations between them.

De Morgan, who had first contacted Hamilton in 1841 to ask about triplets, redoubled his efforts, and in October, 1844, almost exactly one year after Hamilton came upon quaternions, he sent Hamilton an abstract of a paper on triplet systems. He began his abstract with the following acknowledgement, which he repeated in the finished paper: "These imaginaries [of quaternions] are not deductions, but inventions: *their laws of action on each other are assigned.* This idea Mr. De Morgan desires to acknowledge as entirely borrowed from Sir William Hamilton." [14] De Morgan wanted the reader to understand that his triplet systems were entirely different from Hamilton's quaternions, but he also acknowledged that the *idea* of constructing a new algebra came from the quaternions. Hamilton, who had been following just such an idea all along, could not share De Morgan's enthusiasm over the triplets. In all of his early attempts to create triplets he had found cases in which the product of two finite triplets is zero and the quotient of two triplets is indeterminate. [15] Neither of these conditions could he accept. He responded to De Morgan's letter and abstract by saying:

It will surprise me, I confess, if either your theory, or any other person's, of *pure triplets* shall be found to surpass that which I have been led to perceive, as *included* in my theory of quaternions on all, or most, of the three following points:

1st. *Algebraic simplicity*; ... analogy to ordinary algebra, as to the rules of addition and multiplication (the commutative property excepted);
2nd. *Geometrical simplicity*; ... ease of construction; the rule of the diagonal; and, above all, *symmetricity of space*, no one direction being eminent;
3rd. *Determinateness of division*; ... a quotient being never indeterminate or impossible unless the constituents of the divisor all vanish.

Of all these assumed requisites, or things aimed at by me (and I admit that I aimed at others), what *now* appears to me most my own is the SYMMETRICALNESS OF SPACE in my system. If *you* have succeeded in representing this with pure triplets, *eris mihi magnus Apollo.* My *real* is the representative of a sort of *fourth* dimension, inclined equally to all lines in space. [16]

De Morgan's systems (he constructed five) [17] explicitly rejected the requirement of symmetry, and in his paper he wrote, "Sir William Hamilton seems to have passed over triple algebra altogether," and gave the reason as being because Hamilton could not discover a symmetrical modulus. [18] De Morgan's systems also failed to follow the associative law. [19] Of course it is not obvious that a symmetrical modulus of multiplication or the associative law should be more sacred than the commutative law, which Hamilton sacrificed and De Morgan preserved. Hamilton had led the way in departing from the rules of algebra and he could hardly condemn De Morgan for doing the same thing, but his instinct told him that quaternions would lead to greater things than De Morgan's imperfect triplets.

The discovery of quaternions also revived John Graves's enthusiasm for triplets, and he worked on them at the same time that he was developing his system of octaves. In November, 1843, he was again sending Hamilton new formulas for the multiplication of triplets.[20] There was still a possibility that the triplets might amount to something, and Hamilton and Graves were anxious to sort out the priority of their ideas. The matter of priority took on a more serious aspect when James MacCullagh, at a meeting of the Royal Irish Academy, announced that he had anticipated the quaternions in a theorem relative to the ellipse that he had placed on the Junior Fellowship examination in 1842.[21] Hamilton felt himself forced to recall in the minutest detail all the circumstances that might have led to the quaternions, and in particular any hint that he might have received from MacCullagh.[22]

It was an unfortunate time for MacCullagh to bring forth again one of his perennial claims to priority. After a visit to the observatory in October John Graves wrote to his brother Robert that Hamilton's behavior was beginning to exhibit "the morbid activity of brain resulting from overwork" and a "certain nervous irritability of temperament," and went on to say: "It is exceedingly unfortunate that he should lately have been stimulated by other workers in the same field [De Morgan], and worse still, annoyed by unfounded claims [by MacCullagh] to the credit of *suggesting* what is peculiarly his own."[23] Into this sticky situation the third Graves brother, Charles, entered as peacemaker. Charles had been awarded the professorship of mathematics that Hamilton had clearly desired the previous year when a change of professorships had been suggested to the Board of Trinity College, and therefore some lingering shade of resentment probably existed between him and Hamilton. But Charles Graves was successful in persuading MacCullagh to temper his claims, and the potential quarrel was avoided.[24]

Through his discussions with MacCullagh, however, Charles Graves himself became infected with the triplet mania and entered into competition against his brother John and De Morgan. On November 21, 1844, he announced in excited letters to John and to Hamilton that he had at last found the "long sought triplets." He had gotten out of bed at 3:00 A.M. to write them down, and that very day had deposited a sealed note at the Royal Irish Academy to insure his priority.[25] The next day, before he had received Charles's letter, John wrote to Hamilton reassuring him that MacCullagh's claims had no foundation, but also warning him that be believed De Morgan was using Hamilton's ideas without giving proper credit.[26] Confronted with this intense rivalry Hamilton withdrew from the triplet race, telling all involved that it was too late for him to have any claim in the matter.[27] It was not a particularly difficult decision, because he remained convinced that his quaternions would be far more fruitful than any triplet system that might be devised.

For three weeks Charles preserved "a most provoking tranquillity," while John tried desperately to guess what the new discovery might be.[28] When the "veritable triplets" were at last revealed at a general meeting of the Royal Irish Academy the result was one of general disappointment. The new system closely resembled those of John Graves and De Morgan, and differed primarily in the mode of presentation. John Graves's triplets were expressed geometrically, while Charles's were algebraic, but they came to much the same thing.[29] This time it was Hamilton's turn to act as peacemaker, and, speaking from the president's chair at the academy, he gave "testimony to the remarkable fact of two brothers, in two different capitols, having arrived almost at the same moment at the same important conception."[30] While attempting to ease the disappointment of the Graves brothers, Hamilton was also anxious to protect the claims of De Morgan, who had described a similar system in his paper on triple algebra. What might have become a warm controversy rapidly cooled down when it was recognized that the "veritable triplets" of Graves were not so remarkable after all.[31]

The failure of the triplets and octaves to measure up to the quaternions in mathematical importance lessened Hamilton's competition, but did not automatically bring support to the quaternions. Many doubted that the new algebra would have any value. When he was finishing his *Lectures* in 1852, Hamilton admitted to De Morgan that many of the Fellows at Trinity had ridiculed him about his invention, and to his old friend Mortimer O'Sullivan he described the publication of his quaternion views as an "ordeal through which I had to pass, an episode in the battle of life, to know that even candid and friendly people secretly, or, as it might happen, openly, censured or ridiculed me, for what appeared to them my monstrous innovation."[32] There seems to be little evidence of this opposition, and one suspects that the ridicule of which Hamilton speaks was largely in his own imagination. It is true that the publication of the quaternions took courage, particularly since he knew that his metaphysical views were unpopular with other British mathematicians, and it took awhile for mathematicians to understand the implications of the quaternion theory.

Certainly Hamilton did not present his ideas in the clearest possible light. But the only serious opposition came in 1847 at the Oxford meeting of the British Association. Hamilton's paper was a practical one entitled "On Some Applications of the Calculus of Quaternions to the Theory of the Moon," in which he tried to show the advantages of quaternions over coordinates for celestrial mechanics. The *Athenaeum* reported that at the meeting John Herschel praised the quaternions as a "perfect cornucopia, which, turn it on which side you will, something rich and valuable was sure to drop out."[33] What the reporter did not describe was the opposition. When Hamilton rose to give his paper he found himself addressing, among others, Struve, LeVerrier, Herschel, Airy, Adams, Challis, Peacock, and Whewell—a formidable array of mathematical and astronomical talent. Fortunately most of them seemed to be on his side. The first few objectors were minor figures at the meeting who obviously did not understand what Hamilton was talking about. But then Airy, recognizing that the opposition was not presenting its case very well, added his objections. Hamilton described the incident in a letter to Robert Graves:

Mr. Airy, seeing that the subject could not be cushioned, rose then to speak of his own acquaintance with it, which he avowed to be none at all; but gave us to understand that what he did not know could not be worth knowing. He warned all persons, if they should use the method, to do so with the extremest caution; professing to regard me as believing it to be a right one, solely on the ground of the agreement of its results, so far as they have been yet obtained, with those of the older method. What was obscure was to him as if it were erroneous, what was paradoxical was to him as if it were false; and he thought *that* system useless as an algebraic geometry, of which the expressions were so extremely difficult of geometrical interpretation. [34]

In his response to Airy, Hamilton argued that the quaternions were "eminently interpretable" in geometry, and attempted to resume his demonstration, but we can presume that his audience had little success in following him. Peacock spoke favorably of the quaternions in private, and that counted for much. [35]

Airy's complaint about the difficulty of giving the quaternions a geometrical interpretation had its merit. Hamilton himself was not sure what the real part of the quaternion represented in physical terms. It was an "extra-spatial unit," but that name did not define its meaning, and it is possible to see in Airy's objections an anticipation of the future battle between the quaternionists and the vector analysts. Quaternions had more satisfactory algebraic properties than vectors but less satisfactory geometrical properties, which gave Airy grounds for criticism.

To Airy's objection that the quaternion method had not been proved beyond the fact that it usually gave the correct answer, Hamilton replied:

As to my own personal conviction of the correctness of my method, it was doubtless a psychological fact that this conviction had strengthened with practice, and with the satisfactory result of comparison with other methods; but I asserted that its logical ground was the *a priori* examination of principles: and could not consent to let it be supposed that there remained the slightest misgiving in my own mind respecting the soundness of the method of demonstration, any more than there could now be doubt of the value of the method of research. [36]

Of course a psychological conviction, or a derivation from metaphysical first principles, could not prove the consistency of quaternion algebra; and Airy had some grounds for urging caution. On the other hand, he should have tempered his remarks with a bit more tact, considering the obvious sensitivity of Hamilton to criticism of his new invention.

From the moment of his discovery of quaternions the problem of their geometrical representation, which disturbed Airy, had also disturbed Hamilton. He had come to the quaternions through a research for algebraic triplets, but he had always hoped that the algebraic triplets would also provide a three-dimensional corollary to the geometry of the complex plane. He described the geometrical problem of triplets as one of finding a fourth proportional to three lines in space. In the case of complex numbers the line of unit length on the real axis is to the multiplier line as the multiplicand line is to the product line. This proportion holds for the lengths of the lines and also for the angles of inclination of the lines to the real axis.[37] Because there was no general definition for the multiplication of lines in space, it was necessary to invent one. The concept of proportion seemed to be a reasonable approach, because it emphasized geometrical relationships over numerical ones.[38]

Once Hamilton came up with quaternions instead of triplets, however, he saw the problem of proportion in a somewhat different light. Since by analogy to complex numbers the three imaginary parts of the quaternion, **i**, **j**, **k**, could most reasonably represent three mutually perpendicular lines in space, Hamilton now asked how the fourth and real part of the quaternion could be represented as a fourth proportional to the **i**s, **j**s, **k**s. He concluded that this fourth proportional is a line to the extent that one can move along it forwards and backwards, but that it is a line in space of only one dimension:

If we then resolve to retain the assumption of the existence of a fourth proportional ... to three rectangular directions in space, as subject to be reasoned on ... and as determined in direction by its contrast to its own opposite ... we must think of these two opposite directions ... as merely *laid down upon a scale,* but must abstain from attributing to this SCALE any one direction rather than another in tridimensional space ... and the progression *on this scale* ... corresponds less to the conception of *space* itself (though we have seen that considerations of space might have suggested it) than to the conception of *time;* the variety which it admits is not *tri-* but *uni-*dimensional; and it would, in the language of some philosophical systems, be said to appertain rather to the notion of *intensive* than of *extensive* magnitude.[39]

Here is the origin of the term *scalar.* For Hamilton it is a line, but not a line in three-dimensional space. Therefore all directions in space are related to it in the same way, and the quaternion calculus *"selects no one direction in space as eminent* above another, but treats them as all equally related to that *extra-spatial,* or simply SCALAR direction."[40] In this remark Hamilton recognized the major geometrical significance of quaternions—they provide a form of analytical geometry which is coordinate-free. Hamilton was also the first mathematician to use the word *vector* commonly to denote any directed line segment. The expression *radius vector* had been part of the language of mathematics since the early part of the eighteenth century, but Hamilton was the first to give the word *vector* its more general meaning.

The identification of a scalar part and a vector part in the quaternion creates new problems, however, because it is not permitted in geometry to add quantities of different dimensions. The sum of a line and a number is meaningless, and therefore it is not clear what the formula for a quaternion (a + bi + cj + dk) could mean. Hamilton had devised his number couples in response to the same problem as it occurred in complex numbers. A complex number is written as the sum of a real number and an imaginary number. This is again the sum of two unlike things. The number couples removed the necessity for thinking of a complex number as a sum, but how was one to accomplish this revision for quaternions? The apparent impossibility of solving this problem led Hamilton to reconsider the metaphysical foundation of quaternions. In 1846 he suggested that a quaternion might better be called a *grammarithm*, "denoting *partly a number,* and *partly a line,* which two parts are to be conceived as quite *distinct* in kind from each other, although they are *symbolically added,* that is altho' their symbols are written with the sign + interposed." [41]

As he thought more about the problem he concluded that a quaternion is a kind of "symbolical addition" of a scalar and a vector and therefore a synthesis of number and line, and that it could be better understood as an operator than as a number. [42] But the inability of Hamilton to give geometrical *meaning* to the quaternion (or grammarithm) forced him to seek a "symbolical" meaning for his new creation, and this in turn required a partial capitulation to Peacock and the other British advocates of symbolical algebra. In April, 1846, he wrote about his conversion to Robert Graves:

To some extent I have become a convert to the views of those authors [Peacock, Ohm, and Gregory], so far as to admit that there is a sort of symbolical science, or *science of language,* which well deserves to be studied, abstraction being made for awhile of *meaning,* or interpretation; and *forms of expression* being treated themselves as the subject-matter to be studied: in short, I feel an increased sympathy with, and fancy that I better understand that *Philological School,* which was referred to in the introduction to my essay on *Algebra as the Science of Pure Time.* Thus without having renounced my old view . . . I seem to have gained not only a power of reading with increased intelligence and pleasure the works of Peacock, Ohm, and Gregory, but also the possession of a point of view very essential to the metaphysical development of my own intellectual being: because it enables me to see better than before the high functions of language, to trace more distinctly and more generally the influence of signs over thoughts. [43]

In writing to Peacock, Hamilton said that his own views had grown gradually to approximate those of Peacock, but that he was not a total convert, because he still tended to look more habitually beyond the symbols to the things signified than to the symbols themselves. [44]

The product of these reflections was a series of papers in the *Cambridge and Dublin Mathematical Journal,* beginning in 1846 and continuing through 1849, entitled "On Symbolical Geometry." Symbolical geometry, wrote Hamilton, is "analogous in several important respects to what is known as symbolical algebra, but not identical therewith, since it starts from other suggestions and employs, in many cases, other rules of combination of symbols." [45] In the first paper of this series Hamilton pays his compliments to Peacock most generously, but still argues that those aspects of symbolical algebra that he has adopted do not conflict with his older views. [46] These papers by Hamilton are obscurely written and indicate the difficulty that he was having in giving even a symbolical geometrical meaning to his quaternions. In "On Symbolical Geometry," as in the later *Lectures,* a quaternion is defined as a quotient of two lines. He concludes that the quotient of two parallel lines must be a scalar, while the quotient of two perpendicular lines must be a vector. These results require that in general the quotient of any two lines will be the sum of a scalar and a vector. If line *b* is to be divided by line *a* we can always write *b* as the sum of two components, one parallel to the divisor line *a* and the other perpendicular to *a* (figure 23.1). Then $b \div a$ is $(b_{\parallel} + b_{\perp}) \div a$, and, assuming a distributive law for division, we get $(b_{\parallel} \div a) + (b_{\perp} \div a)$, where $(b_{\parallel} \div a)$ is a scalar and $(b_{\perp} \div a)$ is a vector.

In the *Lectures* Hamilton explained much more clearly why the quotient or product of two parallel lines must be a scalar, while the quotient or product of two perpendicular lines must be a vector. It was an argument that he applied to the multiplication of lines, but it could equally well be applied to division. He called the argument a "speculation, of a character partly geometrical, but partly also metaphysical (or *a priori*)" and said that it had occurred to him shortly after the

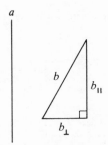

Fig. 23.1. The division of vectors.

discovery of quaternions.[47] He began with the following conditions concerning the multiplication of lines:

a. The direction and magnitude of the product must be determined unambiguously by the two factor lines.
b. The direction and sign of the product line is reversed when *one* of the factor lines is reversed.
c. The relationship of the product and the factor lines must remain the same, independent of any orientation in space. Thus the space is symmetrical and coordinate-free.
d. The distributive law holds for the multiplication of vectors, which may be represented as the sums of components.

First Hamilton asks what should be the product of two parallel vectors α and β. Assume that the product $\alpha\beta$ is a vector parallel to α and β and in the same direction (Figure 23.2a). Then if α and β are both made negative, we have by condition (b) the arrangement shown in figure 23.2b. But this result contradicts condition (c), because vectors α and β could be reoriented in space to coincide with $-\alpha$ and $-\beta$, and the product line $\alpha\beta$ should follow with them.

Assume instead that the product $\alpha\beta$ is parallel to α and β but in the opposite direction (figure 23.3a). Again make α and β both negative and we see that condition (c) is again violated (figure 23.3b).

In the final case assume that the product is a vector inclined at any angle to the parallel factor lines (figure 23.4). But this case is indeterminate. Even if the angle of inclination is given, the collinear factor lines cannot determine the direction of the product in the three-dimensional space, and condition (a) is violated.

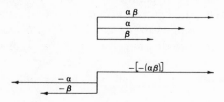

Fig. 23.2. Two collinear vectors with a product vector in the same direction.

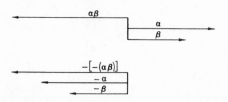

Fig. 23.3. Two collinear vectors with a product vector in the opposite direction.

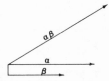

Fig. 23.4. Two collinear vectors with a product vector inclined to them at any angle.

Since these are all the possible cases of vector products of collinear factors, we can conclude that the product of two collinear vectors cannot be a vector; therefore it must be a scalar.

We next consider the product of two mutually perpendicular vectors. Can this product also be a scalar? Not easily, because our conditions suggest contradictory conclusions. Condition (c) suggests that making one factor negative does not change the sign of the product because the arrangements of vectors in figure 23.5 may be reoriented in space to make them coincide. The reversal of one of the vectors does not change the *relative* orientation of the two vectors, and therefore the scalar product should remain unchanged. Condition (b), however, suggests that the change of sign of one of the factors should change the sign of the product. The assumption of a scalar product of two mutually perpendicular vectors seems to imply a contradiction.

But if the product is a vector at right angles to the two factor vectors, then the contradiction is removed. The product is unambiguously determined by the factors (condition [a]); the direction of the product line is reversed when one of the factors is reversed (condition [b]); and the relationship between the factors and product remain the same for all orientations in space (condition [c]). The three configurations in figure 23.6 may be made to coincide. Therefore we may conclude that the product of two mutually perpendicular vectors is a third vector that is perpendicular to them both.[48] There is a problem, however, with the multiplication of perpendicular vectors. Consider two mutually perpendicular *unit* vectors **i** and **j**. Their product is a vector **k** perpendicular to them both (figure 23.7). If we reorient this set of vectors so that **i** and **j** exchange places, the vector **k** will be pointing in the opposite direction. If the product **ij** in the second diagram is a vector pointing downward, then the product **ji** in the first diagram must also be a vector pointing downward, because the relationship of **j** to **i** in the first diagram is exactly the same as the relationship of **i** to **j** in the second diagram. But the product in the first diagram is pointing upward! The only solution is to conclude that **ij** does not equal **ji**. In fact, **ij** $= -($**ji**$)$, because the product of **ij** is exactly opposite to the product **ji**. The factors do not commute.

By similar arguments Hamilton was able to show that the associative

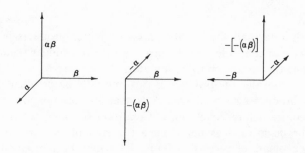

Fig. 23.5. All of these pairs of mutually perpendicular vectors can be made to coincide.

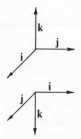

Fig. 23.6. These sets of three mutually perpendicular vectors can be made to coincide.

Fig. 23.7. The multiplication of two mutually perpendicular unit vectors cannot be commutative.

law does hold for the multiplication of vectors, and that the scalar product of two collinear vectors with the same direction should be negative. Once he had demonstrated that the product of collinear vectors must be a scalar and the product of mutually perpendicular vectors must be a vector, he could then apply the distributive law (condition [d]), as he had used it in the papers on symbolical geometry, to conclude that the product of any two vectors is the "symbolical sum" of a scalar and a vector.

Another impulse to a symbolical interpretation of quaternions was Hamilton's discovery of biquaternions in 1844. Biquaternions are quaternions with complex coefficients, and they come up inevitably as roots of real quaternion equations. Hamilton had no satisfactory geometrical interpretation for the "imaginary" parts of these complex coefficients. He had removed them from complex numbers by the introduction of number couples, but they remained imaginary in quaternions and forced him further into a symbolical representation of quaternions. [49]

By such "metaphysical" arguments Hamilton was able to show that any reasonable definition of a product of vectors would be something like the quaternions, and he was able to do this quite independently of the *algebraic* properties of quaternions. Pursuing this beginning he chose to develop quaternions in his *Lectures* from a geometrical point of view.

Hamilton's symbolical geometry, while it provided helpful insights into the nature of quaternions, did not provide an altogether satisfactory geometry of three-dimensional space. In the struggle to obtain such a geometry the quaternions were eventually replaced by pure vectors, but not without bitter protests from Hamilton's quaternionist disciples. Vectors had obvious advantages to the physicist who was looking for the simplest representation of lines in space, but pure vectors in space would have to be represented by triplets, and the search for triplets had failed, as we have seen. Failure for the mathematician, however, may not necessarily mean failure to the physicist, and in vector analysis triplets are multiplied, whatever the consequences. The consequences are an unsatisfactory algebra, which nevertheless turns out to be convenient for the representation of physical phenomena.

If one multiplies only the vector portions of two quaternions, as Hamilton did early in his development of the theory, one obtains a quaternion composed of both a scalar part and a vector part.

$$(0 + x\mathbf{i} + y\mathbf{j} + z\mathbf{k})\,(0 + x'\mathbf{i} + y'\mathbf{j} + z'\mathbf{k}) = -(xx' + yy' + zz')$$
$$+ \mathbf{i}(yz' - zy') + \mathbf{j}(zx' - xz') + \mathbf{k}(xy' - yx').$$

Hamilton denoted the scalar part of the product of two quaternions α and α' as S.$\alpha\alpha'$ and the vector part as V.$\alpha\alpha'$. The scalar part of this product is immediately recognizable as the negative of the dot product in vector analysis, and the vector part is identical to the cross product. The two products employed in vector analysis are therefore immediately recognizable in Hamilton's early work. Even the so-called "del" operator appears early in Hamilton's writings. In 1846 he was already using the expressions

$$\triangleleft = \mathbf{i}\frac{d}{dx} + \mathbf{j}\frac{d}{dy} + \mathbf{k}\frac{d}{dz}$$

and

$$-\triangleleft^2 = \left(\frac{d}{dx}\right)^2 + \left(\frac{d}{dy}\right)^2 + \left(\frac{d}{dz}\right)^2$$

and remarking that "applications to analytical physics must be extensive to a high degree."[50] One wonders, then, why Hamilton did not exploit vector analyis immediately. The answer must be that he had no need for the simplified version of quaternions that vector analysis provided. He could manage the full panoply of quaternion operations with ease; and while lesser minds quailed before their complexity, he felt no reason to sacrifice their beautiful algebraic properties for what P. G. Tait called the "hermaphrodite monster" of vector analysis.[51]

The multiplication of quaternions obeys all the rules of ordinary algebra except the commutative law. The multiplication of vectors, on the other hand, fails miserably on many counts. In the first place there are *two kinds* of multiplication in vector analysis, a condition that would have distressed Hamilton. The dot product of two vectors is not even a vector. Therefore the law of closure fails. The law of the moduli also fails, and therefore division is ambiguous. The cross product of two vectors is a vector, and therefore the property of closure is maintained, but it does not satisfy either the commutative or the associative law, and the law of the moduli fails as well. Also, both products can be zero even when the factors are both finite, a condition that Hamilton especially wished to avoid.[52] It is not surprising, then, that he did not develop vector analysis, although he certainly recognized many of its important properties. Vector analysis proved its worth in physics, and since Hamilton always postponed investigating the applications of quaternions while he worked on the algebra, he was unlikely to develop vector analysis as a practical method.

The inventors of vector analysis were Josiah Willard Gibbs at Yale and Oliver Heaviside in England. Both of these men were led to develop their systems by reading James Clerk Maxwell's *Treatise on Electricity and Magnetism,* published in 1873, eight years after Hamilton's death. Maxwell had learned of quaternions from his classmate and close friend P. G. Tait, who had been Hamilton's chief disciple. So there is a strong intellectual connection between Hamilton's quaternions and the vector analysis of Gibbs and Heaviside.[53]

Maxwell first supported quaternions publicly in a paper called "On the Mathematical Classification of Physical Quantities" (1871), in which he argued that Hamilton's distinction between scalar and vector quantities was of enormous importance for physics. He went on to say that "the invention of the calculus of Quaternions is a step towards the knowledge of quantities related to space which can only be compared, for its importance, with the invention of the triple coordinates by Descartes."[54] His argument is that quite different physical phenomena can have the same or similar mathematical expressions, and that if we pay close attention to these mathematical forms we can reach a much better understanding of the relations between physical phenomena. Hamilton's theory of the characteristic function, which brought together optics and mechanics in the same formal mathematical expression, is a good example of what Maxwell was talking about. (Maxwell's own illustration was William Thomson's analogy between electrical attraction and the steady conduction of heat.)[55] Maxwell then went on to discuss Hamilton's $\lhd$ operator (which Maxwell wrote as ∇) and coin the names "convergence" for the operation that Hamilton wrote as $S\lhd$ and "curl" for the operation that Hamilton wrote as $V\lhd$.[56] Tait had been urging quaternions on Maxwell ever since 1867, arguing that "they go into that $\lhd$ business like greased lightning," and Maxwell, when he went from Aberdeen to Cambridge in 1871, promised to "sow 4nion seed at Cambridge."[57]

Maxwell saw the advantage of quaternions not in their use as a method of calculation, but in their representation of physical phenomena in space: "Now quaternions, or the doctrine of Vectors, is a mathematical method, but it is a method of *thinking* [my emphasis], and not, at least for the present generation, a method of saving thought. ... It calls upon us at every step to form a mental image of the geometrical features represented by the symbols, so that in studying geometry by this method, we have our minds engaged with geometrical ideas, and are not permitted to fancy ourselves geometers when we are only arithmeticians."[58] According to Maxwell, quaternions allow one to express the physics of electricity and magnetism much more directly than is possible with coordinates, so that the mathematics reveals more clearly the nature of the phenomena. This was an important advance, but not one that he felt he could readily exploit in his treatise, because quaternions were so unfamiliar to the average reader. Maxwell's solution was to present his main results in both coordinates and quaternions.

This "bilingual" approach, as he called it, caused several difficulties. Most annoying was the fact that the square of a vector was negative, while in coordinates it seemed that it should be positive. Kinetic energy, for instance, was negative in quaternions. The two methods did not seem to work together well, and Maxwell asked in characteristic language whether one could "plough with an ox and an ass together."[59] Both methods led to the same results, but when used together they produced awkward discrepancies of sign.

In a draft letter written in 1888, Gibbs explained how reading Maxwell's *Treatise on Electricity and Magnetism* led him to devise his system of vector analysis:

My first acquaintance with quaternions was in reading Maxwell's E.&M. [*Electricity and Magnetism*] where Quaternion notations are considerably used. I became convinced that to master those subjects, it was necessary for me to commence by mastering those methods. At the same time I saw, that although the methods were called quaternionic the idea of the quaternion was quite foreign to the subject. In regard to the products of vectors, I saw that there were two important functions (or products) called the vector part & the scalar part of the product, but that the union of the two to form what was called the (whole) product did not advance the theory as an instrument of geom. investigation. Again with respect to the operator ∇ as applied to a vector I saw that the vector part & the scalar part of the result represented important operations, but their union (generally to be separated afterwards) did not seem a valuable idea. This is indeed only a repetition of my first observation, since the operator is defined by means of the multiplication of vectors, & a change in the idea of that multiplication would involve the change in the use to the operator ∇.

I therefore began to work out ab initio, the algebra of the two kinds of multiplication, the three differential operations ∇ applied to a scalar, & the two operations to a vector, & those functions or rather integrating operators wh [which] (under certain limitations) are the inverse of the said differential operators, & wh play the leading roles in many departments of Math. Phys. To these subjects was added that of lin. vec. functions wh is also prominent in Maxwell's E. & M.[60]

Gibbs recognized that the full quaternion algebra was not necessary for the theory of electricity and magnetism. Only the vector portion of the quaternion was involved, and the vector and scalar parts of the product of two vectors were usually employed as independent operations. Thus Gibbs limited his new algebra to vectors, and defined the scalar and vector parts of the quaternion product as two different products defining two different kinds of multiplication. He also made the scalar product of two positive vectors positive, thereby removing the embarrassing minus sign. In the 1888 letter just quoted Gibbs also said that he came to his vector analysis from "some knowledge of Ham's methods," but it is clear that Maxwell's *Treatise on Electricity and Magnetism* and Tait's *Treatise on Quaternions* were the real inspirations for his ideas. These two works deal much more with physical applications of quaternions than did either Hamilton's *Lectures* or his *Elements*, and it was in physical applications that the advantages of pure vectors and the separation of the scalar and vector products were most apparent.

Gibbs had his *Elements of Vector Analysis* printed privately, the first half in 1881 and the second half in 1884, and circulated it to a large number of prominent scientists on the Continent. It was not formally published until 1901, in an expanded and revised form by Gibbs's student Edwin Bidwell Wilson, under the title *Vector Analysis: A Text Book for the Use of Students of Mathematics and Physics Founded upon the Lectures of J. Willard Gibbs*.

At the same time that Gibbs was circulating his treatise Oliver Heaviside was working out his own vector algebra, also under the inspiration of Maxwell's *Treatise on Electricity and Magnetism.* Heaviside first published his method in an 1882 and 1883 paper in the *Electrician* entitled "The Relations between Magnetic Force and Electric Current." Apparently he did not hear of Gibbs's vector analysis until 1888, and his initial work was entirely independent of Gibbs's. Heaviside gave the first detailed treatment of vector analysis in the first volume of his *Electromagnetic Theory,* published in 1893. [61]

By 1893 the battle between the quaternionists and the vector analysts was in full swing. It was really two battles, one of quaternions versus coordinates, and a second one of quaternions versus vectors. The first battle had begun as soon as quaternions appeared, but the second began only in 1890, after vector analysis became widely known. In 1890 Tait entered the controversy on both fronts. In an address to the Physical Society of the University of Edinburgh entitled "On the Importance of Quaternions in Physics" he extolled the "naturalness" of quaternions and claimed that they removed all the artificiality of coordinates and revealed the physical properties of space in the most obvious way. [62] He had already done battle with William Thomson over this very question of quaternions versus coordinates in their coauthorship of the *Treatise on Natural Philosophy,* and in 1894 he was still fighting with Arthur Cayley, who claimed that quaternions were "merely a particular method, or say a theory, in coordinates," Tait replied:

It will be gathered from what precedes that, in my opinion, the term Quaternions means one thing to Prof. Cayley and quite another thing to myself: Thus
To Prof. Cayley Quaternions are mainly a Calculus, a species of Analytical Geometry; and, as such, *essentially* made up of those coordinates which he regards as "the natural and appropriate basis of the science." ... To me Quaternions are primarily a Mode of Representation. ... They *are,* virtually, the thing represented: and are therefore antecedent to, and independent of, co-ordinates: giving, in general, all the main relations, in the problems to which they are applied, without the necessity of appealing to co-ordinates *at all.* [63]

Quaternions, according to Tait, freed the mathematical physicist from the artificial slavery of coordinates and allowed his thoughts to run in their most natural channels. In a marvelous display of Scottish industrialism and Victorian imperialism he compared coordinates

to a steam-hammer, which an expert may employ on any destructive or constructive work of *one general kind,* say the cracking of an egg-shell, or the welding of an anchor. But you must have your expert to manage it, for without him it is useless. He has to toil amid the heat, smoke, grime, grease, and perpetual din of the suffocating engine-room. The work has to be brought to the hammer, for it cannot usually be taken to its work. ... Quaternions, on the other hand, are like the elephant's trunk, ready at *any* moment for *anything*, be it to pick up a crumb or a field gun, to strangle a tiger or to uproot a tree. Portable in the extreme, applicable anywhere ... directed by a little native who requires no special skill or training, and who can be transferred from one elephant to another without much hesitation. Surely this, which adapts itself to its work, is the grander instrument! But then, *it* is the natural, the other the artificial, one. [64]

It can be assumed that Tait knew little about elephants or natives, but his feelings about the superiority of quaternions over coordinates are obvious enough.

In 1890 Tait also opened up a second front against the vector analysts. The occasion was the publication of the third edition of his *Treatise on Quaternions,* the preface of which contained the provocative statement that Gibbs would have to be ranked as "one of the retarders of Quaternion progress" because of his "hermaphrodite monster." Such a challenge could not go without response. Gibbs himself was the model of temperance. Not so Heaviside, who was prepared to respond to Tait in kind. Quaternions, said Heaviside, were generally believed to involve "metaphysical considerations of an abstruse nature, only to be thoroughly understood by consummately profound metaphysicomathematicians, such as Prof. Tait, for example." He admitted that he had himself been burdened by the quaternion approach until he recognized that it was not only unnecessary, but even a "positive evil of no inconsiderable magnitude," whereupon he threw off the quaternion yoke and saw vectors revealed in their true light. "Quaternions," wrote Heaviside, "furnishes a uniquely simple and natural way of treating *quaternions.* Observe the emphasis." [65]

The quaternion-vector controversy was carried on through 1894 in several journals, but most notably in *Nature.* The polemic of Tait and Heaviside brought forth responses from quaternionists all over the globe. One of the most vigorous was Alexander McAuley, a graduate of Cambridge, who became lecturer first at Ormond College in Melbourne, and then at the University of Tasmania. Also supporting the quaternion cause was Alexander Macfarlane, a former student of Tait who was teaching at the University of Texas. Macfarlane became the leading force in the International Association for Promoting the Study of Quaternions and Allied Systems of Mathematics, an association that had its origin in the combined efforts of a Japanese and a Dutch mathematician. From 1900 through 1923 the association published a *Bulletin,* and in 1904 Macfarlane, acting in his position as general secretary, published a voluminous *Bibliography of Quaternions and Allied Systems of Mathematics.* [66] In 1900 the membership of the association included over sixty scientists from fifteen countries, and it revealed the great enthusiasm for quaternions in the United States and the British Commonwealth countries. In 1875, when Hamilton's sister Sydney arrived in Auckland, her first letter back to Ireland said, "Who would imagine Sir George Grey in New Zealand studying Tait and the quaternions."[67] The Governor General had indeed been infected by quaternions, to the extent that he wished to meet Hamilton's sister immediately upon her arrival.

It is difficult to understand from our perspective why the controversy between the quaternionists and the vector analysts became so heated. Mathematicians were free to use whichever method they preferred, and the proof of the pudding would be in the eating, not in any theoretical discussion over the merits of one particular form. In any case, vector analysis came directly from quaternions, and it is hard to see how they could be regarded as rival methods. Vector analysis could best be regarded as "quaternions for the practical man." It provided a simpler and more direct way for approaching physical problems than could be provided by quaternions. It is surprising that Tait, who in so many ways was the epitome of the practical mathematical physicist, could not recognize the value of vector analysis. As the most direct disciple of Hamilton he may have felt responsible for preserving the "quaternion stream pure and undefiled," but in his letters to Hamilton he had repeatedly stated that he was interested only in the physical applications of quaternions and not in their purely mathematical aspects.[68] He was certainly not the "metaphysicomathematician" that Heaviside described. It must be that Heaviside read into Tait's works metaphysical predilections that he had seen in Hamilton's *Lectures*.

Hamilton would not have agreed with all aspects of Tait's defense. In particular Tait criticized a statement by Gibbs to the effect that quaternions were limited to a representation of three-dimensional space. Tait asked: "What have students of physics, as such, to do with space of more than three dimensions?"[69] Hamilton, who invented abstract phase space for his dynamics and who had sought diligently for "polyplets" of ever higher dimensions, would not have insisted on limiting vector spaces to three dimensions. If he had lived to witness the battle between Tait and Gibbs, he might well have sided with Gibbs. For the most part the battle of vectors versus quaternions was an argument without a point.

The extension of vectors to linear spaces with more than three dimensions was being explored by Hermann Günther Grassmann at just the time that Hamilton was developing quaternions.[70] Both Hamilton and Gibbs read Grassmann's work, although not until several years after it first appeared. Grassmann was a schoolteacher in Stettin who had no formal training in mathematics. In an early (1840) unpublished paper on the theory of tides one can already see the beginnings of a vector algebra, and in his *Ausdehnungslehre* of 1844 these ideas were developed in their greatest generality. [71] Vectorial methods appeared independently in several places during the first half of the nineteenth century. The works of August Ferdinand Möbius, Giusto Bellavitis, and Adhémar Barré (Comte de Saint-Venant) are other examples, but the contributions of Hamilton and Grassmann were by far the most important.

Like Hamilton's *Lectures on Quaternions,* Grassmann's book approached the algebra of space from a philosophical point of view. He regarded pure mathematics as "the particular existent which has come to be through thought." Mathematics was "the theory of forms," and Grassmann proceeded to investigate the properties of directed lines in space in the most general and abstract manner. [72] He believed that algebra had reached a much greater degree of abstraction than had geometry, and his book was an attempt to bring the same degree of abstraction to geometry:

It had for a long time been evident to me that geometry can in no way be viewed, like arithmetic or combination theory, as a branch of mathematics; instead geometry relates to something already given in nature, namely, space. I also had realized that there must be a branch of mathematics which yields in a purely abstract way laws similar to those in geometry, which appears bound to space. By means of the new analysis it appeared possible to form such a purely abstract branch of mathematics; indeed this new analysis, developed without the assumption of any principles established outside of its own domain and proceeding purely by abstraction, was itself this science. [73]

Like Hamilton's *Lectures,* the book was unreadable except to the most intrepid mathematicians. A second *Ausdehnungslehre* that was much expanded from the first edition appeared in 1862, but had no more immediate success than the 1844 version.

Hamilton read the 1844 *Ausdehnungslehre* in the years 1852 and 1853, and he was one of the first to appreciate its value. Of course he was most concerned to discover if any of his quaternion work had been anticipated. He concluded that it had not, but admitted that Grassmann "was well worthy to have anticipated me in the discovery of the quaternions; and it appears to me a very remarkable circumstance that he did *not.*" [74] Grassmann did not discover quaternions primarily because his approach was so much more general than Hamilton's. His book includes much of quaternion algebra and vector analysis, but without the concentration on a single algebra.

When Gibbs wrote his *Elements of Vector Analysis* in 1881 he had read Grassmann's *Ausdehnungslehre,* but he claimed that it had only confirmed notions that he already held. [75] Still, he admitted that his methods, "while nearly those of Hamilton, were almost exactly those of Grassmann." In 1888 he was working to obtain republication of Grassmann's works, because, as he wrote to the editor of the *American*

Journal of Mathematics: "I believe that a Kampf ums Dasein is just commencing between the different methods and notations of multiple algebra, especially between the ideas of Grassmann and of Hamilton."[76] It is significant that in 1888 Gibbs saw the imminent controversy as one between the supporters of Hamilton and the supporters of Grassmann.

According to Arthur Cayley, the beginning of multiple algebra dates from Benjamin Peirce's "Linear Associative Algebra" (1870). Before that time it was difficult to draw the line between "ordinary" algebra and other special algebras. Peirce was a professor of astronomy and mathematics at Harvard from 1833 until his death in 1880 and a great quaternion enthusiast. He began teaching quaternions at Harvard as early as 1848, and spoke with great praise of Hamilton in his *Analytical Mechanics* of 1855. This makes him one of Hamilton's earliest supporters, because Tait and Hamilton did not begin to correspond until 1858.[77] Peirce distributed his "Linear Associative Algebra" in lithographed form in 1870. It was finally published in 1881 by his son Charles Sanders Peirce in the *American Journal of Mathematics*. Benjamin Peirce's final evaluation of the quaternions was one that Hamilton would certainly have appreciated. Referring to Hamilton as "the immortal author of quaternions," he said that the greatest value of the square root of minus one as it was used in the quaternions was its "magical power of doubling the actual universe, and placing by its side an ideal universe, its exact counterpart, with which it can be compared and contrasted, and, by means of curiously connecting fibres, form with it an organic whole, from which modern analysis has developed her surpassing geometry."[78] The creation of an ideal or mental universe that mirrored exactly the universe of experience had been Hamilton's primary metaphysical goal.

In his "Linear Associative Algebra" Peirce summarized all the algebras of hypercomplex numbers known at the time and worked out all the possible linear associative algebras composed of elements with less than seven units. It was a mathematical masterpiece, the most original mathematical work to come from the United States, and one from which the field of multiple algebra could readily grow.

It was also a first effort to obtain some order in a chaos of competing systems. By successfully defying the Principle of the Permanence of Form, it seems Hamilton removed all restrictions from algebra. Gibbs later described this state of affairs as follows:

The student of multiple algebra suddenly finds himself freed from various restrictions to which he has been accustomed. To many, doubtless, this liberty seems like an invitation to license. Here is a boundless field in which caprice may riot. It is not strange if some look with distrust for the result of such an experiment. But the farther we advance, the more evident it becomes that this, too, is a realm subject to law. The more we study the subject, the more we find all that is most useful and beautiful attaching itself to a few central principles. We begin by studying *multiple algebras*: we end, I think, by studying MULTIPLE ALGEBRA. [79]

The attempts by Peirce and Gibbs to find the common content in the myriad of new algebras was certainly a step in the right direction. [80]

The passage quoted above is from an 1886 paper by Gibbs entitled "Multiple Algebra." The following year Arthur Cayley published a paper with the same title. [81] Both papers mentioned Hamilton with respect, but devoted most of their attention to Grassmann, whose ideas had a closer connection with the general concept of a "multiple algebra" than did Hamilton's. While Cayley adopted an antigeometrical stance in keeping with his belief that coordinates were superior to vectors or quaternions, [82] Gibbs emphasized the geometrical representation of "multiple numbers." In fact, he went so far as to suggest that the foundations of Cayley's famous 1858 paper "Memoir on the Theory of Matrices" had already been laid by Grassmann in his *Ausdehnungslehre* of 1844. [83] The confusion over priority comes, of course, from the fact that no single person "discovered" linear transformations, and the fact that they can be described symbolically. As we have seen, Gotthold Eisenstein had the idea as early as 1843 when he visited Hamilton at the observatory. [84]

Gibbs in particular recognized that linear transformations were at the root of "multiple algebra":

Our Modern Higher Algebra is especially occupied with the theory of linear transformations. Now what are the first notions which we meet in this theory? We have a set of n variables, say x, y, z, and another set, say x', y', z', which are homogenous linear functions of the first, and therefore expressible in terms of them by means of a block of n^2 coefficients. Here the quantities occur by sets, and invite the notation of multiple algebra. It was in fact shown by Grassmann in his *Ausdehnungslehre* ... that the notations of multiple algebra afford a natural key to the subject of elimination.

Now I do not merely mean that we may save a little time or space by writing perhaps ρ for x, y, and z; ρ' for x', y', and z'; and ϕ for a block of n^2 quantities. But I mean that the subject as usually treated under the title of determinants has a stunted and misdirected development on account of the limitations of single algebra. [85]

In this quotation Gibbs does not call ρ and ρ' *vectors*, but he might just as well have done so, considering the importance that he attributes to Grassmann's geometrical representation of multiple quantities. In fact, Gibbs asked specifically: "If it is a good thing to write in our equation a single letter to represent a matrix of n^2 numerical quantities, why not use a single letter to represent the n quantities operated upon, as Grassmann and Hamilton have done?" [86] Gibbs obviously understood the value of the vector approach for linear algebra, and it is not surprising that he was unwilling to have the concept of a vector confined to three dimensions, as Tait would have wished.

Since the controversies of the 1890s physicists have adopted vector analysis and mathematicians have developed the theory of vector spaces, while quaternions have more or less dropped by the wayside. There have been occasional attempts to revive them, but so far these attempts have failed to stir the scientific community. Sir Edmund Whittaker called for a Hamiltonian revival in 1940 and stated that quaternions might still become "the most natural expression of the new physics."[87] Whittaker may still be right, but the forty years that have passed since he made his appeal have seen a great deal of progress in physics, but little opportunity for quaternions. It seems unlikely that they will become the key to the universe that Hamilton had hoped for, and if by some chance they ever do, it would be a mistake to credit Hamilton with having foreseen their future use. He had insight, but not prevision.

On the other hand, it is also wrong to call the quaternions a failure. E. T. Bell, in his popular *Men of Mathematics,* labeled Hamilton "The Irish Tragedy," because he wasted his magnificent talent on a mathematical dead end.[88] But of course quaternions were *not* a dead end. They opened the way to much of modern algebra. Hamilton and Grassmann showed how new algebras could be constructed that did not necessarily follow all the rules of ordinary algebra. While such a step may appear obvious from the perspective of the twentieth century, it was unquestionably daring in the middle of the nineteenth. Such "obvious" steps are always the work of genius. They are simple to understand once the mind is adequately prepared. They are not simple or obvious at all when the mind must make a radical break with accepted modes of explanation.

In 1859 Hamilton wrote to Tait about quaternions: "*Could* anything be simpler or more satisfactory? Don't you *feel*, as well as think, that we are on a *right track*, and shall be thanked hereafter. Never mind when."[89] Through quaternions Hamilton looked into a new world of mathematical opportunity, and for that insight he does indeed deserve our thanks.

24

The Hodograph
and the Icosian Calculus

FROM THE MOMENT of their discovery until his death, Hamilton spent most of his time on the quaternions, but occasionally he could be drawn away from the subject, and when this occurred he continued to reveal his extraordinary creative power as a mathematician. Two examples of this power were the hodograph and the icosian calculus, which he discovered ten years apart—in 1846 and 1856—and which had nothing to do with the quaternions.

On the night of September 23, 1846, the planet Neptune was discovered at the Berlin Observatory. A Frenchman, Urbain Jean Joseph Leverrier, and an Englishman, John Couch Adams, had both independently predicted the existence of the planet and had calculated its position and motion from the observed perturbations of the neighboring planet Uranus. Only another planet could have caused the irregular motions of Uranus, but solving the inverse problem of perturbation to find its orbit was a mathematical accomplishment of a high order. Hamilton's sometime friend George Biddell Airy had been positively unenthusiastic when Adams had first brought his calculations to Airy's door, and in the subsequent brouhaha of claims and counterclaims, Airy came in for a great deal of abuse from Sir David Brewster and Sir James South, both of them masters of invective. By snubbing Adams, Airy, who as Astronomer Royal should have had the interests of his country at heart, had prevented a great discovery from coming to the British Isles and had given it to the French instead.[1] Airy, whose hide was as thick as a rhinoceros's, remained unperturbed as the storm swept past him, but it became a cause célèbre in the British and French Press. Hamilton wrote a poem in honor of Adams, which he forwarded to him through William Whewell, and several years later Adams married Hamilton's cousin.[2]

The discovery of Neptune and the discovery of conical refraction were often spoken of together as the two best examples of previously unobserved phenomena predicted entirely from theory. In the wake of the discovery of Neptune, Hamilton was naturally attracted to the theory of planetary perturbations on which he had first worked out his system of mechanics twelve years earlier. He chose the subject for his astronomical lectures for 1846, and in the process of working up the lectures he discovered a new geometrical representation of planetary motion that he called the *hodograph*, meaning "to describe the path." He read a paper describing his invention to the Royal Irish Academy on December 14, 1846, and concluded his presentation with an account of another startling calculation, made by the German astronomer Johann Heinrich Mädler, whose book *Die Central-Sonne* (which Hamilton exhibited at the meeting of the academy) claimed that the orbit of our sun revolved about the "central sun," Alcinoe, at a distance thirty-four million times the distance between the earth and our sun.[3] A reporter from the *Dublin Evening Post*, somewhat dazed by all these discoveries and sensing a possible "scoop," confused the discoveries of the hodograph and the central sun, giving credit for both of them to Hamilton, and letters of congratulation began to pour into the observatory. Hamilton wrote to the newspaper and to all of his well-meaning friends dissociating himself from the theory of the central sun and claiming only the hodograph as his discovery. When he later learned that Mädler's theory had not been accepted by sidereal astronomers, he was even more anxious that it not be attributed to him.

The hodograph, on the other hand, remains a valuable way of representing planetary motion. As a planet moves about the sun in its elliptical orbit, both the magnitude and direction of its velocity are constantly changing. If these different velocity vectors are all drawn from a single point rather than from points on the orbit, they will define another curve (the velocity vectors acting as radius vectors in this case), which is the hodograph (figure 24.1).[4]

If the hodograph is drawn so that its origin coincides with the focus of the elliptical orbit, several interesting geometrical properties appear (figure 24.2). The velocity vector of the orbit **v** becomes the radius vector of the hodograph **r**' by definition, and the velocity vector of the moving point on the hodograph **v**' is the acceleration vector of the moving point *A* on the orbit. This can be seen from the fact that by definition

$$\frac{d\mathbf{r}}{dt} = \mathbf{v}$$

on the orbit. On the hodograph the radius vector $\mathbf{r}' = \mathbf{v}$, and therefore $\mathbf{v}' = d\mathbf{r}'/dt = d\mathbf{v}/dt$. The side *BC* of the parallelogram is tangent to the hodograph.

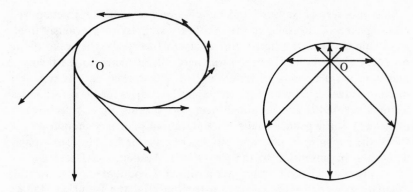

Fig. 24.1. The velocity vectors along the orbit, when drawn from a single origin, define the hodograph.

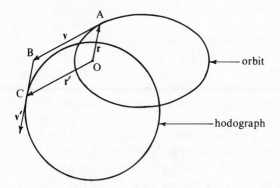

Fig. 24.2. The velocity vector on the hodograph is proportional to the acceleration vector on the orbit.

As the point *A* traces its orbit and the corresponding point *C* traces the hodograph, the parallelogram *ABCO* changes its shape, but retains the same area if the acceleration is always towards a central force at *O*. This can be proved from Kepler's second law (figure 24.3). *AB'* and *B'D'* are infinitesimal elements of the orbit, representing displacements made during equal infinitesimal times. Therefore they are proportional to the velocities at *A* and *B'*. Newton proved (*Principia*, book 1, proposition 1, theorem 1) that Δ*AB'O* and Δ*B'D'O* must have the same area. The parallelograms *AB'C'O* and *B'D'E'O* each have twice the area of the corresponding triangles and are therefore equal. Since the velocity vectors *AB* and *B'D* are proportional to the infinitesimal elements of the orbit *AB'* and *B'D'* respectively, the areas of the large parallelograms *ABCO* and *B'DEO* are equal and will remain the same for every point on the orbit.

Hamilton's most startling discovery was that in the case of motion in an orbit about a central force following Newton's inverse square law, the hodograph is always a *circle* (figure 24.4)! *OA* is the radius vector from the sun (at the focus of the elliptical orbit) to the planet at point *A* on the ellipse. *AB* is the tangent to the ellipse at *A* and is therefore collinear with the velocity vector. Newton proved (*Principia*, book 1, proposition 16, theorem 8) that the line *OB* drawn from the focus *O* perpendicular to the tangent *AB* is *inversely* proportional to the velocity of the planet at *A*. A theorem in the theory of conic sections states that as point *A* moves along the ellipse, point *B* describes a circle with a diameter equal to the major axis of the ellipse.

To find a line *directly* proportional to the velocity vector at *A*, we need only take the line *OC* (see figure 24.4). Euclid proved (*Elements*, book 3,

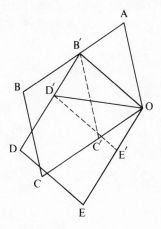

Fig. 24.3. The area of the parallelogram ABCO defined by the orbit and the hodograph remains constant.

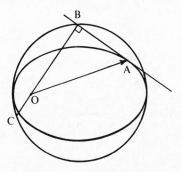

Fig. 24.4. Newton's proposition XVI. The line OB is inversely proportional to the velocity of the planet at A.

proposition 35) that for all chords of a circle passing through any given point within the circle, the products of the lengths of the segments on either side of the given point are equal. In other words, in figure 24.5, $AO \cdot OB = CO \cdot OD = EO \cdot OF$. Therefore in figure 24.4, $OC \cdot OB$ equals some constant k for any point A on the orbit. Since we already know from Newton's theorem that $OB = k'/v_A$ (where k' is a constant and v_A is the velocity at A), we can then write $OC = (k/k')v_A$, and OC is directly proportional to v_A. As point A moves about the ellipse, the line OC gives the magnitude of the velocity vector at A.

Referring again to figure 24.4, we note that the line OC is always at *right angles* to the tangent at A. Therefore while OC has the same magnitude as the velocity vector of the planet, it is at right angles to that vector. Therefore if we turn the circle through 90 degrees, as in figure 24.6, we obtain the hodograph.

Hamilton proved that the hodograph of any path about a central force following Newton's inverse square law is a circle. If the path is an ellipse,

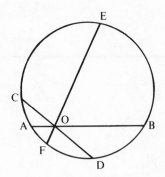

Fig. 24.5. For any line passing through point O the product of the two segments is constant.

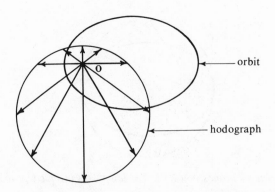

Fig. 24.6. An elliptical orbit and its corresponding hodograph.

the origin is within the circle, as in figure 24.6. If the path is a parabola, the circle passes through the origin, and if the path is a hyperbola, the origin lies outside the circle (figure 24.7).

Hamilton sent descriptions of his hodograph to Airy, Herschel, and Whewell, all of whom returned interested, if less than enthusiastic, replies.[5] William Thomson (later Lord Kelvin) was very enthusiastic, however, and asked Hamilton repeatedly for copies of his papers. He even made the hodograph the subject of a prize essay for his students.[6] Peter Guthrie Tait, an enthusiastic disciple of Hamilton, also wrote on the hodograph, and he and Thomson gave a clever analytical proof of the circular hodograph in their famous *Treatise on Natural Philosophy* (1879). It runs as follows.[7]

The x and y components of the acceleration of the planet are

$$a_x = \frac{d^2x}{dt^2} = \frac{\mu}{r^2}\left(\frac{x}{r}\right) \tag{24.1}$$

and

$$a_y = \frac{d^2y}{dt^2} = \frac{\mu}{r^2}\left(\frac{y}{r}\right), \tag{24.2}$$

where $\mu = Gm$ and $r^2 = x^2 + y^2$. Therefore

$$y\frac{d^2x}{dt^2} = x\frac{d^2y}{dt^2}. \tag{24.3}$$

Integrating (24.3) gives

$$x\frac{dy}{dt} - y\frac{dx}{dt} = C. \tag{24.4}$$

Multiplying (24.1) and (24.4) gives

$$\frac{d^2x}{dt^2} = \frac{\mu x}{Cr^3}\left(x\frac{dy}{dt} - y\frac{dx}{dt}\right)$$

$$\frac{d^2x}{dt^2} = \frac{\mu}{C}\frac{\left(x^2\dfrac{dy}{dt} - xy\dfrac{dx}{dt}\right)}{r^3} = \frac{\mu}{C}\frac{(x^2+y^2)\dfrac{dy}{dt} - y\left(x\dfrac{dx}{dt} + y\dfrac{dy}{dt}\right)}{r^3}$$

$$\frac{d^2x}{dt^2} = \frac{\mu}{C}\frac{r^2\dfrac{dy}{dt} - y\left(r\dfrac{dr}{dt}\right)}{r^3}.$$

Fig. 24.7. Hodographs of parabolic and hyperbolic motion about a central force.

Integration gives

$$\frac{dx}{dt} = \frac{\mu}{C}\left(\frac{y}{r}\right) + A = v_x,$$ (24.5)

and by the same argument

$$\frac{dy}{dt} = \frac{\mu}{C}\left(\frac{x}{r}\right) + B = v_y.$$ (24.6)

Therefore

$$(v_x - A)^2 + (v_y - B)^2 = \frac{\mu^2}{C^2 r^2}(x^2 + y^2) = \frac{\mu^2}{C^2}.$$ (24.7)

In the hodograph v_x and v_y are the x and y coordinates giving the locus of points on the curve. The equation is the equation of a circle with its center displaced from the origin.

If we put the central force at the origin of the x and y axes (point O in figure 24.6), we can see from equation 24.5 that $A = O$, because figure 24.6 shows that when $y = O$, the velocity vector on the orbit has no x component, and if y and v_x are both equal to zero, A must also be equal to zero. If in equation 24.6 we seek the point on the elliptical orbit where $v_y = O$, this occurs at the end of the semiminor axis. At this point the radius vector r is equal to the semimajor axis, and x equals the distance between the focus and the center of the ellipse. Therefore the ratio x/r at this point is equal to the eccentricity ϵ of the ellipse and equation 24.6 becomes $B = -(\mu/C)\epsilon$. Equation 24.7 shows that μ/C is the radius of the circular hodograph and therefore $B = -R\epsilon$ where R equals the radius of the hodograph. B represents the displacement of the center of the hodograph from the origin. This immediately explains the hodographs in figure 24.7. In the case of the parabola, $\epsilon = 1$, and therefore $B = -R$. The origin is on the circumference of the hodograph. In the case of the hyperbola, $\epsilon >$ 1, and therefore $B > -R$. The origin lies outside the hodograph.

Hamilton devised more complex theorems of the hodograph, including his "Law of Hodographic Isochronism," which corresponds to Lambert's central force theorem. This latter theorem seemed important to Hamilton, but when he sent it to Herschel he got the reply: "You are fairly got out of my depth."[8] (This was Herschel's regular response when he did not have the time or inclination to follow Hamilton's long analytical excursions.) Hamilton later discovered that he had been anticipated in the idea of the hodograph, although not in all the theorems, by August Ferdinand Möbius. He acknowledged this anticipation in his *Lectures on Quaternions* of 1853.[9] The hodograph turned out to be an elegant way to describe motion about a central force, but it did not lead to any really new approach. It disappeared for a while from books on rational and celestial mechanics, and has reappeared recently, its discovery attributed to a modern author.[10] The hodograph was scarcely the earth-shaking discovery that the *Dublin Post* reported. It remains as a mathematical curiosity and as a testament to Hamilton's imaginative powers.

An even greater departure from his usual mathematical pursuits than the hodograph was Hamilton's discovery in 1856 of a new "calculus" based on the relationship of the sides of the icosahedron. Because of its origin he called it the *icosian calculus*, although it was not dependent on the geometrical properties of the icosahedron, as the name might imply. Instead it was a study in what is now called graph theory, a field that did not exist as such in the nineteenth century. Graph theory was not entirely new with Hamilton; problems of this kind had been recognized by Leibniz, Euler, Vandermonde, Poinsot, and others, but Hamilton's icosian calculus drew much attention to the field.[11]

A graph, in this mathematical sense, consists of a finite set of points called *vertices*, a finite set of lines called *edges*, and a rule telling which edges join which pairs of vertices.[12] (The use of the words *vertices* and *edges* for the points and connecting lines of the graph indicates that the origins of this science lay in the study of polyhedra.) The icosahedron is one of the five regular polyhedra and it is composed of twenty triangles. The regular polyhedron with the next smaller number of faces is the dodecahedron, which is composed of twelve pentagons. The dodecahedron, however, has twenty vertices, and it is easy to show that the twenty vertices of the dodecahedron stand in the same relationship as the twenty faces of the icosahedron, that is, each of the twenty vertices of the dodecahedron is joined by edges to three other vertices, and each of the twenty sides of the icosahedron is adjacent to three other sides. Thus the relationships of the graph can be studied and displayed equally well on the sides of the icosahedron or on the vertices and edges of the dodecahedron. Hamilton began by studying the former, and concluded by using the latter, because it is easier to display the edges and vertices than the sides. In its final version his new system could be better described as a "dodecahedral calculus."

Hamilton got the idea of his icosian calculus in August, 1856, while attending the British Association Meeting at Cheltenham. He was particularly anxious to go to this meeting because John Graves lived in Cheltenham and had invited Hamilton to be his guest. Lady Hamilton had been seriously ill throughout the summer and Hamilton wrote to Graves that in spite of the fact that she had three devoted children and a niece to look after her, she was dependent on him and might not be able to bring herself to let him go.[13] The illness was a "nervous" one similar to her crisis of 1840 and 1841, and there is some indication that Hamilton feared he might break down himself under the strain of the close confinement and constant attention that his wife required.[14] The trip to Cheltenham would be a welcome break for him, in any case, and he finally got away in time for at least part of the meeting. Unfortunately he was soon striken by a case of gout that immobilized him at John Graves's house, but since there were other scientific guests at the house (W. R. Grove, in addition to John Graves) and since John Graves had probably the best mathematical library in England, Hamilton was able to put his time to good use.

Hamilton was ecstatic about the opportunities that Graves's library gave him: "Conceive me shut up and revelling for a fortnight in John Graves's Paradise of Books! of which he has really an astonishingly extensive collection, especially in the curious and mathematical kinds. Such new works from the Continent as he has picked up! and such rare old ones too!"[15] In addition to the books, Hamilton was stimulated by quaternion problems that Graves posed to him. In particular Graves introduced him to a book by Hermann Scheffler entitled *Der Situationskalkul* that contained a method for finding the greatest common measure of two complex numbers.[16] Graves challenged Hamilton to solve the same problem for quaternions, and in a letter to his son, Hamilton reported: "I dashed off before breakfast (indeed in my shirt) one morning a *general method* for that purpose, though I had never thought of the problem before."[17] It is possible that Scheffler's book started Hamilton thinking about polyhedra, too; more likely, though, it was John Graves, who had still not given up the triplet search and told Hamilton that he was still "hoping to find in trihedral angles natural triplets which would serve as the basis for an imaginary calculus."[18]

Whatever the cause, Hamilton started thinking about trihedral angles and polyhedra, and on October 7, after his return to Dublin, he wrote excitedly to Graves about a new calculus that he had just discovered. He obviously believed it to be important, because he wrote out his letter in a form that might be published (as he had done when he discovered quaternions) and because he was being so scrupulously careful about priority. In fact he invited Graves to write out his own ideas about polyhedra before opening the sealed packet enclosed with his letter.[19] Graves graciously replied that he had not anticipated any of Hamilton's ideas, and that he would not wish to give up his "chief chance in life," which he considered to be the chance of living by Hamilton's fame.[20] Again John Graves had served the useful purpose of sparking a new mathematical inspiration in

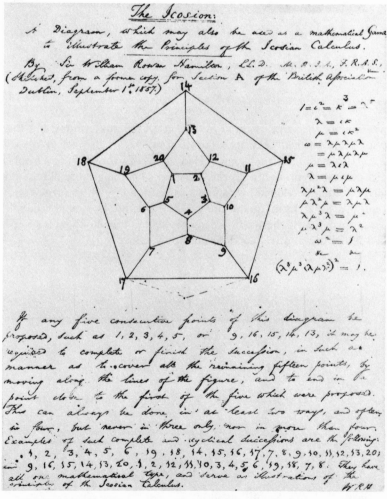

Hamilton's description of the Icosian Calculus and the Icosian Game

Hamilton's mind. With his superior command of the mathematical literature he was in a better position than Hamilton to talk about priority, and
he mentioned other writers on polyhedra, including Poinsot, Euler, and
Kepler, and sent a recent article by Thomas Pennyngton Kirkman that
came dangerously close to Hamilton's ideas and even anticipated some of
his results.

Hamilton begins the description of his calculus by defining three symbols, ι, κ, λ, all of which are roots of unity:

$$\iota^2 = 1 \tag{24.8}$$

$$\kappa^3 = 1 \tag{24.9}$$

$$\lambda^5 = 1 \tag{24.10}$$

He describes one other relation:

$$\lambda = \iota\kappa \tag{24.11}$$

These are not the same as the i, j, k, terms of the quaternions, but there
are some similarities. They are all roots of unity, and they are all associative but not commutative. The way Hamilton talks about his new symbols
indicates that he probably expected to find something more like quaternions, but then developed his calculus as he found it, which turned out to
be quite distinct from quaternions. This calculus is simpler than the
quaternions because the symbols can only be multiplied—there is no addition—so the distributive law does not apply. Since

$$\lambda = \iota\kappa$$

$$\iota\lambda = \iota(\iota\kappa) = \iota^2\kappa = (1)\kappa = \kappa.$$

Hamilton then derives a new symbol μ as follows:

$$1 = \lambda^5 = (\iota\kappa)^5 = \iota\kappa\iota\kappa\iota\kappa\iota\kappa\iota\kappa$$

$$1(\kappa^2) = \lambda^5\kappa^2 = \iota\kappa\iota\kappa\iota\kappa\iota\kappa\iota\kappa(\kappa^2) = \iota\kappa\iota\kappa\iota\kappa\iota\kappa\iota \text{ (because } \kappa^3 = 1)$$

$$\kappa^2\iota = \iota\kappa\iota\kappa\iota\kappa\iota\kappa\iota(\iota) = \iota\kappa\iota\kappa\iota\kappa\iota\kappa \text{ (because } \iota^2 = 1)$$

$$\kappa^2\iota\kappa^2 = \iota\kappa\iota\kappa\iota\kappa\iota$$

He repeats this process until he obtains

$$\kappa^2 \iota \kappa^2 \iota \kappa^2 \iota \kappa^2 \iota = \iota \kappa = (\kappa^2 \iota)^4. \qquad (24.12)$$

Now

$$1 = (\iota^2)(\kappa^3) = \iota(\iota\kappa)\kappa^2 = \iota(\kappa^2\iota)^4\kappa^2 = (\iota\kappa^2)^5.$$

Hamilton now defines $\mu = \iota\kappa^2$, then $\mu^5 = 1$, and μ becomes a fifth root of unity like λ. In any equation involving the λs and μs it is possible to interchange these symbols and have the equation remain valid. He then works out a series of useful transformations such as:

$$\mu = \lambda\iota\lambda, \qquad \lambda = \mu\iota\mu$$

$$\iota = \lambda\iota\mu, \qquad \iota = \mu\iota\lambda$$

$$\lambda\mu^2\lambda = \mu\lambda\mu, \qquad \mu\lambda^2\mu = \lambda\mu\lambda$$

$$\lambda\mu^3\lambda = \mu^2, \qquad \mu\lambda^3\mu = \lambda^2.$$

With these transformations it is possible to transform an operator such as:

$$[\lambda^3\mu^3(\lambda\mu)^2]^2. \qquad (24.13)$$

$$[\lambda^3\mu^3(\lambda\mu)^2]^2 = [\lambda^2(\lambda\mu^3\lambda)\mu\lambda\mu]^2 = [\lambda^2\mu^3\lambda\mu]^2$$

$$= [\lambda(\lambda\mu^3\lambda)\mu]^2 = [\lambda\mu^3]^2 = (\lambda\mu^3\lambda)\mu^3 = \mu^2\mu^3 = \mu^5 = 1.$$

This operator is important because it represents a complete circuit of all the vertices of the dodecahedron, passing through every vertex once and only once (moving only along the edges, not across the faces), and returning to the original vertex. The fact that this operator is equal to unity means that the circuit is complete—that is, the path returns to the original vertex of the dodecahedron from which it started.

But Hamilton developed all this apparatus first, without any mention of its applicability to the dodecahedron. He was primarily interested in the operations of his symbols as a new kind of algebra, and he saw this algebra's relationship to the dodecahedron as only an application of the more general symbolic calculus. This has significance for the question of priority between Hamilton and Thomas Kirkman. After reading Kirkman's papers he wrote to John Graves: "On the geometrical side he [Kirkman] has been dealing with far greater generalities than myself, but I do not observe the slightest trace of his having caught even the notion of my algorithm or *calculus*, such as it is; and my long letter to you contains only *specimens* of such a calculus."[21] Because of the importance of the icosian calculus for the theory of abstract groups, it is significant that Hamilton emphasized the algebraic properties of his new calculus.

In applying the icosian calculus to the dodecahedron he used the projection of the dodecahedron onto one of its faces. In other words, the pentagonal outline of the diagram is one face of the dodecahedron and the other edges are projected onto that face by drawing lines from every vertex to a point below the center of the face (figure 24.8). Because the problem deals only with the connections between the vertices, the actual dimensions of the figure are of no importance and we could just as well draw the diagram another way (figure 24.9).

The most important symbols in Hamilton's calculus are the fifth roots of unity λ and μ. These represent a turn to the right and a turn to the left respectively in moving from any edge to the next. Thus $\lambda(ab) = (bc)$ and $\mu(ab) = (bl)$. (The edge has to be considered as a directed line segment.) It is easy to see why these symbols are fifth roots of unity, because if the operation of a right turn or a left turn is repeated five times, the path comes back to the starting point. The symbol ι represents a reversal of the direction along any edge. Thus $\iota(ab) = (ba)$ and $\iota^2(ab) = \iota(ba) = (ab)$, or $\iota^2 = 1$, which explains why ι is a square root of unity. The symbol κ represents the rotation counterclockwise to the next edge of any edge about its end point.

$$\kappa(ab) = (cb)$$

$$\kappa\kappa(ab) = \kappa(cb) = (lb)$$

$$\kappa\kappa\kappa(ab) = \kappa\kappa(cb) = \kappa(lb) = ab \qquad \therefore \kappa^3 = 1,$$

and so κ is the cube root of unity. The equality $\lambda = \iota\kappa$ can be verified immediately, because a right turn is equivalent to rotating the edge about its end point counterclockwise (κ) and reversing its direction (ι). Thus

$$\lambda(ab) = (bc)$$

$$\kappa(ab) = (cb)$$

$$\iota\kappa(ab) = \iota(cb) = (bc).$$

Obviously these symbols do not commute, because (by comparison to our last equation) $\kappa\iota(ab) = \kappa(ba) = (ta)$, and $\iota\kappa \neq \kappa\iota$.

Hamilton found that his formula (24.13) gave the only possible route that would permit a complete circuit of the dodecahedron and a return to the starting point. The sequence of vertices marked in alphabetical order

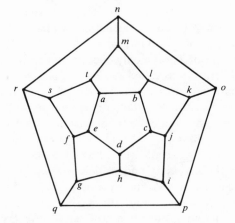

Figure 24.8. The icosian diagram.

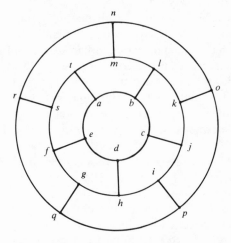

Figure 24.9. An alternate version of the icosian diagram

on figure 24.8 describes such a circuit. The first turn is at vertex *b*, and there is a turn for every vertex, so to complete the circuit of twenty vertices the last turn must be at vertex *a*. The route represented by his formula may be found on the dodecahedron in several ways, because the sequence of right and left turns may be started at any point and may be run in either direction. The circuit that traverses each vertex of a graph once and only once and returns to its starting point is now called a *Hamilton circuit* (Hamilton called it a *complete cyclical succession*). A path through all the vertices that does not return to the starting point is now called a *Hamilton path* (Hamilton called it a *complete noncyclical succession*). It is doubtful that Hamilton rather than Kirkman deserves to have his name used, but what has been hallowed by custom is not easily altered.

A problem comparable to Hamilton's is that of finding a path that traverses all the *edges* of a graph only once (it may traverse a vertex more than once). Such a traverse is called an *Eulerian path* after Leonard Euler and his famous problem of the Seven Bridges of Königsberg. (The graph of the dodecahedron [figure 24.8] does not contain an Eulerian path.)[22] It is relatively easy to demonstrate the conditions that are necessary and sufficient for a graph to have an Eulerian path (it must be connected and have no vertices, or only two vertices, at which an odd number of edges meet), but so far the necessary and sufficient conditions for a graph to have a Hamiltonian circuit have not been found. Some conditions that are necessary but not sufficient, and others that are sufficient but not necessary, have been found, but a complete set of necessary and sufficient conditions still escapes mathematicians.

The question of priority between Hamilton and Kirkman has been raised in retrospect by historians, but it did not seem to have produced any noticeable conflict during Hamilton's lifetime. Thomas Pennyngton Kirkman was rector of a small parish in Lancashire and an enthusiastic mathematical amateur. He was elected fellow of the Royal Society in 1857. Hamilton had known him at least since 1848, when the two men had corresponded about the quaternions.[23] Hamilton was also fond of a book by Kirkman on mathematical mnemonics that set the trigonometric identities into unforgettable verse. It was the only way that Hamilton could remember them.[24]

Kirkman contributed a paper to the Royal Society on August 6, 1855, in which he stated conditions for a complete circuit of the vertices of any polyhedron, but unfortunately these conditions were incorrect. The paper does correctly describe, however, a general class of graphs that does *not* possess a complete circuit, and this was the major significance of the paper.[25] On October 14, 1857, Kirkman submitted another paper that did state that a closed circuit could be drawn through the twenty vertices of the dodecahedron without passing through any vertex more than once, but this was, of course, after Hamilton had already made the same discovery.[26] Hamilton believed that he had not been anticipated by Kirkman's 1856 paper, and he felt that his own interest in a new algebra was quite different from Kirkman's much broader interest in the graphs of polyhedra.

In 1861 Hamilton visited Kirkman on his return from the meeting of the British Association at Manchester and discussed the graphs of polyhedra with him. Subsequent correspondence was friendly, and each man went out of his way to flatter the other, so there does not seem to have been any bitterness at the time.[27] Hamilton especially wanted to send icosian problems to Kirkman's daughter Catherine, and he enjoined Kirkman not to give her the answers, saying, "It would be too great a descent on your part from polyedra to Icosians or Icosian," indicating again that Hamilton recognized Kirkman's work to be more general than his own.[28]

The question of priority arose long after Hamilton's death, when Peter Guthrie Tait, a friend, disciple, and usually a defender of Hamilton, wrote that Hamilton's icosian game "was suggested to him by a remark in Mr. Kirkman's paper on Polyhedra."[29] He later repeated the claim that "Kirkman himself was the first to show, so long ago as 1858, that a 'clear circle of edges' of a unique type passes through all the summits of a pentagonal dodecahedron. Then Hamilton pounced on the result and made it the foundation of his *Icosian Game*, and also of a new calculus of a very singular kind."[30] Kirkman never accused Hamilton of "pouncing" on his results, but he did say: "Attention to these circles was perhaps first called by me in *Phil. Trans.*, vol. 148, 1858, page 160, from which the late Royal Astronomer, Sir W. R. Hamilton, took his idea of the icosian game."[31] It is likely that Hamilton gave a copy of his icosian game to the Kirkman family when he visited them in 1861, because he was sending problems for the game to Kirkman's daughter. The game was marketed in 1859, and that date appeared on the instruction sheet. Kirkman may have assumed that the year on the game was the year of Hamilton's first discovery. In any case the discovery of a Hamiltonian circuit on the dodecahedron was probably made independently by both men, and the assigning of priority at this late date is not a worthwhile undertaking.

From the icosian calculus Hamilton created his "icosian game." He wrote to John Graves:

I have found that some young persons have been much amused by trying a new mathematical game which the Icosian furnishes, one person sticking five pins in any five consecutive points . . . and the other player then aiming to insert, which by the theory in this letter can always be done, fifteen other pins in cyclical succession, so as to cover all the other points, and to end in immediate proximity to the pin wherewith his antagonist had begun. Whatever then may be thought of the utility of these new systems of roots of unity, suggested to me by the study of the ancient solids . . . they will be found to have supplied (a new and innocent) pleasure, not only to algebraists, but even to children: and I am willing to suppose that, on the whole, you will derive some satisfaction from examining them.[32]

Graves and Hamilton continued to discuss the icosian calculus during the early months of 1857, and it was Graves who first had the idea of designing a game board with the icosian diagram engraved upon it and with twenty markers to indicate the route chosen through the graph. Actually the "game" could better be called a "puzzle." One player begins the circuit and challenges his opponent to complete it. There are other challenges that produce Hamilton circuits, Hamilton paths, or in some cases incomplete circuits on the diagram. By the end of February, 1857,

Lady Hamilton's brother-in-law was playing the game on a solitaire board with some of the holes plugged.[33] A year later Hamilton had a round board manufactured from mahogany for the Edgeworth family, and this was apparently the first finished version of the game.[34] Then, early in 1859, a friend of John Graves manufactured a more elegant version with a set of cylindrical markers and a board with legs like a small table.[35] He sent Hamilton one of these sets, much to Hamilton's delight, and in the meantime Graves contacted a London toymaker, John Jaques and Sons, with the suggestion that Hamilton might be willing to sell rights to the game for £30.[36]

Hamilton was pleased with Graves's initiative and was soon in correspondence with Jaques about the game. Jaques offered only £20; after some dickering it was decided that in return for £25 and six copies of the game, Jaques would receive all rights to the game and assistance from Hamilton in composing an instruction sheet and in setting new problems for the game.[37] There was some concern that the game might be too easy; this had been John Graves's criticism, and indeed, it is a simple enough matter to complete a Hamiltonian circuit on the diagram if five vertices in sequence are first specified. Graves said that even children could do it easily, and Hamilton responded: "It piques me a little, I confess it, that the Game should be called 'too easy,' even by *children*, but then you must remember that the lithographs [distributed at the British Association meeting] gave only *one* problem."[38] He set out to invent some problems that would not be so easy. He also urged Jaques to include a description of the mathematics of the game. Jaques agreed, but insisted that the description of the icosian calculus go at the *end* of the instruction sheet rather

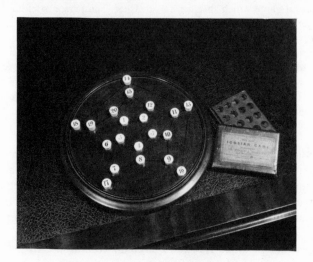

Hamilton's Icosian Game

than at the beginning, where it would tend to discourage nonmathematical readers. Hamilton was willing to put the mathematical description at the end, but he insisted that the mathematics be published as he had written it without any changes.

When Jaques's check for £25 arrived Hamilton was ecstatic. He never considered that any of his mathematical work might have immediate monetary value: "You have made me very happy. When your cheque arrived, I certainly received it with great *glee*; and deposited it, for one night, along with my other medals: for I consider it to be one: with this slight advantage in its form—as regarded convenience of *using* it—that *it* was *intended* to be melted down, which I am not so profane as to do with my gold metals."[39] Hamilton also said that his bankers considered him at this point to be a man of business, since he had obtained *cash* for one of his inventions.[40]

Hamilton had some hopes that the game would sell wildly, but it did not, and it seems that Mr. Jaques never recovered his investment.[41] He marketed two versions of the game, one for the parlor, which was played on a flat board, and another for the "traveler," which was played on an actual dodecahedron with a handle on one face and a large-headed nail at each vertex.[42] The players marked the path by winding a string from vertex to vertex. The great objection to the game was the one raised by John Graves—it was too easy. It is curious that Hamilton, who invented the game, apparently found it more difficult than did his friends. He wrote to Graves that he had no "instinct" for the puzzles set and that they gave him a lot of trouble.[43] One suspects that he made the puzzles more difficult than they needed to be by trying to calculate the solutions rather than by merely trying different paths on the board. In spite of the failure of his game, Hamilton's mathematical instincts were right. The icosian calculus was an important early contribution to graph theory and to the theory of abstract groups.

Curiously enough, Hamilton did *not* recognize the importance of another graph-theoretical problem that came his way at about the same time—the four-color problem. De Morgan posed the problem to Hamilton as follows:

A student of mine asked me to-day to give him a reason for a fact which I did not know was a fact, and do not yet. He says, that if a figure be anyhow divided, and the compartments differently coloured, so that figures with any portion of common boundary *line* are differently coloured—four colours may be wanted, but not more—the following is the case in which four are wanted. Query, cannot a necessity for five or more be invented? As far as I see at this moment, if four *ultimate* compartments have each boundary line in common with one of the other, three of them inclose the fourth, and prevent any fifth from connexion with it. If this be true, four colours will colour any possible map, without any necessity for colour meeting colour except at a point.

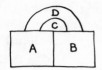

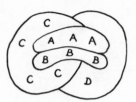

Now it does seem that drawing three compartments with common boundary *ABC* two and two you cannot make a fourth take boundary from all, except by inclosing one. But it is tricky work, and I am not sure of the convolutions—what do you say? And has it, if true, been noticed? My pupil says he guessed it in colouring a map of England. The more I think of it, the more evident it seems. If you retort with some very simple case which makes me out a stupid animal, I think I must do as the Sphynx did.[44]

Of all the problems in graph theory this one has become the most celebrated. There can scarcely be another mathematical problem that is so easy to state and yet so difficult to solve.[45] Since this first statement of the conjecture by De Morgan, untold hours of effort have gone into seeking a solution. When Hamilton received De Morgan's letter he replied: "I am not likely to attempt your 'quaternion of colours' very soon."[46] In fact he never tried it at all, or if he did, he left no record of it, all of which goes to prove that even the most far-seeing of minds may occasionally have a blind spot.

VIII
Last Years

25

Catherine Disney Barlow

THE DISCOVERY OF quaternions had been a high point for Hamilton, a moment when he had been able to lift himself out of the depression of 1842 caused by his wife's illness and by the loneliness of the observatory. He had felt his creative powers return in a dramatic fashion. The quaternions gave him a new direction and commitment to which he was willing to devote the remainder of his life. But the years following the discovery were not easy ones. They included periods of extreme emotional stress caused largely by the return of Catherine Disney, now Catherine Barlow, as a powerful force in his life.

In metaphysics and poetry Hamilton was unquestionably romantic; in his love for Catherine Disney he was hopelessly so. He believed in love at first sight and acted on that belief. From the first time he saw Catherine in 1824 he held her in his heart and mind as a romantic vision never to be tarnished by any events in his subsequent life or in hers. Marriage, children, fame, disaster—nothing dimmed that vision. It kept surging back time and time again, and each time Hamilton was surprised by its power. It was beyond his control. With great effort he could act the perfect Victorian father as far as the outside world was concerned, but no force of will could remove Catherine Disney from his heart.

If Hamilton had been a Shelley or a Byron the intensity of his feelings might have resulted in great creative poetry. He did not have the skill, however, and his creativity was in another direction, as Wordsworth had recognized when he first saw Hamilton's work. His verse and his correspondence are restrained and often stuffy. This stuffiness makes him appear almost silly sometimes—a married man in his fifties mooning over his remembrances of a young girl he had met a quarter century before. But as one reads further one recognizes that the anguish is real, the vision is intense, and the stuffy style is an attempt to reveal and yet to conceal the extent to which he is consumed by his emotions. Hamilton was no Werther, and yet he felt like one inside. In reading his letters about Catherine one first feels the urge to laugh, and then one begins to share his despair.

It was during the year following his discovery of quaternions that Hamilton first began to have trouble with alcohol. The following year, in 1845, his old college friend Thomas Disney paid a visit to the observatory and brought Catherine with him.[1] Hamilton had not seen her since 1830, when he had talked to her at Armagh and had broken the eyepiece of the telescope in his agitation. He left little record of the 1845 visit, but it must have been upsetting in the extreme. In February his problems with alcohol became public knowledge when he collapsed and became violent at the meeting of the Geological Society. In 1847 he suffered the loss of his uncles James and Willey. MacCullagh's suicide occurred in that same year, and Hamilton confessed that it moved him so profoundly that he could scarcely think of anything else. But the year 1848 was the most tumultuous of all, a turmoil that reflected the political shocks of the "year of revolutions."

One of Catherine's sons, James William Barlow, entered Trinity College Dublin in 1842, and in 1848 he competed for a fellowship. Hamilton made Barlow his protégé and coached him in mathematics for the upcoming examination. James did not succeed in winning a fellowship, but it was clear to Catherine that Hamilton had gone out of his way to be of service to her and to her son. In July she wrote to thank him for befriending her son and for helping him with his preparation. According to Hamilton, it was a letter that she would have sent much earlier if she had been able to do so. After their meeting in 1830, Catherine had asked her husband if she might write Hamilton and explain that she had not been responsible for his rejection in 1825 and that she wished to keep his friendship. Her husband refused, probably wisely, considering the strength of feeling between Catherine and Hamilton.

When he received the first letter from Catherine in July, 1848, Hamilton wrote back concealing the fact that she had written first. He was afraid of the difficulties that her husband might make if he intercepted the letter: "I replied very coldly and guardedly at first, and wrote my answer *as if* it has been an *original* letter, and not a reply, and with the intention that it should be shewn by her to her husband, for whom I charged her (icily enough, as I felt at the time, but still I did charge her) with my compliments, if she would take the trouble to deliver them. This she had not the resolution to do—nor have I, knowing more now than I then did, the heart to blame her for it. Letter succeeded letter for awhile, of an increasingly confidential ... character."[2] The exchange continued for six weeks, increasing in intimacy and causing increased distress. Hamilton urged an end to the correspondence; Catherine began to feel that she had to confess it to her husband. Hamilton urged her not to, and the letters continued. It was obvious to both of them that their love had not diminished since 1825. Unhappy marriages on both sides had intensified it. Their mutual love was a sentiment to be savored, but it did nothing to assuage their guilt, and Catherine finally decided that she had to tell her husband. She concluded her last letter: "To the mercy of God in Christ I look alone, for pardon for all my sins."[3]

Before long a letter from Barlow came to the observatory. Hamilton hated Catherine's husband so strongly that he was distressed by his own lack of charity. He expected the letter to be an angry one and he read it as such, but in later years he recalled it as a "curious kind of half-anger" that terminated with a "half-apology."[4] Hamilton replied politely that he was sure Barlow would never expect to hear from him under any circumstances again.

Just as these painful events were drawing to a close Hamilton received an invitation from Lord Rosse to join him at Birr Castle in the company of Airy and Sabine for a series of observations of nebulae to be performed with the giant reflecting telescope. For Hamilton it was a welcome diversion, but not a wise one. Airy as usual was in good spirits, and the party stayed up late observing and socializing. Hamilton's major correspondent during this visit was Catherine's son James, to whom he described the gaiety of the place. He wrote: "For my own part I believe that I had been working rather too hard and too long some time ago, and I felt somewhat languid and fatigued on arriving here; but if I were asked whether I had *enjoyed* myself at Parsonstown, I could safely say that I have done so."[5] Hamilton was putting on a brave front for Barlow and for his host. In the company of Lord Rosse, Airy, and Sabine, he suppressed his feelings, but in later letters he recalled, "My agitation at the end of August of that year was *extreme*, and I remember that I could only find comfort in some . . . *ejaculations* [he tells us that he swore in Greek] while walking by myself thro' the grounds of Parsonstown."[6] He described these same walks to James Barlow as idyllic rambles.

Towards the end of his visit Hamilton received another letter from Catherine that was written in open defiance of her husband. The letter contained a stamped envelope and instructions to Hamilton to mail the letter to her husband in the envelope provided. The contents of the letter and the handwriting showed signs of a complete emotional breakdown. His first impulse was to go to Catherine, but he could not do that with propriety, nor could he trust his own emotions. His purpose would have been innocent, "but who could vouch for the innocence of the results?"[7] While Hamilton stalked through the gardens of Parsonstown trying to decide what to do, Catherine took a heavy dose of laudanum with the expectation that her letter would arrive back at her house after her death. Hamilton, of course, did not return the letter, and Catherine's life was saved, but she had to continue to suffer from her previous guilt with the added burden of having tried to commit the sin of suicide.[8]

Internally Hamilton was in agony, but externally he became even more hearty and jovial. The evenings at Parsonstown were spent in elaborate raillery and literary combat between Hamilton and Airy, each challenging the other to recall classical quotations and fragments of poetry. Unfortunately the challenges extended beyond literature, and Hamilton allowed Airy to goad him into breaking his vow of abstinence. After the first glass of champagne at Parsonstown Hamilton was never again able to mount more than a feeble opposition to the enticement of alcohol. Robert Graves blames Airy for this lapse, but the internal torture of Catherine's suicide attempt must have been the real culprit—if a culprit was needed.

After the events of 1848 Catherine lived almost constantly with her mother or with her sisters and brothers. She saw her husband seldom, although there was no legal separation. Hamilton was able to obtain information about her from her brothers, whom he had known since his college years. His ill-fated correspondence of 1848 was precious, but dangerous to have around. He burned the original letters after copying them out in shorthand, labeling the notebook "Neville and Sydney 1850." He used these code names frequently in his later correspondence when he wished to conceal the real names in talking about the relationship.[9] Through the other Disneys Hamilton sent poems such as "A Prayer for Calm" and "To an Afflicted Friend Suffering under Religious Depression" that he knew would reach Catherine, and a lock of hair to be mingled with hers after his death.[10]

Catherine's son James Barlow continued to work for a fellowship, and Hamilton continued to help him. The fellowship examination in the spring of 1850 was to contain quaternions in the mathematical part. Charles Graves, the professor of mathematics, was lecturing on quaternions using information that Hamilton was sending him, while Hamilton gave the same instruction to Barlow privately and directly.[11] The examination was held in May and this time Barlow was successful.[12] The following August Hamilton and Barlow went to Edinburgh for the meeting of the British Association and shared lodgings, which gave them a chance to become even better acquainted. For Hamilton, Barlow's success and the trip to Edinburgh were gifts for Catherine and revenge against her husband, who could not help James as he had done.[13] While he was in Edinburgh, Hamilton received a note of presentation from Catherine to go with a Bible and Prayer Book that she had already sent him through Dora Disney, Thomas Disney's wife. He kept the inscription separate from the book so that he could use the book without its donor being revealed.[14]

From Scotland he went on to the Lake Country to visit the wife of William Wordsworth. The poet had just died, and Hamilton went to pay his respects and to spend a few days with Robert Graves. The visit to

Rydal Mount was repeated in later years, and it took on symbolic meaning for Hamilton. As part of the pilgrimage in 1854 he went to the churchyard in Grasmere and kissed Dora Wordsworth's gravestone by moonlight.[15] These rituals had deep significance for Hamilton. In April, 1850, while he was coaching Barlow for the fellowship examination, Hamilton had visited his relatives at Trim and had gone over to Summerhill to see once more the place where he first met Catherine: "I ... made pilgrimage ... to the mansion where we first met, now fallen into much decay, and passed into other hands; and ... kissed, in the twilight, alone, the spot whereon I first saw rest the feet of that Beautiful Vision! ... a lady's spaniel, whose mistress I have never seen, but which had seen me kiss the ground, in that large drawing room by twilight, rose with great gravity, and licked my lips."[16]

Another serious blow to Hamilton's psyche was the death of his favorite sister, Eliza, on May 14, 1851. She had been ill for some time, but there had been no indication that her illness would be fatal. She was living in rented lodgings in Dublin, and Hamilton went to stay with her when her condition became critical. Fortunately for Hamilton it was a "beautiful death," so much desired in Victorian Britain. Otherwise it would have been difficult for him to bear. Next to Catherine, Eliza was closest to his heart. "Sydney and I were kneeling at her bedside at the very time of her departure, the *moment* of which we could perfectly fix, although it was as painless as could be conceived, literally a falling asleep, and as we fully believe in the Lord. She had said to Sydney that morning that she was very happy."[17] At the last moment, as Hamilton recalled the event, a shaft of sunlight fell on Eliza, illuminating her body as she died in his arms.[18] He purchased and preserved the pillow on which she died. Four years passed before he had the courage to open, read, and burn her letters.

Hamilton continued to hear about Catherine indirectly through Thomas Disney, but in October, 1853, just after returning from a British Association meeting at Hull, he received a package from Catherine sent through Thomas containing a pencil case and an inscription "From one whom you must never forget, nor think unkindly of, and who would have died more contented if we had once more met."[19] There appeared to be reason to take seriously the premonition of death hinted at in the inscription, and Hamilton immediately determined to try to visit Catherine at whatever risk or cost to propriety. Catherine was staying with her brother Robert in Donnybrook, and Hamilton went there directly to pay a morning call. He did not see her, but received an invitation to return for dinner. This time he saw Catherine,

quite alone, by the firelight, before being summoned to join the family dinner, which she was totally unable to attend; and while she lay, languid and strengthless, but interested and attentive and happy on a sofa to which she had been carried that she might meet me:—kneeling, I offered to her the Book [*Lectures on Quaternions*] which represented the scientific labours of my life. Rising, I received, or took, as my reward, all that she could lawfully give—a kiss, nay many kisses:—for the *known* and *near* approach of *death* made such communion holy. It could not be, indeed, without *agitation* on both sides, that for the first time in our *lives*, our lips then met. . . . Yet dare I to affirm that our affectionate transport, in those few permitted moments, was pure as that of those who in the resurrection neither marry nor are given in marriage, but are as the Angels of God in Heaven.[20]

Looking back over the five years since Catherine's attempted suicide, Hamilton was astonished that he had had the constancy to produce the *Lectures*, his first great work on quaternions, while suffering the anxiety of Catherine's illness and his own separation from her. She was dying without any firm religious belief, which filled Hamilton with horror. He hoped that somehow his love might take the place in her mind of the love of God: "If human affection can thus have a sort of earthly immortality, she ought not to doubt the constancy of that Divine Friend, in whose *love* she certainly *did once* believe, who knows and feels for all our weaknesses and infirmities and who, in his more than earthly love, sticketh closer than a brother."[21] In spite of his agitation, Hamilton asked to see Catherine again and was granted a second interview. She died two weeks later.[22]

When he heard of her death, Hamilton went immediately to the house. He would have liked to have seen Catherine in death, but more importantly he wanted to recover any books, mementos, or incriminating letters that might fall into her husband's hands. It was evening, and the door was opened by Robert Disney: "He went in for and brought out to me, the manuscript book on the Neville and Sydney correspondence, which had been found in my dear departed Catherine's possession, and as I believe, in her bed. Mr. Barlow and perhaps other Barlows were in the house, and I was not invited in, nor did I wish to be so. I saw the upper right hand shutters open but the light was soon afterwards concealed."[23]

From the time of her death Hamilton began to collect what he called "memorials" of Catherine—books, journals, poems, the pencil case, locks of hair (which he mingled with his own hair "as if it were for a joint burial"), and portraits wherever he could find them.[24] All of these he kept locked away in his library desk. He pressed Thomas so hard for a picture that he was rebuked and felt the need to apologize.[25] In March, 1854, Thomas finally loaned Hamilton a miniature portrait for two days; Hamilton arranged to have it copied in Dublin. The copy was tinted unsatisfactorily, but it still had the power to disturb him. "I locked myself up, nearly the whole of that time, to gaze on it alone; but *it* did not soothe."[26] He discovered that the likeness was better when viewed in a mirror, and he preferred to gaze on it that way.[27]

In 1861 he obtained a second and far better portrait of Catherine from her sister Louisa Reid. It also went into the library desk, from which he would take it and look at it "several times a day." It was a private matter at home. He wrote: "I will never show it to my wife."[28]

Hamilton's obsessive involvement with Catherine was not limited to collecting "memorials." He also felt the compulsion to talk about her to any sympathetic listener whose discretion could be relied upon. Augustus De Morgan, with whom he carried on a continuing mathematical correspondence, was the first outside of the Disney family to hear the story. De Morgan accepted the confidence sympathetically, and apparently told Hamilton of similar griefs of his own. For two months Hamilton concluded his letters with the explanation that he could not abstain from writing about the subject for at least a little while longer. In December he wrote that the subject had "continued to agitate [him] to a degree beyond what is rational." He found it difficult to venture out and had just sent his sons with a contribution to a meeting of the Society for Irish Missions Against Popery. He was unshaven and could not go. He said that he himself did not plan to become a Catholic as his friends had done, but added: "I feel just now the strongest possible inclination to pray not *for*, but *to* the deceased lady of my love: and 'Sancta Catharina, ora pro me' is a saying that trembles on my lips."[29]

Another correspondent was Samuel Talbot Hassell, with whom Hamilton had stayed during the British Association meeting at Hull just before Catherine's death. Mrs. Hassell had been particularly friendly, and Hamilton wrote to Mr. Hassell to ask if he had any objection to Hamilton's corresponding with his wife. He said directly that the purpose of the correspondence would be to unburden himself of grief.[30] Mrs. Hassell had loaned him a book that he had given to Catherine to try to reclaim her lost religious conviction, and she was linked to Catherine in his mind by this act. Mr. Hassell, however, thought his wife too sensitive to bear the burden of such confidence and invited Hamilton to write to him instead. Hamilton was embarrassed and even humiliated by revealing so much of himself, but he could not overcome the compulsion. Being a disciple of Coleridge, he recognized in himself the curse of the Ancient Mariner,

> I pass, like night, from land to land;
> I have strange power of speech;
> The moment that his face I see,
> I know the man that must hear me,
> To him my tale I teach.[31]

Thomas Disney and his wife suffered the greatest burden of correspondence, and on Christmas Eve Hamilton wrote: "I think that the frequency and fulness of my letters annoyed you a little, last month."[32] Yet the flow remained unabated; letter followed letter, sometimes more than one letter in a single day to the same person, all full of Catherine. Lady

Campbell was another obvious correspondent, since only she and the Disneys knew the identity of his love.[33]

The Disneys recognized a danger in Hamilton's intense feelings. He hated Catherine's husband and might argue or fight with him in a public way. Hamilton's feelings towards William Barlow became increasingly irrational, partly from hatred, and partly from fear that Barlow would expose him. He believed that Barlow had found some of his poems and possibly some of his letters that might prove incriminating among Catherine's things. In Hamilton's mind Barlow was a criminal for having forced Catherine into marriage, and he threatened Thomas Disney that "at the expense of one minute, and one penny, I can explain to each of those English friends [who knew the story] *who* was the old destroyer of her peace and mine—but I have *no wish* to do so, and shall never do it, except under the highest and most wanton provocation."[34] In more rational moments Hamilton admitted that Barlow had been completely respectable in most ways.

[Catherine's husband] had been *substantially* kind to her. She had a handsome carriage (which Lady Hamilton has never had) and was allowed to receive members of her own family as often (apparently) as she could wish. What I really can in my thoughts *thank* him for, is that he allowed her, especially for the last few years of her life, to reside very much away from him, making *long* visits to her mother, or to brothers and sisters of hers who all adored her: ... I do not imagine that he was ever *rude* to her but she confided to me, almost on her death bed, that she looked forward with *terror* to the bare possibility of her recovering health enough to make it necessary for her to live with him again.

But, in Hamilton's opinion, in 1848 Barlow had "actually preached her into madness, and very nearly into suicide.[35]

There was little chance of an encounter, because Barlow actively avoided Hamilton, and on one occasion "was hiding behind a door, in his son's rooms in ... Trinity College Dublin during a short interview of mine with that son; ... he wanted to listen to our conversation, but ... was afraid to meet me face to face." And yet Barlow was not cowardly about most things. He had been active during the famine and had been absolutely fearless regarding assassination, or so Hamilton had been led to understand.[36] As Hamilton's remarks against Barlow increased in violence, the Disneys intervened and extracted a promise from Hamilton that he would not write to Barlow unless Barlow wrote a "reproachful" letter to him first.[37]

In 1855 Hamilton added three new correspondents, all of whom were told the details of his story. Professor John Pringle Nichol of Glasgow was his host at the 1855 meeting of the British Association at Glasgow. Hamilton became attached to the whole family, and after his return to Dublin wrote to father, son, and daughter. Professor Nichol asked Hamilton to write an article on quaternions for his *Cabinet Cyclopedia,* which Hamilton supplied.[38] He also told Nichol all about Catherine, but concealed her name. In apparent jest, but with more than a hint of seriousness, he wrote to Nichol's daughter: "I am most anxious that in Dublin I should be looked upon as a perfectly prosaic person, with not a bit of the romantic about him, whereas in fact my life has been a romance."[39] This in part explains Hamilton's persistence in telling his tale. His long, drawn-out descriptions of his relationship with Catherine (one can scarcely call it an affair), which may appear overblown to our eyes, were held by him to be highly romantic and poetic. His many correspondents responded with interest and sympathy, which would indicate that they adopted his view of the story rather than the view most modern readers would have.

In 1855 Aubrey De Vere again became a regular correspondent. Travel abroad, his conversion to Catholicism, and then illness had caused him to break off his regular exchange of letters with Hamilton, but when the correspondence was resumed, Hamilton found in Aubrey a most sympathetic listener. He sent Aubrey extracts of the Neville and Sydney correspondence, something that the other confidants had never received, and the most detailed account of all.[40]

The most unusual of his new acquaintances was Lady Jane Francesca Elgee Wilde, wife of Dr. William Wilde and mother of Oscar, who was born only a few months before Hamilton met his mother. In fact, Lady Wilde asked Hamilton to be a sponsor for her "little pagan" Oscar, but for some unexplained reason Hamilton refused.[41] Dr. William Wilde was a member of the Royal Irish Academy, where Hamilton had met him often. A well-known surgeon and archaeologist, he later acquired a scandalous reputation by fathering a natural son (as if fathering the notorious Oscar had not been scandal enough). But at the time Hamilton knew them, the Wildes were respectable. Most of what has been written about them is from a later date. Harry Furniss described them in Dublin as follows:

Lady Wilde, had she been cleaned up and plainly and rationally dressed, would have made a remarkably fine model of the *Grande Dame*, but with all her paint and tinsel and tawdry tragedy-queen get-up she was a walking burlesque of motherhood. Her husband resembled a monkey, a miserable-looking little creature, who apparently unshorn and unkempt, looked as if he had been rolling in the dust. . . .

Opposite to their pretentious dwelling in Dublin were the Turkish Baths, but to all appearances neither Sir William nor his wife walked across the street. At all the public functions these two peculiar objects appeared in their dust and eccentricity. Living caricatures, in evidence, that neither Hogarth nor Dickens in their respective periods had the need to invent characters.[42]

This is biographical debunking of the most irresponsible kind, and certainly does not describe the Wildes as Hamilton knew them. It is true, however, that they were "unusual." Lady Wilde in particular had a romantic background that fascinated Hamilton. In 1845 she had witnessed the funeral of Thomas Davis, the romantic leader of "Young Ireland," and had been quickly caught up in the revolutionary movement of the famine years. She contributed frequently to the *Nation*, the enormously successful journal of "Young Ireland." Her poems and articles appeared under the pseudonym of "Speranza" and took an increasingly violent tone during 1848, climaxing in *Jacta Alea Est* (*The Die Is Cast*), an open call for rebellion. The editor, Gavan Duffy, was imprisoned for sedition, and Speranza might well have followed him into jail if the authorities had not decided that the imprisonment of a woman would probably cause them more trouble than it was worth.[43]

Hamilton often referred to Mrs. Wilde as "the strong-minded lady." She was a poetess and a feminist, two qualities that were sure to attract him. Her views on the status of women were strong indeed: "I have sometimes thought with admiration upon that excellent Chinese custom of drowning half the female children, or that still more excellent Hindu system of burning all the widows. At present myriads of eternal souls come on earth as 'ladies', and, in right of the title, live a life of vacuity, inanity, vanity, absurdity, and idleness. . . . This idle life of ladyhood is indeed the most deplorable thing in the Universe."[44] Associating with such an "original" had just enough danger in it to be exciting. They met for the first time at Colonel Larcom's house in April, 1855, and Lady Wilde made a long visit to the observatory in May. Hamilton reported to Aubrey De Vere:

She is undoubtedly a genius herself, and she won my heart very soon by praising what she had seen of the poems of my deceased sister. . . . She is almost amusingly fearless and original—and *avows* (though in that as in other respects she perhaps exaggerates whatever is unusual about her) that she likes to make a *sensation*. We agreed on many points, and differed on some, but on the whole we got on very well. I think she has a noble nature (though a rebellious one as I told *herself*—she *was* a rebel in 1848 for which I don't like her a bit the worse, though I remained on the Queen's side).[45]

He told Lady Wilde that he found her remarkable, interesting, and even "a very lovable person." He said that he understood her because he had known another female genius, his sister, and that he admired that genius. The genius of a woman, he said, is most striking when she becomes excited and seems to "beat and bruise her breast against the bars of some imprisoning cage of society."[46] Although Hamilton was not likely to do any bruising and beating himself, he was enough of a romantic and a rebel to appreciate it in others.

By the end of June Hamilton had begun to reveal himself more fully in his correspondence with Lady Wilde. He had just read Hawthorne's

Scarlet Letter and it had cost him three nights' sleep: "Ah, I could tell you a story, in real and almost recent life, though stretching back in its extent over fully 30 years, to which that tale has some affinity. ... This 'bearer of the Scarlet Letter' had only sinned in *thought* ... she swallowed poison; and was with difficulty preserved to linger, under unmerited *self-reproach* ... for 5 years longer on earth, bearing, alas! her Scarlet letter with her always." He went on to describe his last visits with Catherine "at the very court of Death" and the "kisses without limit" that he took "when the known and near approach of the Destroying Angel had confounded all earthly distinction."[47]

As the correspondence continued it took on a mildly flirtatious tone, and Hamilton tried to match the style of Lady Wilde, a style that did not come to him naturally. He remarked jokingly that his wife was becoming jealous. In fact it was more than a joke. She was indeed jealous of Lady Wilde, and even more so of Dora Disney, Thomas Disney's wife, who Lady Hamilton thought was fast taking the place of Catherine in Hamilton's affections. In June, 1855, she found a letter from Dora that had dropped out of his pocket. That incident provoked a domestic quarrel, and Hamilton asked Dora to have her letters addressed by Thomas, but he did not stop the correspondence.[48]

Hamilton's unburdening of himself before so many correspondents seemed to have salutary effects. During the next years the references to Catherine became less frequent, although he continued to talk about her with the Disney family. Thomas and Dora were living in nearby Finglass, so Hamilton saw them frequently. Dora obviously liked Hamilton and had been willing to put up with a constant stream of lengthy letters regarding Catherine and numerous appeals for "memorials." She recognized that the letters were expressions of his anguish and she did all that she could to steady his mind and help him through his crisis.

In 1861 Louisa Disney, now Louisa Reid, met Hamilton at Thomas Disney's house and they began to correspond. For a brief period in 1827 he had thought of marrying her. Louisa was fascinated with the story of her sister and her inquiries revived the old feelings in Hamilton. Her letters are teasing and flirtatious, while his are increasingly disturbed. In June Hamilton confessed to Lady Wilde that Louisa had come to visit him at the observatory, where they had last met thirty-four years ago. He had taken her into the meridian room alone. "We did not talk of Astronomy! ... I must admit that our lips met (for the first time in our lives)."[49] The correspondence that ensued from this meeting was semiclandestine. Hamilton insisted that they were doing nothing improper, but he instructed Louisa how to send her letters so that they would come to him directly and not pass through the hands of any servant.[50] Louisa sent him a photograph of Catherine that he mooned over in all sorts of lights—"the morning sun, the noonday sun, the evening sun, the twilight, and now candle-light."[51] He, in turn, sent her a copy of the correspondence of "Neville and Sydney," the only complete copy that he ever let out of his hands.

No person ever replaced Catherine in Hamilton's affections from the moment that he first saw her until his death, and it is difficult to imagine anyone being loved with more constancy over a longer period of time. Perhaps if he had married her the glitter of his romantic vision would have rapidly tarnished, but he did not marry her, and the vision kept all of its luster. His romantic attachment caused him great agony. Not only did he suffer from the loss of Catherine, he also suffered from his inability to express his feelings in a socially acceptable way. Lady Wilde's boldness attracted him because she was fearless in the face of society. He had no such courage. Aubrey De Vere was a philosophical comforter, and the Disneys could stand in some way as a replacement for Catherine herself. His wife was little help, and in this matter could not be. Her feelings are not recorded, but she may have suffered as much as Hamilton. She must have known that his love and attention were directed elsewhere. Certainly she must have known a great deal more about what was going on than he realized. There is no indication that he ever shared his feelings with her. Before they were married he told her that he had once loved Catherine Disney, and he told her when Catherine died; this much was "proper." But he told her nothing more. Whatever she was able to discover or imagine (and it probably was a great deal), she kept it to herself.

26

The End of a Career

When Hamilton presented a copy of his *Lectures on Quaternions* (1853) to Catherine Disney on her deathbed, he was presenting her with the fruits of ten years of labor and what he believed to be the crowning achievement of his life. It was not the end of his interest in the quaternions, however, and with the exception of the icosian calculus, which took much of his attention in 1856, he worked on quaternions almost exclusively until the end of his life in 1865. Hamilton was not a good expositor of his own ideas, and he seemed incapable of acting on the advice of his friends, no matter how good his intentions. The *Lectures on Quaternions* contained over 700 all but impenetrable pages. In 1855, two years after its publication, Hamilton complained to Lady Wilde, "I have not been able to ascertain that a single copy of the Quaternions has been bought here," and two years later he wrote to another correspondent, "When I said that Dublin was not a book-buying place, perhaps I spoke with some little bitterness, on account of its not having bought any copies (that I can hear of) of my own book."[1] Sales were better in England, Scotland, and North America, but the total number of copies sold must have been small. Trinity College ultimately bore the entire expense of the printing and recouped only a small fraction of its investment.[2]

When Hamilton proposed a second book on quaternions, the college was understandably cautious. He began in 1858 with the intention of writing a simple "manual," an easy introduction to the basic theory of quaternions for students and interested beginners. By 1859 he was circulating sample pages. John Herschel warned him constantly to be brief and to the point. Herschel had made a second assault on the

Lectures in 1859 that had carried him through the first three chapters, but as he wrote: "I was again obliged to give it up in despair. Now I pray you to listen to this cry of distress. I feel *certain* that if you pleased you *could* put the whole matter in as clear a light as would make the Calculus itself accessible as an instrument to readers even of less 'penetrating power' than myself, who, having once mastered the *algorithm* and the *conventions* so as to work with it, would then be better prepared to go along with you in your metaphysical explanations."[3] De Morgan said that a manual was just the thing wanted. "I shall be glad to rub up by such a thing. Be brief in explanation—refer fully to the lectures."[4] But when De Morgan asked again about the "manual" in 1864, Hamilton ruefully replied that he had just passed page 700 and that there was no longer any chance for a manual. The book had now become a "reference work."[5]

In 1860 he had promised Dr. Hart, the college bursar, that he had "not the least notion" of extending the book to 400 pages "or anything *like* that." He expected to add only £25 of his own money to the college grant of £100; he stated: "I feel that the *bulk* of the Lectures has been a most serious disadvantage to them and I intend to be much more brief this time."[6]

In fact the "manual" turned into a book entitled the *Elements of Quaternions* and was incomplete at his death. Containing 762 pages of text and 59 pages of introductory materials, it was published posthumously by his son William Edwin. The problems that had plagued the earlier *Lectures* reappeared in the *Elements*. Hamilton sent his manuscript directly to the college printer and made his revisions on the printer's proof. As new ideas came to him he spun them out (just what De Morgan had warned against), and he soon lost control of any organizing structure for the book. He had absolutely promised De Morgan that this time he would not go to press "till the *whole* book is actually in *manuscript* and not merely in embryo, or in my *head*," but all to no avail.[7] The more he wrote the more he felt he had to write, and the greater his debt to the college printer.

As Hamilton labored on the *Elements* he sent the proof sheets to Peter Guthrie Tait, the only real disciple he ever had. Tait was at this time professor of mathematics at Queen's College Belfast. Hamilton received a formal letter of introduction to him from the vice-president of Queens in August, 1858, and soon Tait and Hamilton were in correspondence.[8] Where Herschel had only been able to get through the first three of Hamilton's *Lectures on Quaternions*, Tait claimed in his first letter to have "easily mastered" the first *six* lectures, but he found that the physical applications of the method, which were for him the most important part, gave him a great deal of trouble.

But for your offered assistance, I am afraid I should have had to relinquish all hopes of using quaternions as an instrument of investigation, on account of the time I should have had to spend in acquiring a sufficient knowledge of them. I have all along preferred mixed to pure mathematics, and since I left Cambridge where the former are comparatively little attended to, have been busy at the theories of Heat, Electricity etc. Your remarkable formula for $\{(d/dx)^2 + (d/dy)^2 + (d/dz)^2\}$ as the square of a vector form, and various analogous ones with quaternion operators appear to me to offer the very instrument I seek for some general investigations on Potentials, and it is therefore almost entirely on the subject of "Differentials of Quaternions" that I shall trespass on your kindness.[9]

Hamilton was pleased by Tait's interest and more than a little impressed by the amount that Tait had been able to learn on his own. He wanted to know if Tait had had any assistance from others, for, "If the Lectures on Quaternions have been your *only* teacher, I must consider the result of such a state of things to be not only creditable to *your own* talents and diligence, but also complimentary to, and evidence of, some *didactic capabilities* of my volume."[10]

Hamilton began by sending problems that Tait then worked out and returned to be checked.[11] Soon the pupil was making excellent progress— such good progress, in fact, that he contemplated publishing a brief practical description of the method of quaternions himself. Unlike Hamilton, Tait was eager to get on with the quaternions so that he could turn to other things, and in October, 1860, he wrote that he was urging his publisher, Macmillan, to name a definite time for the appearance of his proposed volume. Hamilton had told Tait that his own book should be ready by January, 1861, and Tait was seeking a provisional publication date for sometime the following summer. It was agreed that he would publish merely a set of "examples in quaternions." This would not be a complete exposition of quaternions, but merely a description of some important applications of them in physics. But in a letter of October, 1860, he referred to his book as an "Introductory Treatise," and it was advertised as such by Macmillan. Tait even asked Hamilton: "If you could favour me occasionally with a sight of some of your proof sheets, I have little doubt of being thereby enabled materially to improve my own."[12]

His letter came as a dash of cold water on a heretofore warm friendship. In 1860 Hamilton still thought he was writing a "manual" on quaternions, and an "introductory treatise" was dangerously like a "manual." He turned to De Morgan for advice, telling him that Tait had been introduced as a "scientific pupil," not as a "rival author." He continued:

The correspondence which ensued was very copious, and confidential. If printed, it would make as I estimate, a not small octavo volume. It was conducted, throughout, on my side (and I supposed upon the other), in a spirit of entire confidence and unreserve.

To my extreme surprise and *disapproval*, after something like a year had thus passed, Mr. Tait advertised, with Messrs MacMillan of Cambridge, about the end of 1859, the publication of "An Introductory Treatise on Quaternions" etc. I ask you, privately, as an honourable man, do you consider this conduct to have been *honourable?*

That *you* (like myself) would have been incapable of it, I am well assured. But perhaps you and I are a pair of fastidious old fools, unfit for the present time. What would London authors say to it?[13]

Hamilton admitted that he had given permission for the publication of a set of "examples," but it seemed to him that Tait was going well beyond that.

In 1860 Tait moved to Edinburgh, where he succeeded J. D. Forbes as professor of natural philosophy. In his inaugural address he went out of his way to praise Hamilton and extol the quaternions. The effect was immediate. Hamilton wrote to Aubrey DeVere that Tait had spoken "in a way that might avail to soften the rugged heart of a Hyrcanian Tiger, or Bear: and *therefore,* let us hope, that of your humble servant! In short, he has completely soothed my *feelings:* as to my *interests* ... I do not care much about *them*—but all is likely to come right. (I had *not* written to him.)"[14] Since Hamilton had not written to Tait, it was probably his chilly silence, possibly combined with a discreet note from De Morgan, that had warned Tait of Hamilton's displeasure. At the end of 1860 Tait wrote two more letters asking for a "perfectly distinct idea" of Hamilton's wishes on the matter.[15] Either Hamilton could not, or would not, speak his mind with sufficient clarity, and so the following June (1861) Tait came to visit him at the observatory. In parting he asked again what Hamilton's wishes were regarding publication, to which Hamilton replied cryptically that they could be summed up in a couple of words, or in a quaternion of figures: *61, 62;* meaning that Hamilton was to publish in 1861 and Tait in 1862.[16] Tait took this to mean that he should wait for Hamilton, no matter how long that might take. Certainly Hamilton was being overly optimistic in suggesting in August that his book might come out that same year.[17] Four more years were insufficient to complete it. Moreover, the quaternions had been public property ever since Hamilton had announced their discovery in 1843, and there was no reason why he should insist on a monopoly on the subject. Tait must have realized all this, but he decided to hold up publication of his book indefinitely rather than risk offending Hamilton further. He responded by saying that he would not be going into print soon and that he was about to begin a "treatise on physics" with William Thomson (later Lord Kelvin), "there being none in English which is worth twopence."[18]

Hamilton could well have allowed Tait to publish his work, because the two men held quite divergent interests. Hamilton, no matter what he might have said to the contrary, was a mathematician first and foremost. In 1860, when his difficulties with Tait began, his enthusiasm was for anharmonic coordinates and the geometry of nets in space. In fact, the *Elements of Quaternions* starts right off with these two rather difficult subjects that have little to do with the quaternions themselves. Quaternions are not even introduced until the second book.[19]

Tait, on the other hand, was primarily a physicist. He told Hamilton that he was "too physical" in his ideas to appreciate an anharmonic equation, and that his primary use of the quaternions was for electromagnetic induction and potentials.[20] In spite of frequent urging from Tait, Hamilton could not bring himself to work on the physical applications of quaternions, and since he was not particularly interested in that aspect, he could have afforded to let Tait publish on it. Tait was able to present some of his views in journal articles, but he continued to hold back his book and urge Hamilton in the direction of physics:

I would ask you ... to reconsider your determination as to the Δ. Δ and φ will be, for perhaps a century to come, the quarter from which extensions of mathematical Physics may most reasonably be looked for; and I think that you *ought* to give, in as brief a form as you please, all you can about them in your new book. ... On this depends the distortion of an elastic solid, the motion of fluids, and a great part of the statical theories of Heat, Electricity, and Magnetism. ... Such things as these would show the enormous value of quaternions to *Physicists,* and that is in my eyes worth any amount of pure analysis or geometrical theorems.[21]

Tait also sent him proof sheets of "Thomson and Tait" containing complimentary references to Hamilton's Principle of Varying Action, but Hamilton confessed his limitation with regard to physics:

The world of Science seems to admit that I had not only read but written to some purpose on *Dynamics* about *thirty years ago;* but a new generation has arisen. *Energy* and *Work,* in the *old* English meaning, are things not unfamiliar to me. But I have only the dimmest views of the modern meanings attached to those terms. From "Thomson and Tait" I hope at last to be instructed on the point: *not,* of course, without patient study on my part. A certain humility and teachableness of the *ethical* kind I can promise. ... Your book when complete, may enable *even me* to understand all this, slight as my *instinct* is for physics.[22]

All of which indicates that Hamilton did not really understand the physical concepts that excited Tait, and he considered himself too old to begin a serious investigation of the subject.

Thomson, who was coauthoring the book with Tait, was not a lover of quaternions, and therefore he successfully prevented their inclusion in the work, but Hamilton was also responsible in part.[23] Tait wrote to him in June, 1863, asking if he might add an appendix giving quaternion solutions for problems that were solved only with difficulty using coordinates. He intended to write only "a dozen or two" pages and would include no explanation of quaternions. They would be used only "as an instrument." Such an appendix would attract people to the quaternions and would draw attention to Hamilton's work.[24] Hamilton replied that he had "perfect confidence" in Tait's sense of propriety, adding that his only wish was not to be anticipated in a book on quaternions.[25] The following August Tait sent a rough draft of the proposed quaternion appendix for volume 1 of "Thomson and Tait," asking: "Would you kindly look over it—and *return it soon*, with answers to the following questions—(1) Is it worth printing? (2) Is its tone satisfactory to you?— If No. 2 is answered in the negative pray suggest any desirable modification."[26] A month later he asked for it back *"within a fortnight"* with Hamilton's opinion as to the advisability of printing it.[27] In December Tait finally gave up. Hamilton had returned half a dozen pages with his comments, but Tait wrote to him that it had been decided not to publish it with their volume. He asked for the return of the remainder of the appendix with a simple imprimatur so that he might publish it as a separate article.[28]

When Graves wrote about these incidents in the third volume of his biography he was mildly critical of Hamilton, suggesting that his treatment of Tait had not been reasonable.[29] Tait, however, disagreed, and criticized Graves in turn for making Hamilton appear unaccountably capricious in their discussions about publication of the quaternion books. He wrote: "I had, of course, no rights in the matter:—and I cheerfully submitted to the restrictions he imposed on me; especially as I understood that he expressly (and most justly) desired to be the first to give to the world his system in its vastly improved form."[30] Hamilton had, however, already given his "system" to the public. By the time of his death he had published 109 papers and an enormous book— all on quaternions.[31] Granted that the *Elements* represented a more refined version of his earlier work (in the *Elements* the coordinate vectors *i, j, k,* appear only incidentally), it is still an exposition of a subject that Hamilton had described in detail over a period of twenty years. It is difficult to see how he could have justified the restrictions he put on Tait, although by nineteenth-century custom it was entirely proper.

Reflecting on their correspondence, Tait concluded that Hamilton was incapable of writing a treatise that would do justice to his own vision of the quaternions.

I do not now think that Hamilton, with the "peculiar turn of mind" of which he speaks, could ever, *in a book*, have conveyed adequately to the world his new conception of the Quaternion. I got it from him by correspondence, and in conversation. When he was pressed to answer a definite question, *and could be kept to it,* he replied in ready and effective terms, and no man could express *viva voce* his opinions on such subjects more clearly and concisely than he could:— but he perpetually planed and repolished his printed work at the risk of attenuating the substance: and he fatigued and often irritated his readers by constant excursions into metaphysics. One of his many letters to me gave, in a few dazzling lines, the whole substance of what afterwards became a Chapter of the Elements; and some of his *shorter* papers in the *Proc. R. I. A.* are veritable gems. But these were dashed off at a sitting, and were not planed and repolished.[32]

On this point Tait was certainly correct. Hamilton could not keep his published works on quaternions within reasonable bounds. During the 1860s he was writing for posterity. What started as a brief "manual" had expanded until it became in his mind his greatest mathematical monument, if not for his generation of mathematicians, then at least for future generations. He put every bit of his physical and mental energy into this one last great effort. He kept a diary of his progress— at first a simple account of work done and lists of pages sent off to the printer.[33] But gradually he began to include in the diary events of his life and brief accounts of his activities. Because he was retiring more and more into the observatory, these activities are not of great interest, except for the way in which he recorded them. One gradually realizes that Hamilton was keeping a record for future biographers. It is at this time that he began to record letters sent, the times of mailing, and sometimes even the name of the servant who took them to the post.

In describing his occasional trips to Dublin he recorded people seen, errands completed, and articles purchased, all in obsessive detail. On one occasion he recorded having purchased gloves at a shop, but because he could not remember the name of the shop, he left a bracketed space that he later filled in with the correct name.[34] He began to boast more than usual about his medals and other honors. Underneath all this show of confidence and self-esteem, however, one can detect his anxiety about his ever-expanding manuscript. It was terribly important to him, but as it grew it put serious burdens on his health and on his bank account.

Hamilton had been fortunate with the publication of his previous *Lectures on Quaternions.* The Board of Trinity College had granted him £200, and another £100 in August, 1854, when he had been confronted with a printers bill that he could not pay.[35] This amount covered all but £20, which came from Hamilton's own pocket. The eventual total cost of printing the *Elements of Quaternions,* however, was close to £500. Hamilton had received a grant of £100 from the board for his proposed "manual," but soon this money was consumed. He was hesitant to go to the board again, but knew that he could never bear the expense of publication by himself. In the summer of 1861 he had an angry exchange with the university printer, Gill, who had apparently made application to the board without his permission. At that point he was prepared to seize the printed pages, blocks of type, and so on from Gill and make his own arrangements with another press.[36] He told Professor Hart that he had put in no less than 10,000 hours of labor on the pages completed thus far and that he believed the book still deserved the college's support.[37] The board bailed him out again with an additional grant of £100, but this time the money was awarded with the stipulation that it not be paid until the work was finished, and that any additional expenses were to come from Hamilton's own pocket.[38] With the immediate financial problem solved by this second grant, Hamilton proceeded to run up the debt again over the next three years. He passed every deadline and every page limit that he set for himself without being able to finish the book. After his death the board assumed the entire expense of the publication except for £50 that Hamilton had already paid. This was an act of great liberality by the board, but coming as it did after his death it could have done nothing to relieve his anxiety in 1864 and 1865.

During the time that Hamilton was working on the *Elements* his life was beginning to change in other ways. One of the ways was a growing concern for his children, whose names appear more frequently in his correspondence. He was probably more attentive than most Victorian fathers, because Lady Hamilton's frequent illnesses caused him to take a greater responsibility for the management of the household than was customary at the time. William Edwin, their eldest son, had been born in 1834, and Hamilton told Herschel that his son had been his "nearly constant companion during the first twelve years of his life."[39] These twelve years came to an end in 1846, when Hamilton reluctantly sent William Edwin to the Reverend Charles Pritchard's boarding school in Clapham, where he would have the sons of the Edgeworths and Herschels as schoolmates. Following his studies at Clapham, William Edwin entered Trinity College Dublin, where he did reasonably well. At times Hamilton urged William into studies that were beyond the capabilities of a child his age, but there is no indication that he drove him the way Uncle James had driven his children. After receiving his son's first report from Pritchard he told Herschel that the grades were "very satisfactory," although he had obviously hoped for more. He admitted feeling pleasure at finding William Edwin at the head of his division in science.[40] In 1852, after William Edwin had entered Trinity College, his verdict to De Morgan was: "he is no prodigy, but a good and sensible and *reasonably* industrious boy, and may yet do very well."[41]

Hamilton's second son, Archibald Henry, nicknamed "Arch," was born a year after William Edwin, on Hamilton's thirtieth birthday. Arch was apparently more serious about his studies than William Edwin had been, because Hamilton talks about taking him through six books of Euclid at a time when Arch was seventeen. We also hear of him avidly reading his Homer, and in 1852 Hamilton called him a "tremendous bookworm," adding the judgment that "he is also a good boy."[42] Arch particularly enjoyed puzzles and ciphers sent by Hamilton's friends De Morgan and Nichol, and he began to make his own investigations into geomagnetic and electrical phenomena while in college. He devised a telegraph system for the observatory and a scheme for a telegraphic connection with the observatory at Greenwich.[43] Arch also seems to have done respectably at Trinity. His only recorded misdemeanor was involvement in a major student riot in 1858.[44] The occasion was the return of the Earl of Eglington as lord lieutenant on a change of ministry. Eglington had been popular, and a holiday spirit reigned in Dublin. Several hundred students milling outside the college threw firecrackers and orange peels at the police and crowd. After the parade had passed an orange struck Colonel Browne of the police in the face. Browne requested that the Scots Greys, who were posted by the Bank of Ireland, charge the crowd of students. When the commander of the Scots Greys laughingly refused to attack "schoolboys," Colonel Brown assembled mounted and foot police. The mounted police charged with sabers, and were followed by the foot police, who carried batons. There were several near-fatal injuries. In the mêlée Arch received a saber stroke from a mounted policeman, but fortunately his wound was from the flat of the sword and not from the edge. Hamilton was of two minds about the affair. He wrote that "no printed report has exaggerated the violence of the police," but at the same time he agreed with the college authorities in "disapproving of the foolish conduct of the boys."[45] Other than this event his sons made it through adolescence without causing him serious distress except for the illnesses that were an inevitable part of family life in the nineteenth century.

His daughter Helen (affectionately called "Moo") was born in 1840, five years after Arch, and it was the birth of this child that precipitated Lady Hamilton's first protracted absence from the observatory. When the boys left for college, little Helen remained at home, and Hamilton became closely attached to her. In a letter to Lady Campbell he gives an insight into what her life must have been: "I have played a few games of draughts this morning with my daughter—dear, sweet, patient child. She has very few sources of amusement, but she enjoys flowers, the garden, and the Bible."[46] Hamilton liked to talk about his daughter with De Morgan, who delighted in sending her puns and conundrums in return.

The reality of serious childhood illnesses is revealed by an exchange of letters in December, 1853. De Morgan's daughter was ill, and Hamilton tried to lift his spirits with a light-hearted quotation from the Beggars' Opera: "I wonder any man alive would ever rear a daughter," and then went on to say, "I am apt to estimate the happiness of any friend of mine by the circumstance of his or her having a daughter. Of course the most exquisite *un*happiness might be derived from such a source: but that alternative we may dismiss from thought." De Morgan replied: "I think, if I remember right, that I wrote you a short note, telling you how completely the illness of my eldest daughter prevented my even reading your notes. On the 23rd of last month she died in her sixteenth year ... I am just beginning to recover—I will not say my spirits—but power enough to attend to things absolutely essential."[47]

Hamilton's turn came in August, 1858, when little Helen became seriously ill while visiting her relatives at Trim. He gave up his plans to attend the British Association meeting and hurried to Trim, where he stayed for three weeks seeing her through her crisis. Finally, when she was able to get out of bed, he reported to his sister Sydney that she was better, but added that the burden on him had been heavy, both financially and emotionally: "This has, of course, been a *very* expensive business to me, but at least the child's *life* appears to have been saved."[48] Little Helen recovered slowly. She was still an invalid the following July when Hamilton took her to Rhyl in North Wales to stay for a while with her mother's relatives. In September he returned to take her to Fulneck, where she stayed with his Uncle Willey at the Moravian settlement while he attended the British Association meeting at Aberdeen. Hamilton also visited for a while at Fulneck and then brought Helen back to the observatory, fully recovered, but only after almost a year of convalescence.[49]

Occasionally Hamilton would take one of his children along on a business or vacation trip. His wife was unable to travel at all, and he enjoyed the children's companionship. Arch went with him to the 1854 British Association meeting at Liverpool and then on to the Lake Country, where they visited old friends in the vicinity.[50] In 1858 he took William Edwin to Edgeworthstown, where a formal welcome was made to William Edgeworth, Francis Edgeworth's son, who was returning from military duty in India.[51] And then in 1864 he took his daughter Helen on an excursion into Wicklow.[52] These occasional trips drew him closer to his children and gave them some sense of what his life was like outside the confines of the observatory.

The letters of 1856 give the first indication that the expenses of his children and of his quaternion publications might be causing him financial difficulties. William Edwin had shown an interest in engineering, and Hamilton arranged to have him work in the engineering firm of George Willoughby Hemans, son of the poetess Felicia Hemans, whom Hamilton had known thirty years earlier. Hemans agreed to take William Edwin into his employ for £400, an enormous sum, but one that Hamilton seemed to regard as entirely reasonable. Hamilton could not raise the entire amount, and finally settled with Hemans for £350 and the stipulation that it be paid immediately. In order to raise the money he borrowed against his life insurance at 5 percent interest, thereby incurring a debt that rolled around with unnerving regularity.[53] He warned William Edwin to be careful of his financial affairs:

You know that I have not *much* money to spare, but still I hope that in any little distress you will always apply to *me, first, and not* to any comrade of yours. Even the accepting of 5^s loan, without the *clear* understanding of its being repaid on the following day, or at all events *very soon,* has a lowering tendency, not merely in other people's opinion, but in one's own. ... tell me freely if you are [in want of money], and you know that it is not out of an inquiring habit that I recommend you to keep a *most minute* journal of your expenses and in such a form, that if I should ever wish to see it, I would be able to understand it fully.[54]

Hamilton must have felt relieved at having his oldest son gainfully employed, although his employment came about through a substantial cost to himself. William Edwin lived in London, and for part of the time his lodgings were next-door to De Morgan's, although they seldom saw each other.[55]

In spite of Hamilton's investment in his education, William Edwin does not seem to have been able to find a permanent position as an engineer, and in 1862, at the age of twenty eight, he was still unemployed. In May of that year events took a sudden turn. Hamilton's sister Sydney wrote with enthusiasm about a letter in the *Athenaeum* written by Dr. Edward Cullen describing a project for cutting a canal across the Isthmus of Darien in Central America.[56] The letter announced that a French expedition was sailing within a month to survey the route that he had discovered in 1849. Cullen claimed the acquaintanceship and support of Emperor Napoleon III, with whom he had discussed the project in 1853, 1857, and 1859. He admitted that an expedition of 1854 had been unsuccessful in finding the route, but blamed that failure on the engineer in charge of the expedition, Lionel Gisborne, who had been so eager to find the transit himself that he had left Cullen locked up on shipboard.[57] In a letter to the *Irish Times,* Cullen also described an observatory at Bogota built in 1803. Since it was at the highest elevation and at the lowest latitude of any observatory in the world, it should have the best viewing conditions, and Sydney, who had collected a substantial amount of astronomical knowledge during her years at Dunsink, wrote to Cullen to ask about the observatory and about the opportunities for immigration.[58]

Sydney Margaret Hamilton

In August she wrote to Hamilton again to say she had decided to go with Cullen to South America and that she wanted to take William Edwin with her. "Being literally devoid of money here he may be extremely glad of such an opportunity of making the acquaintance of South American Merchants and of seeing the world. I always loved him from his infancy, and I cannot and will not easily be put off doing something for him now, and he certainly has no excuse for refusing my invitation."[59] From her own savings Sydney proposed to pay William Edwin's travel expenses and his maintenance for one year. She pointed out the uncomfortable fact that his prospects were getting worse and worse in Ireland and that he had no means of support. South America was a rich country. She had heard that even the snail shells could bring as much as £50 each in London, and that the land yielded its return in only four months.[60] Surprisingly, Hamilton gave his consent the next day. Granted that his only other alternative would have been to try to find a job for William Edwin at the observatory, it is still surprising that he agreed to the expedition without looking further into Cullen's reputation and the accuracy of his claims. He did check with T. Romney Robinson, the astronomer at Armagh, about the Bogota observatory, and learned that it had already been considered by the English astronomers as a possible location for the southern reflector, but the idea had been given up because "among those hybrid Spaniards, there is no real security of life or property." They would resent the arrival of a British astronomer; besides, the society had no funds for such an undertaking.[61] Sydney remained undaunted and hoped to be able to reach Bogota to look at the observatory herself, but there is no indication that she ever succeeded.

In arranging the expenses of the journey Hamilton had to admit some financial embarrassment: "My insurance policy is already burdened to a somewhat dangerous extent . . . in consequence of the loan which I incurred for WEH, altho' I have scarcely ever (I think) reminded him of it. . . . But I fully purpose to contribute what I can, to Will's share of the expedition."[62] The fact of the matter was that Cullen was a fraud— one of the more colorful frauds to propose a canal across Central America. It is true that Napoleon III had a passionate interest in an isthmian canal (although he favored the Nicaragua route over the Darien route that Cullen proposed). Napoleon had even organized a company, La Canale Napoléone de Nicaragua, while a prisoner at the fortress of Ham, and upon his escape in 1846 had written a pamphlet on the subject.[63] It is possible that he had actually received Cullen on the three occasions that Cullen claimed, but the rest of Cullen's letter strayed a good way from the truth. The expedition of 1854 to Darien had failed, not because of Lionel Giscombe's perfidy, but because Cullen's supposed transit did not exist. He had claimed to have discovered in 1849 a gap in the mountains of Darien at an elevation no higher than 150 feet above sea level, and to have walked easily across the isthmus several times, each journey taking only a few hours.[64]

A measure of Cullen's persuasiveness was the size of the expedition that he had been able to instigate in 1854. It was composed of ships and engineers from Great Britain, France, the United States, and New Granada (the name of the government seated at Bogota that controlled the territory). A British party starting from the Pacific side of the isthmus could find no gap and lost three men to hostile Indians before they gave up. An American party under Lieutenant Isaac Strain set out from the Atlantic side and soon found themselves lost in what was, and still is, one of the most impenetrable jungles of the entire world. Darien is the one unfinished gap in the Pan American Highway, and it remains as formidable now as it was in 1854. Lieutenant Strain and his crew fought the jungle for forty nine days before a starving remnant of his crew finally reached the Pacific Ocean. Cullen's story of an "easy walk" across the isthmus was a lie.[65] Instead of an elevation of 150 feet, the actual minimum elevation for the mountains is approximately 1000 feet—making a canal at this location entirely impractical.

After the failure of the expedition of 1854 Cullen disappeared from public view and served as an army doctor in the Crimea for several years. In 1861 he was back in Dublin with new tales of Central America and its advantages for unemployed Irishmen. The discovery of gold in California in 1848 had greatly increased the traffic across the isthmus, and by 1855 there was a railroad linking the Atlantic and Pacific Oceans at Panama, but it had been built at an enormous cost. The mortality rate among the workers, many of whom were Irish, led to the rumor that there had been one man buried for every tie laid on the railroad, which should have led Sydney and William Edwin to think twice before emigrating.[66]

The other usual route for travelers crossing the isthmus was through Nicaragua, from Greytown (San Juan del Norte) up the San Juan River, across Lake Nicaragua, and down the other side by coach to San Juan del Sur. Napoleon's dream of a canal along this route was never fulfilled, but Cornelius Vanderbilt set up a transit that became a lucrative rival to the Panama route for several years.[67] The opening of the Nicaragua route suddenly made Greytown an important harbor, and it was to Greytown that Sydney and William went in 1863.

Once Hamilton had given his consent, Sydney arranged with Cullen to have William Edwin come along.[68] The emigrants departed from Liverpool in September, 1862. Sydney reported that there had been some trouble at Liverpool with the shipping company, but when Cullen received a supporting letter from the ambassador, the authorities "saw that Dr. Cullen was no imposter and that [they] might be *somebodies*," and there was no further trouble.[69] Sydney probably expected to receive a grant of land, because Cullen claimed to have received 200,000 acres and a concession for a canal from the government of New Granada in 1852. But since the government of New Granada was prepared to grant Cullen's land and concession to almost any European or American who asked for it, Cullen's claims were insecure to say the least, and there is no indication that Sydney or William Edwin ever acquired any land.[70]

The ship first stopped at Jamaica. From there Sydney and William Edwin wrote that they intended to sail to Santa Marta in New Granada. Possibly they hoped to reach Bogota and take up Cullen's claim.[71] But this plan fell through and they went to Greytown instead, in the spring of 1863. Greytown was already on the decline economically by the time the Hamiltons got there. William Walker and his "filibustering" associates had taken advantage of the political turmoil in Nicaragua to seize control of the country and the transit in 1855.[72] Walker finally ended his career before a Nicaraguan firing squad in 1860, and the transit was opened again just before the Hamiltons arrived, but traffic was much reduced and the company never again turned a profit.

Sydney was enthusiastic about Greytown in spite of the difficulties of life there. She had hoped to find a job as a schoolteacher, but as the population of the town was less than a thousand, and that largely Spanish, there did not seem to be a great demand for her services.[73] The food was terrible (William Edwin reported eating turtle, sea cow, monkey chops, alligators' eggs, guava, and other unappetizing peculiarities) and terribly expensive, but his first letters were also optimistic.[74] He ran into trouble with the authorities over his passport, however, and soon his father was receiving requests for money and help in obtaining papers of identification.[75] In June William Edwin wrote a frightened letter:

I now find that the long residence in the tropics is telling seriously and rapidly on my health. (About a fortnight ago the thermometer was 97° in the *shade* and the glare of the sun was so strong that from looking suddenly on *the grass* I was so affected that I could not read or write for about 10 days afterwards the optic *nerve* was so excited.) For the last fortnight I have been almost every day confined to bed with dry burning fever and chills with violent headache and difficulty of sleep. Even Sydney [who] always said that I would get over the heat, and get accustomed to the climate admits now that she wished I could get some change of air.[76]

The letter ends with an appeal for money to return home before his health was completely ruined. Hamilton replied that he could send no more money, but that if William Edwin would spend some of the money already sent to try to reach a place with a better climate he would try to replace it.[77] He recommended against his immediate return to Ireland, because that

William Edwin Hamilton

"would have too much the air of an admitted *failure.*"[78] William Edwin decided to go to Canada, while Sydney with her naive optimism stayed on in Greytown. He later recalled the horrible food in Nicaragua, adding: "Of course SMH [Sydney] is not to blame as she had nothing better herself, but she might have given me some idea of the sort of miserable diet there. She is greatly to be envied in one way, she is so sanguine and believes so firmly, in the fortune that she is *going* to make, but she's as easily imposed on as ever."[79] Sydney collected and sold boxes of natural history specimens and gave lessons in Spanish to a Spaniard (who almost persuaded her to go into partnership with him in buying an island in Lake Nicaragua). William Edwin scoffed at her helplessness in financial matters. The lessons to the Spaniard were a financial loss because she fed him tea ("my tea by the way," mentions William Edwin) and bread and butter, all of which were hideously expensive. And yet Sydney remained eternally optimistic.

In Canada William Edwin went to stay with the Walter Keatings, the same relatives of Lady Hamilton's with whom Helen had stayed in Wales in 1859. They had since emigrated to Barrie in Ontario and now were boarding William Edwin. The family connection did not persuade them to reduce their charges, however, and after several months Hamilton wrote an anguished letter to Edwin about an unexpectedly large bill that he had received:

It is totally out of the question that I should undertake to pay Mr. Keating in Canada for you in 1864, anything *like* what I did at Rhyl ... for Helen in 1859.

Any debt which you may unadvisably contract, I will do all in my power to assist you in repaying. I have no doubt that you had no option, in the circumstance, but to go to Barrie in Canada. But you now understand, that I can by no means undertake to comply with Mr. Keating's terms. It will be strange if, after all the money that has been spent on your education and subsequent professional advancement including £350 paid down to Mr. Hemans, you cannot find means to support yourself.[80]

There follow complaints about Gill, the university printer, and the expense of his book. The matter was obviously weighing on Hamilton's mind, and in April, 1864, he wrote to his other son: "The more I think of it, the less use there appears, in your brother remaining in Canada. As regards the Keatings, while paying an extravagant sum, I am put in the position of *seeming* to be a beggar on them."[81]

William Edwin continued to look for a job in Ottawa and Toronto, but he had no luck, not even when he offered to work as a common laborer. In a defensive letter he admitted that the emigrants' dream was severely tarnished in Canada:

Of course I know that people in the old country (who have got an idea of Canada as a wild country where a person can live for almost nothing and dress as he pleases) will say that no one can fail to do well in Canada except through his own fault, but although of course a certain deference is due to the opinions of acquaintances, that deference may be carried too far when it obliges one to remain in a poor country [such as] Canada, where the great opens are for capitalists large and small, mechanics and bona fide laborers and which is too little advanced in civilization to encourage literary efforts or highly educated men without sufficient interest to get appointments before leaving home.[82]

In the United States he could probably get a job. In fact, he had worked briefly in New York on the way from Nicaragua, but the Civil War made that impossible now. Hamilton sent a formal letter offering to pay £20 with 6 percent interest to anyone who would loan that amount to his son for his passage home.[83] William Edwin was back in Dublin by the end of the year.[84]

While William Edwin was on his expedition to America, Archy, who had been ordained in 1860, began his duties as a clergyman, first as an unpaid assistant curate at Castleknock, their own parish, and then at Dunlavin, where he assisted Hamilton's friend John O'Regan.[85] O'Regan was particularly fond of Arch, and as archdeacon of the diocese he was in a position to find him a suitable living.[86] Charles Graves, the professor of mathematics of Trinity, was also a man of considerable influence, and Hamilton worked through these two men to try to get a place for Arch. Graves's reply was encouraging: "I am on the lookout for an appointment that would suit Archibald, and I have two or three kind and sensible friends helping me. It will be strange if we do not succeed."[87] Hamilton worked hard trying to get Arch situated, and of course he had to pay his son's expenses during the times that Arch was unemployed. Arch held brief temporary appointments at various parishes until April, 1864, when he was finally offered the senior curacy of Clogher by Dean Ogle Moore, an old friend of John O'Regan's.[88]

Hamilton must have breathed a sigh of relief to have at least one of his children placed in the world. The first reports on Arch from Clogher indicated that he was settling in well, and a year after his appointment the dean wrote: "A. Hamilton satisfactory, tho' at times a little dreamy."[89] But the dean was having trouble with his parish. He wrote to O'Regan that he had been "publicly held up to scorn and indignation and reviled as a papist, and spy, and informer against orangemen."[90] Unfortunately Arch became embroiled in the battle between the dean and his congregation. In July the dean's report was not good:

An unfavourable change has come over [Archibald] of late. I have noticed symptoms of it occasionally which gave me uneasiness—but yesterday it was revealed most painfully.

He has latterly kept very much away from me—after his return from Dublin he did not come to me for a week. I now find that, flattered and cajoled by the malcontents, he has adopted their views and taken their part. Separating himself from me in the parish *pro tanto*.

He very plainly and honestly enough, tells me that his independent thought and calculations and opinions (as he believes them to be) ought to be the guide of his actions. If he remains under this delusion, God help him. . . .

Perhaps I have spoiled him . . . I can no otherwise account for his waywardness. But from whatever cause it has sprung, tis evident he and I cannot go on together unless he lays it aside, and becomes sensible that in practical things he is a mere child.[91]

Here was new trouble to burden Hamilton, but there was little he could do about it. By the summer of 1865 he was seriously ill.

Hamilton had suffered from occasional attacks of gout over a good many years. He had had a severe attack in 1856 at Cheltenham when he was visiting John Graves, and other, less severe, bouts had followed. In the summer of 1865, however, his condition became alarming. He had gone to the opening of the International Exhibition in Dublin on May 9, but had suffered an attack the following week.[92] He seemed to recover, but another attack came in early June, this time accompanied by convulsions and followed by bronchitis. William Edwin, who was now at home, wrote to John O'Regan: "We have all been in a state of nervous tension last week, hope and fear succeeding each other every hour— now we are a little easier and that last week seems like some horrid dream separating life into two halves."[93] It was only with great difficulty that William Edwin was able to "screw anything out of the doctors" in the way of a diagnosis. They finally told him that the illness was caused by "suppressed gout" that flew first to the head and then to the heart.[94] William Edwin reported that it was "one of the most difficult and extraordinary cases which ever occurred in medical practice," which probably means the doctors did not know exactly what was wrong.

The greatest problem was keeping Hamilton quiet and away from his work and worries about money. With the realization that death might soon be upon him, he labored mightily on the *Elements* with every ounce of his ebbing strength. William Edwin wrote to John O'Regan:

We hope to get him to go somewhere for change of air . . . and if possible rest from mathematical work, which at present is very exhausting and injurious to him. . . . *He* feels in many ways that he is quite equal to continue his most difficult Quaternion investigations—but when he makes the attempt he then finds either that physical exhaustion soon stops him or that, if he does go on working, a worse reaction follows. He feels this and most painfully, though he will not admit it, or rest from mathematical work for an hour of his own free will.[95]

The travel for a change of air was never more than one short drive down the lane in his phaeton. Even this brief expedition caused extreme fatigue and a relapse. A letter from Hamilton to Robert Graves reporting this event begins firmly enough, but the handwriting becomes shaky halfway through, indicating that the exertion of writing a letter was too much for him.

Money worries would not leave him. The university printer presented his bill at the end of May, and there was a balance of £145 chargeable against him that he could not pay.[96] At the end of July he finally sent in his statement of income tax, apologizing for the delay in a handwriting that was constantly deteriorating.[97] Hamilton declared an income of £607 in 1865. It should have been sufficient for his needs, and one suspects that many of his worries about money were imagined, but a single charge of £145 was almost a quarter of his annual income. It must have come as a shock. In his illness he became increasingly reluctant to sign checks, and finally was unable to do so. The grocer refused to give any further credit. In fact, all the Hamiltons' creditors began knocking on the door. There was not even money to pay the doctor. On one occasion Hamilton tried to write a check, but had a terrible attack while writing it, and money became a forbidden subject in a house that had to be kept absolutely quiet. His daughter Helen was ill, too, and could not be of much help. She went to stay with Arch at Clogher. Lady Hamilton was home, but was also ill. The burden fell on William Edwin, who managed matters well under difficult circumstances. Hamilton's Aunt Jane Willey was at the observatory, and the Rathbornes, who lived next door at Dunsinea, also helped. It was John Rathborne who eased the family's way through the financial crisis with a loan.[98]

By the end of August it was becoming clear that Hamilton would not live much longer. Robert Graves was in Dublin at the time and was called to the observatory on September 2, 1865, by William Edwin. When he arrived he found Hamilton calm and alert, but conscious that he was dying. Hamilton turned their conversation to religious matters and asked particularly if Graves thought God could love him. Once reassured, he asked for prayer, a reading of Psalm 145, and communion from his parish clergyman the next day. He would have preferred to receive the sacrament from Graves or from Arch, but being punctilious to the last in matters of religion, he preferred to follow the rubrical direction of the Prayer Book. But death caught him before the appointed hour. Around 2:00 in the afternoon, when he felt the end coming, he "solemnly stretched himself at his full length upon his bed, and symmetrically disposed his arms and hands, thus calmly to await his death." It was a beautiful death, of the kind much admired by the Victorians.[99]

Graves's account is belied at least in part by a scrap of paper containing what were probably Hamilton's last written words. The note is dated September 2, 1865, and must have been written only a few hours before his death. It is written in an extremely feeble hand and it states: "Sir W. R. Hamilton wishes his letters to be given to Mr. W. E. Hamilton by the *postman* and *not* to *any other* person here. The postman is *not* to give any letters for me to anyone *but* Mr. WEH who *alone* is appointed to receive them. WRH."[100] Hamilton's last hours were less serene than Graves would have us believe.

When Arch and Helen arrived from Clogher, Helen wrote her reactions down, also in a shaky hand, reflecting her emotion and her recent illness.

I felt it wrong to go into the room till I was nicely dressed and my hair settled, because of a presence. The world seemed on Saturday about 2 o'clock full of silence. As we walked about the roads ... I could not be impatient for the train, though I said to my heart "God will not let him die till we come." Still a feeling of the inevitable was so strong it filled my mind. I felt I could not change the decree. At or very near 2 o'clock I said and could not help saying it "Of all the easy and short journeys we will ever take, the journey from this world to the next will be the shortest: see the trouble it is to get to Dublin, but it will be no trouble to leave this world. We make too much of death." Still I was gathering flowers for him, but said to myself, perhaps he will not be able to look at them. ... The feelings most vivid were the Presence of God, the fact that my Father was not there and the prevailing sense of rest. I never felt afraid, only awed. ... I realized death when I touched his forehead with my lips and felt that chill which is unlike all else. ... This visitation has utterly abolished the fear of death. I wondered what it was that I had been afraid of all my life. I said to Jane [Willey] while I looked upon him, "Who would ever think of praying for the soul of that man. We are I think not to insult God by asking for what he has so obviously granted."[101]

The funeral was held on September 7, and the fellows, scholars, and students, in their academic dress, accompanied by the council and members of the Royal Irish Academy, walked in procession behind the hearse from the college gate to the grave at Mt. Jerome Cemetery.

Soon after the funeral Charles Graves wrote about Hamilton: "He was a wonderful creature: and now that he is removed from amongst us people judge him more tenderly, and more fairly estimate his greatness. For infirmities such as he possessed there are excuses the force of which can be felt only by those who have had some large experience of human nature and its frailties, or who have themselves felt the craving for stimulus after the excessive exertion of mind."[102] Graves also obtained an official report on Hamilton's death from the physician who had been in attendance. The most significant thing about the report was the fact that the physician felt compelled to add a statement that he had called on Hamilton "very unexpectedly early and late ... to watch the progress of his health under such mental pressures and I take leave solemnly to declare that I never found him under the influence of alcoholic stimulants—Moreover I am of opinion the human mind would be rendered quite incapable of executing such a mighty task if the functions of the human brain were tainted by intemperate habits, to which it has been alleged Sir William was subject."[103] The obvious intent of the report was to lay to rest rumors about Hamilton's alcoholism, but they were not easily dispelled. About the same time James Disney wrote to Robert Graves, who was working on the biography: "I hear not without regret from my brother Thomas that you did not mean to shrink from exhibiting his melancholy, or deplorable failing to public view."[104] The "deplorable failing" could only be drinking to excess. The general opinion around Dublin, however, was that Hamilton had worked himself to death.[105]

After the funeral Arch returned to Clogher to do battle with his dean. Little Helen accompanied him, and William Edwin stayed at the observatory to settle his father's estate. Arch asked his mother to join them at Clogher, but she chose not to come.[106] She was in failing health and rapidly losing her eyesight. Fortunately the government voted to extend Hamilton's pension of £200 per year for the benefit of his widow and daughter. Lady Hamilton lived in complete seclusion until her death in 1869.[107]

John O'Regan and Robert Graves became the mainstays of the family in the years following Hamilton's death. Graves helped William Edwin arrange for the publication of the incomplete *Elements,* and through his brother Charles persuaded Trinity College to bear the entire cost of its publication. William Edwin had hoped to realize some profit from the book, but this seemed unlikely. The college left the copyright with Hamilton's heirs, but asked that his scientific manuscript notebooks be given to the college library as part compensation for what was certain to be a financial loss.[108]

After settling the estate and reviewing again his prospects for employment, William Edwin went back to Canada, returning only once to Dublin, in 1871, to sell the property that remained to him. His letters from Canada became increasingly flamboyant. He talked of writing for newspapers, managing immigration schemes, and investing money at huge profits.[109] The profits were not always realized, however, because in November, 1871, Arch warned John O'Regan not to pay anything further to William Edwin, and listed for O'Regan the amounts of money that he himself had already given him.[110] The following June his complaints to O'Regan became more bitter:

In consequence of what William Edwin wrote to [Aunt Sydney], I shall certainly recommend her strongly to withdraw the money from whatever investment he put it in, and to invest in some other way. ... I also think she ought not to answer any letters on the subject which William Edwin may write. I am sorry that I entrusted him with the matter at all but it seemed natural and I thought would gratify him in a harmless way. I fear he is totally unfit to be trusted in money matters of that kind. He fancied himself a man of business and has entirely lost the property which my father left him; besides a good deal of my sister's and the greater part of mine, which I am now extremely sorry that I ever gave up to him. My only excuse for doing so is that I really hoped that he would turn over a new leaf after my father's death, and my idea was to accumulate all the available capital that I could into his hands. I think now that the sooner and the more plainly he is made to feel how contemptible as well as wicked his conduct has been, the better. It is possible his eyes may yet be opened in some degree to see himself as others see him.[111]

As a result of his financial debacles, William Edwin had trouble getting answers to his letters—nobody wanted to get enmeshed any further in his financial affairs—and he gradually disappeared from view. He settled at Chatham in Ontario, which had a large Irish population, and wrote editorials and local news for the Chatham *Planet*.[112] The last item in the manuscripts from William Edwin is a copy of the *Chatham Market Guide* from 1891.[113] It advertises "Personal Saturday distribution by W. E. Hamilton, Editor and Proprietor." Among the advertisements for trusses and binder twine is one for "Specialty in Albums! enquire of W. E. Hamilton." Among the albums offered are "Indian Horrors or Massacres by Red Men," "Tragic Death of Sitting Bull," and, for the local Irish immigrants, "Portraits of all the Popes from St. Peter to Leo XIII with beautiful picture of the Vatican, the Pope's palace, with eleven thousand rooms in centre $2. Apply to W. E. Hamilton." William Edwin was fifty-seven by this time, only three years younger than his father was at the time of his death. He seemed destined to live with grand schemes that never quite succeeded, and one feels that he must have inherited his father's naiveté but none of his brilliance.

Arch continued to have trouble at Clogher after his father's death, and finally left in 1867. It took the combined efforts of Helen and Robert Graves to persuade him to go back to his responsibilities.[114] Subsequent letters from Arch became increasingly odd, until a letter to O'Regan in June, 1870, became absolutely incoherent—a jumble of odd sketches, Latin expletives, wild stories in Irish dialect, and the like.[115] Other letters from the same year seem completely reasonable, but the general impression is one of growing emotional instability. Sydney wrote from Greytown: "The account of Archy is too terrible. God grant that he may not destroy himself or anyone else, for no one can say how mental derangement might end."[116]

Helen's future seemed more secure. Even before Hamilton's death she had fallen in love with John O'Regan; she wrote in her diary in 1862: "I understand that glorious nature. I could play on his heart as on a harp, it should answer to my least touch and he loves me—I am his *friend*, his eye rests with pleasure on my face when he looks at me. Oh how glorious and how gentle should *I* be to be worthy of the love of *such* a man."[117] John O'Regan had had severe family and financial problems of his own, which prevented him from marrying Helen until

1869, when she was twenty-nine and he was fifty-two. Both he and Helen felt particularly blessed by the marriage, which allowed them to lay aside past worries and start a new life together.

In many ways Helen was more like her father than either of her brothers. In a commonplace book she began to write a novel, probably when she was about eighteen. It is romantic in the extreme and continued over many pages through many years. It was broken off, probably at her father's death, because it resumed with a statement dated July 8, 1866: "This book begun in affluence of love and time I dedicate to the absent and the dead. It seems to be my *duty* to return to writing this novel as the only means of relieving an almost uncontrollable depression of spirits. I gave up the task because it interfered with my *home* duties—having *no home* now *that* reason does not hold. Therefore I take up the task partly as a duty and also to give my mind some object of deep interest."[118] In spite of her resolve to write daily, the book contains only occasional entries. After her marriage her life became much busier. She became pregnant within the year and delivered a son christened John R. H. O'Regan. There was a complication with the

Helen Eliza Amelia Hamilton (Mrs. John O'Regan)

delivery, however, and she did not recover. A desperate letter from Robert Graves suggested a blood transfusion, but nothing seemed to help.[119] In her illness she turned again to her manuscript book and wrote the words: "Something of me is imaged here. Sketches might be selected from this book for publication. I leave to my friends farewell. What my novel might have been I know. The world never will. 'All things come to an end but thy commandment is exceeding broad.' Farewell!"[120] She died on June 21, 1870.

While working on his biography of Hamilton Robert Graves entered into correspondence with Ellen De Vere O'Brien, Hamilton's old love from the 1830s. In response to Ellen's request for information about Hamilton's children, he gave the following report as of 1873:

The eldest son, William Edwin, passed creditably through College and became a Civil Engineer. His intellectual powers are of a high order, *but* he has no moral principle, has lived a most irregular life, and is now near Toronto, living from hand to mouth teaching and lecturing. The second son, Archibald Henry, also passed through College creditably, and became a Clergyman. In some respects his intellect is also of a high order, and he is a *good* man both morally and religiously; the *but* in his case is that eccentricity in him seems fast ripening into insanity. He commuted and compounded, as a Curate, and has given everything away—mostly to the poor, and is now almost entirely dependent on his brother-in-law, Archdeacon O'Regan, who married in 1869 the only daughter of Sir W., a sweet bright shy and rather eccentric girl, but also religious and good. She was as happy as a woman could be for about a year and died after giving birth to a son who lives and for an infant promises to be a remarkable man. ... His daughter was very dear to Sir W. and perhaps gave him the only sweetness of affection tasted by him in his latter years.[121]

In his biography Graves made his discussion of Hamilton's children as brief as possible.

Sydney, the eternal adventurer, stayed in Nicaragua until 1873 or 1874, when she returned to Dublin and then emigrated again the following year, this time to New Zealand, where she became the matron of an asylum for the insane.[122] She lived in Auckland until her death in 1888 at age of seventy-eight.[123]

Arch held a series of curacies and lived until 1914. After 1895 he held no regular appointment, but assisted where he could, mostly on locum tenens for sick clergy. The O'Regans supported him, first the archdeacon, and then his son, John. He became an increasingly eccentric figure. John's widow still remembers "Uncle Arch," an old man with the courtly manners and clerical garb of the nineteenth-century Anglican minister. He was "not of this world," but greatly beloved by children, whom he would entertain at length with stories of old Ireland.[124]

It was Arch who revealed better than any other commentator why Hamilton's children did not succeed in the world in the ways in which

Archibald Henry Hamilton

the world usually measures success. When Robert Graves asked for his judgment about his father's intellectual powers, Arch responded:

With regard to his intellectual ability, I really do not pretend to judge. ... I know I would be esteemed a mere butterfly in *that* generation; or at least in that coterie in which Brougham, Whately, Newman were lights in the firmament. ... They all appeared to me to move in such an awful separate region that ordinary mortals like myself didn't really belong in the same world—and this feeling ... had the effect of driving *me* at least quite out of the common beaten track; in which it seemed quite *hopeless* to excel, or even to obtain any seat at all—as if a commoner should desire to sit in the House of Peers.

In fact I can truly say for myself that I have taken more pains and trouble to get out of the common groove, than anyone that I know has done to get into it. Of his intellectual abilities then I do not pretend to judge, ... having neither part nor lot in the matter. But there is *another* department in which *everyone* with a *soul* is not only *entitled,* but is *bound* to judge. I mean of course practical Religion and Morality.

It is *this,* and not in the Intellectual department that I proudly recognize some stir of blood or lineage, however faintly—and once or twice in my life I have felt inclined to claim (in my own mind I mean) relationship with so truly *great* a character.[125]

It is difficult to grow up the child of a genius, especially if you do not share that genius yourself. What Hamilton's children could share with him was not his knowledge and intellect, but his human qualities of affection and enthusiasm. That enthusiasm, however, was combined with a driving ambition that must have made the observatory an uncomfortable place at times. One could not say that the Hamiltons had a happy home life. Lady Hamilton's constant illness, her timidity, and her anxiety, combined with Hamilton's dependence on alcohol, must have aggravated the stresses that are part of any household.

Hamilton was also too much beset with what might have been for him ever to find happiness in what he had. His ideal but passionate love for Catherine Disney and his relentless correspondence with her family after her death created a dream of an ideal life against which his own sad and troublesome situation at home was only too easily juxtaposed. Much of his life had been a disappointment, whatever other face he might have presented to the world. Such is the frequent fate of romantics. If he had not chosen Helen Bayly he would have married someone equally sensitive and delicate. It is hard to see how a marriage with Catherine would have been significantly better. The romantic ideal is almost by definition other-worldly, and to find happiness in common society is a betrayal of that ideal. One part of Hamilton's personality sought a strictly conventional home life, but the other part, the romantic part, could not be satisfied with the mundane. He sought truth, for himself and for humanity. His mathematical achievements were to be both his monument and a gift to mankind. The pursuit of mathematical truth was, in his mind, a truly religious undertaking. Under these circumstances it is not surprising that writing the *Elements* became an all-consuming effort in his last years, so much so that he continued to work on the book as long as he could hold a pen. It is also not surprising that his sons were unable to find in him a model on which they could easily pattern themselves.

27

Hamilton and Nineteenth-Century Science

THE IDEALIST VIEW of science that Hamilton represented is not one that the modern scientist can easily share, nor was it the predominant view of British science in the nineteenth century. British science was always strongly inductionist, and after the appearance of John Stuart Mill's *System of Logic* in 1843 it became increasingly so. One may well ask, then, how Hamilton came to hold his idealist views and how typical his position was. Can we make any generalizations about British science based on his life and career, or was he a special case? A possible source for the origin of his ideas is in the Scottish school of common-sense philosophers, particularly Dugald Stewart. Hamilton read Stewart's *Philosophical Essays* in a copy he borrowed from the Edgeworths, and it was probably Stewart's essay "The Idealism of Berkeley" that caused him first to think about point forces in space and to undertake a more thorough study of Berkeley's works. He also obtained hints about Kant from reading Stewart, which led him to study the *Critique of Pure Reason* in the original German.

Scottish philosophy also had an indirect influence on Hamilton. At a time when the university curricula at Dublin and Cambridge were becoming more mathematical, the Scots continued to emphasize metaphysics. The conflict between the "mathematical" curriculum in England and the "metaphysical" curriculum in Scotland created a continuing debate over the relationship between metaphysics, logic, and mathematics. William Whewell and Augustus De Morgan were both very much involved in the debate. Whewell carried on a famous controversy with Sir William Hamilton of Edinburgh (no relation to William Rowan

Hamilton) on the relative merits of mathematics and metaphysics, and De Morgan debated the same Edinburgh Hamilton on the logical problem of the quantification of the predicate. William Rowan Hamilton did not take an active part in these controversies, but he was certainly aware of them. They indicate that logical and metaphysical questions were still capable of arousing controversy, in spite of the moribund state of British philosophy in the first half of the nineteenth century.[1]

The influence of Scottish philosophy on Hamilton was small, however, compared to that exercised by Samuel Taylor Coleridge. Coleridge gave Hamilton the foundation for his metaphysics and also led him to Kant. He personally urged Hamilton to read Kant and loaned him his own copies of Kant's works. Coleridge was unquestionably Hamilton's major philosophical source, the person he recognized as his principal teacher in metaphysics. But Coleridge's philosophy would never have seized Hamilton so strongly if it had not also been closely associated with romantic poetry. Coleridge and Wordsworth were primarily poets, and their philosophy was a large part of their poetry. Hamilton insisted that science and poetry were the two clearest expressions of truth, and he expended much effort to show that truths expressed in these two different forms were fundamentally the same. His enthusiasm for poetry actually preceded his enthusiasm for mathematics, and his love of the ideal and the abstract was confirmed in his poetry before it was confirmed in his science. The style of his life and his modes of thinking were closely tied with his poetry.

Hamilton's poetic style, however, was much like his scientific style—abstract, ideal, and with an avoidance of concrete images. This emphasis on the abstract separated him from Wordsworth and Coleridge more than he realized. Hamilton shared their desire to unify and harmonize human experience, but he did not recognize the need for a concrete expression of his idealism. His poetry lacked vigor precisely because it always moved on such an elevated plane. It was the power of abstraction that made him a creative mathematician, but the same power of abstraction made him an inadequate poet.

Probably the most important concept Hamilton obtained from Coleridge was the meaning of the word *science*. *Science* was, for Hamilton, "strict, pure, and independent; deduced by valid reasonings from its own intuitive principles."[2] For Coleridge, *Science* was "any chain of truths which are either absolutely certain, or necessarily true for the human mind, from the laws and constitution of the mind itself."[3] A *Science* for both Coleridge and Hamilton was created by the reason alone acting on intuitive principles without any empirical content. The fact that *Science* as Hamilton and Coleridge defined it could not be derived from experience might suggest that it would have no practical value, but this was not the case. Hamilton believed that any science could be applied to the world of experience even though it was not derived from that world. The reason for this is that the forms of our thought that give us a true *Science* are mirrored closely by the actual events in the physical world. The organization of our minds and the organization of the physical world are in happy accord. This mirroring is, according to Hamilton, the work of God.

With his view of science, it is not surprising that Hamilton placed mathematics on such a lofty plane. For him the search for triplets and quaternions was not merely an exercise in abstract mathematics; it was also an inquiry into the workings of the mind and at the same time an inquiry into the order of nature. The characteristic function revealed a heretofore undiscovered unity between the sciences of optics and mechanics, and the prediction of conical refraction confirmed its value in physics. The quaternions were numbers constructed from the primitive intuition of time, and therefore they were derived from an "intuitive principle" or "form of thought," and, as Hamilton expected, they also had value in physical science, because they revealed more clearly than previous algebras had the quantitative relationships between physical phenomena. Thus any suggestion that his science was too mathematical and not sufficiently empirical would have brought the rejoinder from Hamilton that progress in mathematics must inevitably produce progress in physics. The intentional infusion of the empirical into his work would have been an unnecessary encumbrance that would have removed it from the realm of *Science.*

It is easy to criticize Hamilton by arguing that his definition of *science* was a misuse of language and, that by rejecting the empirical content of science, he was defining himself out of the company of practicing scientists in the nineteenth century. But before we raise this criticism, we should remember that it was just at this time—during the second quarter of the nineteenth century—that the modern meaning of *science* arose. The *Oxford English Dictionary* is helpful on this point. It reminds us that the word *scientist* did not exist in the English language until 1840, when it was consciously coined by Hamilton's friend William Whewell. Whewell wrote: "We need very much a name to describe a cultivator of science in general. I should incline to call him a Scientist." Wordsworth faced the same problem of terminology in 1829 when he wrote, probably as a result of his conversations with Hamilton at Dunsink Observatory, "a *savant* who is not also a poet in soul and a religionist at heart is a feeble and unhappy creature."[4] Wordsworth was referring here specifically to the natural scientist, but because the word did not yet exist, he used the French word *savant* instead. He might have used the expression "natural philosopher," but that expression obviously no longer served Wordsworth's purpose. The meaning of the word *science* and the kind of person practicing science were changing. New words were needed to express new ideas.

In 1867, two years after Hamilton's death, William G. Ward, Newman's supporter in the Oxford Movement, wrote in the *Dublin Review*: "We shall . . . use the word 'science' in the sense which Englishmen so commonly give to it; as expressing physical and experimental science, to the exclusion of theological and metaphysical," indicating that the word *science* had recently changed its meaning and that the change required notice. *Science* was also the name for the portions of ancient and modern philosophy, logic, and cognate subjects included in the course of study at Oxford University for a degree in the school of Literae Humaniores. The *Oxford English Dictionary* quotes E. A. Freeman writing in 1884 about science at the middle of the century: "I don't mean . . . cutting up cats, but what science meant then, Ethics, Butler, and such like." Thus Hamilton's view of *science* may not have been so unusual at a time when the meaning of the word was rapidly changing. It is true that he felt intellectually isolated; he complained in 1831 to Aubrey De Vere: "I differ from my great contemporaries, my 'brother-band', not in transient or accidental, but in essential and permanent things: in the whole spirit and view with which I study Science,"[5] and yet his isolation did not prevent him from carrying on a sympathetic exchange of ideas with John Herschel, William Whewell, Augustus De Morgan, and Peter Guthrie Tait. If these stalwarts of British science could find value in his metaphysical position, then it could not have been entirely atypical.

Of course it was part of the romantic movement to place oneself on the "outside" of society, separated from the humdrum affairs of the world, and Hamilton's feeling of isolation was an essential part of his poetical idealism. It was necessary for him to separate himself from the common herd, whatever philosophical position the herd might take, and therefore his isolation was partly his own choosing. He exaggerated the differences between himself and his "brother-band" of scientists in order to emphasize his own uniqueness.

Hamilton claimed to have drawn the inspiration for his science from metaphysical idealism, and to a large extent that claim was true. His work in optics, mechanics, and algebra were not logically dependent on his metaphysics, but his metaphysics did set the style and establish the direction of his research. It is difficult to find other British scientists who drew directly from the same source of inspiration, but we should not expect other scientists to be so consciously metaphysical as Hamilton. In fact a scientist rarely formulates a complete philosophy to support his working methods. Usually his assumptions, whether physical or metaphysical, are largely unconscious. Therefore a scientist, even though he may reject a certain philosophical position, will take account of that position in his thought. Metaphysical idealism was not a popular approach to science in nineteenth-century Britain, but it was not unknown, particularly to William Whewell, John Herschel, and Michael Faraday. It did not have to become a dominant philosophy in order to have its effect, and that effect was to counteract the strictly mechanical view of nature that had characterized science in the eighteenth century.

The abstract concepts of electromagnetic fields, energy, and light waves, which emerged in the nineteenth century, were significant departures from the mechanical view of the eighteenth century. The creators of these concepts were not all metaphysicians, but they all found that "forces" in space were a useful substitute for the interactions of hard material atoms. How much of this change should be attributed to the introduction of German metaphysical idealism into Britain is difficult to say; if Hamilton serves as an example, it was important.

Idealism became popular in Britain only in the last quarter of the nineteenth century, but as John Passmore, in his *A Hundred Years of Philosophy,* reminds us, it had flourished twice before, once in the seventeenth century with the Cambridge Platonists, and once in the eighteenth century with Berkeley.[6] In both cases it had been brought forward as a defense against materialism. While there was no continuing tradition of idealism in Britain, Passmore mentions that "literary philosophers" such as Coleridge and Carlyle helped to keep the idealist tradition alive even while it was being ignored by academic philosophers. Hamilton's career would indicate that the "literary philosophers" may have had as great an impact on the natural sciences as did the professional academic philosophers. The turn in British physics away from mechanistic explanations comes at one of those times when the literary philosophers were enjoying their greatest popularity; in fact, it follows right on the heels of the romantic movement in England. Without insisting on a direct causal relationship, we can recognize a philosophical mood in Britain that made nonmechanistic explanations in natural science more acceptable. Moreover, Hamilton lived in an age of universal genius. Later on in the century, as the meaning of *science* and the profession of the scientist were narrowed, this universality was lost. But Hamilton and his scientific associates still had an education and a range of interests that was broad, including literature, poetry, history, and philosophy, both classical and modern. A literary current as powerful as English romanticism could determine their approach to the world just as surely as the teachings of an academic philosopher. In Hamilton's case this is undoubtedly what happened. Granted that he moved from Coleridge to Kant for his rigorous metaphysics, it was still Coleridge and the other romantic poets who set the direction of his philosophical journey.

Hamilton hated mechanism, and yet he made the most important contribution to the science of mechanics in the nineteenth century. This paradox serves to warn us that as the concepts of physics became more abstract in the nineteenth century, so did the mathematical formulation of those concepts. It is a process that has continued on into the twentieth century, until much physical law has now become so abstract that it can be stated *only* mathematically. Hamilton fostered this trend as much as he could. He believed that mathematics revealed the real unity of physical phenomena in a way that experiment could not. The same thing is true of his algebra. Through a metaphysical examination of the concepts of mathematics he arrived at a more general abstract meaning for number and for the operations of algebra. The interaction between his intuitive approach and the formal approach of De Morgan, Boole, and the Cambridge algebrists was extremely productive in removing the restrictions on algebra that had limited it to the concepts and operations of arithmetic. Thus his idealism had its value in pure mathematics as well as in physics.

Hamilton's idealism was also reflected in his relationships with other people, in his religious beliefs, and in his general view of the world. It caused him untold amounts of pain. In some cases it was pain that he welcomed, but it was pain nevertheless. Before the advent of Freud it was possible to believe that ideal relationships such as the one that he imagined between himself and Catherine Disney were actually possible. He must have believed it sincerely, or he would not have tortured himself with this dream for forty years. One could explain it, no doubt, by sublimated sexual impulses, latent incestual love for his sister Eliza, and early separation from his mother, but it is more helpful to see it as Hamilton saw it, as an ideal love, "pure as that of those who in the resurrection neither marry nor are given in marriage, but are as the Angels of God in Heaven."[7] In his science, in his poetry, and even in his love of women, Hamilton constantly tried to elevate himself to an ultramundane world. In many ways it was an admirable effort, but because his body had to live in this world—wherever his mind might be—he suffered the disappointments of all idealists. The men and women who contribute the most to our world are likely to be uncomfortable in it, and Hamilton was no exception, but in spite of his human failings he was, as Charles Graves said, a "wonderful creature."[8]

Appendix

The Equations of Conjugation and Logarithms of Number Couples*

HAMILTON BELIEVED that the common way of writing a complex number as the sum of a real and an imaginary part $(x + y\sqrt{-1})$ was misleading, because it expressed the sum of unlike things. He sought another way to express complex numbers. A complex function expressed in the common way,

$$u + v\sqrt{-1} = \Phi(x + y\sqrt{-1}),$$

where u, v and x, y are real quantities and where u, v depend on x, y, is differentiable at x_0, y_0 if and only if it has a total differential at x_0, y_0, and if the following relations hold:

$$\frac{\partial u}{\partial x} = \frac{\partial v}{\partial y}, \qquad \frac{\partial u}{\partial y} = -\frac{\partial v}{\partial x}. \tag{A.1}$$

These are now called the Cauchy-Riemann equations. Hamilton called them his *Equations of Conjugation*. They are derived from the properties of complex numbers, but they express a relationship between *real* numbers. Hamilton believed they might be taken as primitive expressions and used to *define* complex numbers without the introduction of imaginary quantities. In particular they might help define a relationship between couples of real numbers,

$$(u, v) = \Phi(x, y),$$

*Adapted from *British Association Report* (1834), pp. 519–23, *Math. Papers*, 3:97–100.

which would replace complex functions written with imaginaries. He called these couples *algebraic couples* or *conjugate functions*.

In developing his theory of couples Hamilton finds that he must first define the operations of addition and multiplication for algebraic couples, and he does this in a fashion completely analogous to the same operations on complex numbers.

$$(x, y) + (a, b) = (x + a, y + b) \tag{A.2}$$

$$(x, y) \times (a, b) = (xa - yb, xb + ya) \tag{A.3}$$

Now arises the question of how to define an algebraic couple raised to the power of another couple:

$$(u_{x,y}, v_{x,y}) = (a, b)^{(x,y)} \tag{A.4}$$

where $u_{x,y}$ means that u is a function of x, y. The major properties of the exponential function of single real numbers are:

$$a^x a^y = a^{x+y}$$

$$a^1 = a.$$

Hamilton establishes the same properties for number couples by *definition*:

$$(a, b)^{(x,y)}(a, b)^{(\xi,\eta)} = (a, b)^{(x+\xi, y+\eta)} \tag{A.5}$$

$$(a, b)^{(1,0)} = (a, b). \tag{A.6}$$

Taking $(u_{x,y}, v_{x,y})$ as the exponential function for number couples (A.4), then from (A.3):

$$(u_{x,y}, v_{x,y})(u_{\xi,\eta}, v_{\xi,\eta}) = (u_{x,y}u_{\xi,\eta} - v_{x,y}v_{\xi,\eta}, u_{x,y}v_{\xi,\eta} + v_{x,y}u_{\xi,\eta}).$$

From (A.5):

$$(u_{x,y}, v_{x,y})(u_{\xi,\eta}, v_{\xi,\eta}) = (u_{x+\xi, y+\eta}, v_{x+\xi, y+\eta}).$$

Combining these two equations and taking advantage of the fact that if two couples are equal, the corresponding parts of those couples are also equal,

$$u_{x,y} u_{\xi,\eta} - v_{x,y} v_{\xi,\eta} = u_{x+\xi,y+\eta}$$

$$\text{(A.7)}$$

$$u_{x,y} v_{\xi,\eta} + v_{x,y} u_{\xi,\eta} = v_{x+\xi,y+\eta},$$

and from equations (A.4) and (A.6),

$$(u_{1,0}, v_{1,0}) = (a, b)^{(1,0)} = (a, b)$$

$$u_{1,0} = a, \qquad v_{1,0} = b. \qquad \text{(A.8)}$$

In solving the pair of equations (A.7) Hamilton notices that they stand in the same relationship as the trigonometrical identities:

$$\cos a \cos b - \sin a \sin b = \cos (a + b)$$

$$\sin a \cos b + \cos a \sin b = \sin (a + b).$$

Writing $a = \rho_{x,y}$, $b = \rho_{\xi,\eta}$, the trigonometric formulas give:

$$\cos \rho_{x,y} \cos \rho_{\xi,\eta} - \sin \rho_{x,y} \sin \rho_{\xi,\eta} = \cos (\rho_{x,y} + \rho_{\xi,\eta})$$

$$\cos \rho_{x,y} \sin \rho_{\xi,\eta} + \sin \rho_{x,y} \cos \rho_{\xi,\eta} = \sin (\rho_{x,y} + \rho_{\xi,\eta}).$$

Substituting $\cos \rho_{x,y} = u_{x,y}$ and $\sin \rho_{x,y} = v_{x,y}$ into equations (A.7) gives the same left members of the above equations, but to equate the right members we must make

$$\rho_{x,y} + \rho_{\xi,\eta} = \rho_{x+\xi,y+\eta}.$$

Let $\rho_{x,y} = \alpha y + \beta x$, then $(\alpha y + \beta x) + (\alpha \eta + \beta \xi) = \alpha(y + \eta) + \beta(x + \xi)$, which is the desired solution.

Also, since

$$e^{\alpha' y + \beta' x} e^{\alpha' \eta + \beta' \xi} = e^{[\alpha'(y+\eta)+\beta'(x+\xi)]},$$

Hamilton writes the most general solution of equations (A.7) in the form

$$u_{x,y} = e^{(\alpha' y + \beta' x)} \cos (\alpha y + \beta x)$$

$$v_{x,y} = e^{(\alpha' y + \beta' x)} \sin (\alpha y + \beta x).$$

Hamilton had already proved that if functions of the form

$$u_{x,y} = e^{\rho_{x,y}} \cos \phi_{x,y}$$

$$v_{x,y} = e^{\rho_{x,y}} \sin \phi_{x,y} \tag{A.9}$$

are conjugate functions (that is, if they obey the Cauchy-Riemann equations), then functions $\rho_{x,y}$ and $\phi_{x,y}$ are also conjugate functions (this is easily confirmed by finding the Cauchy-Riemann equations for $u_{x,y}$, $v_{x,y}$) and

$$\frac{\partial \rho}{\partial x} = \frac{\partial \theta}{\partial y}, \qquad \frac{\partial \rho}{\partial y} = -\frac{\partial \theta}{\partial x}.$$

Then

$$\frac{\partial(\alpha'y + \beta'x)}{\partial x} = \frac{\partial(\alpha y + \beta x)}{\partial y}, \qquad \frac{\partial(\alpha'y + \beta'x)}{\partial y} = -\frac{\partial(\alpha y + \beta x)}{\partial x},$$

from which we get $\beta' = \alpha$, $\alpha' = -\beta$. This is the only place the Equations of Conjugation enter into Hamilton's argument. The general solution for the exponential function (A.4) then becomes:

$$u_{x,y} = e^{(\alpha x - \beta y)} \cos (\alpha y + \beta x)$$

$$v_{x,y} = e^{(\alpha x - \beta y)} \sin (\alpha y + \beta x) \tag{A.10}$$

What remains is to evaluate α and β in terms of the base couple (a, b). From (A.8) and (A.10) we get:

$$u_{1,0} = a = e^\alpha \cos \beta$$

$$v_{1,0} = b = e^\alpha \sin \beta$$

$$a^2 + b^2 = e^{2\alpha}(\sin^2 \beta + \cos^2 \beta) = e^{2\alpha}$$

$$e^\alpha = \sqrt{a^2 + b^2}$$

$$\alpha = \int_1^{\sqrt{a^2+b^2}} \frac{dr}{r}. \tag{A.11a}$$

The quantity β can have many values for the same value of $\sin \beta$ and $\cos \beta$. Therefore $\beta = \beta_0 + 2\pi i$ where $-\pi < \beta_0 \le \pi$ and $i = 1, 2, 3, \ldots$ and

$$\sin \beta_0 = \frac{b}{\sqrt{a^2 + b^2}}, \qquad \cos \beta_0 = \frac{a}{\sqrt{a^2 + b^2}}. \tag{A.11b}$$

We now have in equations (A.10) and (A.11a and b) the complete solution of (A.4). The solution introduced one arbitrary integer, $i = 1, 2, 3, \ldots$
The inverse of (A.4) is

$$(x_{u,v}, y_{u,v}) = \log_{(a,b)} (u, v) \qquad \text{(A.12)}$$

From (A.9) we have

$$u = e^{\rho} \cos \theta, \qquad v = e^{\rho} \sin \theta$$

$$u^2 + v^2 = e^{2\rho}(\cos^2 \theta + \sin^2 \theta) = e^{2\rho}$$

$$e^{\rho} = \sqrt{u^2 + v^2}$$

$$\rho_{u,v} = \int_1^{\sqrt{u^2+v^2}} \frac{dr}{r} \qquad \text{(A.13a)}$$

$$u = \sqrt{u^2 + v^2} \cos \theta, \qquad v = \sqrt{u^2 + v^2} \sin \theta.$$

But as was the case for β, θ is also cyclical in $2\pi k$, so that

$$\theta = \theta_0 + 2\pi k, \qquad -\pi < \theta_0 \leq \pi, \qquad k = 1, 2, 3, \ldots$$

We have introduced a second arbitrary integer, $k = 1, 2, 3, \ldots$, into the expression for the logarithm of a number couple, and

$$\cos \theta_0 = \frac{u}{\sqrt{u^2 + v^2}}, \qquad \sin \theta_0 = \frac{v}{\sqrt{u^2 + v^2}}. \qquad \text{(A.13b)}$$

From (A.9) and (A.10) we have

$$\rho = \alpha x - \beta y$$

$$\theta = \alpha y + \beta x.$$

Solving simultaneously for x

$$\alpha\rho = \alpha^2 x - \alpha\beta y$$

$$\beta\theta = \beta^2 x + \alpha\beta y$$

$$\alpha\rho + \beta\theta = (\alpha^2 + \beta^2)x$$

Solving simultaneously for y

$$\beta\rho = \alpha\beta x - \beta^2 y$$

$$\alpha\theta = \alpha^2 y + \alpha\beta x$$

$$\beta\rho - \alpha\theta = -y(\alpha^2 + \beta^2)$$

$$x = \frac{\alpha\rho + \beta\theta}{\alpha^2 + \beta^2}, \qquad y = \frac{\alpha\theta - \beta\rho}{\alpha^2 + \beta^2}. \qquad \text{(A.14)}$$

Equations (A.14) give x, y in terms of α, β and ρ, θ.
Equations (A.13 a and b) give ρ, θ in terms of u, v.
Equations (A.11 a and b) give α, β in terms of a, b.
We have now found the couple x, y that is the logarithm of the couple u, v taken to the base a, b.

Hamilton has confirmed the argument of his friend John T. Graves that the logarithm of a complex number to a complex base involves two arbitrary constants. In Hamilton's expression for the logarithm of a number couple to a base couple the expressions θ and β each introduce an arbitrary integral constant. Hamilton notices that by equations (A.14) and (A.3),

$$(x, y)(\alpha, \beta) = (\rho, \theta).$$

This suggests a new expression for the logarithm in terms of number couples.

$$(x, y) = \frac{(\rho, \theta)}{(\alpha, \beta)} \tag{A.15}$$

John Graves had written

$$\mathop{\text{Log}}_{e} 1 = \frac{2k\pi\sqrt{-1}}{1 + 2i\pi\sqrt{-1}}.$$

In number couples the logarithm of 1 would be written

$$(x, y) = \mathop{\log}_{(a,b)} (u, v) = \mathop{\log}_{(e,0)} (1, 0),$$

from which we obtain the following values:

$$u = 1, v = 0, a = e, b = 0,$$

and solving for (x, y) in (A.15) using (A.11) and (A.13),

$$(x, y) = \frac{(\rho, \theta)}{(\alpha, \beta)} = \frac{(0, 2k\pi)}{(1, 2i\pi)},$$

which agrees with Graves's expression.

Abbreviations Used in the Notes

Graves   Robert Perceval Graves, *Life of Sir William Rowan Hamilton, Including Selections from His Poems, Correspondence, and Miscellaneous Writings,* 3 vols. (Dublin: Hodges, Figgis & Co.), (London: Longmans, Green & Co.), vol. 1, 1882; vol. 2, 1885; vol. 3, 1889; with an addendum, 1891.

JG   Trinity College Library, Dublin, MS 4015, "Photocopies of that part of Hamilton's correspondence in the possession of John Graves, Holme Grange."

Math. Papers   William Rowan Hamilton, *The Mathematical Papers,* 3 vols. (Cambridge: At the University Press), vol. 1, "Geometrical Optics," ed. A. W. Conway and J. L. Synge, Cunningham Memoir no. 13, 1931; vol. 2, "Dynamics," ed. A. W. Conway and A. J. McConnell, Cunningham Memoir no. 14, 1940; vol. 3, "Algebra," ed. H. Halberstam and R. E. Ingram, Cunningham Memoir no. 15, 1967.

Misc.   Trinity College Library, Dublin, MS 1492, "William Rowan Hamilton miscellaneous papers," in nine boxes, uncataloged. Reference is to box number.

NLD   National Library of Ireland, Dublin

OR   Trinity College Library, Dublin, MSS 5123-33, "Papers of Sir William Rowan Hamilton and of his son-in-law Archdeacon O'Regan listed and indexed at Trinity College Library, Dublin, Oct. 1970–April 1971."

ORSUP   Trinity College Library, Dublin, MSS 7773-76, "Additional Papers of Sir William Rowan Hamilton and his son-in-law Archdeacon John O'Regan." Manuscript numbers refer to John O'Regan's letter list.

OS   Trinity College Library, Dublin, MS 1493, "Correspondence of Sir William Rowan Hamilton."

RIA   Royal Irish Academy, Dublin.

TCD   Trinity College Dublin

TCD notebook   Trinity College Library, Dublin, MS 1492, "Mathematical papers of Sir William Rowan Hamilton, Manuscript notebooks." Reference is to the notebook number.

WRH   William Rowan Hamilton

Notes

Introduction

1. B. A. Gould to WRH, May 17, 1865, OR 1655; Graves, 3:205, Gould to WRH, May 20, 1865, OR 1657; Graves 3:205-6, WRH to A. S. Hart, June 13, 1865, Graves, 3:207.
2. Augustus De Morgan, "Sir W. R. Hamilton," *Gentleman's Magazine* 220 (1866): 133, quoted from Michael J. Crowe, *A History of Vector Analysis: The Evolution of the Idea of a Vectorial System* (Notre Dame, Ind.: University of Notre Dame Press, 1967), pp. 21-22.
3. TCD notebook 17, not paginated.
4. William Whewell to WRH, Mar. 27, 1834, Graves, 2:81, and WRH to Whewell, Mar. 18, 1834, Graves, 2:80.
5. I have discussed this problem at greater length in Thomas L. Hankins, "In Defense of Biography: The Use of Biography in the History of Science," *History of Science* 17 (1979): 1-16.
6. This criticism is stated by Cornelius Lanczos, "William Rowan Hamilton—An Appreciation," *University Review* (National University of Ireland) 4 (1967): 156, and Eric Temple Bell, *Men of Mathematics* (New York: Simon and Schuster, 1965), p. 358.
7. Hamilton often filled his notebooks by starting at both ends! He would write from the front filling only the right-hand pages. Then he would turn the book over and write from the back, again using only the right-hand page. The writing starting from the back appears upside down when reading from the front. But then he would usually give up this system half-way through and scribble wherever he could find a blank space.
8. Often inspiration came on one of his frequent walks. He describes trying to scratch equations on a leaf of ivy with a sharp rock. Accustomed as he was to such bizarre writing equipment, the famous story of his cutting quaternions into the stones of Broome Bridge is not so surprising.
9. WRH to Thomas [Disney], June 22, 1852, TCD notebook 103.5, p. 168.
10. The story of the chop bones originated with Sir Robert Ball, Hamilton's successor at the observatory (see OR 2294).
11. Graves, 1:vi-vii.

Chapter 1

1. Harold Nicolson, *The Desire to Please: A Story of Hamilton Rowan and the United Irishmen* (New York: Harcourt, Brace, and Co., 1943), p. 112.
2. Archibald Hamilton to his wife, Sarah, July 27, 1806, ORSUP 1.
3. Nicolson, *Desire to Please,* p. 176.
4. Ibid., p. 177.
5. Sydney Hamilton to Sarah Hamilton, June 8, 1807, ORSUP 10.

6. Archibald Hamilton to his wife, Sarah, Sept. 29, 1814, Graves, 1:44. Nicolson says that Archibald Hamilton had been guilty of improper financial speculation and brought his problems upon himself (Nicolson, *Desire to Please*, p. 176), and Aunt Sydney wrote in 1807 that he had gotten into financial difficulty arranging his mother's business before his difficulties with Rowan began (Sydney Hamilton to Sarah Hamilton, Apr. 2, 1807, ORSUP 10). When Archibald died in 1820 WRH was concerned about his affairs: "Uncle was joined in a Bond of very large amount with my Father and it is not yet possible to say how seriously he may be affected in consequence" (WRH to sister Eliza, Jan. 20, 1820, ORSUP 19).

7. The connection between Archibald Hamilton and Archibald Rowan is so complex as to defy description. Archibald Rowan was not a Rowan at all, but a Hamilton! (For this reason he is usually referred to in Ireland as "Hamilton Rowan." His godson, the mathematician, is just as frequently called "Rowan Hamilton," which provides almost unlimited opportunity for confusion.) Rowan's father was Gawen Hamilton of Killyleagh, County Down. He took the name of his wealthy maternal grandfather, William Rowan, in order to become his heir.

Sometime in the second half of the eighteenth century Archibald Rowan's parents, the Gawen Hamilton's, were shipwrecked off the coast of Scotland at Kirkmaiden in Galloway. They were rescued by the Reverend James McFerrand, minister of the parish. Mrs. Gawen Hamilton became attached to the McFerrand's daughter Grace and brought her to Killyleagh at age fifteen, where she gave her the advantages of an education and a tour of the Continent. When Grace was satisfactorily polished, she was introduced to another Hamilton, William Hamilton, "a very eminent apothecary of Dublin," and was married with a dowry of £500. (Graves, 1:2-5. The main source for Graves's information was a memorandum by WRH's sister Eliza [OR 1922]. Other documents on family history are OR 81, 1891, 1927, 1929, 1930, 2135, 2136, 2307, ORSUP 10, OS 573, 905, and TCD notebook 114.5, pp. 64ff.) William and Grace McFerrand Hamilton were Archibald Hamilton's parents, and William Rowan Hamilton's paternal grandparents.

Although there is no known tie of kinship between the Hamiltons of Dublin and those of Killyleagh, there was a close tie of friendship through Grace. Her son Archibald Hamilton was a frequent visitor at Killyleagh, a supporter of Rowan's revolutionary politics, and an accomplice in his escape from prison in 1794. It is not surprising, then, that Archibald Hamilton served as Rowan's estate agent.

The parentage of William Hamilton, the apothecary and grandfather of William Rowan Hamilton, is unknown. It was the cause of a vigorous debate in the 1890s between William Rowan Hamilton's first biographer, Robert Perceval Graves, and the physicist Peter Guthrie Tait. Tait claimed Hamilton as a Scot, and the Scottish claims were pursued by R. E. Anderson in his article for the *Dictionary of National Biography*. Graves restated the Irish claims in an addendum to his biography of Hamilton, and the battle raged back and forth through several issues of the *Athenaeum* in the spring of 1891. Tait made his claims in biographical notices in the *North British Review* 45 (1866); 37-74, and the *Encyclopedia Britannica*. Graves's response was in *Addendum to the Life of Sir William Rowan Hamilton: On Sir William Rowan Hamilton's Irish Descent: on the Calculus of Quaternions*. Other articles were in the *Athenaeum*, no. 3308 (March 21, 1891), pp. 380-81; no. 3310 (April 4, 1891), p. 444; and no. 3313 (April 25, 1891), pp. 539-40. There can be no doubt that William Hamilton was apothecary in Dublin, because there exists a diploma stating that "at Michaelmas Assembly 1836 Sir William R. Hamilton—apothecary was admitted into the liberties and Franchises of the City of Dublin by birth" (RIA MS 23.0.49). He was actually Astronomer Royal at the time, but was registered under the occupation of his grandfather. Hamilton occasionally referred to his Scottish connections, but there is nothing in the manuscripts to support Tait's claims. The family tradition held that his particular branch of the Hamiltons came to Ireland at the time of James I (WRH to Augustus De Morgan, July 26, 1852, OS 626; Graves, 3:393).

8. Graves, 1:218.

9. There are serious problems with this story. I give the account as Graves gives it. The letter from Rowan is dated Nov. 6, 1835, and is OR 501. The date must be approxi-

mately correct because it is congratulating WRH on his knighthood. It is almost certainly from Rowan himself, because it mentions his ten children, which is, indeed, the number of children Rowan had. Yet Nicolson, *Desire to Please,* and the *Dictionary of National Biography* give Rowan's death date as Nov. 1, 1834, while Hamilton's knighthood was conferred on Aug. 15, 1835.

10. Graves, 1:24-26.
11. Patronage was extremely important in Ireland under the Union. It was more important in Ireland than in England, and the English were amazed at how pertinacious the Irish were in struggling for office. At least half of all the mail received by the chief secretary dealt with requests for office, and Hamilton's correspondence, from the time he became Astronomer Royal in 1827, contains letter after letter requesting testimonials (R. B. McDowell, *Public Opinion and Government Policy in Ireland, 1801-1846* [London: Faber and Faber, 1952, pp. 47-48]).
12. John Talbot of Hallonshire, later Earl Furnival, Earl of Shrewsbury, mentioned by Shakespeare in *Henry IV* (part I, act II) as "the scourge of France." On the monuments of Trim see Reverend Joseph P. Kelly, "The Porchfields," *Touring Trim,* no. 2 (1968).
13. The Duke of Wellington had studied at the Diocesan School of Trim. In the eighteenth century Talbot's Castle had been owned briefly by Jonathan Swift and by "Stella" before him.
14. WRH to Augustus De Morgan, July 26, 1852, OS 626; Graves, 3:392.
15. J. C. Beckett, *The Making of Modern Ireland, 1603-1923* (London: Faber and Faber, 1966), pp. 289-91.
16. R. B. McDowell, ed., *Social Life in Ireland, 1800-1845* ("Irish Life and Culture" 12, The Cultural Relations Committee of Ireland [Dublin: Three Candles, 1957]), pp. 15-18.
17. T. W. Freeman, *Pre-Famine Ireland: A Study in Historical Geography* (Manchester: Manchester University Press, 1957), p. 13. Freeman gives a population figure of 2,845,932 for 1785 (p. 15). K. H. Connell, in his *Population of Ireland, 1750-1845* (Oxford: At the Clarendon Press, 1950) warns that the precensus figures are much too small. He estimates a population growth that, though still very large, is more comparable to that of England (p. 2).
18. K. H. Connell, "Population," in McDowell, *Social Life in Ireland,* p. 93.
19. Freeman, *Pre-Famine Ireland,* p. 168.
20. Sydney Hamilton to Sarah Hamilton, Sept. 2, 1809, ORSUP 10.
21. Sydney Hamilton to Sarah Hamilton, Apr. 17, 1812, ORSUP 10.
22. WRH to sister Eliza, May 22, 1822, ORSUP 46.
23. WRH to sister Grace, June 7, 1823, OR 65. In 1822 William had been even more concerned about insurrection in the south (WRH to Eliza Hamilton, Jan. 7, 1822, ORSUP 40).
24. WRH to Arthur Hamilton, May 12, 1823, OR 61.
25. William Wordsworth to WRH, July 24, 1829, Graves, 1:333.
26. Cousin Arthur was James Hamilton's cousin, and therefore WRH's cousin once removed.
27. Graves, 1:141.
28. Ibid., 30.
29. Sydney Hamilton to Sarah Hamilton, Oct. 17, 1808, Graves, 1:31.
30. RIA MS 23.0.49.
31. Sydney Hamilton to Sarah Hamilton [1810], Graves, 1:40.
32. Archibald Hamilton to Mr. Beilby, May 18, 1815, Graves, 1:45.
33. Archibald Hamilton to his daughter Grace, Jan. 30, 1815, OR 17; Graves, 1:44, and the same to the same May 23 [1815], Graves, 1:45-46.
34. Sydney Hamilton to Sarah Hamilton, May 15, 1812, Graves, 1:42.
35. Sydney Hamilton to Sarah Hamilton, May 1810, OR 2139.
36. Sydney Hamilton to Sarah Hamilton, no date, Graves, 1:33.
37. Sydney Hamilton to Sarah Hamilton, Aug. 16, 1809, Graves, 1:34.
38. "To this day I remember mourning over the cryings which were caused to her, an almost infant, by her father's determination (happily since frustrated, tho' I remember him with sincere gratitude and affection), to make of *her* a prodigy. She is just a very sensible

and a very agreeable woman now" (WRH to Thomas Disney, Jan. 26, 1854, TCD notebook 123.5, p. 51).
39. Sydney Hamilton to Sarah Hamilton, Jan. 11, 1811, Graves, 1:41.
40. WRH mentions his "Compendious Treatise of Algebra" in 1818 (OR 26; Graves, 1:54), but I have not been able to find it among the manuscripts.
41. William and his sisters were also extremely impressed by Zerah's sixth finger (WRH to sister Eliza, May 15, 1820, ORSUP 26). The Colburns were a polydactylous family (*National Cyclopaedia of American Biography*, 57 vols. [New York: James T. White & Co., 1892-1977], 7:74).
42. WRH to sister Eliza, May 15, 1820, ORSUP 26, and WRH to sister Eliza, June 5, 1820, OS 3; Graves, 1:81.
43. "Letter to the Editor of the 'Irish Ecclesiastical Journal,'" Graves, 2:380-83. Friedrich Gottlob Röber, Professor of Architecture at Dresden, discovered a method for constructing the regular heptagon that he believed was employed by the Egyptians in constructing the Temple of Edfu on the Nile. Although the construction is not exact, it is such a good approximation that it amazed Hamilton and he devoted a substantial amount of time to its study (Graves, 3:141-46, and WRH, "On Röber's Construction of the Heptagon," *Philosophical Magazine*, 4th ser. 27 [Feb., 1864]: 124-32).
44. WRH to Augustus De Morgan, Feb. 5, 1852, Graves, 1:53.
45. Archibald Hamilton to his wife, Sept. 4, 1805, OR 6. Archibald also enjoyed his drink (Sydney Hamilton to Sarah Hamilton 1812, ORSUP 10).
46. WRH to sister Sydney, Mar. 24, 1821, ORSUP 31; Graves, 1:50.
47. Archibald Hamilton to WRH, May 16, 1819, OR 28.
48. WRH to sister Eliza, Sept. 15, 1819, ORSUP 15.
49. Archibald Hamilton to WRH, May 20, 1819, ORSUP 11.
50. Archibald Hamilton to WRH, May 29, 1819, OR 32.
51. There are many letters and passages of notebooks in shorthand among the manuscripts. WRH's shorthand was apparently a variant or combination of several shorthands popular at the time. Graves had an amanuensis who helped him transcribe the shorthand manuscripts. I am especially grateful to June Z. Fullmer for her assistance with Hamilton's shorthand.
52. The document is dated July 17, 1819, OR 36.
53. WRH to sister Eliza, June 5, 1820, ORSUP 27; Graves, 1:81.
54. WRH to his father, Archibald, July 9, 1819, OR 35.
55. Archibald Hamilton to his daughter Grace, June 25, 1819, ORSUP 12.
56. Archibald Hamilton to his daughter Grace, Aug. 2, 1819, OR 39.
57. Graves, 1:64.
58. WRH to sister Grace, Feb. 2, 1820, OS 2.
59. Graves, 1:73-74.
60. WRH to sister Eliza, Jan. 20, 1820, OS 1; Graves, 1:75.
61. WRH to sister Eliza, May 15, 1820, ORSUP 26.
62. "Eliza appears a little angry at my having advised her in my last letter" (WRH to sister Grace, June 8, 1819, OR 33).
63. WRH to sister Eliza, Oct. 8, 1823, OR 74; Graves, 1:152.

Chapter 2

1. WRH to Cousin Arthur Hamilton, Sept. 4, 1822, Graves, 1:111.
2. WRH to Cousin Arthur, Sept. 4, 1822, Graves, 1:111.
3. WRH to Cousin Arthur, Sept. 4, 1822, Graves, 1:112.
4. WRH to Cousin Hannah Hutton, Oct. 9, 1821, Graves, 1:92.
5. WRH to Cousin Arthur, Sept. 4, 1822, Graves, 1:112.
6. WRH to Cousin Arthur, Apr. 14, 1822, ORSUP 42.
7. WRH to Cousin Arthur, Apr. 21, 1822, ORSUP 43; Graves, 1:99-100, and WRH to sister Eliza, May 10, 1822, ORSUP 45; Graves, 1:100.
8. WRH to Cousin Arthur, May 5, 1822, ORSUP 44.

9. Graves publishes William's correction as an appendix to vol. 1, on pages 661-62. The error of Laplace came in his discussion of the parallelogram of forces.
10. WRH to Cousin Arthur, Oct. 31, 1822, ORSUP 58; Graves, 1:119-20.
11. WRH had attended the fellowship examinations in the Spring of 1820 and 1821. Boyton had been beaten in 1820, but he won the examination in 1821 (Graves, 1:81, 91).
12. WRH to sister Eliza, July 18, 1822, Graves, 1:108. Eliza later claimed that Boyton gave WRH his entire library when Hamilton moved to the observatory (Graves, 1:238).
13. For information on the new mathematical course at Trinity College Dublin, see A. J. McConnell, "The Dublin Mathematical School in the First Half of the Nineteenth Century" (Quaternion Centenary Celebration), Royal Irish Academy, *Proceedings* 50, sect. A, no. 6 (Feb., 1945): 75-88.
14. John Brinkley, *Elements of Astronomy* (Dublin: Graisberry & Campbell, 1813).
15. It was a school of geometricians, however, and Hamilton's interests turned more and more to algebra. Augustus De Morgan complained in letters to WRH that the Dublin mathematicians limited themselves too exclusively to geometry (De Morgan to WRH, Feb. 2, 1852, and WRH to De Morgan, Feb. 9, 1852, OS 541, 543; Graves, 3:333-35, and De Morgan to WRH, May 3, 1852, OS 580; Graves, 3:353).
16. The paper on "Developments" was sent later because Brinkley asked to see it in a more developed form.
17. WRH later added the following note to the paper: "This curious old paper, found by me today in settling my study, must have been written at least as early as 1822. It contains the germ of my investigations respecting Systems of Rays, begun in 1823. W. R. H. February 27, 1834" (OR 52; Graves, 1:115).
18. WRH to Cousin Arthur, May 31, 1823, OR 64; Graves, 1:141.
19. WRH to sister Eliza, Sept. 23, 1822, Graves, 1:114-5. When Hamilton wrote this letter he had just finished reading David Brewster's *Life of Newton*.
20. WRH to Aunt Mary Hutton, Aug. 26, 1822, Graves, 1:110-11.
21. "To Eliza," Oct. 30, 1824, Graves, 1:168.
22. WRH to Arabella Lawrence, 1825, OR 112; Graves, 1:192-95.
23. Sydney Hamilton to Sarah Hamilton, n.d., OR 1931.
24. WRH to sister Grace, Feb. 2, 1820, OS 2; Graves, 1:75.
25. Graves, 1:62.
26. Hamilton attempted to get Eliza's poems published in *Blackwoods' Magazine* with Wordsworth's support (Graves, 2:28-29; also Graves, 1:682-85).
27. WRH to sister Eliza, July 18, 1822, Graves, 1:103, 109-10.
28. Hamilton was particularly fond of one passage from the "The Minstrel," which he later associated with the quaternions. (TCD notebooks 24 and 103.5, and Graves, 3:238-39.) It reads as follows:

> And Reason now thro' Number, Time, and Space,
> Darts the keen lustre of her serious eye,
> And learns from facts compared the laws to trace,
> Whose long progression leads to Deity.
> Can mortal strength presume to soar so high!
> Can mortal sight, so oft bedimmed with tears,
> Such glory bear!—for lo, the shadows fly
> From nature's face; confusion disappears,
> And order charms the eyes, to harmony the ears.

29. WRH to sister Eliza, July 14, 1821, Graves, 1:91.
30. WRH to Cousin Arthur, Oct. 12, 1822, Graves, 1:117-18.
31. WRH to sister Eliza, Jan. 28, 1832, OR 55; Graves, 1:127-28.
32. Eliza Hamilton to WRH, Mar. 23, 1823, OR 58.
33. OR 2152 (1832).
34. Cousin Arthur to WRH, Oct. 4, 1822, ORSUP 54. In a letter to Eliza dated July 18, 1822 (ORSUP 49; Graves, 1:108), William writes: "I am going to enter college in October," and he tells her of the decision on Oct. 9, 1822 (ORSUP 55; Graves, 1:115-16).

35. WRH to sister Eliza, Jan. 28, 1823, OR 55; Graves, 1:126.
36. Graves, 1:142.
37. For a more detailed account see Graves, 1:154-55. The actual examination records are among the TCD muniments (Trinity Examinations V. Mun. 27.6).
38. *The Book of Trinity College Dublin, 1591-1891* (Dublin: Hodges, Figgis & Co., 1892), p. 119. The reform of the classical course had been accomplished in 1793.
39. The gold medals at degree examinations had been introduced in 1793, so they were not among the oldest traditions of the college. See Constantia Elizabeth Maxwell, *A History of Trinity College Dublin, 1591-1892* (Dublin: The University Press, Trinity College, 1946), p. 154.
40. The college records show him as a pensioner (Graves, 1:235), but in submitting Hamilton's name for the astronomical professorship, Charles Boyton writes: "I do not know the form of making application for sizars" (TCD muniments WB × 10/22).
41. Maxwell, *History of Trinity College*, p. 141.
42. The exchange of letters is OR 1954. On Hamilton's challenge, see Graves, 3:236.
43. Eliza to Aunt Susanna Willey, Feb. 4, 1824, OR 76.
44. WRH to Cousin Arthur, May 2, 1824, OR 84, WRH to sister Grace, Oct. 14, 1825, OR 109; Graves, 1:190, and WRH to sister Grace, June 10, 1825, OR 105.
45. WRH to Cousin Arthur, Sept. 28, 1823, Graves, 1:148-49.
46. Maxwell, *History of Trinity College*, p. 154. Hamilton was surprised in 1852 to discover the fellowship examination being conducted in English. Some of the fellows automatically fell back into speaking Latin, much to Hamilton's amusement. WRH to De Morgan, June 2, 1852, OS 603; Graves, 3:371-72.
47. WRH to sister Eliza, June 5, 1820, ORSUP 27; Graves, 1:81.
48. "From 1800, when she published *Castle Rackrent*, to 1814, when Scott published *Waverley*, Maria Edgeworth was easily the most celebrated and successful of practising English novelists" (Marilyn Butler, *Maria Edgeworth: A Literary Biography* [Oxford: At the Clarendon Press, 1972], p. 1).
49. Butler, *Maria Edgeworth*, p. 394.
50. Maria Edgeworth to Miss Honora Edgeworth, Aug. 28, 1824, Graves, 1:161.
51. WRH to sister Grace, Aug. 27, 1824, Graves, 1:162.
52. Herbert Butler, "The Country House—the Life of the Gentry," in *Social Life in Ireland, 1800-1845*, ("Irish Life and Culture" 12, The Cultural Relations Committee of Ireland [Dublin: Three Candles, 1957]), ed. R. B. McDowell, p. 28. Butler calls Edgeworthstown the most illustrious of all the country houses.
53. Hamilton communicated his famous papers "On a General Method of Dynamics" to the Royal Society of London through Francis Beaufort (see *Math. Papers*, 2:103), and Beaufort distributed books and offprints in England for him.
54. Eliza Hamilton to WRH, Sept. 1, 1824, OR 89.
55. WRH to sister Eliza, Sept. 7, 1824, OR 90.
56. Maria Edgeworth to WRH, Jan. 4, 1828, OR 148.
57. James Hamilton to WRH, July 1, 1828, JG 20. This reads as if it were a reference to Butler's own marriage to Harriet, but then it should have been written in 1826.
58. Butler, *Maria Edgeworth*, p. 405.
59. TCD notebook 16, f. 27.
60. Samuel Lewis, *A Topographical Dictionary of Ireland*, 2 vols. (London: S. Lewis, 1837), 2:244-45. Hamilton's Aunt Mary Hutton lived in the village of Summerhill, and Hamilton had occasion to go there from Trim, occasionally even on foot.
61. The Disney sons entered Trinity College as follows: Thomas (Nov., 1815), Robert (Nov., 1817), Edward (July, 1821), Henry (July, 1822), James (July, 1823), and Lambert (July, 1825) (*Alumni Dublinenses: A Registry of the Students, Graduates, Professors and Provosts of Trinity College in the University of Dublin, 1593-1860*, ed. George Burtcaeli and Ulrich Sadlier [Dublin: Alex Thom & Co., 1935]). The daughters' names were Catherine, Louisa, Sarah, and Anne.
62. "Wonderful hour! of my sitting, irregularly, from the very first,—beside her: when, without a word said of love, we gave away our *lives* to each other. *She* was, as you know, *beautiful;* I was only *clever* and (already) celebrated" (WRH to Louisa Disney Reid, Aug. 16, 1861, TCD notebook 123.5, pp. 213ff).

63. WRH to Uncle James, Jan. 11, 1825, OR 94; Graves, 1:170.
64. WRH to Uncle James, Jan. 11, 1825, OR 94; Graves, 1:173.
65. WRH to Uncle James, Jan. 11, 1825, OR 94; Graves, 1:173.
66. WRH to Uncle James, Jan. 11, 1825, OR 94; Graves 1:172.
67. ORSUP 72.
68. WRH to Louisa Disney Reid, July 13-14, 1861, and another letter of the same date, July 13, 1861, TCD notebook 123-25, pp. 34ff, 42ff. When Hamilton was caught up in remembrances of Catherine it was not unusual for him to write two letters to the same person on one day.
69. WRH to Louisa Disney Reid, Aug. 10, 1861, TCD notebook 123.5 pp. 211ff.
70. In 1861 Hamilton recalled an evening at the Stanleys' when they all sat around the fire and listened to Catherine play the harp. "Alas, there was another person in the room (Mr. William Barlow), whose presence or absence seemed then to me a matter of supreme indifference" (WRH to Louisa Disney Reid, July 13 and 14, 1861, TCD notebook 123.5 pp. 34ff).
71. Graves, who probably knew Barlow, refers to him in his biography as an "elder suitor" (Graves, 1:182).
72. WRH to Aubrey De Vere, Aug. 24, 1855, NLD 905, pp. 109ff.
73. Hamilton said that it was the only *serious* illness he had ever suffered (WRH to De Morgan, Dec. 14, 1853, Graves, 1:182).
74. WRH to Peter Guthrie Tait, Oct. 6, 1858, TCD notebook 146, p. 59.
75. WRH to De Vere, Aug. 29, 1848, NLD 5765, fol. 423.
76. The poem is dated Jan. 1, 1826 (OR 2157; Graves, 1:183-85).
77. Uncle James to WRH, Apr. 26, 1825, OR 100; Graves, 1:181-82.
78. Graves, 1:186. The committee was composed of Henry Harte, Dionysius Lardner, and Doctor MacDonnell.
79. WRH to Uncle James, Jan. 11, 1825, OR 94; Graves, 1:170-73.

Chapter 3

1. WRH to Uncle James, Oct. 24, 1826, Graves, 1:220.
2. WRH to Uncle James, Oct. 24, 1826, Graves, 1:220.
3. Uncle James to WRH, May 10, 1827, OR 117.
4. This information comes from the TCD muniments WB × 10/22 "Letters etc. respecting the Election of a Professor of Astronomy."
5. George Biddell Airy to Robert Phipps, Registrar, April 20, 1829, TCD muniments WB × 10/22.
6. Charles Boyton to WRH, June 8, 1827, OR 118. Boyton's information turned out not to be entirely correct. While the board was prepared to raise the salary for Airy, they apparently were not prepared to do the same for Hamilton, because on Oct. 9, 1827, Hamilton wrote to Airy that the salary had *not* been increased and that he was receiving less than he would have received as a fellow (Graves, 1:275). His salary was raised in 1831 to £700 with the added stipulation that he take no more private pupils. The £700, however, *included* the salary for an assistant (£100) and for a gardener (£20) (*Registry of TCD*, vol. 1830, p. 25; Graves, 1:434).
7. TCD muniments WB × 10/22.
8. Although Hamilton was elected unanimously, the election was preceded by a long debate (WRH to sister Grace, June 14, 1827, OR 120). Ten days after the election the board gathered at the observatory to take stock of the grounds and equipment (*Registry of TCD*, vol. 1827, p. 402). The board ordered a new house built for the assistant, Thompson, and ordered Hamilton to go to Cloyne to consult with Brinkley, supplying an additional £10 for travel.
9. John Brinkley, Bishop of Cloyne, to Rev. C. Boyton and fellows, June 14, 1827, TCD muniments WB × 10/ 22.
10. Brinkley to WRH, June 26, 1827, OR 127; Graves, 1:240.
11. WRH memorandum, July 2, 1827, OR 128; Graves 1:241.

12. WRH to Boyton, June 11, 1827, TCD muniments WB × 10/22.
13. T. Romney Robinson to WRH, June 21 [1827], OR 125; Graves 1:245.
14. William Macneile Dixon, *Trinity College Dublin* (London: F. E. Robinson & Co., 1902), pp. 33 and 112; and *The Book of Trinity College Dublin, 1591-1891* (Dublin: Hodges, Figgis & Co., 1892), p. 93.
15. *Book of Trinity College*, p. 121.
16. Records of Observations at Dunsink Observatory, Transit Observations, Circle Observations. These are large manuscript notebooks in the Observatory Library.
17. WRH to Robinson, May 12, 1831, OS, 71; Graves 1:431.
18. WRH to sister Sydney, July 29, 1827, OR 132, Sydney Hamilton to WRH, Aug. 1, 1827, OR 134, and WRH to sister Sydney, Aug. 2, 1827, OR 135; Graves, 1:249-50.
19. WRH to sister Sydney, Aug. 2, 1827, OR 135; Graves, 1:249.
20. WRH to Aunt Mary Hutton, Aug. 25, 1827, OR 140; Graves 1:251.
21. WRH to sister Eliza, Aug. 21, 1827, OR 137.
22. WRH to Cousin Arthur, Aug. 15, 1827, OR 136.
23. WRH to Aunt Mary Hutton, Aug. 25, 1827, OR 140; Graves 1:250-52.
24. WRH to sister Eliza, Aug. 30, 1827, OR 141; Graves, 1:253.
25. WRH to Alexander Nimmo, Aug. 31, 1827, Graves, 1:258. Hamilton stated that he had previously looked upon nature only with the eye of a poet; but now he viewed it also with the eye of a painter and had begun to make sketches. In Killarney they had traveled with the painter John Glover (1767-1849) (Graves, 1:252).
26. WRH to sister Eliza, Sept. 16, 1827, Graves, 1:262.
27. William Wordsworth to WRH, Sept. 24, 1827, Graves, 1:268.
28. Robert Southey to Thomas Digges La Touche, Dec. 9, 1827, Graves, 1:270. Hamilton had distant family connections to Southey through La Touche, to whom this letter was addressed. His Aunt Elizabeth, James's wife, was from the La Touche family. Nimmo was definitely a Scot, but a Scot with a playful sense of humor. From the Lake Country Hamilton and Nimmo went to Edinburgh. The following letter refers to their adventures there: "My Dear Sir Wm. Hamilton. I have heard it reported (and the report said to rest on good authority) that Mr. George Combe of Edinburgh [pronounced] your head to indicate the possession of intellectual powers but little better than those belonging to an idiot; and that this mistake (for a slight one it must be confessed to be) was made on the occasion of your being presented to him by Mr. Nimmo, an Engineer, as a young man whom he was about to take as an apprentice and whose fitness for this pursuit he wished to have judged of from the form of his head" (R. J. Evanson to WRH, April 24 [no year], OR 1890).
29. Graves, 1:264-66. William Wordsworth to WRH, Sept. 24, 1827, Graves, 1:266, and Arabella Lawrence to WRH, June 18, 1827, ORSUP 68. Eliza had sent him a similar poem recounting his romance with Catherine Disney, which she claimed to have found somewhere; actually she wrote it herself (Eliza to WRH, Nov. 22, 1825, OS 29).
30. William Wordsworth to WRH, Sept. 24, 1827, Graves, 1:266.
31. Arabella Lawrence to WRH, June 18, 1827, ORSUP 68.
32. Lawrence to WRH, Oct. 22, 1827, OR 144.
33. Lawrence to WRH, Sept. 2, 1828, JG 23.
34. JG 3; Graves, 1:273.
35. Reminiscences by sister Eliza on Charles Boyton, Graves, 1:238, and Susanna Willey to WRH and sister Sydney, June 23 and 25, 1827, OR 126.
36. Graves, 1:216-17.
37. Memorandum by William Edwin Hamilton on his father, Graves, 3:240. See also Graves, 1:210.
38. WRH to sister Eliza, Aug. 30, 1827, OR 141; Graves, 1:253.
39. WRH to sister Eliza, Sept. 9, 1827, OR 142; Graves, 1:260.
40. WRH to sister Eliza, Sept. 26, 1827, OR 143; Graves, 1:271.
41. WRH to sister Sydney, Mar. 30, 1828, OR 152; Graves, 1:292.
42. WRH to Maria Edgeworth, Apr. 15, 1828, OR 153; Graves, 1:292.
43. Graves, 1:286.
44. WRH to sister Eliza, Apr. 2, 1828, JG 9.
45. WRH to sister Eliza, Apr. 2, 1828, JG 9.

46. Marilyn Butler, *Maria Edgeworth: A Literary Biography* (Oxford: At the Clarendon Press, 1972), pp. 423–24.
47. WRH to Louisa Disney Reid, July 22, 1861, TCD notebook 123.5, pp. 87ff.
48. WRH to Lady Wilde, July 2, 1861, TCD notebook 123.5, pp. 18–20.
49. WRH to Lady Wilde, June 30, 1861, TCD notebook 123.5, pp. 15–16.
50. Dated July 13, 1827, Graves, 1:246–47.
51. WRH to sister Sydney, Nov. 12, 1828, TCD notebook 25; Graves, 1:305.
52. WRH to sister Sydney, Nov. 19, 1828, TCD notebook 26.

Chapter 4

1. WRH to Samuel Taylor Coleridge, Oct. 3, 1832, Graves, 1:592. This letter is a draft that was never sent. Hamilton revised it considerably and finally sent it in Feb., 1833 (Graves, 2:36).
2. "On a General Method in Dynamics," Introductory remarks, *Math. Papers*, 2:105.
3. See chapter 2.
4. *Math. Papers*, 2:104.
5. Heinrich Hertz, *The Principles of Mechanics Presented in a New Form* (New York: Dover, 1956), p. 32.
6. Translated from the third German edition by Henry L. Brose (London: Methuen and Co., 1923), pp. 555–56.
7. Erwin Schrödinger, "The Hamilton Postage Stamp: An Announcement by the Irish Minister of Posts and Telegraphs," in "A Collection of Papers in Memory of Sir William Rowan Hamilton," ed. David Eugene Smith, *Scripta Mathematica Studies*, no. 2 (New York: Scripta Mathematica 1945), p. 82.
8. Schrödinger was speaking in 1945 in Dublin at a centenary celebration of Hamilton's discovery of quaternions shortly after Schrödinger had fled Germany and found refuge at the Dublin Institute for Advanced Studies.
9. S.-F. Lacroix, *Traité du calcul différentiel et du calcul intégral*, 3 vols. (Paris: J.-B.-M. Duprat, 1797–1800), 1:xxv–xxvi, quoted in René Taton, *L'oeuvre scientifique de Monge*, 1st ed. (Paris: Presses Universitaires de France, 1951), p. 133.
10. Lacroix's book on analytical geometry was enormously popular, going through twenty-five editions in the course of a century (Carl Boyer, "Cartesian Geometry from Fermat to Lacroix," *Scripta Mathematica* 13 [1947]: 134).
11. Taton, *L'oeuvre scientifique*, pp. 132–33.
12. References to Hamilton's mathematical reading appear mostly in his study journals, where he keeps a record of his readings in preparation for examinations.
13. WRH to John F. W. Herschel, Sept. 1, 1847, OS 408; Graves, 2:591.
14. WRH to Cousin Arthur, July 19, 1823, OR 70; Graves, 1:142.
15. WRH to Cousin Arthur, May 31, 1823, OR 64; Graves, 1:141.
16. Graves says that he believes Hamilton was referring to his characteristic function (Graves 1:142), but the characteristic function does not appear in the paper on caustics, nor is it even hinted at, which leads me to believe that Hamilton was considering a new theory of rectilinear congruences.
17. The *Application de l'analyse à la géométrie* appeared first in 1795 as *Feuilles d'analyse appliquée à la géométrie à l'usage de l'Ecole Polytechnique*. It went through several changes of title. The edition used by Hamilton was probably the fourth edition of 1809.
18. Assuming a homogeneous medium.
19. Etienne Malus, "Traité d'optique," *Mémoires présentés à l'Institut des sciences, lettres et arts par divers savans ... Sciences mathématiques et physiques* 2 (1811): 214–302.
20. Gaspard Monge, "Mémoire sur la théorie des déblais et des remblais," *Mémoires de l'Académie Royale des Sciences* (1781), pp. 686–704.
21. "Malus ne se referant jamais au travail antérieur de Monge, de nombreux mathématiciens ont cru à tort, à la suite de Hamilton, Kummer et Mannheim, que la première

mention des congruences de droites se trouvait dans son traité, alors que Monge est incontestablement le créateur de cette notion" (Taton, *L'oeuvre scientifique*, p. 203).

22. E. E. Kummer, "Allgemeine Theorie der gradlinigen Strahlen-Systeme," *Journal für die reine und angewandte Mathematik* 57 (1860):189-230. "Diese von *Hamilton* zuerst behandelte Theorie der allgemeinen gradlinigen Strahlensysteme, durch eine neue Begrundung, der analytischen Geometrie des Raumes anzueignen und sie zugleich in mehreren wesentlichen Punkten zu vervollstandigen, soll der Zweck der gegenwartigen Abhandlung sein" (quoted from *Math. Papers*, 1: xxii n).

23. Carl Boyer has called the analytical geometry of Monge an "analytical revolution" comparable in importance to the "chemical revolution" that was taking place at the same time (Taton, *L'oeuvre scientifique*, p. 147).

24. *Math. Papers*, 1:88-106.

25. Misc. 312/1-40. The table of contents that Hamilton published with part one indicates that part three should begin at paragraph 110 and end at paragraph 161. This manuscript begins at paragraph 110 and ends at paragraph 160, so it is evidently incomplete.

26. Sir Joseph Larmor, *Mathematical and Physical Papers*, 2 vols. (Cambridge: At the University Press, 1929), 1:640. The best exposition of Hamilton's optical theory is John L. Synge, *Geometrical Optics: An Introduction to Hamilton's Method* ("Cambridge Tracts in Mathematics and Mathematical Physics," no. 37 [Cambridge: At the University Press, 1937]). The exposition that follows is close to the one given by Synge. Synge wrote this book after editing Hamilton's mathematical papers. He studied the papers in more detail than anyone before him had done. The editors' appendixes to the *Mathematical Papers*, 1:461-517, give a more advanced mathematical commentary. Two books that treat Hamilton's optics in conjunction with his mechanics are Cornelius Lanczos, *The Variational Principles of Mechanics*, 4th ed. (Toronto: University of Toronto Press, 1970); and Wolfgang Yourgrau and Stanley Mandelstam, *Variational Principles in Dynamics and Quantum Theory*, 3d ed. (Philadelphia: W. B. Saunders Co., 1968).

27. This assumes that the medium is isotropic. The theorem does not hold for nonisotropic media.

28. The fact that the wave front is normal to the direction of propagation can be proved directly from Huygens's construction.

29. *Math. Papers*, 1:14.

30. Ibid., 463, editors' note 2, "History of the Theorem of Malus."

31. The geometrical significance of this argument is easier to see in vector notation (which is, of course, entirely anachronistic in this context). Let $\mathbf{n} = (\alpha, \beta, \gamma)$, then the condition for integrability of $\alpha dx + \beta dy + \gamma dz$ is $\mathbf{n} \cdot (\nabla \times \mathbf{n}) = 0$.

32. Hamilton writes: "I have called the *characteristic function* of the initial and final coordinates ... the *variation of the action, or the time, expended by light of any one colour, in going from one variable point to another*" (*Math. Papers*, 1:168).

33. Monge, *Application de l'analyse*, pt. 2, p. 29.

34. Hamilton mentioned this work in his "Theory of Systems of Rays", pt. 2 (*Math. Papers*, 1:104).

35. Hamilton originally intended to make his theory of curved rays into a study of atmospheric refraction, but in keeping with the theoretical character of his paper he later changed it to a simple study "On Systems of Curved Rays."

Chapter 5

1. *Math. Papers*, 1:1.

2. Uncle James to WRH, Apr. 30, 1827, ORSUP 67.

3. WRH to T. Romney Robinson, Feb. 1, 1828, Graves, 1:290.

4. WRH to "My Lord" [Brinkley], Jan. 20, 1828, OS 18. In other notes Hamilton mentions that he had part two almost finished in February, 1826, more than two years before part one appeared in print (TCD notebook 16, p. 24).

5. TCD notebook 11, unpaginated.
6. Misc. I, "Characteristic function." In this manuscript the characteristic function is denoted by i, which indicates that the manuscript was written at the same time as the last part of the table of contents of the "Theory of Systems of Rays," where the same notation appears (*Math. Papers,* 1:8-9). Throughout the published papers the characteristic function is consistently denoted by *V.*
7. Graves, 1:201; TCD notebook 16, p. 24.
8. Graves, 1:211. Graves saw this letter, which he did not publish because it was "unnecessary." It has since been lost. If it ever comes to light it will be helpful in establishing the chronology of Hamilton's ideas.
9. TCD notebook 17, p. 19.
10. Ibid., pp. 22, 27.
11. Ibid., p. 22.
12. WRH to Cousin Arthur, Sept. 14, 1826, Graves, 1:218.
13. TCD notebook 17, pp. 22, 23.
14. Ibid., p. 24.
15. Ibid., p. 25.
16. Ibid., p. 26.
17. This second interpretation explains why Hamilton did not immediately define the characteristic function as the integral of time.
18. This last paragraph (number 161) is not in the manuscript version (which ends with paragraph number 160), and is only mentioned in the printed table of contents. It was probably added at the last minute. (See *Math. Papers,* 1:9 for the "analogous principle".) At these two places in the "Theory of Systems of Rays" Hamilton expresses the variation of the action integral as δI or $\delta i = \int v ds$, repeating the notation first used in his studies of 1826. The fact that he uses the same notation indicates that these brief but important paragraphs were related to his studies of September, 1826. Everywhere else in the "Theory of Systems of Rays" and in all the supplements the characteristic function appears as *V.* Hamilton probably changed his notation in 1826, because he had still not fixed on a final definition for *V.* In part one, *V* stood for the length of the ray; in part two it represented the time of transmission or "optical length" of the ray; by 1828, when he published the entire theory, he wished to define the characteristic function by the action integral. It was only in the first supplement that he finally decided on a single definition of the characteristic function as the action integral, and a single notation *V.*
19. The *W* function and the *T* function from the third supplement are explained by J. L. Synge in his *Geometrical Optics: An Introduction to Hamilton's Method* ("Cambridge Tracts in Mathematics and Mathematical Physics," no. 37 [Cambridge: At the University Press, 1937]), chaps. 2, 3.
20. Graves, 2:345; *Math. Papers,* 1:506.
21. WRH to William Wordsworth, Jan. 17, 1832, Graves, 1:513-14, and WRH to the Countess of Dunraven, Jan. 17, 1832, Graves 1:514.
22. WRH to Aubrey De Vere, Jan. 6, 1832, OS 78; NLD 5764, fol. 7; Graves, 1:516-17.
23. Hamilton considered publishing a fourth supplement on the practical application of his theory, which would probably have contained much of his work on aberrations and on the design of optical instruments. It is unfortunate that he decided against publishing it, for much of the labor was necessarily repeated by later mathematicians trying to apply his methods. See *Math. Papers.* 1:xxiii-xxiv (editors' comments).
24. WRH to Viscount Adare, Sept. 12, 1832, Graves, 1: 591.
25. Royal Irish Academy, *Transactions* 17 pt. 1 (1837): 1-144. The complete volume bears the year 1837, although the number in which the third supplement appears was out in 1833.
26. Uncle James to WRH, Apr. 30, 1827, ORSUP 67.
27. On the method of Bruns see *Math. Papers,* 1:xxiii, 488, and 493. J. L. Synge and M. Herzberger carried on a controversy over the respective merits of the work done by Hamilton and Bruns. J. L. Synge, "Hamilton's Method in Geometrical Optics," *Journal of the Optical Society of America* 27 (1937): 75-82, and idem, "Hamilton's Characteristic Function and the Bruns Eiconal," *Journal of the Optical Society of America* 27 (1937): 138-44.

Chapter 6

1. WRH to John F. W. Herschel, Dec. 18, 1832, Graves, 1:627.
2. The nineteenth-century term was *biaxal*; modern texts usually use *biaxial*.
3. Graves, 1:623; *Math. Papers,* 1:284. Hamilton closed his third supplement with a section entitled "Combination of the Foregoing View of Optics with the Undulatory Theory of Light" in which he announced conical refraction. *Math. Papers,* 1:277.
4. *Math. Papers,* 1:291.
5. The correct angle is the one that will cause a plane wave incident on the crystal to refract into a plane wave normal to the optic axis of the crystal.
6. A good explanation of conical refraction is in Arthur Schuster, *An Introduction to the Theory of Optics* (London: Edward Arnold, 1920), pp. 179–92. In internal refraction the refracted wave front is normal to the optic axis; in external refraction it is normal to the "direction of single ray velocity" or the "cusp ray," as Hamilton called it. The optic axis and the cusp ray are not the same lines in the crystal, although they are close enough to cause confusion.
7. *Math. Papers,* 1:277–78. This result comes directly from the fact that the index of refraction is *inversely* proportional to the speed of light in the medium. The component parallel to the surface is unchanged by reflection or refraction (*Math. Papers,* 1:301). Hamilton's first treatment of anisotropic media is in his first supplement (*Math. Papers,* 1:110–11).
8. For a series of media that may be anisotropic and heterogeneous, Hamilton wrote two equations:

$$0 = \Omega\,(\sigma,\, \tau,\, \upsilon,\, \text{x},\, \text{y},\, \text{z},\, \chi)$$

$$0 = \Omega'\,(\sigma',\, \tau',\, \upsilon',\, \text{x}',\, \text{y}',\, \text{z}',\, \chi).$$

The primed quantities are for the initial rays, the χ is the chromatic index. In deducing the wave surface, the initial point is fixed, the medium is homogeneous (but not isotropic), and he considers only one color of light, making the chromatic index χ unnecessary. The expression

$$\left(\frac{\partial \text{V}}{\partial x}\right)^2 + \left(\frac{\partial \text{V}}{\partial y}\right)^2 + \left(\frac{\partial \text{V}}{\partial z}\right)^2 = \upsilon$$

for homogeneous isotropic media is a special case of his function Ω. In dynamics the analogous function to Ω is the Hamilton-Jacobi equation.
9. *Math. Papers,* 1:291–93.
10. Graves, 1:623. Hamilton's poet-friend Aubrey De Vere was present as a visitor at the meeting. On the discovery of conical refraction see George Sarton, "Discovery of Conical Refraction by William Rowan Hamilton and Humphrey Lloyd (1833)," *Isis* 17 (1932):154–70.
11. Lloyd's first experiment was an attempt to observe exterior conical refraction. Rather than use a full cone of rays, he covered a lens, leaving three small holes for the light to pass through, making these convergent rays incident on the crystal. The three rays emerged from the other side of the crystal indicating a cone, but, to his chagrin, they also appeared when he used the *wrong* angle of incidence.
12. WRH to George Biddell Airy, Oct. 25, 1832, Graves, 1:625, and Airy to WRH, Nov. 4, 1832, OR 294; Graves, 1:625.
13. WRH to Humphrey Lloyd, Nov. 10, 1832, OR 299; Graves, 1:625–26.
14. Lloyd to WRH, undated, OR 300; Graves, 1:626.
15. Lloyd to WRH, Dec. 14, 1832, OR 305; Graves, 1:626.
16. Lloyd to WRH, Dec. 18, 1832, OR 307; Graves, 1:628.
17. See Max Born and Emil Wolf, *Principles of Optics: Electromagnetic Theory of Propagation, Interference and Diffraction of Light,* 2d ed. (New York: The Macmillan Co., 1964), pp. 688–90.
18. OR 308. Lloyd worried about *legal* objections that the Royal Irish Academy might raise

against his papers if he published them first in the *Philosophical Magazine*. Lloyd to WRH, Feb. 4, 1833, OR 340.

19. Airy to WRH, Dec. 23, 1832, OR 309; Graves, 1:628.
20. Lloyd writes: "On examining this curious phaenomenon more attentively, I discovered the remarkable law etc." in "On the Phaenomena Presented by Light in Its Passage along the Axes of Biaxal Crystals," *Philosophical Magazine*, 3d ser. 2 (Feb. 1833):116-17. Graves quotes Airy's letter of Dec. 23, 1832, in its entirety *except* for the passage describing conical polarization.
21. WRH to Lloyd, Jan. 1 and 2, 1833, OR 312; Graves, 1:629-31.
22. WRH to Lloyd, Jan. 3, 1833, OR 313, another of Jan. 3, 1833, OR 315, Jan. 4, 1833, OR 316, Jan. 5, 1833, OR 317, Jan. 6, 1933, OR 318. On Jan. 15, 1833, Herschel wrote in response to a letter from Hamilton that he had heard of conical polarization from Airy, but not that it had been verified (OR 321). On Jan. 1, 1833, Hamilton wrote to Lloyd stating precisely what he believed his claim to be and asking Lloyd to include it in his forthcoming article in the *Philosophical Magazine*.
23. Airy to WRH, Jan. 15, 1833, OR 322; Graves, 1:631.
24. Airy to WRH, Jan. 28, 1833, OR 333; Graves, 1:632.
25. Lloyd's papers on conical refraction appeared in two consecutive numbers of the *Philosophical Magazine*, Feb. and Mar., 1833. The second was "Further Experiments on the Phaenomena Presented by Light in Its Passage along the Axes of Biaxal Crystals," *Philosophical Magazine*, 3d ser. 2 (Mar., 1833): 207-10.
26. MacCullagh's paper was "The Double Refraction of Light in a Crystallized Medium According to the Principles of Fresnel," Royal Irish Academy, *Transactions* 16, pt. 2 (1830): 65-78. Hamilton wrote "Review of Two Scientific Memoirs of James MacCullagh, B. A., *National Magazine*, Dublin 1 (Aug., 1830): 145-49.
27. WRH to Lloyd, Jan. 1, 1833, Graves, 1:629.
28. WRH to Airy, Jan. 4, 1833, Graves, 1:631. MacCullagh's construction is the one commonly given. It is much simpler than Hamilton's. See George Salmon, *A Treatise on the Analytic Geometry of Three Dimensions*, 5th ed., 2 vols. (Dublin: Hodges, Figgis & Co., 1912-15), 2:128-41.
29. James MacCullagh, "Note on the Subject of Conical Refraction," *Philosophical Magazine*, 3d ser. 3 (July, 1833): 114-15.
30. James MacCullagh, "Additional Note on Conical Refraction," *Philosophical Magazine*, 3d ser. 3 (Sept., 1833): 197. A carefully documented account of the close of this controversy is in Graves, 1:685-92. Apparently MacCullagh had been urged to claim priority by an "inconsiderate friend" (WRH to Lord Adare, July 30, 1834, Graves, 2:99).
31. WRH to George Salmon, Aug. 22, 1857, TCD notebook 140, pp. 149-50.
32. *Math. Papers*, 1:288-90.
33. WRH to S. T. Coleridge, Feb. 3, 1833, Graves, 2:37, and George G. Stokes, "Report on Double Refraction," *British Association Report*, 1862, p. 270.
34. Graves, 1:637.
35. When Lubbock, the secretary of the Royal Society, wrote to Hamilton informing him of the award of the Royal Medal, he did not mention Hamilton's "General Method in Dynamics," which the society had published in its *Philosophical Transactions*. The medal was awarded "for your discoveries in Optics, and particularly that of Conical Refraction" (Lubbock to WRH, Nov. 30, 1835, Graves, 2:170).

Chapter 7

1. Hamilton was a poor rider, but he did not lack courage. He named his first horse Comet because he was such a wild and errant creature. After a bad fall Hamilton relegated Comet to the carriage and replaced him by the more sedate Planet.
2. Graves, 1:368.
3. Maria Edgeworth to WRH, Apr. 27, 1828, Graves, 1:294. It is not surprising that Francis was having trouble with science at Cambridge. In 1830 he wrote to Hamilton describ-

ing a pendulum clock invented by his father that would keep time for forty years because it operated entirely without friction and was driven by a weight that never descended! (Francis Beaufort Edgeworth to WRH, Jan. 21, 1830, NLD 11132, fol. 1.)

4. Hamilton confessed that the boys were a constant anxiety. Special accommodations had to be arranged for them at the observatory, and their health was a matter of concern (WRH to T. Romney Robinson, Jan. 30, 1829, Graves, 1:323-24, and Uncle James to WRH, Jan. 22, 1829, OR 170). The problem was soon solved, however, when Anglesey was recalled in January, 1829. Hamilton wrote: "I cannot regret his removal, although my temper disposes me to think as little as possible about state affairs" (same correspondence).

5. WRH to Lord George Paget, Jan. 15, 1830, JG 59.

6. F. B. Edgeworth to WRH, Aug. 7, 1829, Graves, 1:337-40.

7. WRH to William Wordsworth, May 14, 1829, Graves, 1:331, and WRH to Robinson, Aug. 3, 1829, Graves, 1:335-36.

8. WRH to F. B. Edgeworth, Nov. 20, 1829, JG 49; Graves, 1:348-49.

9. WRH to Wordsworth, May 14, 1829, Graves, 1:331.

10. Wordsworth to WRH, July 24, 1829, Graves, 1:333.

11. Sister Eliza to Cousin Arthur, Aug. 13, 1829, OR 186.

12. Alfred North Whitehead, *Science and the Modern World* (New York: Mentor Books, 1958), p. 84.

13. William Wordsworth, *The Complete Poetical Works,* ed. Andrew J. George (Boston: Houghton, 1932), *Excursion,* bk. 4, line 1153.

14. Ibid., lines 1144-45.

15. Ibid., lines 1251-63.

16. Hamilton's sister Eliza left an account of Wordsworth's visit to the observatory (Graves, 1:313). On Wordsworth's attitude towards science, see Meyer H. Abrams, *The Mirror and the Lamp: Romantic Theory and the Critical Tradition* (New York: W. W. Norton Co., 1956), pp. 309-10.

17. WRH to F. B. Edgeworth, Oct. 31, 1829, Graves, 1:346-47.

18. Samuel Taylor Coleridge, *Biographia Literaria,* 2 vols., ed. J. Shawcross (Oxford: Oxford University Press, 1939), 2:221.

19. Ibid., 2:12, and Samuel Taylor Coleridge, *Aids to Reflection, and the Confessions of an Inquiring Spirit* (London: George Bell & Sons, 1901), p. xviii.

20. Coleridge placed the intellect in the understanding, not in the reason. However, he claimed that the understanding, as a mediate faculty, had two extremities or poles, the sensual and the intellectual. The intellectual pole was turned towards the reason and was illuminated by its light (lumen), while the sensual pole remained in darkness (Charles R. Sanders, *Coleridge and the Broad Church Movement* [Durham, N.C.: Duke University Press, 1942], pp. 45-46).

21. Graves, 1:500-501.

22. Ibid., 501.

23. Ibid., 652.

24. Helena Pycior, "The Role of Sir William Rowan Hamilton in the Development of British Modern Algebra" (Ph.D. diss., Cornell University, 1976), pp. 39-41.

25. Graves, 1:314.

26. Although Hamilton never converted Wordsworth to his own ideas about science, there is evidence that he budged him a little. In a note of 1829, probably written soon after his visit to Hamilton, Wordsworth conceded that "analysing, decomposing, and anatomising" are not necessarily unfavorable to the perception of beauty, if they lead to admiration and love and are done by a man of real genius. But Wordsworth concludes with the warning that "a *savant* who is not also a poet in soul and a religionist in heart is a feeble and unhappy creature" (Quoted from Geoffrey Durrant, *Wordsworth and the Great System: A Study of Wordsworth's Poetic Genius* [Cambridge: At the University Press, 1970], p. 5). Hamilton could easily escape this criticism. Even more revealing is a change that Wordsworth made in book 5 of the *Prelude.* In 1850 he added the lines:

> Poetry and geometric truth together serve the human spirit.
> The one that had acquaintance with the stars,
> And wedded soul to soul in purest bond

Of reason, undisturbed by space and time;
The other that was a god, yea many gods,
Had voices more than all the winds . . .
(Durrant, *Wordsworth and the Great System,* p. 20)

Here poetry and mathematics are joined in just the way that Hamilton would have wanted.

27. Aubrey De Vere originated the derogatory remark about Wordsworth's "pedlars and spades and *id genus omne.*" He attacked Wordsworth for "want of Ideality, and therefore of completeness" (Aubrey De Vere to WRH, June 22, 1832, NLD 5764, fol. 53; Graves, 1:579-83).

28. Melvin Miller Rader, *Wordsworth: A Philosophical Approach* (Oxford: At the Clarendon Press, 1967), pp. 66-71.

29. Graves, 1:503. This quotation first appeared as an isolated jotting in one of Hamilton's many manuscript notebooks and was later incorporated into his lecture. See TCD notebook 26.

30. The enthusiasm of the Irish aristocracy for astronomy is not easy to explain. Lord Rosse built the largest reflector in the world at Birr Castle. Hamilton was present when it was erected in 1835. Edward J. Cooper also built an excellent observatory at Markree Castle in Sligo, and Hamilton's pupil built his own observatory at Adare.

31. Francis Goold to [WRH], May 18, 1829, OR 180, and Adare to WRH, Aug. 9, 1829, JG 38, and again on Sept. 9, 1829, JG 41.

32. Graves, 1:358.

33. WRH to Cousin Arthur, Mar. 26, 1830, OR 202.

34. WRH to Mrs. Thomas Disney, May 3, 1854, TCD notebook 103.5, pp. 277ff.

35. Graves, 1:361.

36. WRH to Lady Campbell, Apr. 8, 1830, OS 65; Graves, 1:362-63.

37. Graves, 1:604, and Eliza Hamilton to Dora Wordsworth, Nov. 14, 1831, OS 77. Felicia Hemans died in 1835.

38. Hamilton had to write to Cousin Arthur for money. WRH to Cousin Arthur, Aug. 14, 1830, ORSUP 79.

39. Hamilton's pupil was Edwin Richard Windham Wyndham Quin, third earl of Dunraven and mount-earl in the peerage of Ireland. He held the title Viscount Adare until he assumed the title of earl upon the death of his father. The Quins were an ancient Irish family, but much of the family wealth came through Adare's mother, who was daughter and heiress of Thomas Wyndham of Dunraven Castle, Glamorganshire. Adare took his B.A. degree at Trinity College Dublin in 1833 and was a member of Parliament representing Glamorganshire from 1837 through 1851. He was a Conservative, but became a convert to Catholicism and devoted most of his efforts to safeguarding religious education in Ireland. His early interest in astronomy later gave way to archaeology, and he did much to support the work of George Petrie and other Irish archaeologists. See also John Cornforth, "Adare Manor, Co. Limerick," *Country Life* 145 (1969): 1230-34, 1302-6, and 1366-69.

40. WRH to sister Grace, Sept. 17-18, 1830, Graves, 1:391-92.

41. WRH to sister Grace, Sept. 9, 1831, OR 242; Graves, 1:452-53. Graves deletes the account of the dance. Apparently it was not sufficiently dignified for his view of Hamilton.

42. Wordsworth to WRH, Jan. 24, 1831, Graves, 1:424-25, WRH to Wordsworth, Feb. 2, 1831, Graves, 1:425-28, and WRH to Wordsworth, Feb. 2, 1831, Graves, 1:425-28.

43. WRH to Mrs. W. Rathborne, Aug. 20, 1831, ORSUP 88; Graves, 1:442-43.

44. WRH to Wordsworth, Oct. 14, 1831, Graves, 1:269-71.

45. WRH to sister Grace, Sept. 14, 1831, Graves, 1:454-55.

46. Dora Wordsworth to Eliza Hamilton, July 28, 1833, NLD 905, p. 77.

47. WRH to Wordsworth, Oct. 14, 1831, Graves, 1:469-71.

48. Wordsworth to WRH, Oct. 27, 1831, Graves, 1:473-76.

49. WRH to sister Eliza, Dec. 11, 1831, OR 349.

50. R. P. Graves to Ellen De Vere O'Brien, June 2, 1874, TCD MS 5020. See also R. P. Graves to Ellen, May 3, July 25, June 2, July 29, Nov. 28, 1874, and Apr. 27, June 8, and June 16, 1875, TCD MS 5020.

51. WRH to Ellen De Vere, Dec. 11, 1831, TCD MS 5020.

52. Eliza Hamilton to WRH, Dec. 16, 1831, OR 250.
53. WRH to sister Eliza, Dec. 19, 1831, OR 252.
54. Ellen De Vere to Eliza, May 29 [1832], TCD, notebook 129.5, p. 330.
55. Lady Dunraven to WRH, May 8, 1832, Public Record Office of Northern Ireland, Belfast, Dunraven Papers, D3196.
56. While Hamilton courted Ellen De Vere, Francis Edgeworth considered marrying one of Hamilton's sisters, either Eliza or Sydney. They were duly invited to Edgeworthstown for inspection, and Francis vacillated between them, unable to make up his mind. In a letter describing the visit Maria called Sydney "intellect" because of her talent for mathematics, but Francis feared that she might have "too hard a character." Eliza was called "sensibility" because of her taste for poetry, but her weakness was a lack of "repose" and a tendency to become agitated too easily. Later in the year Francis passed through Dublin on his way to London and came within "one quarter of an inch" of proposing to Eliza. Once in London he went to his old lodgings, where he met a group of Spanish refugees and promptly married one of them named Rosa Eroles. She was seventeen years old, without fortune, and, as he wrote to his mother, she was "fat, voluptuous and made for love." So much for "intellect" and "sensibility." Francis concluded that in matters of marriage philosophical idealism was not an appropriate guide. Maria was not entirely enthusiastic about the marriage, but she preferred Rosa to Hamilton's sisters, because, as she wrote: "I for my part did not feel that I could have *loved* either of these sisters—though I believe them perfectly good—and clever, too." This entire episode is described in a highly amusing letter from Maria Edgeworth to her cousin Sophy Ruxton, Dec. 27, 1831. I am extremely grateful to Mrs. Christina Colvin for providing me with a copy of this letter and for copies of the correspondence between WRH and Francis Edgeworth.
57. WRH to Viscount Adare, Sept. 23, 1831, OR 244; Graves, 1:458-60.
58. WRH to De Vere, Dec. 31, NLD 5764, fol. 5, and WRH to De Vere, Dec. 14, 1831, NLD 5764, fol. 4.
59. WRH to De Vere, Feb. 9, 1832, NLD 5764, fol. 19.
60. De Vere to WRH, Feb. 3, 1832, NLD 5764, fol. 15.
61. De Vere to WRH, n.d., NLD 5764, fol. 25; Graves, 1:528.
62. Graves, 1:564.
63. WRH to sister Eliza, June 25, 1832, OR 276; Graves, 1:575-76, and WRH to De Vere, Oct. 13, 1832, OS 86, NLD 5764, fol. 115; Graves, 1:619-20.
64. WRH to De Vere, Oct. 30, 1832, TCD notebook 86, NLD 5764, fol. 127; Graves, 1:621.
65. Dated Dec. 21, 1832—two months after the dream (Graves, 1:620).
66. WRH to De Vere, Nov. 7, 1832, NLD 5764, fol. 131; Graves, 1:622.

Chapter 8

1. Graves, 2:2, 61.
2. Robert P. Graves to Ellen De Vere, July 25, 1873, TCD MS 5020.
3. Nov. 12, 1832, Graves, 2:4.
4. Dora Wordsworth to Eliza Hamilton, Apr. 17, [1833], TCD MS 2504, fol. 48.
5. WRH to Helen Bayly, Dec. 29, 1832, ORSUP (o).
6. WRH to Aubrey De Vere, Nov. 12, 1832, NLD 5764, fol. 135; Graves 2:3.
7. De Vere to WRH, Nov. 16, 1832, NLD 5764, fol. 137; Graves 2:5.
8. WRH to Mrs. Bayly, Nov. 17, 1832, OR 301.
9. WRH to Mrs. Bayly, Nov. 21, 1832, NLD 5764, fol. 139. Sending Aubrey's letters dwelling on his sister's relationship with Hamilton was not intelligent. One would think the matter could better have been ignored.
10. WRH to Mrs. Bayly, Nov. 22, 1832, ORSUP (a): ". . . your letter has produced in my favour a great and visible change."
11. WRH to Helen Bayly, Nov. 24, 1832, ORSUP (c).
12. WRH to Mrs. Rathborne, Nov. 24, 1832, ORSUP (d). Hamilton wrote two additional

letters on that day, one to Mrs. Rathborne and one to Helen (f, g, and e). He was obviously in earnest.

13. WRH to Mrs. Rathborne, Nov. 27, 1832 (i), and WRH to Helen Bayly Nov. 28, 1832 (j) and Dec. 3, 1832 (l).
14. WRH to Helen Bayly, Nov. 26, 1832, OR 303; Graves, 2:11-12.
15. WRH to Mrs. Rathborne, with note from Mrs. Rathborne to Mrs. Bayly, Dec. 16, 1832, ORSUP (n).
16. WRH to Helen Bayly, Dec. 10, 1832, ORSUP (m). Other parts of this "correspondence course" are Jan. 25, 1833, OR 330, and Jan. 30, 1833, OR 336.
17. Helen Bayly to WRH, Jan. 19, Jan. 24, and Feb. 3, 1833, OR 325, 329, and 339.
18. WRH to Helen Bayly, Jan. 22, 1833, OR 327.
19. WRH to Mrs. Bayly, Apr. 3, 1833, OR 356.
20. Helen Hamilton to Mrs. Bayly, Oct. 17, 1833, OR 382.
21. WRH to Helen Bayly, Jan. 16, 1833, OR 323.
22. Personal communication from Mrs. Phoebe R. O'Regan. Mrs. O'Regan is the wife of John R. H. O'Regan, Hamilton's grandson.
23. WRH to Helen Bayly, Jan. 26, Feb. 1, Feb. 6, and Feb. 9, 1833, Graves, 2:19-26.
24. Helen Bayly to WRH, Jan. 24, 1833, OR 329.
25. WRH to Helen Bayly, Jan. 22, 1833, OR 327.
26. Graves, 2:523.
27. WRH to sister Eliza, May 10, 1834, OR 393.
28. WRH to Helen Bayly, Mar. 14, 1833, OR 352.
29. Helen Bayly to WRH, Jan. 24, 1833, OR 329.
30. WRH to Mrs. Bayly, July 13, 1833, OR 372.
31. WRH to Mrs. Bayly, Sept. 16, 1833, ORSUP 123. Treatment consisted of shower baths, which were installed at the observatory, and the raising of a "small blister" on the stomach, which was in a state of inflammation, and calomel.
32. WRH to Mrs. Bayly, Sept. 16, 1833, ORSUP 123, and Helen Hamilton to Mrs. Bayly, Oct. 5, 1833, OR 380; Graves 2:61.
33. Helen Hamilton to Mrs. Bayly, Oct. 5, 1833, OR 380.
34. Helen Hamilton to Mrs. Bayly, Oct. 5, 1833, OR 380.
35. Hamilton wanted Mrs. Bayly to bring a mattress with her from Bayly Farm (WRH to Mrs. Bayly, Oct. 17, 1831, OR 382).
36. WRH to William Wordsworth, July 20, 1834, Graves, 2:97-99.
37. WRH to Aubrey De Vere, Aug. 29, 1834, NLD 5764, fol. 191; OS 95; Graves, 2:105.
38. WRH to Helen Hamilton, Feb. 22, 1835, OR 446.
39. Helen Hamilton to WRH, Sept. 19, 1834, OR 415.
40. WRH to Helen Hamilton, Feb. 22, 1835, OR 446, and Mar. 14, 1835, OR 457.
41. WRH to De Vere, Jan. 30, 1835, NLD 5764, fol. 199.
42. Archibald was born on the anniversary, to the hour, of Hamilton's birth.
43. Mrs. Bayly to WRH, Apr. 24, 1837, ORSUP 118a.
44. Mrs. Bayly to WRH, Aug. 22, 1837, OR 579.
45. Mrs. Bayly to WRH, Sept. 2, [1837], OR 1865.
46. Hamilton may have felt compelled to visit him. Adare's first son had died the previous May.
47. WRH to Cousin Arthur, Oct. 25, 1838, OR 665.
48. WRH to the Marquess of Northampton, Sept. 3, 1838, Graves, 2:270-74.
49. WRH to Captain William Bayly, Apr. 28, 1838, OR 636.
50. The O'Regan collection contains a large number of letters related to Bayly family affairs. See also Hamilton's memorandum on Cousin Arthur's estate, Oct. 12, 1843, ORSUP 8.
51. Graves, 2:310.
52. Eliza Hamilton, notebook, ORSUP 74.
53. Graves, 2:320.
54. WRH to De Vere, Aug. 30, 1840, NLD 5764, fol. 265.
55. WRH to sister Grace, Oct. 7, 1840, OR 764, and WRH to De Vere, Aug. 30, 1840, NLD 5764, fol. 265.
56. WRH to Charles Boyton, May 22, 1841, Graves, 2:332-34.
57. Helen Hamilton to WRH, Nov. 29, 1841, OR 831.

58. Graves, 2:354.
59. Sydney stayed on after Helen's return for at least another five months, and the other sisters were also frequent visitors (WRH to Robert Graves, May 4, 1842, OS 204).
60. Graves, 3:233.

Chapter 9

1. *Edinburgh Review* 1 (1802–3): 450–56, 457–60; *Edinburgh Review* 5 (1804–5): 97–103.
2. Laplace, *Oeuvres complètes*, 14 vols. (Paris: Gauthier-Villars et Fils, 1878–1912), 12: 288.
3. Ibid., p. 271.
4. *Math. Papers*, 1: 10.
5. *The Quarterly Review* 2 (1809): 337, quoted from R. H. Silliman, "Augustin Fresnel (1788–1827) and the Establishment of the Wave Theory of Light" (Ph.D. diss., Princeton University, 1968), pp. 136–37, and Thomas Young to David Brewster, Sept. 13, 1815, *Miscellaneous Works*, 1: 360, quoted from Silliman, "Augustin Fresnel," p. 138.
6. Eugene Frankel, "Jean Baptiste Biot: The Career of a Physicist in Nineteenth-Century France" (Ph.D. diss., Princeton University, 1972), pp. 250–51.
7. Ibid., pp. 259–60.
8. Silliman, "Augustin Fresnel," p. 158 and Frankel, "Jean Baptiste Biot," p. 302.
9. Silliman, "Augustin Fresnel," p. 198.
10. Poisson showed that according to Fresnel's theory there should be a white spot in the center of the shadow cast by a circular screen. It made Fresnel's theory appear ridiculous, but when Fresnel performed the experiment, he found the white spot just where his theory predicted it would be (Silliman, "Augustin Fresnel," p. 198).
11. Ibid., pp. 206–7. The report was made on June 4, 1821.
12. Ibid., pp. 211–12 and 215–27.
13. Ibid., p. 220.
14. Silliman says that Biot forced the issue by calling for Arago's report, now five years overdue, on Fresnel's memoir on chromatic polarization (Ibid., pp. 233–37). Frankel places the blame largely on Arago ("Jean Baptiste Biot," p. 318).
15. Silliman, "Augustin Fresnel," p. 262.
16. Ibid., p. 259.
17. Henry Coddington, *An Elementary Treatise on Optics,* 2d ed. (Cambridge: J. Deighton & Sons, 1825); James Wood, *The Elements of Optics: Designed for the Use of Students in the University,* 2d ed. (Cambridge: J. Burges, 1801); and John Stack, *A Short System of Optics,* 3d ed. (Dublin: Printed at the University by D. Graisberry, 1820).
18. Humphrey Lloyd, *Treatise on Light and Vision* (London: Longman, Rees, Orme, Brown, and Green, 1831), p. 7, quoted from D. B. Wilson, "The Reception of the Wave Theory of Light by Cambridge Physicists (1820–1850): A Case Study in the Nineteenth-Century Mechanical Philosophy" (Ph.D. diss., The Johns Hopkins University, 1968), p. 28.
19. Wilson, "Reception of the Wave Theory," p. 36.
20. Ibid., p. 37. The report was completed in Dec. 1827.
21. John F. W. Herschel, "Light," in *Encyclopaedia Metropolitana,* 29 vols. (London: B. Fellowes; F. & J. Rivington, etc., 1817–45), 4: 533.
22. Ibid., p. 450, and Wilson, "Reception of the Wave Theory," p. 39.
23. Herschel, "Light," p. 475, quoted from Wilson, "Reception of the Wave Theory," p. 75.
24. Herschel, "Light," p. 456, quoted from Wilson, "Reception of the Wave Theory," p. 67.
25. Herschel, "Light," p. 537.
26. Ibid., p. 538.
27. Ibid., p. 582.
28. Airy was Lucasian Professor of Mathematics at the time. The following year he was appointed Plumian Professor of Astronomy and director of the Cambridge Observatory.

("Airy," *Dictionary of Scientific Biography*, 15 vols. [New York: Scribner, 1970-78], 1: 85). His papers on eyepieces were: "On the Principles and Construction of the Achromatic Eye-Pieces of Telescopes and on the Achromatism of Microscopes," Cambridge Philosophical Society, *Transactions* 2 (1826): 227-52, dated Apr. 26, 1824; "On the Use of Silvered Glass for the Mirrors of Reflecting Telescopes," Cambridge Philosophical Society, *Transactions* 2 (1826): 105-18, dated Nov. 25, 1822; "On the Spherical Aberration of the Eye-pieces of Telescopes," Cambridge Philosophical Society, *Transactions* 3 (1830): 1-63.

29. George Biddell Airy, *Mathematical Tracts on the Lunar and Planetary Theories, the Figure of the Earth, Precession and Nutation, the Calculus of Variations, and the Undulatory Theory of Optics: Designed for the Use of Students in the University*, 3d ed. (Cambridge: J. J. Deighton, 1842), pp. v-vi, quoted from Wilson, "Reception of the Wave Theory," p. 80.

30. George Biddell Airy, "On a Remarkable Modification of Newton's Rings," Cambridge Philosophical Society, *Transactions* 4 (1833): 279-88, dated June 21, 1831.

31. George Biddell Airy, "On the Phaenomena of Newton's Rings when Formed between Two Transparent Substances of Different Refractive Powers," Cambridge Philosophical Society, *Transactions* 4 (1833): 409-24, dated Feb. 16, 1832.

32. Almost certainly Bartholomew Lloyd (Graves, 1: 233).

33. James Hamilton to WRH, Apr. 30, 1827, ORSUP 67.

34. Quoted from "Airy," *Dictionary of Scientific Biography*, p. 85.

35. Hamilton's first difficulty came in 1843 when the provost and Senior Fellows of Trinity College, who were responsible for overseeing the observatory, sent to Hamilton a copy of Airy's annual report and asked that he produce a comparable document (Joseph H. Singer to WRH, Mar. 23, 1843, ORSUP 159; see also Graves, 2: 410). In 1853 a royal commission investigating Trinity College and Dunsink Observatory published a report stating that the evidence "does not show sufficient diligence on the part of the observers." (Great Britain, *Parliamentary Papers*, vol. 45, no. 1637 [1852/53], "Report of Her Majesty's Commissioners," p. 82). Hamilton found that the commission had miscalculated the number of observations, and he defended himself (WRH to Augustus De Morgan, June 30, 1853, OS 728; Graves, 3: 452-53), but he had to admit that he was not a successful observational astronomer.

36. George Biddell Airy to WRH, July 23, 1827, Graves, 1: 273-74, WRH to Airy, Oct. 9, 1827, Graves, 1: 274-76, Airy to WRH, Oct. 31, 1827, Graves, 1: 276, and WRH to Airy, Nov. 7, 1827, Graves, 1: 276-77.

37. Graves, 1: 550.

38. WRH to Cousin Arthur, Apr. 18, 1832, OR 267; Graves, 1: 553.

Chapter 10

1. The term *master spirits* was used by Vernon Harcourt in his address to the British Association. (*Philosophical Magazine*, 3d ser. 7 [1835]: 291).

2. On the founding of the British Association, see A. D. Orange, "The Origins of the British Association for the Advancement of Science," *British Journal for the History of Science* 6 (1972): 152-76, and idem, "The Idols of the Theatre: The British Association and Its Early Critics," *Annals of Science* 32 (1975): 277-94. Also, L. Pearce Williams, "The Royal Society and the Founding of the British Association for the Advancement of Science," *Notes and Records of the Royal Society* 16 (1961): 221-33.

3. John Herschel, "Sound," in *Encyclopaedia Metropolitana*, 29 vols. (London: B. Fellowes, F. & J. Rivington, etc., 1817-45), 4:810n.

4. Charles Babbage, *Reflections on the Decline of Science in England and On Some of Its Causes* (London: B. Fellowes, 1830).

5. David Brewster, "Charles Babbage, *Reflections on the Decline of Science in England*," *Quarterly Review* 43 (1830): 305-42.

6. David Brewster to [ ], Nov. 15, 1830, JG 115.

7. Orange, "Origins," p. 153.

8. WRH to John Herschel, Dec. 3, 1830, OS 67; Graves, 1:408. Williams argues that the

contested election of 1830 at the Royal Society led directly to the foundation of the British Association ("Royal Society," pp. 221-33).

9. William Whewell to James D. Forbes, July 14, 1831, quoted from Orange, "Origins," p. 161.
10. David Brewster to John Phillips, Nov. 19, 1831, Orange, "Origins," p. 171.
11. William Wordsworth to WRH, Jan. 24, 1831, Graves, 1:424.
12. Orange, "Origins," p. 167.
13. Graves, 1:571.
14. WRH to Viscount Adare, July 20, 1832, Graves 1:573.
15. Graves, 1:573.
16. Wordsworth to WRH, May 8, 1833, Graves, 2:46.
17. Graves, 2:50-52.
18. Orange, "Idols," pp. 279-81.
19. Ibid., p. 282.
20. Ibid., p. 288. A thorough study of the continuing conflict between Brewster and Whewell is John Hadley Brooke's "Natural Theology and the Plurality of Worlds: Observations on the Brewster-Whewell Debate," *Annals of Science* 34 (1977): 221-86.
21. The Brewster review of "Memoir and Correspondence of the Late Sir James Smith," *Edinburgh Review* 57 (1833): 41.
22. William Whewell, *Astronomy and General Physics Considered with Reference to Natural Theology* (London: W. Pickering, 1833). Brewster's review was in the *Edinburgh Review* 58 (1834): 422-57.
23. Brewster edited the *Edinburgh Magazine,* the *Scots Magazine,* the *Edinburgh Encyclopaedia,* and the *Philosophical Magazine.* He probably had strong feelings about what a proper editing job should entail.
24. Brewster's review in *Edinburgh Review* 58 (1834): 455. He quotes Hamilton from the third supplement to the "Theory of Systems of Rays" (*Math. Papers,* 1:165).
25. Brewster's review in *Edinburgh Review* 58 (1834): 456-57.
26. WRH to William Whewell, Mar. 18, 1834, Graves, 2:80.
27. Graves, 2:108-9.
28. Brewster to Vernon Harcourt, Apr. 28, 1832, quoted from Orange, "Idols," p. 282.
29. David Brewster, review in *Edinburgh Review,* 60 (1834-35): 363-94. This quotation is from p. 390.
30. Ibid., p. 391.
31. Ibid., p. 392.
32. Ibid., p. 374.
33. "In the last Edinburgh Review is an article on the British Association, apparently written by Brewster, and containing many expressions of soreness some sneers which I take to myself" (WRH to Viscount Adare, Feb. 27, 1835, Graves, 2:124).
34. WRH to Whewell, Apr. 6, 1835, Graves, 2:148, and Whewell to WRH, Apr. 12, 1835, Graves 2:148-49.
35. Brewster's review in *Edinburgh Review* 60 (1834-35): 389.
36. Ibid., p. 383.
37. The *Athenaeum,* Aug. 8, 1835, p. 632, quoted from Orange, "Origins," p. 172.
38. *British Association Reports* (1835), p. liv.
39. Ibid., p. lvi.
40. David Brewster, "Report on the Recent Progress of Optics," *British Association Report* (1831-32), pp. 308-22, and Humphrey Lloyd, "Report on the Progress and Present State of Optics," *British Association Report* (1834), pp. 295-413.
41. Brewster, "Recent Progress of Optics," p. 316.
42. Ibid., pp. 319-322.
43. Ibid., p. 317.
44. The Oxford meeting of the British Association was held in June, 1832. Hamilton announced his prediction of conical refraction in October, 1832. He was studying Cauchy's theory by the end of the year. WRH to Humphrey Lloyd, Jan. 1, 1833, Graves, 1:629.
45. Lloyd, "Present State of Optics," pp. 295-96.
46. Ibid., p. 296.
47. *Dictionary of National Biography,* 63 vols., with supplementary volumes (New York: Macmillan, 1908-), 16:219.

48. *Philosophical Magazine,* 3d ser. 2, no. 8 (Feb., 1833): 81–94.
49. Richard Potter, "A Reply to the Remarks of Professors Airy and Hamilton," *Philosophical Magazine,* 3d ser. 2 (Apr. 1833): 281.
50. WRH, "On the Effect of Aberration in Prismatic Interference," *Philosophical Magazine,* 3d ser. 2 (Mar., 1833): 191–94, dated Feb. 12, 1833.
51. Ibid., p. 191.
52. Potter, "Reply to the Remarks," p. 281.
53. WRH to Baden Powell, Apr. 5, 1833, OR 360; Graves, 2:40.
54. Whewell to WRH, May 17, 1833, OR 362; Graves, 2:47.
55. Richard Potter, "A Reply to the Remarks," pp. 276–81.
56. WRH to Adare, Apr. 22, 1833, Graves 2:23–44.
57. *Philosophical Magazine,* 3d ser. 2 (June, 1833): 451.
58. Potter, "Reply to the Remarks," p. 278.
59. David Brewster, "Observations on the Absorption of Specific Rays, in Reference to the Undulatory Theory of Light," *Philosophical Magazine,* 3d ser. 2 (May, 1833): 360–61, dated Apr. 13, 1833.
60. Ibid., p. 361.
61. George Biddell Airy, "Remarks on Sir David Brewster's Paper 'On the Absorption of Specific Rays, etc.,'" *Philosophical Magazine,* 3d ser. 2 (June, 1833): 420, dated May 7, 1833.
62. Ibid., p. 422.
63. Ibid., p. 423.
64. John Herschel, "On the Absorption of Light by Coloured Media, Viewed in Connexion with the Undulatory Theory," *Philosophical Magazine,* 3d ser. 2 (Dec., 1833): 401–12, dated Oct. 19, 1833. The quotation is from p. 401.
65. See Michael Sutton, "Sir John Herschel and the Development of Spectroscopy in Britain," *British Journal for the History of Science* 7 (1974): 42–60.
66. Brewster, "Observations on the Absorption of Specific Rays," p. 360.
67. Little is known about Barton. The information that he was controller of the mint comes from J. C. Poggendorff, *Biographisch—literarisches Handwörterbuch zur geschichte der exacten wissenschaften.* (Leipzig: J. A. Barth. 1863–1973), 1:111.
68. John Barton, "On the Inflexion of Light," *Philosophical Magazine,* 3d ser. 2 (Apr., 1833): 263–69; and idem, "On the Inflection of Light, in Reply to Professor Powell," *Philosophical Magazine.,* 3d ser. 3 (Sept., 1833): 172–78. Also an earlier paper in the *Philosophical Transactions,* abstracted in the *Philosophical Magazine,* 2d ser. 10 (1831): 300–301, noticed by Powell in 2d ser. 11 (1832): 2.
69. Baden Powell, "Remarks on Mr. Barton's Paper 'On the Inflexion of Light'," *Philosophical Magazine,* 3d ser. 2 (June, 1833): 424–34; and idem, "Remarks on Mr. Barton's Reply, Respecting the Inflexion of Light," *Philosophical Magazine,* 3d ser. 3 (Dec., 1833): 412–16.
70. Powell, "Remarks on Mr. Barton's Paper," p. 425.
71. Powell, "Remarks on Mr. Barton's Reply," p. 416.
72. Barton, "On the Inflection of Light," pp. 172–78.

Chapter 11

1. Sir Edmund Whittaker, *A History of the Theories of Aether and Electricity,* 2 vols. (New York: Harper Torch Books, 1960), 1:137.
2. Peter Guthrie Tait to WRH, Oct. 19, 1858, quoted from Cargill Gilston Knott, *Life and Scientific Works of Peter Guthrie Tait* (Cambridge: At the University Press, 1911), pp. 122–23.
3. WRH to Samuel Taylor Coleridge, Oct. 3, 1832, OS 85; Graves, 1:593.
4. WRH, "On the Application to Dynamics of a General Mathematical Method Previously Applied to Optics," *British Association Report* (1834), pp. 513–18, quoted from *Math. Papers,* 2:212. It is true that in June Hamilton had expressed to H.F.C. Logan some dissatisfaction with Boscovich's atoms on the grounds that they introduced discontinuities that Hamilton believed might better be replaced by a "*plenum* of energies," and at the

British Association meeting in which he declared his belief in Boscovichean points he also argued that Poisson's conclusions respecting the atomic constitution of matter were not necessary for a mathematical investigation of physical laws. So there seemed to be some ambivalence in his stance. In his actual calculations, however, he always assumed mass points with forces between them (WRH to H.F.C. Logan, June 27, 1834, Graves, 2:88).

5. Baden Powell to WRH, Mar. 29, 1834, Graves, 2:78.
6. Graves, 2:108.
7. See reference to Baden Powell in *Philosophical Magazine* 3d ser. 6 (1835): 266.
8. Humphrey Lloyd to WRH, Aug., 1834, Graves, 2:107.
9. Baden Powell, "M. Cauchy on the Undulatory Theory of Light," *Philosophical Magazine*, 3d ser. 4 (May, 1834): 396–97.
10. Baden Powell, "An abstract of the Essential Principles of M. Cauchy's View of the Undulatory Theory, leading to an Explanation of the Dispersion of Light; with Remarks," *Philosophical Magazine*, 3d ser. 6 (1835): 16–25, 107–13, 189–93, 262–67. Cauchy's first work on dispersion was in 1830 (*Bulletin des Sciences Mathématiques* 14 [1830]: 9), but a much more detailed study appeared in 1836 ("Mémoire sur la dispersion de la lumière," *Nouveaux Exercises de Mathématiques* [1836], in *Oeuvres complètes, 26 vols., 2d ser.) (Paris: Gauthier-Villars et Fils, 1882-1938), 10:185–464.
11. Baden Powell, "Further Observations on M. Cauchy's Theory of Dispersion of Light," *Philosophical Magazine*, 3d ser. 8 (1833): 24–28, 112–14, 204–11, 305–09; and vol. 9 (1836): 116–19.
12. Powell first stated that Hamilton had "taken up" the subject and therefore he, Powell, would say nothing more about Hamilton's research ("Further Observations," pp. 25-28). But soon afterwards he said that his mention of Hamilton's having "taken up" the problem did not refer to any separate publication, and in subsequent articles he presented Hamilton's "exact method," obviously with Hamilton's blessing. Hamilton wrote to Baden Powell that the formula that Powell had derived from Cauchy's theory was only approximate because it restricted the aether particles to act only on their immediate neighbors. Hamilton's improvement came in extending the range of action to more distant particles (see WRH to Baden Powell, Oct. 24, 1835, *Math. Papers,* 2:583-95).
13. *Math. Papers, 2:446-50, British Association Report* (1838), pp. 2-6, read Aug. 1838.
14. He favored a rather complex rule proposed by Poisson, but concluded (in the third person): "Sir William Hamilton did not, however, intend to exclude the hypothesis, that the function may contain several alternations of such repulsive and attractive terms,—much less did he deny that at great distances it may reduce itself to the law of the inverse square" (*Math. Papers, 2:449). Both of these conditions had been first suggested by Boscovich. Hamilton rejected the inverse proportion to the fourth power in letters to Lloyd, Dec. 24, 1836, Graves, 2:192; and to Adare, Mar. 8, 1837, Graves, 2:195.
15. *British Association Report* (1838), p. 6. John Herschel may have suggested this idea to Hamilton. Herschel was particularly enthusiastic about MacCullagh's 1837 paper on crystalline reflection and refraction, and in writing to Hamilton shortly before the meeting, he proposed a model that might account for dispersion:

> As regards the distribution of the ether in crystals etc. it has always appeared to me that a uniform distribution of its molecules through space whether occupied or unoccupied by gross matter [one of MacCullagh's assumptions] might account for the phaenomena provided we admit that its molecules may to a certain degree *adhere* to the gross corporeal particles so as not to move without some slight degree setting them in motion and dragging them to and fro. Their vibrations would therefore be retarded as those of a loaded string and produce that retardation of its waves which we know does take place in refraction (J. Herschel to WRH, June 15, 1838, JG 96).

Here was a possible model for dispersion since the coupling between the particles of the aether and the molecules of the crystal would probably depend on the frequency of the light. Humphrey Lloyd took up Herschel's theory and declared at the British Association meeting of that year that "the complete solution of the problem of refraction could not be attained until we took into consideration the action of the particles of the body, as well as those of the ether," a theory that Lloyd related directly to Herschel's work on absorption. (*Athenaeum* [1838], p. 626.) On the development of this idea see R. T. Glazebrook, "Report on Optical Theories," *British Association Reports* (1885), pp. 202, 212-51.

16. WRH to George Biddell Airy, Sept. 23, 1836, Graves, 2:191; Whittaker, *History of the Theories*, 1:159-60. MacCullagh's first contribution to the theory of light was in 1830, when he published his geometrical construction of the Fresnel wave surface. Hamilton wrote a review of this paper. In 1836 MacCullagh published papers on optical activity and on metallic reflection.

17. Royal Irish Academy, *Transactions* 18 (1839): 31-74, read Jan. 9, 1837. MacCullagh had first described his theory at the Dublin meeting of the British Association in 1835 (*British Association Report* [1835], pt. 2, pp. 7-8).

18. John F. W. Herschel to WRH, June 15, 1838, JG 96; Graves, 2:262.

19. WRH to Airy, Sept. 23, 1836, Graves, 2:191.

20. "Address by the President," Royal Irish Academy, *Proceedings* 1 (1836-39):215.

21. Ibid.

22. Ibid., pp. 220-21.

23. Royal Irish Academy, *Transactions* 18 (1839): 68.

24. James MacCullagh, "An Essay towards a Dynamical Theory of Crystalline Reflexion and Refraction," Royal Irish Academy, *Transactions* 21 (1848): 17-50, read Dec. 9, 1839.

25. Ibid., p. 50.

26. Whittaker, *History of the Theories*, 1:143.

27. Stokes brought this criticism against MacCullagh's aether in 1862. See Kenneth F. Schaffner, *Nineteenth-Century Aether Theories* (Oxford: Pergamon Press, 1972), p. 64.

28. Hamilton studied Green's papers in 1839. Graves, 2:286-87.

29. Schaffner, *Aether Theories*, p. 84.

30. Ibid., pp. 91-98.

31. *Math. Papers* 2:xiv-xv.

32. The discovery of the distinction between wave and group velocity is frequently ascribed to J. W. Strutt (Lord Rayleigh). He mentions it in his *Theory of Sound*, 2 vols. (1877; New York: Dover, 1945), 1:301-02, and in an article, "On Progressive Waves" (*Proceedings of the London Mathematical Society* 9 [1877]: 21), which was appended to later editions. In the article he claims Stokes was the first person to make the distinction. There is no mention of Hamilton.

33. Herschel to WRH, Feb. 13, 1839, Graves, 2:290-91.

34. WRH to Herschel, Feb. 8, 1839, *Math. Papers*, 2:599-607.

35. *Math. Papers*, 2:582, 605.

36. "Researches on the Dynamics of Light," Royal Irish Academy, *Proceedings*, 1 (1836-40): 267-70, read Feb. 11, 1839, and "Researches Respecting Vibration Connected with the Theory of Light," Royal Irish Academy, *Proceedings* 1 (1836-40): 341-49, read June 24, 1839, in *Math. Papers*, 2:576-82. Hamilton announced this memoir in his second paper in the *Proceedings, Math. Papers*, 2:582.

37. *Math. Papers*, 2:451-575. Hamilton's unpublished studies of the dynamics of darkness are in TCD notebooks 39, 40, 52, and 53.

38. WRH to Herschel, Mar. 19, 1841, OS 193. Graves publishes this letter in Graves, 2:337, but omits the part in which Hamilton says he is sending his old papers on vibrations, which he never completed.

39. Whittaker, *History of the Theories*, 1:261-65.

40. This solution was suggested by Stokes, Sellmeier, Helmholtz, and Maxwell. There is a good chapter on dispersion in Francis A. Jenkins and Harvey E. White, *Fundamentals of Optics*, 3d ed. (New York: McGraw-Hill Book Co. 1957), chap. 23.

41. Richard Potter, "On a Method of Performing the Simple Experiment of Interferences with Two Mirrors Slightly Inclined, So as to Afford an *Experimentum Crucis* as to the Nature of Light," *Philosophical Magazine*, 16 (1840): 380-87, and idem, "On the Phenomenon of Diffraction in the Centre of the Shadow of a Circular Disk, Placed before a Luminous Point, as Exhibited by Experiment," *Philosophical Magazine* 19 (1841): 155. See also Geoffrey Cantor, "The Reception of the Wave Theory of Light in

Britain: A Case Study Illustrating the Role of Methodology in Scientific Debate," *Historical Studies in the Physical Sciences* 6 (1975): 118. Potter claimed that under certain circumstances he was able to obtain a dark central band in the interference pattern where the wave theory predicted a bright band.

42. Cantor, "Reception of the Wave Theory," p. 128.
43. *Athenaeum*, no. 796 (July 23, 1842), pp. 662-63.
44. Hamilton never published a description of his theory, nor does one remain in the manuscripts. There are, however, hints at a theory in his letters to Herschel. In a letter of 1839 Herschel asked Hamilton, "Have you yet found the three axes of the universe?" and again in 1841, "I hope you have not forgotten the *three axes of the universe"* (John Herschel to WRH, Dec. 11, 1839, OS 182; Graves, 2:312-13 and another, Mar. 5, 1841, OS 192; Graves, 2:337). In one of these letters the "three axes of the universe" were associated with fixed points in space, which would indicate that they were coordinates of an absolute space.

The origins of Hamilton's theory are of an early date. The following manuscript scrap entitled "Principles of Theoretical Mechanics" comes from Jan. 1830, a time when Hamilton had not yet begun any serious study of metaphysics:

> When we find or believe that a certain set of properties and powers belongs to or is centrally connected with a determined point of space at a determined point of time; we call that point at that moment a material point or atom; and we say that this material point moves when we find or believe that this set of properties and powers, or a set which from resemblance and regularity of change we consider as being the same, is centrally connected with other points of space at other moments of time, by a gradual and unbroken transition (TCD misc II).

Hamilton is saying that an atom of matter is identified by the perception of a set of powers or properties associated with a point in space, and if that set of powers or properties is subsequently associated with a neighboring place rather than the place where it was initially perceived, we say that the atom has "moved." Thus *motion* is the perception of a set of powers at different places at different times. There is no need to assume the existence of a material object actually transferred from one place to another. Herschel came up with the same idea eleven years later and wrote to Hamilton: "I have a metaphysical theory ... that *force* as well as matter consists of indivisible units, and that motion of matter is only a successive excitement of active forces in consecutive molecules of the Ether. What think you of such a doctrine?" (J. Herschel to WRH, Mar. 5, 1841, OS 192; Graves 2:337.) Hamilton responded that something of the sort had occurred to him but "had not been followed up."

45. *Manchester Guardian*, Aug. 3, 1842, no. 1415, no pagination. The description of the debate begins in the number for July 20, 1842.
46. *Manchester Guardian*, Aug. 3, 1842.
47. John F. W. Herschel, *Popular Lectures on Scientific Subjects* (New York: George Routledge and Sons, 1880), p. 284. Although Herschel does not attribute this theory to Hamilton directly, the few hints in their correspondence indicate that this is almost certainly the theory that Hamilton presented at the Manchester British Association meeting.
48. *Athenaeum*, July 23, 1842, no. 769, pp. 662-63.
49. *Manchester Guardian*, Aug. 3, 1842.
50. *Athenaeum*, July 23, 1842, no. 769, p. 663.
51. *Athenaeum*, July 30, 1842, no. 770, p. 687.
52. WRH to R. P. Graves, Oct. 1, 1842, Graves, 2:391-92.
53. In 1837 Hamilton became president of the Royal Irish Academy against the determined opposition of MacCullagh. One of his first acts was to revise the procedure for granting medals and other honorary awards to members of the academy. There was one medal to be awarded in June of 1838 for the most important paper in pure or mixed mathematics communicated during the three years that ended in March, 1837, that was already printed. The *only* papers falling into this category were by MacCullagh and by Hamilton himself. Hamilton arranged for the medal to be awarded to MacCullagh for his paper on *Crystalline Reflection and Refraction.* (Graves, 2:256-65.)
54. WRH to Augustus De Morgan, Jan. 31, 1852, OS 538; Graves, 2:331-32.

55. "MacCullagh, James," *Encylopaedia Britannica*, 9th ed., 24 vols. (New York: Scribners, 1885), 15:133.
56. For a later comparison of the work of Hamilton and MacCullagh see the article "Light: Nature of" by H. A. Lorentz in the eleventh edition of the *Encyclopaedia Britannica* (29 vols. [Cambridge: At the University Press, 1910-11], 16:608-23). Lorentz does not give MacCullagh his due.
57. *Abstracts of the Papers Communicated to the Royal Society of London* (later titled *Proceedings*) 5 (1834-50): 712-18. On Hamilton's authorship see Graves, 2:597.
58. Whittaker, *History of the Theories*, 1:142, 144; and Schaffner, *Aether Theories*, p. viii.
59. Faraday made use of Boscovichean point atoms in some of his theorizing on electromagnetism. He occasionally imagined polar particles similar to Hamilton's (particular cases being his researches on induction and diamagnetism), but Faraday always directed his conjectures towards the explanation of physical phenomena, while Hamilton thought more in mathematical and metaphysical terms. See L. Pearce Williams, *Michael Faraday: A Biography* (New York: Basic Books, 1965), pp. 298, 378-81.

   Hamilton first met Faraday in 1834. In 1837 they joined with Whewell in attacking atoms at the Liverpool meeting of the British Association (*Athenaeum* [1837], p. 747, and Williams, *Faraday*, p. 356). This was the occasion of Faraday's first public avowal of Boscovich's theory. Hamilton had made a special effort to join Faraday in the debate. Maria Edgeworth's sister Sophy Fox, who was at the Liverpool meeting, wrote: "At the chemical Section he [Hamilton] went to quarrel with the atomic theory, for he wished the world to be resolved into a series of mathematical points, remarking that the nearer all the Sciences approached Section A (mathematics and physics) the nearer they will be to perfection" (Journal of Caroline Fox, Sept. 15, 1837, OR 2139). At the Cambridge meeting of the British Association in 1845, two years after Hamilton's discovery of quaternions, he again met Faraday and discussed with him the analogy between the multiplication of quaternions and the laws of electrical currents (Graves, 3:482). After this discussion Hamilton requested that a note be placed in the records of the association. It appeared as follows: "Sir W. Hamilton said that he wished to have placed on the records the following conjecture as to a future application of quaternions: is there not an analogy between the fundamental pair of equations, $ij = k, ji = -k$, and the facts of opposite currents of electricity corresponding to opposite rotations?" (*British Association Report* (1845), p. 3.) In 1854 he again took up the "electromagnetic quaternion" briefly, with the urging of Humphrey Lloyd, and in 1858 he received the same urging from P. G. Tait (P. G. Tait to WRH, Aug. 19, 1858, OS 1007), but nothing came of it. Hamilton's mathematical and metaphysical instincts were just the right ones for the creation of an electromagnetic theory, but he lacked the physical insight. He later confessed to Tait that his instinct for physics was slight (WRH to P. G. Tait, Aug. 28, 1863), OR 1586; Graves, 3:150).
60. WRH to Herschel, Mar. 19, 1841, OS 193; Graves, 2:338.

Chapter 12

1. James Clerk Maxwell to Peter Guthrie Tait, July 14, 1871, quoted in Cargill Gilston Knott, *Life and Scientific Works of Peter Guthrie Tait* (Cambridge: At the University Press, 1911), pp. 99-100.
2. WRH to Humphrey Lloyd, Feb. 9, 1833, OR 345.
3. WRH to Lloyd, Sept. 2, 1833, OR 379.
4. WRH to Viscount Adare, July 22, 1833, Graves, 2:54.
5. *Math. Papers*, 2:311-32.
6. WRH, "Problem of Three Bodies by My Characteristic Function," *Math. Papers* 2:1-102. The first entry in the notebook is dated Dec. 16, 1833.
7. It is often difficult to determine which parts of a book were new and which were borrowed from a previous publication. Todhunter says this was one of the major objections to Whewell's books (I. Todhunter, *William Whewell, D.D.; An Account of His Writings*, 2 vols. [London: Macmillan and Co., 1876], 1:20).

8. Whewell was already planning to write a history of the sciences in 1819. See J. C. Hare to William Whewell, Dec. 21, 1819, Todhunter, *William Whewell,* 1:101. Hamilton referred to this discussion on May 25, 1833 (WRH to William Whewell, Graves, 2:49).

9. WRH to Aubrey De Vere, May 7, 1832, Graves, 1:553, and WRH to Samuel Taylor Coleridge, June 14, 1832, Graves, 1:559.

10. Whewell also sent his Bridgewater Treatise (Whewell to WRH, Mar. 18, 1833, OR 353; Graves, 2:40-41). Two months later, in a letter formally inviting Hamilton to stay at Trinity College during the upcoming meeting, Whewell mentioned that he had directed his London publisher to send "a little book on the laws of motion" that he had written "for our 'non-reading men' who have a very limited taste for Metaphysics." He added: "I think my notions of the distinction [between statics and dynamics] are more clear than when we talked on the subject before, and I shall be glad to discuss it further" (Whewell to WRH, May 17, 1833, OR 362; Graves, 2:47). Hamilton replied that he had received Whewell's "Tract on the Use of Definitions" and his *First Principles of Mechanics, With Historical and Practical Illustrations* (1832). As for the *First Principles of Mechanics,* he said that he wished to study it in conjunction with a treatise that Whewell had given him in Cambridge in 1832 in order to see how his views might differ from those of Whewell.

11. WRH to Whewell, May 25, 1833, Graves, 2:48-49.

12. *Math. Papers,* 1:314.

13. Graves, 2:68.

14. Ibid., 71.

15. In 1837 Lloyd discovered that the Royal Irish Academy still did not send its *Transactions* to the Institute of France (Lloyd to WRH, Feb. 21, 1837, OR 556).

16. WRH to Uncle James, Mar. 12, 1834, Graves, 2:74.

17. WRH to Whewell, Mar. 18, 1834, Graves, 2:81.

18. WRH to Whewell, Mar. 31, 1834, Graves, 2:82-83.

19. Cambridge Philosophical Society, *Transactions* 5, pt. 2 (1833-35): 149-72.

20. WRH to Whewell, Mar. 31, 1834, Graves 2:82, and WRH to Viscount Adare, Apr. 4, 1834, Graves 2:83.

21. Whewell has the most trouble with the third law of motion, largely because he did not carefully distinguish between the concepts of mass and weight. The same confusion appears in the *Philosophy of the Inductive Sciences* and was the basis of a criticism by Maxwell ("Whewell's Writings and Correspondence," in *The Scientific Papers of James Clerk Maxwell,* ed. W. D. Niven, 2 vols. [Cambridge: At the University Press, 1890] 2:531-32).

22. Whewell argued that the first axiom of causation requires that an object in motion continue moving in the same direction if no force is exerted on it, but it is *not* proved by that axiom that the speed of the object will remain constant. Therefore Newton's first law of motion is, according to Whewell, a combination of the a priori first axiom of causation and the empirical information that inertial motion takes place at uniform speed.

23. Note in the preface to the *Mechanical Euclid* (1837), mentioned in Todhunter, *William Whewell,* 1:24.

24. This was the position of Jean d'Alembert, whom Whewell names as the person who first "fully achieved" the analytical expressions of the laws of motion (William Whewell, *History of the Inductive Sciences from the Earliest Time to the Present,* 3d ed., 2 vols. [New York: D. Appleton and Co., 1901], 1:355.)

25. Whewell, *History of Scientific Ideas,* pt. 3 of the *Philosophy of the Inductive Sciences,* 2 vols. (London: John W. Parker and Son, 1858), 1:180-92. I discuss Hamilton's use of the second analogy in chapter 19.

26. "Sir William R. Hamilton, Our Portrait Gallery no. XXVI," *Dublin University Magazine* 19 (Jan.-June, 1842): 94-110. The "Portrait Gallery" was a regular feature of the magazine.

27. Ibid., p. 107.

28. OR 2106.

29. WRH to Lloyd, Oct. 6, 1834, OR 419; Graves, 2:110-11.

30. *Math. Papers,* 2:162. The paper was received Oct. 29, 1834, read Jan. 16, 1835, and published in the volume of *Transactions* for that year (Royal Society, *Philosophical Transactions* 1 [1835]:95-144).

Chapter 13

1. William Whewell to WRH, Mar. 27, 1834, Graves, 2:81.
2. The following account of Hamilton's mechanics is an expansion of the terse presentation that he gives in his published papers. Other accounts of Hamiltonian dynamics are listed in the bibliographical essay.
3. *Math. Papers,* 1:316.
4. In his biography Graves gives a popular "Account of a Theory of Systems of Rays" written by Hamilton himself, which Graves dates April 23, 1827 (Graves, 1:228-31). But this date is almost certainly incorrect. It contains a reference to the Law of Varying Action, which did not appear elsewhere until 1833; a reference to algebra as the science of pure time, a notion that Hamilton adopted only in 1832; and the same arguments about induction and deduction as those made in his paper "On the Paths of Light and the Planets." Judging by this internal evidence, the manuscript quoted by Graves was probably written some time after September, 1833, and may have been an early draft of "On the Paths of Light and the Planets."
5. *Math. Papers,* 1:168.
6. Ibid., 107. In the first supplement Hamilton remarked that he had previously called this fundamental equation the *Principle of Constant Action,* "partly because it gives immediately the differential equation of that important class of surfaces, which, on the hypothesis of undulation are called *waves,* and which, on the hypothesis of molecular emission may be named *surfaces of constant action.* But in the present Supplement, it is proposed to designate the fundamental formula by the less hypothetical name of the *Equation of the Characteristic Function.*"
7. *Math. Papers,* 1:9.
8. In manuscripts from 1826 and in his published table of contents, Hamilton first wrote *I* or *A* for the action, distinguishing it from the characteristic function *V,* which he had formerly defined as the length of the ray. In the first supplement he went back to *V,* but defined it as the action integral.
9. See Cornelius Lanczos, *The Variational Principles of Mechanics,* 4th ed. (Toronto: Toronto University Press, 1970), pp. 56-57, on the commutative properties of δ and d.
10. *Math. Papers,* 2:160.
11. WRH to John F. W. Herschel, Oct. 17, 1834, OS 99; Graves, 2:112-17.
12. Herschel to WRH, June 13, 1835, OS 109; Graves, 2:127.
13. This derivation has since become standard. Wolfgang Yourgrau and Stanley Mandelstam, *Variational Principles in Dynamics and Quantum Theory,* 3d ed. (Philadelphia: W. B. Saunders Co., 1968), pp. 34-36, and Herbert Goldstein, *Classical Mechanics* (Reading, Mass.: Addison-Wesley Publishing Co., 1950), pp. 16-18.
14. $$L = T + U = 2T - H = \Sigma \dot{\eta} \frac{\partial T}{\partial \eta} - H = \Sigma \dot{\eta} \omega - H = \Sigma \omega \frac{\partial H}{\partial \omega} - H.$$

15. WRH, "On the Application to Dynamics of a General Mathematical Method Previously Applied to Optics," *British Association Report* (1834); *Math. Papers,* 2:213; and WRH to John Herschel, Oct. 17, 1834, Graves, 2:114.
16. *Math. Papers,* 2:167.
17. C.G.J. Jacobi, "Über die Reduction der Integration der partiellen Differentialgleichungen erster Ordnung zwischen irgend einer Zahl Variabeln auf die Integration eines einzigen Systemes gewöhnlicher Differentialgleichungen," *Journal für die reine und angewandte Mathematik* 17 (1837): 97-162.
18. Goldstein, *Classical Mechanics,* chap. 9.
19. *Math. Papers,* 2:179. For a discussion of the relative contributions of Hamilton and Jacobi to the theory, see *Math. Papers,* 2:613-21; Lanczos, *Variational Principles,* pp. 254-64; and Arthur Cayley, "Report on the Recent Progress of Theoretical Dynamics," *British Association Reports* (1857), pp. 9-26.
20. *Math. Papers,* 2:179.
21. Cayley, "Recent Progress of Theoretical Dynamics," p. 10.
22. H.F.C. Logan to WRH [1837], Graves 2:85.
23. Logan to WRH, Oct. 16, 1837, Graves 2:206.

24. WRH to Logan, Apr. 14, 1838, Graves 2:247. Hamilton complained bitterly to J. W. Lubbock about the difficulty of getting foreign journals and asked Lubbock to send him Jacobi's paper. WRH to Lubbock, Oct. 1837, *Math. Papers,* 2:283.
25. *Math. Papers,* 2:163.
26. WRH to Humphrey Lloyd, Jan. 16, 1836, and WRH to Viscount Adare, Jan. 29, 1836, Graves, 2:177-78. These manuscripts are now published in *Math. Papers,* 2:297-407.
27. WRH to Lloyd, Jan. 16, 1836, Graves, 2:177.
28. WRH, "Calculus of Principal Relations," *British Association Reports* (1836), in *Math. Papers,* 2:408-10.
29. *Math. Papers,* 2:217-83; Graves, 2:189.
30. WRH to Adare, July 5, 1842, Graves, 2:390-91.
31. Graves, 2:388.

Chapter 14

1. Hamilton's geometrical optics could be used to describe wave fronts by the Principle of Constant Action. To this extent it could be used in conjunction with the undulatory theory. But the theory of the characteristic function does not give the wave equation describing the actual vibration of the medium. To pursue this problem Hamilton had to resort to other methods.
2. Erwin Schrödinger, "Quantisierung als Eigenwertproblem (Zweite Mitteilung)," *Annalen der Physik,* 4th ser. 79 (1926). The translation is from the *Collected Papers on Wave Mechanics,* trans. from the 2d German ed. of the *Abhandlungen* by J. F. Shearer and W. M. Deans (London: Blackie and Son, 1928), p. 18. I have altered an occasional word in the translation in order to make it more faithful to the German.
3. Erwin Schrödinger, "The Fundamental Idea of Wave Mechanics," published in *Science and the Human Temperament,* trans. James Murphy and W. H. Johnston (New York: W. W. Norton, 1935), p. 192. The wave-particle duality of quantum mechanics leads one naturally to ask if there is any real difference between electrons and light, or is it just a historical accident that electrons were first observed as particles and light as waves? An examination of the theory indicates that it was not an accident, that there are good reasons physics developed as it did. See R. E. Peierls et al., "A Survey of Field Theory," *Reports on Progress in Physics* 18 (1955): 471-73.
4. Schrödinger, *Collected Papers on Wave Mechanics,* p. 13.
5. William Thomson and Peter Guthrie Tait, *Treatise on Natural Philosophy,* retitled *Principles of Mechanics and Dynamics* (1867; New York: Dover Publications, 1962), 1:354.
6. E. T. Whittaker, *Treatise on the Analytical Dynamics of Particles and Rigid Bodies,* 2d ed. (1904; Cambridge: At the University Press, 1917), pp. v, 288.
7. Ibid., p. 288.
8. Schrödinger, *Collected Papers on Wave Mechanics,* p. 13.
9. Felix Klein, *Vorlesungen über die Entwicklung der Mathematik im 19. jahrhundert,* 2 vols. (Berlin: Verlag von Julius Springer, 1926-27), 1:198.
10. Ibid.
11. Eduard Study, "Sir William Rowan Hamilton," *Jahresbericht der deutschen Mathe-matiker-Vereinigung* 14 (1905): 421-24, and idem, "Über Hamilton's geometrische Optik und deren Beziehung zur Theorie der Berührungstransformationen," *Jahresbericht der deutschen Mathematiker-Vereinigung* 14 (1905): 424-38.
12. Georg Prange, "W. R. Hamilton's Bedeutung für die geometrische Optik. Habilitationsrede, gehalten am 26. Februar 1921" (in Halle), *Jahresbericht der deutschen Mathematiker-Vereinigung* 30 (1921): 69-82. In 1933 Prange gave one of the most faithful accounts of Hamilton's theory: "Die allgemeinen Integrationsmethoden der analytischen Mechanik," *Encyklopädie der mathematischen Wissenschaften* 4, no. 4: 505-804.
13. Klein, *Vorlesungen* . . . , p. 199. Hamilton had rejected the notion that his principle had any teleological implications. He wrote: "Although the law of least action has thus

attained a rank among the highest theorems of physics, yet its pretensions to a cosmological necessity, on the ground of economy in the universe, are now generally rejected. And the rejection appears just" (*Math. Papers*, 1:317).

14. Leo Koenigsberger, *Hermann von Helmholtz*, trans. Frances A. Welby (New York: Dover Publications, 1906), p. 350. See also Wolfgang Yourgrau and Stanley Mandelstam, *Variational Principles in Dynamics and Quantum Theory*, 3d ed. (Philadelphia: W. B. Saunders Co., 1968), p. 163.

15. Heinrich Hertz, *The Principles of Mechanics Presented in a New Form* (New York: Dover, 1956), p. 32.

16. Ibid., p. 23.

17. Max Planck, *Scientific Autobiography and Other Papers*, trans. Frank Gaynor (New York: Philosophical Library, 1949), pp. 179 and 181. Planck's reply to Schrödinger's letter describing the new discoveries in 1926 is characteristic: "I find it extremely congenial that such a prominent role is played by the action function ... I have always been convinced that its significance was far from exhausted." Planck to Schrödinger, Apr. 2, 1926, in Erwin Schrödinger, *Letters on Wave Mechanics*, ed. K. Przibram, trans. Martin J. Klein (New York: Philosophical Library, 1967), p. 3.

18. Yourgrau and Mandelstam, *Variational Principles*, p. 165.

19. Sir Joseph Larmor, *Mathematical and Physical Papers*, 2 vols. (Cambridge: At the University Press, 1929), 1:31 and 70.

20. Ibid., 1:640.

21. Schrödinger, *Collected Papers on Wave Mechanics*, p. 14.

22. Constance Reid, *Hilbert: With an Appreciation of Hilbert's Mathematical Works by Hermann Weyl* (New York: Springer Verlag, 1970), pp. 47 and 103.

23. A. Sommerfeld and J. Runge, "Anwendung der Vektorrechnung auf die Grundlagen der geometrischen Optik," *Annalen der Physik*, 4th ser. 35 (1911): 289–93. Cited in Schrödinger, *Collected Papers on Wave Mechanics*, p. 18.

24. Max Jammer, *The Conceptual Development of Quantum Mechanics* (New York: McGraw-Hill Book Co., 1966), pp. 255–58; Armin Hermann, "Schrödinger," *Dictionary of Scientific Biography*, vol. 12, pp. 217–23; V. Raman and Paul Forman, "Why Was It Schrödinger Who Developed de Broglie's Ideas?" *Historical Studies in the Physical Sciences* 1 (1969): 291–314; and J. Gerber, "Geschichte der Wellenmechanik," *Archive for History of Exact Science* 5 (1968-69): 349–416.

25. Schrödinger to Wilhelm Wien, Dec. 27, 1925, in Hermann, "Schrödinger," p. 219.

26. Jammer, *Conceptual Development*, pp. 257-58.

27. Schrödinger, *Collected Papers on Wave Mechanics*, p. 20.

28. Louis de Broglie, "Recherches sur la théorie des quanta," *Annales de physique* 3 (1925): "Fermat's principle applied to the wave, becomes identical to the principle of least action applied to the particle. The rays of the wave are identical to the trajectories of the particle (p. 22). ... Guided by the idea of a profound identity between the principle of least action and Fermat's principle, I have been led, from the beginnings of my research on this subject, to admit that . . . the possible dynamic trajectories of the [particles] coincide with the possible rays of the [waves]" (p. 45).

29. Raman and Forman, "Why Was It Schrödinger?" pp. 293-303. Raman and Forman (pp. 303-10) see a preparation for, if not an anticipation of, de Broglie's ideas in an earlier paper by Schrödinger, "Über eine bemerkenswerte Eigenschaft der Quantenbahnen eines einzelnen Elektrons" (1922).

30. Jammer, *Conceptual Development*, p. 260; Erwin Schrödinger, *Abhandlungen zur Wellenmechanik* (Leipzig: J. A. Barth, 1928), p. 12.

31. Erwin Schrödinger, *Les Prix Nobel en 1933* (Stockholm: Imprimerie Royale, 1935), p. 87.

32. Cornelius Lanczos, "William Rowan Hamilton—An Appreciation," *University Review* (National University of Ireland) 4 (1967):155. Lanczos's statement carries weight because he was for many years a colleague of Schrödinger's at the Dublin Institute for Advanced Study.

33. Schrödinger, *Les Prix Nobel en 1933*, p. 87; William T. Scott, *Erwin Schrödinger: An Introduction to His Writings* (Amherst: University of Massachusetts Press, 1967), p. 2; and Stefan Meyer, "Friedrich Hasenöhrl," a eulogy in *Physikalische Zeitschrift*

16, no. 23 (1915):432. According to Armin Hermann, Schrödinger began to attend Hasenöhrl's lectures in 1907, during his third semester at the University of Vienna. He was greatly impressed by Hasenöhrl's inaugural lecture on the work of his predecessor, Ludwig Boltzmann ("Schrödinger," p. 217-18).

34. Friedrich Hasenöhrl, "Über die Anwendbarkeit der Hamiltonschen partiellen Differentialgleichung in der Dynamik kontinuierlich verbreiteter Massen," *Festschrift Ludwig Boltzmann* (Leipzig: J. A. Barth, 1904), pp. 642-46.

35. Thomas Kuhn, John L. Heilbron, Paul Forman, and Lini Allen, *Sources for History of Quantum Physics: An Inventory and Report* (Philadelphia: American Philosophical Society, 1967), p. 124. Microfilm 39, E. Schrödinger Papers, to c. 1920. These manuscripts appear to be lecture notes. In some cases it is possible to find brief jottings that are later written out in full, as if Schrödinger had later expanded on notes taken hurriedly during lecture. This would tend to indicate that the notes are from his student days. But a reference in the notes to Schwarzschild's papers of 1917 proves that at least a portion of the notes come from the postwar years, 1918 to 1920, when Schrödinger returned to Vienna. The blank notebooks in which Schrödinger wrote these notes were printed in Vienna, another indication that the notes were probably taken in Vienna.

36. Scott, *Erwin Schrödinger*, pp. 2-3.

Chapter 15

1. WRH to Aunt Mary Hutton, June 7, 1842, OS 5.
2. Graves, 1:93.
3. Grace Hamilton to Eliza Hamilton, July 14, 1824, OS 6, and WRH to Sister Eliza, Oct. 15, 1821, ORSUP 37.
4. Graves, 2:642.
5. Ibid., 687.
6. Ibid., 447-49.
7. WRH to Sister Eliza, Aug. 28, 1820, ORSUP 29.
8. Hamilton describes the voting process in a letter to Augustus De Morgan, July 12, 1852, OS 618; Graves, 3:383.
9. Maria Edgeworth to Honora Edgeworth, Aug. 12, 1831, quoted from Michael Hurst, *Maria Edgeworth and the Public Scene: Intellect, Fine Feeling, and Landlordism in the Age of Reform* (Coral Gables, Fla.: University of Miami Press, 1969), p. 65.
10. She called in the "hanging gale," a portion of the rent that the tenants had traditionally been allowed to hold (Ibid., pp. 77-80).
11. Ibid., p. 71.
12. WRH to De Morgan, July 20, 1852, OS 622; Graves, 3:388. Hamilton was referring to an election in the late 1830s.
13. Quoted from J. C. Beckett, *The Making of Modern Ireland, 1603-1923* (London: Faber and Faber, 1966), p. 304.
14. Dora Wordsworth to Eliza Hamilton, Oct. 26, 1831, Graves, 1:471-73.
15. William Wordsworth to WRH, Jan. 24, 1831, Graves, 1:424, and June 13, 1831, Graves, 1:429.
16. WRH to Wordsworth, Oct. 29, 1831, Graves 1:476-78.
17. Wordsworth to WRH, Nov. 22, 1831, Graves, 1:491-93.
18. WRH to Sister Eliza, Mar. 21, 1832, Graves, 1:536-38.
19. WRH to Wordsworth, June 15, 1832, Graves, 1:566-68.
20. In particular it maintained the disenfranchisement of the forty-shilling freeholder. Sir Llewellyn Woodward, *The Age of Reform, 1815-1870* (Oxford: at the Clarendon Press, 1962), pp. 80-87.
21. As reported in the Dublin *Evening Mail,* Aug. 20, 1834, and quoted from R. B. McDowell, *Public Opinion and Government Policy in Ireland, 1801-1846* (London: Faber and Faber, [1952]), pp. 113-14. See also WRH to Viscount Adare, Aug. 20, 1834, Graves, 2:100-101.

22. Norman Gash, *Sir Robert Peel: The Life of Sir Robert Peel after 1830* (London: Longman, 1972), p. 96.
23. Aubrey De Vere to WRH, Mar. 24, 1835, NLD 5764, fol. 207; Graves, 2:131.
24. WRH to Uncle James Hamilton, Jan. 9, 1832, OR 254.
25. WRH to His Grace, Lord John George Beresford, Mar. 4, 1836, OR 518, and reply, Mar. 24, 1836, OR 522.
26. It was properly called the Royal College of St. Patrick at Maynooth. The Orange Society was founded the same year and set the stage for subsequent sectarian battles.
27. Maria Edgeworth to Honora Edgeworth, Aug. 12, 1831, quoted from Hurst, *Maria Edgeworth*, p. 66.
28. WRH to De Morgan, July 26, 1852, Graves, 3:393.
29. Galen Broeker, *Rural Disorder and Police Reform in Ireland, 1812-1836* (London: Routledge and Kegan Paul, 1970), p. 219.
30. Thomas Drummond to WRH, Feb. 12, 1838, OR 628. See also Graves, 1:288, 303; 2:180; 3:388-89, 471.
31. Gash, *Sir Robert Peel*, p. 217.
32. Wordsworth to WRH, Jan. 20, 1839, Graves, 2:292.
33. WRH to Lord Northampton, Jan. 20, 1839, OR 683; Graves, 2:294.
34. Marquess of Northampton to WRH, Jan. 30, 1839, OR 683; Graves, 2:294-96.
35. Beckett, *Making of Modern Ireland*, pp. 312-13. The destruction of the Irish language was probably due as much to the dislocations of the Great Famine as it was due to the national schools.
36. Eliza Hamilton to WRH, Mar. 23, 1832, OR 263. On the national school system see Norman Atkinson, *Irish Education: A History of Educational Institutions* (Dublin: Allen Figgis, 1969), chap. 5.
37. De Vere to WRH, Sept. 17 [1840], NLD 5764, fol. 267; Graves 2:321.
38. WRH to William Sewell, Nov. 2, 1840, Graves 2:326-27. On the foundation of Saint Columba's, see Atkinson, *Irish Education*, pp. 86-89.
39. Instead, William Edwin went to a school in Clapham, England, founded by Charles Pritchard (later Savilian Professor of Astronomy at Oxford). Pritchard offered to take William Edwin free of charge, probably because he felt that it would be advantageous to have the son of a prominent mathematician in his school. The sons of John Herschel, John Graves, and one of the Edgeworths were also at the school. The distinguished company and the advantageous financial arrangements persuaded Hamilton to accept the offer. There is a long series of letters involving the negotiation with Pritchard extending from July 17, 1846, to Aug. 17, 1846, OR 356, 358-361, and 363.
40. WRH to De Vere, Feb. 6, 1847, NLD 5765, fol. 401; Graves, 2:558. Aubrey's brother Stephen De Vere was a major philanthropist who took emigrants to Canada to escape the famine, and there housed them and cared for them when they were ill. On the work of the De Veres see S. M. Paraclita Reilly, *Aubrey De Vere: Victorian Observer* (Dublin: Clonmore and Reynolds, 1956), pp. 24-28, and Wilfrid Philip Ward, *Aubrey De Vere: A Memoir Based on His Unpublished Diaries and Correspondence* (London: Longmans, Green and Co., 1904), pp. 97-143.
41. De Vere to WRH, Feb. 23, 1847, NLD 5765, fol. 407; Graves, 2:558-60.
42. WRH to John F. W. Herschel, Apr. 1, 1847, OS 384; Graves, 2:565.
43. John T. Graves to R. P. Graves, Nov. 25, 1844, OS 271; Graves, 2:462-63.
44. Graves, 2:491.
45. WRH to Robert Mallet, Mar. 2, 1846, Graves, 2:507-9.
46. Graves, 2:507.
47. WRH journal entry, May 11, 1846, Graves, 2:522.
48. WRH to R. P. Graves, Apr. 30, 1846, Graves, 2:520-22.
49. WRH to George Peacock, Oct. 13, 1846, Graves, 2:527-29.
50. WRH to Humphrey Lloyd, Nov. 23, 1841, OR 824, Lloyd to WRH, Nov. 24, 1841, OR 827, WRH to Lloyd, Nov. 25, 1841, OR 828, and WRH to Adare, Dec. 31, 1841, Graves, 2:351-56.
51. Thomas Hutton to WRH, Nov. 18, OR 823.
52. WRH to Adare, Mar., 1842, Graves, 2:361-62.

53. WRH to Lloyd, Nov. 1, 1845, and WRH to the Members of the Council of the Royal Irish Academy, Nov. 17, 1845, Graves, 2:496-99.
54. OR 996; Graves, 2:511.
55. De Morgan to WRH, Feb. 15, 1846, OS 352; Graves, 2:504-5 and 3:263.
56. Graves, 2:245.
57. "The Royal Irish Academy and Its Library; A Brief Description" (Dublin, 1971).
58. William Betham to WRH, Apr. 23, 1839, OR 699; Graves, 2:281.
59. Graves, 2:320.
60. WRH to William Smith O'Brien, Apr. 20, 1836, OS 528; Graves, 2:180-82.
61. Graves, 2:339-40.
62. Graves, 3:160. The controversy generated a great deal of correspondence; see OR 517, 527, 528, 531, 543, 769, 771, 791, 792, 794, 798, 799, 800, 802, 804, 809, 812, 1921, and 1944.
63. WRH to Adare, Aug. 31, 1844, OR 936.
64. WRH to Adare, Sept., 1844, OR 940.
65. James Hamilton to WRH, July 26, 1848, ORSUP 9.
66. WRH to John Nichol, Nov. 8, 1855, TCD notebook, 129.5, pp. 51ff.
67. WRH to Madame Ranke, Dec. 14, 1849, Graves, 2:645-47. Robert Perceval Graves's sister married the German historian Leopold von Ranke. Hamilton was godfather to their son.
68. WRH to De Vere, Sept. 5, 1848, NLD 5765, fol. 439; Graves, 2:627.

Chapter 16

1. Walter Alison Phillips, ed., *History of the Church of Ireland from the Earliest Times to the Present Day,* 3 vols. (Oxford: Oxford University Press, 1933), 3:305-6.
2. Phillips, *History of the Church,* 3:334.
3. Phillips, *History of the Church,* 3:291.
4. R. W. Church, *The Oxford Movement: Twelve Years, 1833-1845* (1891; Hamden, Conn.: Archon Books, 1966).
5. The Hibernian Bible Society had as one of its founders the Rev. James H. Singer, Regius Professor of Divinity at Trinity College and later bishop of Meath. Singer was a colleague of Hamilton's. He joined with the Rev. Caesar Otway in founding the *Christian Examiner.* Hamilton knew Otway, too, because he mentions meeting him at the foot of the mountain of Helvellyn during his first visit to Wordsworth in 1827, and traveling with him during part of that tour (WRH to sister Eliza, Sept. 16, 1827, Graves, 1:262).
6. Graves, 2:312, and Archbishop Whately to WRH, Dec. 24, 1853, OR 1257.
7. Graves, 2:306, and WRH to Prof. Hennessey, May 28, 1858, OR 1362.
8. Wilfrid Philip Ward, *Aubrey De Vere: A Memoir Based on His Unpublished Diaries and Correspondence* (London: Longmans, Green and Co., 1904), pp. 14-15.
9. WRH to Humphrey Lloyd, Nov. 29, 1837, OR 594, WRH to Lloyd, Dec. 4, 1837, OR 595, WRH to Viscount Adare, Dec. 14, 1837, OR 612 (all in Graves, 2:212-21).
10. WRH to sister Eliza, Jan. 10, 1843, OR 878.
11. E. Jane Whately, *Life and Correspondence of Richard Whately, D.D., Late Archbishop of Dublin,* 2 vols. (London: Longmans, Green and Co., 1866), 2:26.
12. This distinction between different groups of Broad Churchmen is made by Charles Richard Sanders in his book *Coleridge and the Broad Church Movement* (Durham, N.C.: Duke University Press, 1942), p. 14.
13. Samuel Taylor Coleridge to WRH [Apr. 6, 1832], Graves, 1:545. Their one joint effort was an attempt to convert Arabella Lawrence away from her Unitarian belief. Hamilton had met her through the Edgeworths and she had given him his introduction to Coleridge. Both Hamilton and Coleridge tried, without success, to prove to her that Unitarianism was a shallow creed. Coleridge called it "that noiseless sand-shoal and wrecking shallow of Infra-Socinianism, yclept most calumniously and insolently, Unitarianism: as if Tri-unitarian were not as necessarily Unitarian as an apple pie

must be pie." Hamilton had sent Arabella his objections to Unitarianism the previous year, even before he met Coleridge (Graves, 1:464-66, and Coleridge to WRH, Apr., 1832, Graves, 1:542-43, Coleridge to Arabella Lawrence, Mar., 1832, Graves, 1:543-45).
14. Sanders, *Coleridge and the Broad Church,* pp. 73 and 77.
15. Ibid., p. 88.
16. Ibid., p. 32.
17. Ibid., p. 49.
18. See Alice D. Snyder, *Coleridge on Logic and Learning, with Selections from the Unpublished Manuscripts* (New Haven, Conn.: Yale University Press, 1929). Snyder admits that Coleridge's *Logic* is largely incomprehensible at first reading, and that full comprehension requires considerable familiarity with Kant's philosophy.
19. John Henry Cardinal Newman, *Apologia Pro Vita Sua, Being a History of His Religious Opinions* (New York: Longmans, Green and Co., 1947), p. 88.
20. Whately, *Richard Whately,* 2:154-55.
21. WRH to Uncle James Hamilton, Aug. 7, 1824, OR 88; Graves, 1:159.
22. WRH to Lord Northampton, Sept. 3, 1838, Graves 2:270-74. When Peter LaTouche died in 1827, Knox returned to his lodgings on Dawson Street, Dublin, where he lived until 1831. Like Coleridge, he would hold forth on a wide variety of subjects before visitors who came to witness his power as a conversationalist. Hamilton does not mention visiting Knox in Dawson Street. By 1827 he was already becoming a disciple of Coleridge.
23. Newman, *Apologia Pro Vita Sua,* p. 88.
24. WRH to Rev. Mortimer O'Sullivan, Feb. 25, 1859, Graves, 3:111.
25. For example, WRH to Charles Hort, Dec. 9, 1839, OR 726; Graves, 2:306-7.
26. Adare to WRH, Feb. 25, 1840, Graves, 2:317. He was commenting on Newman's Tract 85. The famous Tract 90 came in 1841. Adare expanded on his concern in another letter, Mar. 28, 1840, Graves, 2:318-19.
27. WRH to Adare, June 3, 1840, Graves, 2:319.
28. See especially Aubrey De Vere to WRH, July 10, 1837, NLD 5764, fol. 243; Graves, 2:200.
29. WRH to Robert P. Graves, Aug. 6, 1841, Graves 2:344.
30. WRH to the Rev. George Montgomery, Mar. 26, 1845, Graves, 2:481. Montgomery replied that he was leaving Dublin to collect his thoughts and to think things through (Montgomery to WRH, Mar. 31, 1845, OR 951). On their earlier friendship, see WRH to De Vere, Jan. 3, 1843, NLD 5764, fol. 313.
31. John T. Graves to WRH, Feb. 10, 1846, OS 351.
32. WRH to J. T. Graves, Mar. 3, 1846, OS 353.
33. WRH to Rev. William Sewell, Nov. 2, 1840, Graves, 2:326.
34. WRH to Rev. William Sewell, Nov. 2, 1840, Graves, 2:326.
35. Graves, 3:236.
36. Ibid., 2:612.
37. ORSUP 158.
38. WRH to sister Eliza, Nov. 19, 1843, Graves, 2:450-51.
39. Graves, 2:380-83.
40. De Vere to WRH, Sept. 2, 1848, NLD 5765, fol. 425; Graves, 2:615.
41. The Hamilton manuscripts contain a letter from Aubrey De Vere to Archbishop Whately (Jan. 11, 1811, OR 1061) that accompanied a copy of his book. Aubrey says that he realizes that "some of the views put forward in it will not meet your Grace's approbation," but he hopes that it will be of interest to him nonetheless.
42. De Vere to WRH, Feb. 27, 1851, NLD 5765 fol. 455; Graves, 2:667.
43. WRH to De Vere, Mar. 8, 1851, NLD 5765, fol. 461; Graves, 2:667.
44. The correspondence is OR 1140-1150. See also Graves, 2:655-66.
45. De Morgan held unorthodox religious views himself and was not a member of the Anglican Church, but he was in a good position to judge Hamilton's actions. He was a strong theist who was inclined towards Unitarianism, but he had a deep hatred of all sectarianism. He refused to join any church and destroyed all chances of obtaining his Master of Arts degree and a fellowship at Cambridge. Having been involved in

religious controversy throughout much of his life, he was well versed in doctrinal matters.
46. WRH to Augustus De Morgan, Sept. 1, 1852, OS 650; Graves, 3:407-9, and De Morgan to WRH, Sept. 3, 1852, OS 653; Graves, 3:412-13.
47. Charles Thomas Longley to WRH, Aug. 7, 1850, OR 1126. Hamilton's interest in St. Chrysostom probably came in part from the influence of Alexander Knox, who had argued that the Anglican Church was infinitely closer to the Eastern Church than it was to the Roman Church.
48. TCD notebook 103.5, pp. 201-2, and Elizabeth Whately to WRH, Jan. 3, 1854, OR 1260.
49. In December, 1853, Whately had another conflict with the rector of Castleknock, this time over a criticism of the Society for the Propagation of the Gospel that the rector had published without consulting Whately, his Diocesan, or the President of the Society for the Province. Hamilton, who was president of the local chapter, seemed to be inclined to support his bishop against his rector. Whately made much of the danger of "Tractite errors" in the Castleknock parish (Richard Whately to WRH, Dec. 24, 1853, OR 1257).
50. WRH to De Morgan, July 26, 1852, OS 626; Graves, 3:392, De Morgan to WRH, July 27, 1852, OS 627; Graves, 3:394, WRH to De Morgan, July 28-29, 1852, OS 629; Graves, 3:395.
51. WRH to De Morgan, July 28-29, 1852, OS 629; Graves, 3:395.
52. WRH to De Vere, Aug. 4, 1855, NLD 5765, fol. 523; Graves, 3:31-33.
53. WRH to De Vere, May 14, 1856, TCD notebook 129.5, pp. 331ff.
54. De Vere to WRH, Apr. 10, 1856, NLD 5765, fol. 675.
55. De Vere to WRH, Feb. 1, 1856, NLD 5765, fol. 653, Graves, 3:61-62.
56. Graves, 3:64.
57. WRH to Lady Wilde, May 13, 1858, TCD notebook 140.7, p. 163; Graves, 3:99-100.
58. De Vere to WRH, July 29, 1857, NLD 5765, fol. 963, Graves, 3:79.
59. "Memoir of the Rev. John O'Regan by His Son" (unpublished). Hamilton's two sons never married. His descendants are all in the O'Regan line.
60. WRH to De Vere, July 30, 1857, NLD 5675, fol. 795; Graves, 3:80.
61. WRH to De Vere, Apr. 25, 1856, NLD 5765, fol. 681.

Chapter 17

1. *Math. Papers*, 3:117n.
2. WRH to the Rev. Richard Townsend, May 14, 1855, TCD notebook 126, pp. 14-15. These two lines came from Hamilton's sonnet "The Tetractys," written in Oct., 1846, three years after the discovery of quaternions. In his later years Coleridge made much of the "divine Tetractys" of Pythagoras, which he believed would provide a philosophic ground and principle of unity for the "three" of the Trinity. (See Walter Jackson Bate, *Coleridge* [New York: The Macmillan Co., 1968], pp. 215-18.) Hamilton apparently shared some of this vision. The poem ends:

And when my eager and reverted ear,
Caught some faint echoes of an ancient strain,
Some shadowy outline of old thoughts sublime,
Gently He smiled to mark revive again,
In later age, and occidental clime,
A dimly traced Pythagorean lore;
A westward floating, mystic dream of FOUR.

(Quoted in Graves, 2:525).

3. Ernest Nagel, "'Impossible Numbers': A Chapter in the History of Modern Logic," *Studies in the History of Ideas*, 3 vols., ed. the Department of Philosophy of Columbia University (New York: Columbia University Press, 1918-35), 3:429-74.
4. WRH to John T. Graves, Oct. 20, 1828, Graves, 1:304.

5. WRH to Aubrey De Vere, Feb. 9, 1831, NLD 5764, fol. 19; Graves, 1:519.
6. WRH to George Peacock, Oct. 13, 1846, Graves, 2:528.
7. "It is this law which secures the complete identity of the two sciences as far as those results exist in common, and without which the latter science would degenerate into a science expressing the arbitrary combinations of symbols only" (George Peacock, *Treatise on Algebra,* 2 vols., 2d ed. [1840–1845; New York; Scripta Mathematica, 1940], 1:v).
8. Nagel, "'Impossible Numbers,'" p. 454. Peacock, *Treatise on Algebra,* 2:59, gives a slightly different statement.
9. Cornelius Lanczos, "William Rowan Hamilton—An Appreciation," *University Review* (National University of Ireland) 4 (1967):156.
10. Eric Temple Bell, *Men of Mathematics* (New York: Simon and Schuster, 1937), p. 359. Bell is referring to L.E.J. Brouwer and the Dutch school of "intuitionists."
11. G. J. Whitrow, *The Natural Philosophy of Time* (New York: Harper and Row, 1963), p. 118.
12. Ernest Nagel also tries to put Hamilton into the same school with De Morgan and company. (Nagel, "'Impossible Numbers,'" p. 461.)
13. Dan Griffin to WRH, Apr. 13, 1835, OR 472.
14. I cannot understand why Hamilton never took a greater interest in the writings of Boole. In 1852 De Morgan asked Hamilton to help find a publisher for Boole's book, saying: "I shall be very glad to see his [Boole's] work out, for he has, I think, got hold of the true connexion of algebra and logic" (De Morgan to WRH, Oct. 1852, OS 666; Graves, 3:421-22). In 1855 Boole asked Hamilton to write a testimonial recommending him for a position as mathematical examiner and sent a copy of his now-famous book, *An Investigation of the Laws of Thought* (George Boole to WRH, Jan. 29, 1855, OR 1297). Hamilton probably never read the book, because he wrote in his reply that he found it "an able work" and did not elaborate any further. He excused himself from writing the testimonial by saying that John Graves had paid more especial attention to Boole's speculations on the mathematical analysis of logic than he had (WRH to George Boole, Feb. 3, 1855, TCD notebook 126, pp. 3-4). Boole's reply, dated Feb. 4, 1855 (OR 1298), states that he fully understands Hamilton's hesitation: "The field of analysis is so wide that our tracks of investigation have widely diverged." Hamilton either could not recognize the value of mathematical logic or he never read far enough in Boole's work to understand the argument.
15. L. Pearce Williams, "Kant, Naturphilosophie, and Scientific Method," in *Foundations of Scientific Method: The Nineteenth Century,* ed. Ronald N. Giere and Richard Westfall (Bloomington: Indiana University Press, 1973), pp. 5-6.
16. De Vere to WRH, Dec. 5, 1846, NLD 5765, fol. 377; Graves, 2:541, and De Vere to WRH, July 25, 1855, NLD 5765, fol. 493; Graves, 3:30.
17. Quoted from Meyer Howard Abrams, "Coleridge's 'A Light in Sound': Science, Meta-Science, and Poetic Imagination," *Proceedings of the American Philosophical Society* 116, no. 6 (Dec., 1972):461.
18. Among Coleridge's poems Hamilton was especially fond of "Christabel." He told his sister Eliza, "My love of the supernatural is one cause, doubtless, of my singular fondness for Christabel" (WRH to sister Eliza, Sep. 5, 1831, Graves, 1:448).
19. WRH to De Vere, May 13, 1835, OS 108; NLD 5764, fol. 225; Graves, 2:142.
20. *Math Papers,* 3:3.
21. Ibid., 3:5.
22. Samuel Taylor Coleridge, *Aids to Reflection, and the Confessions of an Inquiring Spirit* (London: George Bell & Sons, 1901), p. 224.
23. Immanuel Kant, *Critique of Pure Reason,* trans. Norman Kemp Smith (London: Macmillan and Co., 1961), A34-B50. Page numbers are to the first and second editions as they are given in Smith's translation.
24. Kant, *Critique of Pure Reason,* A38-39—B55-56.
25. Thomas McFarland, *Coleridge and the Pantheistic Tradition* (Oxford: At the Clarendon Press, 1969), p. 215.
26. WRH memorandum, Graves, 1:437.
27. McFarland, *Coleridge,* p. 216, quoted from *The Friend.*

28. Coleridge, *Aids to Reflection*, p. 161.
29. Samuel Taylor Coleridge, *Collected Works*, 6 vols. in 9, ed. Barbara Rooke (Princeton, N.J.: Princeton University Press, 1969), vol. 4, *The Friend*, pt. 1, p. 158.
30. Coleridge, *The Friend*, 1:159.
31. Coleridge, *Aids to Reflection*, p. 161.
32. Sanders, *Coleridge and the Broad Church*, pp. 35-45.
33. Coleridge, *Aids to Reflection*, p. 166.
34. Charles Richard Sanders, *Coleridge and the Broad Church Movement* (Durham, N.C.: Duke University Press, 1942), p. 24.
35. Ibid.
36. Coleridge, *Aids to Reflection*, p. xlvi.
37. WRH to Lord Adare, Aug. 23, 1831, Graves, 1:444.
38. Coleridge occasionally skirted close to the neo-Pythagorean belief that the world was actually composed of number. An example occurs in *The Friend* where he asks whether the hypothetical atoms of physics were not merely symbols of algebraic relations "representing powers essentially united with proportions of dynamic ratios—ratios not of powers, but that are powers (quoted from John Muirhead, *Coleridge as Philosopher* [New York: Humanities Press, 1954], p. 98n). How an algebraic ratio can become a dynamic power is not clear.

Chapter 18

1. Misc. papers, box 6.
2. TCD notebook 24.5, fol. 49.
3. Ibid. Hamilton incorporated some of these remarks in his introductory lecture on astronomy of Nov. 8, 1832 (Graves, 1:643).
4. Graves, 1:208-9, TCD notebook 16, fol. 49, and WRH to Arthur Hamilton, Apr. 14, 1824, OR 79.
5. JG 37, copied from the *Foreign Review and Continental Miscellany* 4 (1830): 97ff.
6. *The Foreign Review and Continental Miscellany* 4 (1830):117.
7. William Wordsworth, Sr., to WRH, Sept. 9, 1830, Graves, 1:393, and Nov. 26, 1830, Graves, 1:401.
8. WRH to Wordsworth, Oct. 29, 1831, Graves, 1:478.
9. Of these six attempted translations, four are in the uncatalogued miscellaneous papers, one in TCD MS notebook 24.5, fol. 101, and another is in notebook 70, not foliated. Hamilton had the first edition in 1831; he read the second edition in 1834.
10. WRH to Aubrey De Vere, July 3, 1832, Graves, 1:585, and WRH to sister Eliza, Mar. 15, 1832, Graves, 1:535.
11. WRH to De Vere, Aug. 29, 1834, Graves, 2:103.
12. Wordsworth to WRH, Jan. 24, 1831, Graves, 1:425.
13. Alice D. Snyder, *Coleridge on Logic and Learning, with Selections from the Unpublished Manuscripts* (New Haven, Conn.: Yale University Press, 1929), p. 68.
14. Samuel Taylor Coleridge to WRH, Apr. 4, 1832, Graves, 1:545.
15. WRH to Viscount Adare, July 19, 1834, Graves, 2:96, and Aug. 20, 1834, Graves, 2:100. WRH to De Vere, Aug. 29, 1834, Graves, 2:103-5. WRH to Adare, Apr. 19, 1842, Graves, 2:364. WRH to Rev. William Lee, July 2, 1859, ·Graves, 3:115-17.
16. Graves, 1:437.
17. WRH to De Vere, Mar. 27, 1832, Graves, 1:541.
18. Samuel Taylor Coleridge, *Complete Poetical Works.* 2 vols., ed. Ernest Hartley Coleridge, (Oxford: Oxford University Press, 1912), 1:419-20.
19. Samuel Taylor Coleridge, *Collected Works*, ed. Barbara Rooke, (Princeton, N.J.: Princeton University Press, 1969), vol. 4, *The Friend*, pt. 1, p. 440.
20. OS 10.
21. A good history of this controversy is Florian Cajori, "History of the Exponential and Logarithmic Concepts," *American Mathematical Monthly* 20 (1913):5, 35, 75, 107, 148, 173, 205.
22. WRH to John T. Graves, Oct. 20, 1828, Graves 1:303-4.

23. Graves, 1:307-8. J.F.W. Herschel to J. T. Graves, undated, JG 114.

24. Herschel to J. T. Graves, undated, JG 113.

25. Herschel to J. T. Graves, Jan. 24, 1829, JG 112.

26. Hamilton to Herschel, Feb. 25, 1829, OS 35. The paper appeared as "An Attempt to Rectify the Inaccuracy of Some Logarithmic Formulae. By John Thomas Graves, of the Inner Temple, Esq. V. P., Read December 18, 1828," *Philosophical Transactions of the Royal Society* 209 (1829):pt. 1, 171-86.

27. George Peacock, "Report on the Recent Progress and Present State of Certain Branches of Analysis," *British Association Reports* (1834), p. 266n. See Cajori, "Exponential and Logarithmic Concepts," p. 178.

28. WRH to J. T. Graves, June 11, 1829, OS 54. Graves writes: "Though I may seem inclined to the opinion that imaginary quantities—the poetry of algebra—are, like Virtue, desirable for their own sakes; in reality I shudder to reflect that the addition of one string to the lyre of numbers might, like the introduction of Chinese soldiers into an Ursuline convent give birth to prodigious rubbish" (J. T. Graves to WRH, Aug. 8, 1829, OS 60, and Jan. 1, 1830, OS 63).

29. Augustus De Morgan, "On the Foundation of Algebra," Cambridge Philosophical Society, *Transactions* 8 (1844-49):141.

30. John Warren, *A Treatise on the Geometrical Representation of the Square Roots of Negative Quantities* (Cambridge: At the University Press, 1828). On Hamilton's knowledge of the geometrical representation of complex numbers, see his "Preface" to *Lectures on Quaternions, Math. Papers*, 3:135n. See also J. T. Graves to WRH [1829], OS 1349. This is the letter in which Graves tells Hamilton about Warren's book. Unfortunately it is undated. Graves was in correspondence with Warren. J. T. Graves to WRH, Aug. 8, 1829, OS 60, and Aug. 15 [1829], OS 1352.

31. Hamilton to De Morgan, June 25, 1864, Graves, 3:189.

32. Augustin-Louis Cauchy, *Oeuvres complètes*, 26 vols., 2d ser. (Paris: Gauthier-Villars et Fils, 1882-1938), 2:3.

33. Preface to *Lectures on Quaternions, Math. Papers*, 3:123.

34. Morris Kline, *Mathematical Thought from Ancient to Modern Times* (New York: Oxford University Press, 1972), p. 626.

35. WRH to J. T. Graves, June 11, 1829, OS 54.

36. WRH to J. T. Graves, Aug. 5, 1829, OS 58.

37. Hamilton believed erroneously that the Cauchy-Riemann equations held for *all* complex functions, while in fact they are only valid when the first partial derivatives exist and are continuous. See Helena Pycior, "The Role of Sir William Rowan Hamilton in the Development of British Modern Algebra" (Ph.D diss., Cornell University, 1976), p. 93.

38. For instance, C. C. MacDuffee, "Algebra's Debt to Hamilton," *Scripta Mathematica* 10 (1944):25. The number couples have stood the test of time. See Ross A. Beaumont and Richard S. Pierce, *The Algebraic Foundations of Mathematics* (Reading, Mass.: Addison-Wesley, 1963), pp. 286-98. Hamilton also called the number couples *conjugate functions* because of their relationship to his Equations of Conjugation.

39. See my Appendix for an outline of Hamilton's argument.

40. Graves, 1:639-54.

41. Ibid., 2:32.

42. WRH to Adare, July 19, 1834, Graves, 2:96-97.

43. WRH to J. T. Graves, Sept. 11, 1834, OS 96, and Sept. 13, 1834, OS 97.

44. Graves, 2:138, and WRH to De Vere, May 13, 1835, Graves, 2:141. The entire series of letters was later published by Robert P. Graves, "Sir W. Rowan Hamilton on the Elementary Conceptions of Mathematics," *Hermathena* 6 (1879):469-89.

45. WRH to De Vere, Oct. 4, 1835, Graves, 2:164.

46. Royal Irish Academy, *Transactions* 17 (1837):293-422.

47. *Math. Papers*, 3:15-16.

48. Ibid., xv.

49. Ibid., 233. I learned of Hamilton's discovery of the associative law from an undergraduate thesis by Barbara Underwood Cohen, "From Algebra to Algebras: A Study of Changing Conceptions of Mathematics" (Harvard University, Apr., 1966).

50. *Math. Papers*, 3:96.

51. See Eric Temple Bell, *Development of Mathematics*, 2d. ed. (New York: McGraw-Hill, 1945), p. 189.
52. WRH, "Researches on Quaternions," read Nov. 13, 1843, *Math. Papers*, 3:159. Hamilton was even more explicit in a letter to De Morgan dated May 7, 1847 (Graves, 2:574).

Chapter 19

1. *Math. Papers*, 3:7.
2. Immanuel Kant, *Critique of Pure Reason*, trans. Norman Kemp Smith (London: MacMillan and Co., 1961), B35.
3. *Math. Papers*, 3:7.
4. Ibid.
5. Hamilton to Aubrey De Vere, May 13, 1835, Graves, 2:142.
6. "Arithmetic produces its concepts of number through successive addition of units in time" (quoted from Norman Kemp Smith, *A Commentary to Kant's "Critique of Pure Reason,"* 2d. ed. (London: MacMillan and Co., 1923), p. 129.
7. Kant, *Critique* (A142-B181).
8. Ibid., (A145-B184).
9. Ibid., (A142-43—B182).
10. Augustus De Morgan, "On the Foundation of Algebra," Cambridge Philosophical Society, *Transactions* 7 (1839-42):174.
11. Hamilton to Augustus De Morgan, May 8, 1841, Graves, 2:342. He quotes Kant's *Critique* (A162-63—B293-94).
12. Kant's use of these notions in arithmetic is a current topic of debate. See Kaarlo Jaakko Hintikka, "Kant on the Mathematical Method," *Monist* 51 (1967):352-75; and Charles Parsons, "Kant's Philosophy of Arithmetic," in *Philosophy, Science, and Method*, ed. Sidney Morgenbesser, Patrick Suppes, Morton White (New York: St. Martins's Press, 1969), pp. 568-94.
13. Kant, *Critique*, (A713-B741, A724-B752).
14. Hintikka believes that Kant was thinking specifically about the Euclidean method of proof, where the construction or "display" is called the *ecthesis* or *setting-out* (Hintikka, "Mathematical Method," p. 361).
15. Parsons, "Kant's Philosophy," pp. 581-82.
16. Kant is not even consistent in his claim that only the inner sense of time is involved in arithmetic construction. In the schematism he derives number from ordered succession in time; but in the Transcendental Doctrine of Method he generates number from an intuition of "the universal element in the synthesis of one and the same thing in time and space" (A724-B753). Hintikka argues that for Kant the important aspect of intuition is not its immediacy, but the fact that all intuitions are individual and singular. Both aspects are closely related, because Kant realizes that in order for intuitions to be immediate, they must be of singular objects (Hintikka, "Mathematical Method," pp. 354ff.).
17. Smith, *Commentary*, p. 131.
18. For L.E.J. Brouwer and other intuitionists the existence of mathematical entities is synonymous with the possibility of their construction, while their opponents, the formalists, emphasize the static and timeless character of mathematical objects. Brouwer announced his intuitionist position by rejecting Kant's apriority of space, but adhering the more resolutely to the apriority of time (L.E.J. Brouwer, "Intuitionism and Formalism," in *Philosophy of Mathematics*, ed. Paul Benacerraf and Hilary Putnam [Englewood Cliffs, N.J.: Prentice-Hall, 1964], p. 69). He constructs the natural numbers from "the intuition of the bare two-oneness," which he calls the basal intuition of mathematics. This bare two-oneness is intuited from the division of time, which he in turn claims is "the fundamental phenomenon of the human intellect" (p. 69).
19. *Math. Papers*, 3:9.

20. Ibid., 29. Hamilton speaks specifically of counting as an ordinal relation in a letter to the Viscount Adare dated Aug. 9, 1843 (Graves, 2:417).
21. Hamilton to Adare, May 16, 1833, Graves, 2:46. Misc. papers, box 6, Feb. 16, 1830 "Triads," Dec., 1830 "Triads," and undated "Fundamental Idea."
22. This is one of the weakest points in Hamilton's essay. An evaluation is in Jerold Mathews, "William Rowan Hamilton's Paper of 1837 on the Arithmetization of Analysis," *Archive for History of Exact Science* 19 (1978):177-200.
23. *Math. Papers*, 3:76-77.
24. The definition of multiplication is not wholly arbitrary, because Hamilton has already defined addition of number couples in the obvious way: $(a_1, a_2) + (b_1, b_2) = (a_1 + b_1, a_2 + b_2)$, and he wants the following distributive law to hold: $(a_1, a_2) [(b_1, b_2) + (c_1, c_2)] = (a_1, a_2) (b_1, b_2) + (a_1, a_2) (c_1, c_2)$. A second restriction is that division of number couples should be determinate as long as the divisor is not $(0, 0)$. These two restrictions do not completely determine the process of multiplication, but they limit the possible choices (*Math. Papers*, 3:80-83).
25. Misc. papers, box 6, Apr. 21, 1832.
26. TCD notebook 35.5, unpaginated, underlining by Hamilton.
27. Kant, *Critique* (A189-B234).
28. Ibid., (A190-B235, A193-B238).
29. Ibid., (A194-B239).
30. Ibid., (A193-B238).
31. Ibid., (A31-B46).
32. N. K. Smith says in his *Commentary* that they are not consistent. "There are for Kant, two orders of time, subjective and objective. Recognition of the latter (emphasized and developed in the *Analytic*) is, however, irreconcilable with his contention that time is merely the form of inner sense" (p. 137).
33. *Math. Papers*, 3:7.

Chapter 20

1. Paolo Ruffini, *Teoria generale delle Equazioni, in cui si dimostra impossibile la soluzione algebraica delle equazioni generali di grado superiore al quarto* (Bologna: Nella stamperia di S. T. d'Aquino, 1799). Histories of the attempts to solve the equation of the fifth degree are: J. Pierpont, "Zur Geschichte de Gleichung des V. Grades (bis 1858)," *Monatshefte für Mathematik und Physik* 6 (1895):15-68; H. O. Foulkes, "The Algebraic Solution of Equations," *Science Progress in the Twentieth Century* 26 (1932):601-8; L. E. Dickson, *Modern Algebraic Theories* (Chicago: Sanborn, 1926), chap. 10; Luboš Nový, *Origins of Modern Algebra* (Prague: Academia Publishing House of the Czechoslovak Academy of Sciences, 1973), pp. 20-72; and Felix Klein, *Lectures on the Icosahedron and the Solution of Equations of the Fifth Degree*, trans. George Gavin Morrice (1884; New York: Dover Publishing, 1956), pp. 153-59.
2. Morris Kline, *Mathematical Thought from Ancient to Modern Times* (New York: Oxford University Press, 1972), pp. 597-608, 752-70.
3. Niels Henrik Abel, *Mémoire, sur les équations algébriques, où on démontre l'impossibilité de la résolution de l'équation générale du cinquième degré* (Christiania: Impr. de Groendahl, 1824), and idem, "Démonstration de l'impossibilité de la résolution algébrique des équations générales qui passent le quatrième degré," *Journal für reine und angewandte Mathematik* 1 (1826):65-84.
4. Graves 2:155.
5. George Birch Jerrard, *Mathematical Researches*. 3 vols. (Bristol, 1832-35). Jerrard also published an *Essay on the Resolution of Equations*, (London: Taylor and Francis, 1859, plus numerous articles in the *Philosophical Magazine* (see the *Royal Society Catalogue of Scientific Papers, 1800-1863* 3 (1869):547-48, for a complete list).

6. Graves, 2:155.
7. John T. Graves to WRH, Oct. 22, 1835, OR 500.
8. WRH to J. W. Lubbock, Dec. 2, 1835, OS 113, WRH to Lubbock, Apr. 20, 1836, OS 123, Lubbock to [WRH], Apr. 13, 1836, OS 120, Lubbock to G. B. Jerrard, Apr. 18, 1836, OS 121, WRH to Lubbock, May 7, 1836, OS 124, Jerrard to WRH, May 9, 1836, OS 126, Lubbock to WRH, May 12, 1836, OS 127, Jerrard to WRH, May 16, 1836, OS 128.
9. WRH, "Theorem Respecting Algebraic Elimination Connected with the Question of the Possibility of Resolving in Finite Terms the General Equation of the Fifth Degree," *Philosophical Magazine*, 8 (1836):538–43, and *Philosophical Magazine* 9 (1836)28–32; *Math. Papers*, 3:471–77.
10. WRH to Lubbock, May 30, 1836, OS 129. See also WRH to Jerrard, June 1, 1836, OS 130.
11. WRH to Lubbock, July 1836, OS 133, and John Phillips to WRH, July 19, 1836, OS 132.
12. WRH to Jerrard, July 28, 1836, OS 136.
13. Jerrard to WRH, July 25, 1836, OS 135; Graves, 2:185.
14. Jerrard to WRH, Aug. 12, 1836, OS 138.
15. WRH to Jerrard, Aug. 18, 1836, OS 139.
16. *Math. Papers*, 3:xx–xxi, and Klein, *Icosahedron*, pp. 156–59.
17. WRH to sister Sydney, Oct. 31, 1836, OR 547. Hamilton wrote a testimonial in favor of Jerrard in 1838 that confirms the fact that he continued to hold Jerrard's work in high regard (WRH to Rev. Dr. MacDonnell, June 6, 1838, OR 640).
18. Further objections were voiced by Cockle and Cayley in the *Philosophical Magazine* (vols. 17–24, 1859–62). See Klein, *Icosahedron*, p. 158.
19. WRH, "Inquiry into the Validity of a Method Recently Proposed by G. B. Jerrard Esq., for Transforming and Resolving Equations of Elevated Degrees," *British Association Report* (1836), pp. 295–348. *Math. Papers*, 3:481–516.
20. Royal Irish Academy, *Transactions* 17 (1837):293–422, and *Math. Papers*, 3:517–69. Also, "Exposition of the Argument of Abel," *British Association Report* (1837), pt. 2, p. 1, and "Investigations Respecting Equations of the Fifth Degree (May 22, 1837)," Royal Irish Academy, *Proceedings* 1 (1841):76–80, *Math. Papers*, 3:478–80.
21. Dickson, *Modern Algebraic Theories*, p. 178n, and Kline, *Mathematical Thought*, p. 755.
22. One such claim was transmitted by James MacCullagh (James MacCullagh to WRH, Mar. 9 [no year], OR 1971). The other was from P. Gerolamo Badano, *Nuove Ricerche sulla Risoluzione Generale delle Equazioni Algebriche* (Genova: Tipografia Ponthenier, 1840). Hamilton replied in two papers: "On a Method Proposed by Professor Badano for the Solution of Algebraic Equations," Royal Irish Academy, *Proceedings* 2 (1844):275–76, *Math. Papers*, 3:570–71; and "On Equations of the Fifth Degree: And Especially on a Certain System of Expressions Connected with Those Equations, Which Professor Badano Has Lately Proposed," Royal Irish Academy, *Transactions* 19 (1843):329–76, *Math. Papers*, 3:572–602.

Chapter 21

1. *Math Papers*, 3:96.
2. "Thirteen years after Hamilton's death G. Frobenius proved that there exist precisely three associative division algebras over the reals, namely, the real numbers themselves, the complex numbers and the real quaternions" (*Math. Papers*, 3:xvi). For more detail see Kenneth O. May, "The Impossibility of a Division Algebra of Vectors in Three Dimensional Space," *American Mathematical Monthly* 73 (1966):289–91.
3. Misc. papers, box 2 (uncatalogued). See also WRH to Viscount Adare, Jan. 4, 1831, Graves, 1:418–19.
4. Preface to the *Lectures on Quaternions* in *Math. Papers*, 3:137. When Hamilton wrote this preface he was almost morbidly concerned about questions of priority, as appears from his correspondence with De Morgan, and therefore he spared no

detail. His account agrees with the evidence of the manuscripts and supplements them on several points.

5. In Jan., 1831, Hamilton was using Laplace's equation of continuity to try to obtain a triplet system. See Helena Pycior, "The Role of Sir William Rowan Hamilton in the Development of British Modern Algebra" (Ph.D. diss., Cornell University, 1976). p. 96.
6. *Math. Papers*, 3:128.
7. Ibid., 126-31.
8. WRH to John T. Graves, Apr. 13, 1836, OS 119. Hamilton wrote that he was sending a copy of the essay with Humphrey Lloyd, who was going to England.
9. J. T. Graves to WRH, Aug. 8-13, 1836, OS 137, and *Math. Papers*, 3:138-40. This letter and the preface to the *Lectures on Quaternions* indicate that Graves and Hamilton had discussed the triplets long before 1836, although the subject does not appear in the extant correspondence before this date.
10. J. T. Graves to WRH, Aug. 13, 1836, OS 137.
11. J. T. Graves to WRH, Aug. 13, 1836, OS 137.
12. Samuel Taylor Coleridge, *Collected Works*, ed. Barbara Rooke (Princeton, N.J.: Princeton University Press, 1969), vol. 4, *The Friend*, pt. 1, p. 158.
13. Graves, 1:439.
14. G.N.G. Orsini, *Coleridge and German Idealism: A Study in the History of Philosophy with Unpublished Material from Coleridge's Manuscripts* (Carbondale: Southern Illinois University Press, 1969), pp. 109-10; see also Alice D. Snyder, *Coleridge on Logic and Learning, with Selections from the Unpulished Manuscripts* (New Haven, Conn.: Yale University Press, 1929), pp. 126 and 129n.
15. The noetic Pentad appears in numerous places in Coleridge's work. See Snyder, *Coleridge on Logic and Learning*, p. 71n.
16. John H. Muirhead, *Coleridge as Philosopher* (New York: Humanities Press, 1954), p. 86n.
17. Orsini, *Coleridge and German Idealism*, p. 243.
18. Aubrey De Vere to WRH, Apr. 7, 1834, Graves, 2:90, and WRH to De Vere, May 9, 1834, Graves, 2:93.
19. WRH to De Vere, May 13, 1835, Graves, 2:142.
20. J. T. Graves to WRH, Oct. 17, 1840, OS 186, Oct. 18, 1840, OS 187, and Oct. 23, 1840, OS 188; WRH to J. T. Graves, Oct. 24, 1840, OS 189; Graves, 2:328; and J. T. Graves to WRH, Oct. 30, 1840, OS 190. These are described in *Math. Papers*, 3:140-41.
21. J. T. Graves to WRH, Oct. 23, 1840, OS 188.
22. J. T. Graves to WRH, Oct. 30, 1840, OS 190.
23. WRH to J. T. Graves, Oct. 24, 1840, Graves, 2:328. WRH to Charles Boyton, May 22, 1841, Graves, 2:333.
24. Graves, 2:329-30.
25. Ibid., 1:547. I have not been able to find this particular triad of categories anywhere in Coleridge's published papers.
26. WRH to Viscount Adare, Apr. 19, 1842, Graves, 2:363-76.
27. Graves, 2:376-79.
28. In the *Critique of Judgment* Kant wrote: "It has been thought somewhat suspicious that my divisions in pure philosophy should almost always come out threefold. But it is due to the nature of the case." If the divisions were to be analytic according to the law of contradiction, then the arrangements of categories would be dichotomous, but in order "to meet the requirements of synthetic unity in general, namely (1) a condition, (2) a conditioned, (3) the concept arising from the union of the conditioned with its condition, the division must of necessity be trichotomous." (*Critique of Judgment*, trans. James Creed Meredith [Oxford: At the Clarendon Press, 1957], pt. 1, p. 39n.).
29. Kant speaks directly to Hamilton's view that there is aesthetic beauty in science. "Something, then, that makes us attentive in our estimate of nature to its finality for our understanding—an endeavour to bring, where possible, its heterogeneous laws under higher, though still always empirical, laws—is required, in order that, on meeting with success, pleasure may be felt in this their accord with our cognitive

faculty, which accord is regarded by us as purely contingent" (*Critique of Judgment*, pt. 1, p. 28).
30. Graves, 2:365. In the *Critique of Judgment* (pt. 1, p. 24) Kant writes: "There is in nature a subordination of genera and species comprehensible by us: Each of these genera again approximates to the others on a common principle, so that a transition may be possible from one to the other, and thereby to a higher genus." This passage may have suggested to Hamilton a hierarchy of triads.
31. Kant, *Critique of Judgment*, pt. 1, p. 14.
32. Ibid., p. 25. For the implications of these remarks for science, see L. Pearce Williams, "Kant, Naturphilosophie, and Scientific Method," in *Foundations of Scientific Method: the Nineteenth Century*, ed. Ronald N. Giere and Richard Westfall (Bloomington: Indiana University Press, 1973), pp. 3-22.
33. Kant, *Critique of Judgment*, pt. 1, pp. 24-25.
34. Graves, 2:366.
35. Kant, *Critique of Judgment*, pt. 2, p. 33. On Kant's "highest intelligences" see the *Critique of Pure Reason*, trans. Norman Kemp Smith (London: Macmillan and Co., 1961), A670-71—B698-99. Kant, however, would never call the teleological judgment a matter of faith (*Critique of Judgment*, pt. 2, pp. 143-44).
36. John Graves responded enthusiastically to Hamilton's triads, and asked specifically if they had the property of polarity. Hamilton replied that they did indeed have that property (Graves, 2:378-79).

Chapter 22

1. WRH to son Archibald, Aug. 5, 1865, Graves, 2:434-35.
2. He spoke on the calculus of differences, the calculus of probabilities, and the equation of the fifth degree (Graves, 2:415).
3. Graves, 2:408.
4. Thomas Hawkins has shown in several articles that the idea of matrices was understood by mathematicians long before Cayley gave them their modern name and notation: Thomas Hawkins, "Another Look at Cayley and the Theory of Matrices," *Archives internationales d'histoire des sciences* 27 (1977): 82-112, and "Cauchy and the Spectral Theory of Matrices," *Historia Mathematica* 2 (1975): 1-29. Eisenstein's paper was "Allgemeine Untersuchungen über die Formen dritten Grades mit drei Variabeln, welche der Kreistheilung ihre Enstehung verdanken," *Journal für die reine und ungewandte Mathematik* 28 (1844): 289-374.
5. Quoted from Hawkins, "Another look at Cayley," p. 86.
6. Helena Pycior points out in her thesis that there were mathematicians read by Hamilton (John Walker, John Bonnycastle, and L. B. Francoeur) who felt the need to *prove* the commutative law, and Hamilton himself felt the need to demonstrate its validity in ordinary algebra. (Pycior, "The Role of Sir William Rowan Hamilton in the Development of British Modern Algebra" [Ph.D. diss., Cornell University, 1976], pp. 124-26.)
7. 2 vols. (Berlin: T. H. Riemann, 1828-29). The best discussion of Ohm's work is in Luboš Nový, *Origins of Modern Algebra* (Prague: Academia Publishing House of the Czechoslovak Academy of Sciences, 1973), pp. 83-90.
8. WRH to Arthur Moore, May 31, 1834, OR 396; Graves, 2:84. WRH to John T. Graves, Feb. 28, 1835, OS 106. Hamilton found that Ohm had preceded Graves in some of his conclusions regarding logarithms of complex numbers. George Peacock also read and appreciated the work of Ohm.
9. The translation was published as *The Spirit of Mathematical Analysis* (London: Parker, 1843.) The letters referred to are Alexander John Ellis to WRH, May 15, 1843, OS 219; Graves, 2:416, WRH to J. T. Graves, May 17, 1843, OS 220, and Lord Adare to WRH, Aug. 4, 1843, OS 221.
10. Graves, 2:410-13.
11. WRH to son Archibald, Aug. 5, 1865, Graves, 2:434-35.
12. WRH to Peter Guthrie Tait, Oct. 15, 1858, Graves, 2:435-36.

13. WRH, "On a New Species of Imaginary Quantities Connected with the Theory of Quaternions," *Math. Papers*, 3:111-16.
14. WRH to J. T. Graves, Oct. 17, 1843, *Math. Papers*, 3:106-10.
15. There have also been two modern reconstructions of the discovery: E. T. Whittaker, "The Sequence of Ideas in the Discovery of Quaternions," Royal Irish Academy, *Proceedings* 50, sect. A, no. 6 (Feb., 1945), pp. 93-98; and B. L. van der Waerden, "Hamilton's Discovery of Quaternions," *Mathematics Magazine* 49 (1976): 227-34.
16. *Math. Papers*, 3:107.
17. TCD in notebook 24.5, p. 119. The entry is dated Aug., 1842. In a system of triplets, $x + iy + jz$, where x, y, z, are real numbers and $i^2 = j^2 = -1$, he sets $ji = i$ and $ij = j$. I learned of this early experiment with noncommutative systems from Pycior, "Role of Sir William Rowan Hamilton," p. 147.
18. *Math. Papers*, 3:103.
19. Ibid., 107.
20. Hamilton claimed to have written $i^2 = j^2 = k^2 = ijk = -1$ on the bridge, but it seems unlikely that the inspiration came to him in such a neat and condensed form (WRH to son Archibald, Aug. 5, 1865, Graves, 2:435). This notebook has been preserved. The pertinent pages are reproduced in Graves, 2:436-37.
21. WRH memo of July 18, 1848, Graves, 2:439.
22. J. T. Graves to WRH, Oct. 31, 1843, Graves, 2:443.
23. Augustus De Morgan to WRH, Dec. 16, 1844, Graves, 2:475.
24. *Math. Papers*, 3:110.
25. WRH to Humphrey Lloyd, Dec. 3, 1844, Graves, 2:475-77.
26. WRH to J. T. Graves, Oct. 16, 1843, *Math. Papers*, 3:108, and memorandum of Oct. 16, 1843, Graves, 2:439-40.

Chapter 23

1. For material in this section I am especially indebted to Michael J. Crowe's *A History of Vector Analysis: The Evolution of the Idea of a Vectorial System* (Notre Dame, Ind.: University of Notre Dame Press, 1967), and Helena Pycior's "The Role of Sir William Rowan Hamilton in the Development of British Modern Algebra" (Ph.D. diss., Cornell University, 1976).
2. "On Quaternions: Or on a New System of Imaginaries in Algebra," *Philosophical Magazine* 25 (1844): 10-13, 241-46; 26 (1845): 220-24; 29 (1846): 26-31, 113-22, 326-28; 30 (1847): 458-61; 31 (1847): 214-19, 278-93, 511-19; 32 (1848): 367-74; 33 (1848): 58-60; 34 (1849): 294-97, 340-43, 425-39; 35 (1849): 133-37, 200-204; 36 (1850): 305-6. These are run together in *Math. Papers*, 3:227-97. "On Symbolical Geometry," *Cambridge and Dublin Mathematical Journal* 1 (1846): 45-57, 137-54, 256-63; 2 (1847): 47-52, 130-33, 204-9; 3 (1848): 68-84, 220-25; 4 (1849): 84-89, 105-18. This series will appear in the fourth volume of *Math. Papers*. Another important early paper is "Researches Respecting Quaternions," Royal Irish Academy, *Transactions* 21 (1848): 199-296, *Math. Papers*, 3:159.
3. The correspondence is described in considerable detail in *Math. Papers*, vol. 3, app. 3, "Four and Eight Square Theorems," pp. 648-56. See also L. E. Dickson, "On Quaternions and Their Generalization and the History of the Eight Square Theorem," *Annals of Mathematics*, 2d ser. 20 (1919): 155-71, for the subsequent history.
4. WRH to John T. Graves, July 8, 1844, OS 251; *Math. Papers*, 3:650.
5. Hamilton was prepared to notice this weakness, because in his first published paper on the quaternions he stated explicitly that quaternion multiplication preserves an important property of regular algebra, "that which may be called the *associative* character of the operation" ("On a New Species of Imaginary Quantities Connected with the Theory of Quaternions," Royal Irish Academy, *Proceedings* 2 [1844]: 424-34, in *Math. Papers*, 3:114). This paper is marked as having been communicated on Nov. 13, 1843, but it closes with a note that it had been abstracted from a larger paper to appear in the *Transactions* of the academy (*Math. Papers*, 3:116). That longer paper,

also dated Nov. 13, 1843, was not published until 1848, and when it did appear it concluded with a note mentioning works written as late as June, 1847 ("Researches Respecting Quaternions: First Series," Royal Irish Academy, *Transactions* 21, [1848]: 199-296, in *Math. Papers,* 3:159-216, "note A," pp. 217-26). In the note Hamilton stated that his presentation on Nov. 13, 1843, had been "in great part oral" (*Math. Papers,* 3:225); therefore it is possible that his statement of and naming of the associative law was added later when his original communication was to be printed in 1844. It is thus possible that he first recognized the importance of the associative law in 1844, when he began work on Graves's octaves. On this point see Pycior, "Role of Sir William Rowan Hamilton," p. 152, and Crowe, *History of Vector Analysis,* p. 16, n. 26.

6. WRH to Robert P. Graves, Aug. 16, 1844, OS 254.
7. Graves, 2:461.
8. Arthur Cayley, "On Jacobi's Elliptic Functions and on Quaternions," *Philosophical Magazine* 26 (1845): 208-11.
9. J. T. Graves to WRH, Mar. 3, 1845, OS 330.
10. J. T. Graves, "On the Theory of Couples," *Philosophical Magazine* 26 (1845): 315-20; *Math. Papers,* 3:651.
11. J. T. Graves to WRH, Jan. 13, 1844, *Math. Papers,* 3:648-49.
12. *Math. Papers,* 3:651-55. Hamilton and Young exchanged many letters in the summer of 1847 (OS 390-97, 400, 403-5, 407, 414, 416, 418, 420-21). Young lost his position in 1849 when the new Queen's Colleges were founded. The government removed its support from the Belfast Institution, which was then dissolved, and Young became absolutely destitute. De Morgan considered this to be a great injustice. In 1852 Hamilton was sending money to Young, but as he told De Morgan: "It required management to get him to accept *anything,* and . . . I could not induce him to receive so much as I wished to give" (WRH to Augustus De Morgan, May 26, 1852, OS 597; Graves, 3:367).
13. Royal Irish Academy, *Transactions* 21, pt. 2 (1848): 311-38. See also Royal Irish Academy, *Proceedings* 3 (1845-47): 526-27, and *Proceedings* 4 (1847-50): 19-20.
14. Graves, 3:251. De Morgan to WRH, Oct. 11, 1844, OS 256; Graves, 2:472. The abstract is given completely in Graves, 3:251-52. De Morgan's paper on triplets is "On the Foundation of Algebra," pt. 4, completed Oct. 9 and read Oct. 28, 1844, Cambridge Philosophical Society, *Transactions* 8, pt. 3 (1844-49): 241-54.
15. WRH to J. T. Graves, Nov. 29, 1844, OS 276, and preface to the *Lectures on Quaternions* (*Math. Papers,* 3:129-31). Although the law of the modulus insures the possibility of division, it does not assure that the division is completely unambiguous. The ambiguity in quaternion division is caused by the fact that multiplication is not commutative.

We can describe the problem of division of quaternions as one of finding the quotient r of two quaternions p and q, thus $p/q = r$. But this can mean finding an r such that $p = qr$, or it can mean finding an r such that $p = rq$. Because the multiplication of quaternions is not commutative, the value of r may be different in the two cases. The ambiguity may be removed in the following way:

Consider the quaternion $q = a + b\mathbf{i} + c\mathbf{j} + d\mathbf{k}$.

We define a *conjugate* of q as $q' = a - b\mathbf{i} - c\mathbf{j} - d\mathbf{k}$.

Then the *norm* of q is $N(q) = qq' = q'q = a^2 + b^2 + c^2 + d^2$.

(The modulus, a measure of length, is merely the square root of the norm.)

We define the *inverse* of q as $q^{-1} = q'/N(q)$, $N(q) \neq 0$, and $qq^{-1} = q^{-1}q = 1$.

To find the r such that $p = qr$, we write $q^{-1}p = q^{-1}qr$ and $q^{-1}p = r$.

To find the r such that $p = rq$, we write $pq^{-1} = rqq^{-1}$ and $pq^{-1} = r$.

See Morris Kline, *Mathematical Thought from Ancient to Modern Times* (New York: Oxford University Press, 1972), p. 780.

16. WRH to De Morgan, Dec. 9, 1844, OS 283; *Graves,* 3:253-54.
17. See his abstract in Graves, 3:252.
18. De Morgan, "On the Foundation of Algebra," p. 242.

19. Ibid., p. 253, and his abstract in Graves, 3:252.
20. J. T. Graves to WRH, Nov., 1843, OS 226A. In a long letter to Hamilton of Nov. 4, 1844, John Graves reviewed the history of their correspondence over triplet systems (OS 259).
21. Graves, 2:463-72.
22. WRH to J. T. Graves, Nov. 20, 1844, OS 265, and J. T. Graves to WRH, Nov. 22, 1844, OS 268; Graves, 2:465-70. In the preface to his *Lectures* Hamilton mentioned MacCullagh's theorem, but stated that "it did not happen to supply me with any suggestion" (*Math. Papers*, 3:142n.).
23. J. T. Graves to R. P. Graves, Nov. 25, 1844, OS 271; Graves, 2:462-63.
24. WRH to R. P. Graves, Feb. 3, 1844, Graves, 2:431-32, and WRH to J. T. Graves, Nov. 27, 1844, OS 274.
25. Charles Graves to J. T. Graves, Nov. 21, 1844, OS 266, and Charles Graves to WRH, Nov. 21, 1844, OS 267.
26. J. T. Graves to WRH, Nov. 22, 1844, OS 268; Graves, 2:467-70.
27. WRH to De Morgan, Dec. 9, 1844, OS 283; Graves, 2:473-74.
28. WRH to R. P. Graves, Dec. 7, 1884, OS 282. The correspondence between Hamilton and the three Graves brothers about Charles's triplets is contained in OS 266-76, 280, 282, 287, 289, 291-99, 301-2, 312, 315-17, 327, and 332.
29. Charles Graves, "On Algebraic Triplets," Royal Irish Academy, *Proceedings* 3 (1847): 51-54, 57-64, 80-84, 105-8.
30. Quoted by R. P. Graves in a letter to J. T. Graves, Feb. 23, 1845, OS 327; Graves, 2:479, and American Philosophical Society MS, uncatalogued.
31. In 1878 F. Georg Frobenius proved that the only linear associative algebras that had real coefficients, a finite number of primary units, and a unit element for multiplication, and that obeyed the product law (that finite factors must have a finite product) are those of the real numbers, the complex numbers, and real quaternions. In 1898 Adolph Hurwitz proved that real numbers, complex numbers, real quaternions, and biquaternions (invented by William Kingdon Clifford in 1871) are the only linear associative algebras satisfying the product law (Kline, *Mathematical Thought*, p. 793).
32. WRH to De Morgan, Feb. 9, 1852, OS 543; Graves, 3:336, and WRH to Mortimer O'Sullivan, Aug. 4, 1853, OR 1219; Graves, 2:683.
33. *Athenaeum*, no. 1027 (July, 1847), p. 711.
34. WRH to R. P. Graves, July 2, 1847, Graves, 2:585.
35. WRH to J. R. Young, July 10, 1847, Graves, 2:579.
36. Graves 2:587.
37. $\alpha:\beta::\gamma:\delta$ where α is of unit length.

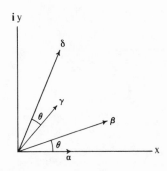

The proportion holds for length because $\alpha\delta = \beta\gamma$ and it holds for *angle* because the angular relation between α and β and between γ and δ is Θ.
38. *Math. Papers*, 3:149-52, 356.
39. Ibid., 358. Hamilton's argument is obscure. The best presentation is *Math. Papers*, 3:148-52.
40. WRH, "On Quaternions," communicated Nov. 11, 1844, *Math. Papers*, 3:359.

41. TCD notebook 50, p. 36, quoted from Pycior, "Role of Sir William Rowan Hamilton," pp. 187-188. Hamilton coined the word *grammarithm* from the Greek words for line and number (*Math. Papers*, 3:147, 367).

42. WRH to Humphrey Lloyd, July 20, 1848, OS 429.

43. WRH to R. P. Graves, Apr. 30, 1846, Graves, 2:521-22.

44. WRH to George Peacock, Oct. 13, 1846, Graves, 2:527-28. See also WRH to J. R. Young, June 19, 1847, Graves, 2:578-79.

45. WRH, "On Symbolical Geometry," *Cambridge and Dublin Mathematical Journal* 1 (1846): 45. Hamilton dates his introduction to the first paper Oct. 16, 1845.

46. "On Symbolical Geometry" 1 (1846): 46.

47. *Math. Papers*, 3:145.

48. The product of two perpendicular vectors might be a vector at some angle other than perpendicular to the plane determined by the factor vectors, but Hamilton shows in a somewhat more complex argument that this cannot be the case. The product must be perpendicular to both of the factors. (*Math. Papers*, 3:148).

49. Helena Pycior's thesis called my attention to the significance of biquaternions (Pycior, "Role of Sir William Rowan Hamilton," pp. 196-98. See also *Math. Papers*, 3:126). Hamilton's biquaternions are *not* the same thing as Clifford's biquaternions, mentioned in note 31 to this chapter. There was an unfortunate duplication of terms. Benjamin Peirce wrote in his famous "Linear Associative Algebra" (1870): "Hamilton's total exclusion of the imaginary of ordinary algebra from the calculus as well as from the interpretation of quaternions will not probably be accepted in the future development of this algebra. It evinces the resources of his genius that he was able to accomplish his investigations under these trammels. But like the restrictions of ancient geometry, they are inconsistent with the generalizations and broad philosophy of modern science. With the restoration of the ordinary imaginary, quaternions become Hamilton's biquaternions." (*American Journal of Mathematics* 4 (1881): 105n.

50. *Math. Papers*, 3:263, 377.

51. Peter Guthrie Tait, *An Elementary Treatise on Quaternions*, 3d ed. (Cambridge: at the University Press, 1890), p. vi.

52. Crowe, *History of Vector Analysis*, p. 28.

53. Crowe confirms this connection with a great deal of additional evidence (*History of Vector Analysis*, chap. 5).

54. London Mathematical Society, *Proceedings* 3 (1871): 259. Quoted from Crowe, *History of Vector Analysis*, p. 131.

55. Crowe, *History of Vector Analysis*, p. 130.

56. Let $\sigma = \mathbf{i}t + \mathbf{j}u + \mathbf{k}v$ be a vector, then

$$\nabla \sigma = \left(\mathbf{i}\frac{\partial}{\partial x} + \mathbf{j}\frac{\partial}{\partial y} + \mathbf{k}\frac{\partial}{\partial z} \right) \left(\mathbf{i}t + \mathbf{j}u + \mathbf{k}v \right),$$

and by quaternion multiplication

$$\nabla \sigma = -\left(\frac{\partial t}{\partial x} + \frac{\partial u}{\partial y} + \frac{\partial v}{\partial z} \right) + \mathbf{i}\left(\frac{\partial v}{\partial y} - \frac{\partial u}{\partial z} \right)$$
$$+ \mathbf{j}\left(\frac{\partial t}{\partial z} - \frac{\partial v}{\partial x} \right) + \mathbf{k}\left(\frac{\partial u}{\partial x} - \frac{\partial t}{\partial y} \right),$$

where "Convergence of σ" $= S \nabla \sigma = -\left(\frac{\partial t}{\partial x} + \frac{\partial u}{\partial y} + \frac{\partial v}{\partial z} \right)$

and "Curl of σ" $= V \nabla \sigma = \mathbf{i}\left(\frac{\partial v}{\partial y} - \frac{\partial u}{\partial z} \right) + \mathbf{j}\left(\frac{\partial t}{\partial z} - \frac{\partial v}{\partial x} \right) + \mathbf{k}\left(\frac{\partial u}{\partial x} - \frac{\partial t}{\partial y} \right).$

57. P. G. Tait to J. C. Maxwell, Dec. 13, 1867, and Maxwell to Tait, Oct. 9, 1872, both quoted from Crowe, *History of Vector Analysis*, pp. 132-33.

58. "Quaternions," published anonymously in *Nature* 9 (1873):137-38, quoted from Crowe, *History of Vector Analysis*, pp. 133-34.

59. Crowe, *History of Vector Analysis*, pp. 137-38.

60. J. W. Gibbs to Victor Schlegel, in draft, 1888, quoted from Crowe, *History of Vector Analysis*, p. 152.

61. Crowe, *History of Vector Analysis,* p. 163.
62. Peter Guthrie Tait, "On the Importance of Quaternions in Physics," *Philosophical Magazine,* 5th ser. 29 (Jan., 1890): 84-97.
63. Peter Guthrie Tait, "On the Intrinsic Nature of the Quaternion Method," Royal Society of Edinburgh, *Proceedings* 20 (1893-94): 277-78, quoted from Crowe, *History of Vector Analysis,* pp. 212-13.
64. Quoted from Crowe, *History of Vector Analysis,* p. 214.
65. Ibid., pp. 170-73.
66. Ibid., p. 218.
67. Sydney Hamilton to John O'Regan, July 21, 1875, ORSUP 13.
68. The quotation is from Heaviside. See Crowe, *History of Vector Analysis,* p. 173.
69. Peter Guthrie Tait, "The Role of Quaternions in the Algebra of Vectors," *Nature* 43 (1891): 608.
70. The use of the word *linear* always entailed the idea of linear independence, although it was expressed in different ways by different writers. In his very first letter to De Morgan, written before the discovery of quaternions, Hamilton wrote: "As to triplets, I must acknowledge that though I fancied myself at one time to be in possession of something worth publishing about them, I never could resolve the problem which you have justly signalized as the most important in this branch of (future) Algebra; to *assign* two sympols Ω and ω, such that the one symbolical equation $a + b\Omega + c\omega = a_1 + b_1\Omega + c_1\omega$ shall give the three equations $a = a_1, b = b_1, c = c_1$." (WRH to De Morgan, May 12, 1841, OS 196; Graves, 3:247). This is a demand for linear independence of Ω and ω. Benjamin Peirce gave the following definition: "An algebra in which every expression is reducible to the form of an algebraic sum of terms, each of which consists of a single *letter* with a quantitative coefficient, is called *a linear algebra*" (Peirce, "Linear Associative Algebra," p. 107). The meaning here is that the product of any two units gives one of the units, as, for example $\mathbf{i\,j} = \mathbf{k}$ in quaternions. The modern criteria for a linear algebra are that the algebra be associative $i(jk) = (ij)k$ and bilinear $i(cj + dk) = c(ij) + d(ik)$ and $(ci + dj)k = c(ik) + d(jk)$ (Garrett Birkhoff and Saunders MacLane, *A Survey of Modern Algebra* [New York: Macmillan Co., 1953], p. 239).
71. Hermann Günther Grassmann, *Die lineale Ausdehnungslehre, ein neuer Zweig der Mathematik, dargestellt und durch Anwendungen auf die übrigen Zweige der Mathematik, wie auch auf die Statik, Mechanik, die Lehre vom Magnetismus und die Krystallonomie erläutert* (Leipzig: O. Wigand, 1844).
72. Grassmann, *Ausdehnungslehre,* quoted from Crowe, *History of Vector Analysis,* p. 65.
73. Ibid., p. 64.
74. WRH to J. T. Graves, Sept. 30, 1856, OS 862; Graves, 3:70. Hamilton discusses the *Ausdehnungslehre* in the preface to his *Lectures* (*Math. Papers,* 3:153-54).
75. Crowe, *History of Vector Analysis,* p. 153.
76. J. W. Gibbs to Thomas Craig, quoted from Crowe, *History of Vector Analysis,* p. 161.
77. Peirce's *Analytical Mechanics* does not *use* quaternions. It merely states: "Much must soon become antiquated and obsolete as the science advances, and especially when we shall have viewed the full benefit of the remarkable machinery of HAMILTON's *Quaternions*" (*A System of Analytic Mechanics* [Boston: Little, Brown and Co., 1855], p. 476). In 1858 Hamilton wrote to Aubrey DeVere: "A very handsome American book, Peirce's *Analytic Mechanics,* which has just reached me, combines (as I find) those older and newer researches of mine, which I had been in some degree *contrasting,* in our recent conversations here. You will be pleased to see the figure which your old friend makes in that book" (Apr. 21, 1858, TCD notebook 140.3, p. 137).
78. Peirce, "Linear Associative Algebra," pp. 216-17.
79. J. W. Gibbs, "Multiple Algebra," *Proceedings of the American Association for the Advancement of Science* 35 (1886): 66.
80. James Byrnie Shaw, in his *Synopsis of Linear Associative Algebra: A Report on Its Natural Development and Results Reached Up to the Present Time,* Carnegie Institution Publication no. 78 (Washington, D.C.: Carnegie Institution, 1907), said that there had always been two philosophical approaches to what he called "complex algebra." One regarded the number in such an algebra as a mere "multiplex" of

units; the other considered it to be a single entity. Hamilton was definitely in the second camp.

81. Gibbs, "Multiple Algebra," pp. 37-66. Arthur Cayley, "On Multiple Algebra," *Quarterly Journal of Pure and Applied Mathematics* 12 (1887): 270-308; idem, *Collected Mathematical Papers,* 14 vols. (Cambridge: At the University Press, 1897), 12:459-89.

82. Cayley, "On Multiple Algebra," p. 460.

83. Gibbs, "Multiple Algebra," p. 41. While Gibbs gave the priority for the idea of matrices to Grassmann, Henry Taber, in an important treatise on matrices in 1889, gave the honor of their discovery to Hamilton! "Before Cayley's memoir appeared, Hamilton had investigated the theory of such a symbol of operation as would convert these vectors into three linear functions of those vectors, which he called a linear vector operator. Such an operator is essentially identical with a matrix as defined by Cayley. . . . Hamilton must be regarded as the originator of the theory of matrices, as he was the first to show that the symbol of a linear transformation might be made the subject-matter of a calculus" (Henry Taber, "On the Theory of Matrices," *American Journal of Mathematics* 11 (1889): 337). Taber may have been recalling the Cayley-Hamilton theorem, which states that every square matrix satisfies its characteristic equation. Hamilton gave this theorem for quaternions in his *Lectures* (1853). Cayley gave it in his "A Memoir on the Theory of Matrices" (1858). The background is given in *Math. Papers,* 3:xviii.

84. See chapter 22.

85. Gibbs, "Multiple Algebra," pp. 42-43.

86. Ibid., p. 52. Gibbs goes on to say that describing linear transformations by determinants instead of matrices is "very like the play of Hamlet with Hamlet's part left out" (p. 44).

87. E. T. Whittaker, "The Hamiltonian Revival," *Mathematical Gazette* 24 (1940): 158.

88. E. T. Bell, *Men of Mathematics* (New York: Simon and Shuster, 1937).

89. Tait, *Elementary Treatise on Quaternions,* p. viii.

Chapter 24

1. The story is well told in Morton Grosser, *The Discovery of Neptune* (Cambridge: Harvard University Press, 1962).

2. Graves, 2:551-52.

3. WRH, "The Hodograph, Or a New Method of Expressing in Symbolical Language the Newtonian Law of Attraction," *Math. Papers,* 2:287-92; Graves, 2:542-46.

4. Hamilton wrote three papers on the hodograph, "The Hodograph, Or a New Method of Expressing in Symbolical Language the Newtonian Law of Attraction," *Math. Papers,* 2:287-92; "On a Theorem of Hodographic Isochronism," *Math. Papers,* 2:293; and "On a Theorem of Anthodographic Isochronism," *Math. Papers,* 2:293-94, the last two being statements of theorems. The elementary properties of the hodograph can be demonstrated by simple geometric arguments, but Hamilton gave no diagrams, which makes his description hard to follow. The diagrams are provided by J. M. Child in "Hamilton's Hodograph," *The Monist* 25 (1915):615-24.

5. WRH to G. B. Airy, Jan. 2, 1847, OS 377, Whewell to WRH, Feb. 8, 1847, Graves, 2:561, and WRH to John Herschel, Jan. 6, 1847, OS 378.

6. William Thomson to WRH, Apr. 17, 1847, OR 1036; May 1, 1847, OR 1037; Oct. 19, 1847, OR 1048; Oct. 26, 1847, OR 1050; Nov. 7, 1847, OR 1054; Nov. 20, 1847, OR 1055; Dec. 6, 1847, OR 1057; Dec. 2, 1848, OR 1087.

7. William Thomson and Peter Guthrie Tait, *Treatise on Natural Philosophy,* retitled *Principles of Mechanics and Dynamics,* 2 vols. (1879; New York: Dover Publications, 1962), 1:26-28. Tait included several theorems on the hodograph in Peter Guthrie Tait and William John Steele, *A Treatise on Dynamics of a Particle with Numerous Examples,* 5th ed. (London: Macmillan and Co., 1882), pp. 298-309, 316-19, and in Tait, "Note on the Hodograph," Royal Society of Edinburgh, *Proceedings* (Dec. 16,

1867), in Tait's *Scientific Papers*, 2 vols. (Cambridge: At the University Press, 1898–1900), 1:78–82.

8. John Herschel to WRH, Apr. 20, 1847, OS 387; Graves, 2:565–66. Hamilton stated his Law of Hodographic Isochronism as follows: "If two circular hodographs, having a common chord, which passes through or tends towards a common centre of force, be cut perpendicularly by a third circle, the times of hodographically describing the intercepted arcs will be equal." *(Math. Papers,* 2:293). Lambert's theorem states: "In elliptic orbits of equal major axes, described in the same periodic time, under the action of a central force to a focus, the times of describing any sectors are the same, if the chords of the sectors and also the sum of the bounding radii vectores are the same" (Child, "Hamilton's Hodograph," p. 621). Hamilton gave no proof of this theorem, but one is included in the article by Child and another is given in *Math. Papers,* 2:630.

9. *Lectures on Quaternions* (Dublin: Hodges and Smith, 1853), p. 614.

10. Emil Jan Konopinski, *Classical Descriptions of Motion* (San Francisco: W. H. Freeman, 1969), p. 101. Problem 4.11 states: "Find the 'Kepler orbit in momentum space' as in a diagram having p_x, p_y as orthogonal axes. Show that the 'orbit' is a circle centered on a point with $|p| = Lc/b^2$ and having radius La/b^2. [This curious result was brought to the author's attention in a lecture by Dr. M. Gutzwiller.]" The bracketed statement is by Konopinski. L is the angular momentum, and a and b are the semimajor and semiminor axes of the elliptical orbit, respectively. Another "rediscovery" of the hodograph, this time attributed to Hamilton, appears in a recent series of articles in the *American Journal of Physics.* See Herbert Goldstein, "Prehistory of the 'Runge-Lenz' Vector," *American Journal of Physics.* 43 (1975): 737–38, and idem, "More on the Prehistory of the Laplace or Runge-Lenz Vector," *American Journal of Physics* 44 (1976):1123–24; and J. Stickforth, "The Classical Kepler Problem in Momentum Space," *American Journal of Physics* 46 (1978):74–75.

11. On the icosian calculus and graph theory see Norman L. Biggs, E. Keith Lloyd, and Robin J. Wilson, *Graph Theory, 1736–1936* (Oxford: At the Clarendon Press, 1976); J. Riverdale Colthurst, "The Icosian Calculus," Royal Irish Academy, *Proceedings* 50, sect. A (1945):112–21; Ross Honsberger, *Mathematical Gems* (2 vols. [Washington, D.C.: Mathematical Association of America, 1973–76]); O. Ore, "Note on Hamilton Circuits," *American Mathematical Monthly* 67 (1960):55.

On the importance of the Icosian calculus for group theory see G. A. Miller, "Note on William R. Hamilton's Place in the History of Abstract Group Theory," *Bibliotheca Mathematica,* ser. 3, 11–12 (1910–12):314–15, and Stella Mills, "British Group Theory, 1850–1890" (Ph.D. diss., University of Birmingham, 1976). Hamilton's published papers on the Icosian calculus and a selection of his unpublished manuscripts are in *Math. Papers,* vol. 3.

12. Biggs, Lloyd, Wilson, *Graph Theory,* p. 9.

13. WRH to John T. Graves, July 7, 1856, OS 851; Aug. 5, 1856, OS 853; and Aug. 8, 1856, OS 854.

14. Graves, 3:51.

15. WRH to Humphrey Lloyd, Sept. 9, 1856, Graves 3:67. See Alison R. Dorling, "The Graves Mathematical Collection in University College London," *Annals of Science* 33 (1976):307–10.

16. The full title is *Der Situationskalkul. Versuch einer arithmetischen Darstellung der niederen und höheren Geometrie auf Grund einer abstracten Auffassung der räumlichen Grössen, Formen und Bewegungen, etc.* (Braunschweig, 1851).

17. WRH to son William, Aug. 21, 1856, OS 855.

18. J. T. Graves to WRH, Oct. 13, 1856, OS 865. See also J. T. Graves to WRH, Sept. 23, 1856, OS 860.

19. Hamilton wrote about his new calculus:

[It] was suggested to me by a recent consideration of some of the polyhedra to which consideration you had encouraged me; and as you mentioned, without explanation, in the Mathematical Section lately that the subject had suggested to you "a new imaginary" words which I had forgotten until yesterday, and which even now I may not accurately remember: —if it shall turn out that I have been merely reproducing what you did, I do not want to claim any part of the merit. You, on the other hand, may choose to inform me, at least generally, what *your* view was: for you will find, in the enclosed memorandum, a sketch of *mine* sufficient to identify it. I thought that you might like to have the option of reading the memorandum, or not reading it, (and you are most free to choose) before you write to me. . . . If you have only found *one* new imaginary, and not (as I have done) a system of three non-commutative symbols ($\iota\kappa\lambda$) entirely distinct from *ijk*, we are so little likely to have clashed that you may as well open my letter (or memorandum at once). It is only on the *chance* of our results being *very like each other* that I thought you might *choose* to answer before reading (WRH to J. T. Graves, Oct. 7, 1856, OS 863; Graves, 3:71).

20. J. T. Graves to WRH, Oct. 27, 1856, OS 868.
21. WRH to J. T. Graves, Nov. 1, 1856, OS 870; Graves, 3:73-74.
22. Biggs, Lloyd, Wilson, *Graph Theory*, pp. 1-12.
23. Graves 2:635. Kirkman wrote one of the early papers on quaternions: "On Pluquaternions and Homoid Products of Sums of *n* Squares," *Philosophical Magazine* 33 (1848):447-59, 494-509.
24. Thomas P. Kirkman, *First Mnemonical Lessons to Geometry, Algebra and Trigonometry* (London: J. Weale, 1852). Kirkman asked to have this book returned in 1850 (T. P. Kirkman to WRH, Dec. 1, 1850, OR 1138), which means that Hamilton must have had the book in manuscript. See also WRH to Helen Hamilton, Aug. [1861], OR 1934.
25. Biggs, Lloyd, Wilson, *Graph Theory*, pp. 28-31. Thomas P. Kirkman, "On the Representation of Polyedra," Royal Society, *Philosophical Transactions* 146 (1856): 413-18.
26. Thomas P. Kirkman, "On the Partitions of the *r*-Pyramid, Being the First Class of *r*-Gonous *x*-edra," Royal Society, *Philosophical Transactions* (1858), pp. 145-62.
27. Graves, 3:135-36.
28. WRH to T. P. Kirkman, May 23, 1862, OR 1568. Kirkman insisted on using the spelling *polyedra* instead of *polyhedra*. Hamilton complied with this rather strange demand in his correspondence with Kirkman.
29. Royal Society, *Philosophical Transactions* (1858), p. 160.
30. Peter Guthrie Tait, "Listing's *Topologie*," *Philosophical Magazine*, 5th ser. 17 (1884): 30-46.
31. Thomas P. Kirkman, "Problem 6610 and Solution," *Educational Times Reprints* 35 (1881):112-16.
32. WRH to J. T. Graves, Oct. 17, 1856, OS 866; Graves, 3:72.
33. WRH to J. T. Graves, Feb. 26, 1857, OS 899.
34. WRH to Helen Hamilton, July 25, 1858, OR 1367, and WRH to J. T. Graves, Mar. 25-26, 1859, OS 1060.
35. J. T. Graves to WRH, Jan. 30, 1859, OS 1039, WRH to J. T. Graves, Apr. 1, 1859, OS 1069, and WRH to J. T. Graves, Apr. 4, 1859, OS 1073.
36. Graves describes his version of the game as follows: "The men were cylinders, numbered from 1 to 20, each cylinder being numbered on both of its flat sides. They fitted into holes in the Board, not numbered nor lettered, as the men are numbered. . . . The men are held in playing in a small wooden frame kept in the left hand" (J. T. Graves to WRH, Jan. 30, 1859, OS 1039).
37. WRH to John Jaques, Jan. 27, 1859, TCD notebook 151, p. 7; Jaques to WRH, Feb. 1, 1859, OS 1040; WRH to Jaques, Feb. 3, 1859, TCD notebook 151, p. 9; WRH to J. T. Graves, Feb. 5, 1859, TCD notebook 151, p. 8; Jaques to WRH, Feb. 5, 1859, OS 1043; Jaques to WRH, Feb. 14, 1859, OS 1045; Jaques to WRH, Feb. 21, 1859, OS 1046; WRH to Jaques, Mar. 5, 1859, TCD notebook 151, p. 53; and WRH to J. T. Graves, Mar. 25-26, 1859, OS 1060.
38. WRH to J. T. Graves, Mar. 25-26, 1859, OS 1060.
39. WRH to Jaques, Apr. 20, 1859, TCD notebook 151, p. 71.
40. WRH to Jaques, Mar. 5, 1859, TCD notebook 151, p. 53.
41. Graves, 3:55.

42. Edouard Lucas, *Récréations mathématiques,* 4 vols. (Paris: Gauthier-Villars, 1882-94), 2:201.
43. WRH to J. T. Graves, Apr. 1, 1859, OS 1069.
44. Augustus De Morgan to WRH, Oct. 23, 1852, Graves, 3:422-23, and Biggs, Lloyd, Wilson, pp. 90-91. The sphynx killed herself in disgust when Oedipus gave the answer to her puzzle.
45. In 1976 a computer-aided proof indicated that the four-color conjecture is true, but there is still some question about the validity of the proof. Kenneth Appel and Wolfgang Haken, "The Solution of the Four-Color-Map Problem" *Scientific American* 237, no. 4 (Oct., 1977):108-21. See also Kenneth O. May, "The Origin of the Four-Color Conjecture," *Isis* 56 (1965):346-48; and John Wilson, "New Light on the Origin of the Four-Color Conjecture," *Historia Mathematica* 3 (1976):329-30.
46. WRH to De Morgan, Oct. 26, 1852, OS 669; Graves, 3:423.

Chapter 25

1. WRH to Robert Disney, Nov. 6, 1853, TCD notebook 103.5, p. 183, and WRH to Thomas Disney, Nov. 8, 1853, TCD notebook 103.5, p. 187.
2. WRH to Aubrey De Vere, Aug. 7, 1855, NLD 905, p. 85.
3. WRH to De Vere, Aug. 7, 1855, NLD 905, p. 86.
4. WRH to J. P. Nichol, Nov. 26, 1855, TCD notebook 129.5, pp. 123ff.
5. WRH to James W. Barlow, Aug. 30, 1848, Graves, 2:620. Barlow's letters to WRH are Sept. 7 and Sept. 13, 1848, ORSUP 9.
6. WRH to De Vere, Sept. 9, 1855, NLD 5765, fol. 555.
7. WRH to De Vere, Sept. 9, 1855, NLD 5765, fol. 555.
8. In recounting this story Hamilton said Catherine's last *rational* letter was written on Oct. 5, 1848, when she had "that terrible attack of mental disease" (WRH to Mrs. [Thomas] Disney, Nov. 10, 1853, TCD notebook 103.5, p. 203). Elsewhere he recalled that he received the letter while still at Parsonstown (WRH to De Vere, Sept. 9, 1855, NLD 5765, fol. 555). But in a letter written Sept. 11, 1848, he said this was his *last day* at Parsonstown (WRH to De Vere, Sept. 11, 1848, NLD 5765, fol. 499; Graves, 2:629). Hamilton must have confused the two months in his remembrances.
9. This notebook was almost certainly destroyed by Robert Graves. Graves publishes brief extracts of a correspondence of July, 1848, between Hamilton and an "old friend." These could only be parts of his letters to Catherine. Of course they are edited to remove any personal references and they dwell largely on religious matters. If the rest of the correspondence was like these extracts, it was certainly harmless (see Graves, 2:610-12).
10. WRH to T. Disney, Apr. 17, 1850, TCD notebook 114.5, pp. 161ff.
11. WRH to Barlow, Jan. 15, 1850, TCD notebook 114.5, p. 17, Mar. 1, 1850, OS 470, Mar. 7, 1850, TCD notebook 114.5, p. 106, Mar. 7, 1850, TCD notebook 114.5, p. 108, Apr. 22, 1850, TCD notebook 114.5, p. 168, Apr. 13-May 11, 1850, OS 474, and May 13, 1850, OS 479.
12. Samuel Sullivan to WRH, June 1, 1850, ORSUP 9 (congratulating Hamilton on Barlow's success).
13. WRH to Nichol, October 10, 1855, NLD 905, pp. 151ff, and WRH to Helen Hamilton, Aug. 5, 1850, OR 1125. See also WRH to sister Eliza, Aug. 19, 1850, OR 1129, in which Hamilton writes: "James Barlow (son of Mrs. Wm Barlow) who very recently obtained a Fellowship in Dublin, chiefly (as it is supposed) through his knowledge of the quaternions (acquired through my assistance), was at Edinburgh too, and we lodged in the same house, and had one common sitting room, so that we saw a good deal of each other."
14. WRH to Mrs. T. Disney, Sept. 12, 1854, TCD notebook 103.5, pp. 291-92.
15. WRH to De Vere, July 19, 1855, NLD 5765, fol. 491, WRH to Lady Campbell, Oct. 6, 1854, TCD notebook 123.9, p. 205; Graves 3:14-15, WRH to Countess Dowager of Dunraven, May 19, 1855, OS 828; Graves, 3:26-28.

16. WRH to De Vere, Oct. 18, 1855, NLD 5765, fol. 597; Graves, 2:648. In October, 1850, Hamilton learned to his dismay that the Disney estate at Rock Lodge was listed among the notices of encumbered estates and was to be sold. His old friend Thomas Disney was obviously in financial difficulty. At first there was little Hamilton could do to help, but two years later the Disneys moved to Finglass near the observatory and Hamilton saw much more of them. In 1854 Hamilton used all the influence he could muster to try to find a job for Thomas (WRH to his Cousin Bessy Hamilton, Oct. 24, 1850, TCD notebook 114.8, p. 19, T. Disney to WRH, Feb. 9, 1852, OR 1178; the WRH letterbook, OR 2286, has many letters on this subject between the dates of Nov. 23, 1854, and May 11, 1856).
17. WRH to R. P. Graves, June 16, 1851, OR 1158; Graves, 2:669.
18. WRH to Lady Wilde, May 28, 1855, NLD 905, p. 5.
19. WRH to Samuel Talbot Hassell, Dec. 9, 1853, TCD notebook 123.5, p. 159.
20. WRH to De Vere, Aug. 7, 1855, NLD 905, pp. 88–89.
21. WRH to R. Disney, Oct. 25, 1853, TCD notebook 103.5, pp. 177–78.
22. R. Disney to WRH, Nov., 1853, TCD notebook 103.5, p. 180.
23. Memorandum added to a copy of a letter from WRH to R. Disney, TCD notebook 103.5, pp. 181–82.
24. WRH to Mrs. T. Disney ("Dora"), Nov. 10, 1853, TCD notebook 103.5, pp. 203ff, and July 6, 1861, TCD notebook 123.5, pp. 21ff.
25. WRH to T. Disney, Nov. 11, 1853, TCD notebook 103.5, pp. 206–7, and to the same, Nov. 13, 1853, and Nov. 14, 1853, TCD notebook 103.5, p. 194.
26. WRH to Louisa Reid, July 6, 1861, TCD notebook 123.5, pp. 21ff.
27. WRH to T. Disney, May 19, 1854, TCD notebook 103.5, pp. 280–81.
28. WRH to Lady Wilde, June 30, 1861, TCD notebook 123.5, pp. 15–16.
29. WRH to Augustus De Morgan, Dec. 11, 1853, TCD notebook 103.5, p. 199, and also Nov. 8, 1853, OS 761, Nov. 10, 1853, OS 762, and Dec. 12, 1853, OS 763.
30. WRH to Hassell, Nov. 29, 1853, TCD notebook 123.5, p. 151. Copies of other letters from this correspondence are in the same notebook.
31. WRH to Hassell, Dec. 30, 1853, TCD notebook 123.5, p. 167.
32. WRH to T. Disney, Dec. 24, 1853, TCD notebook 103.5, pp. 233–34. Copies of the extensive correspondence with the Disneys are in TCD notebook 103.5.
33. The letters to Lady Campbell are in TCD notebook 123.5.
34. WRH to T. Disney, Jan. 20, 1854, TCD notebook 103.5, p. 247.
35. WRH to Nichol, Nov. 26, 1855, TCD notebook 129.5, pp. 123ff.
36. WRH to Nichol, Nov. 14–15, 1855, TCD notebook 129.5, pp. 106ff.
37. WRH to T. Disney, Feb. 3, 1854, TCD notebook 103.5, p. 250ff.
38. WRH to Nichol, Nov. 13, 1855, TCD notebook 129.5, pp. 54–55. Copies of the rest of the Nichol correspondence are in this same volume.
39. WRH to Miss Nichol, July 23, 1856, TCD notebook 129.5, pp. 135ff.
40. WRH to De Vere, Aug. 7, 1855, NLD 905, pp. 78–97.
41. WRH to De Morgan, May 4, 1855, OS 825; Graves 3:496–98.
42. Harry Furniss, *Some Victorian Women,* quoted in Terence de Vere White, *The Parents of Oscar Wilde: Sir William and Lady Wilde* (London: Hodder and Stoughton, 1967), p. 127.
43. "The Hour of Destiny," which appeared in the issue before the one containing "Jacta Alea Est," contained these words: "Ireland! Ireland! It is no petty insurrection—no local quarrel—no party triumph that summons you to the field. The destinies of the world—the advancement of the human race—depends now on your courage and success; for, if you have *courage* success must follow. ... It is a death struggle between the oppressor and the slave—between the murderer and his victim. Strike!—Strike! Another instant, and his foot will be upon your neck—his dagger at your heart" (Quoted from White, *Parents of Oscar Wilde,* p. 104). The rebellion, when it came in August under the leadership of William Smith O'Brien, was nothing more than a skirmish with the police.
44. Quoted from White, *Parents of Oscar Wilde,* p. 104.
45. WRH to De Vere, Aug. 7, 1855, quoted from White, *Parents of Oscar Wilde,* p. 129.
46. WRH to Lady Wilde, June 30, 1855, NLD 905, pp. 27ff.

47. WRH to Lady Wilde, June 29, 1855, NLD 905, pp. 25ff.
48. WRH to Mrs. T. Disney, June 14, 1855, TCD notebook 123.5, pp. 139ff.
49. WRH to Lady Wilde, June 30, 1861, TCD notebook 123.5, pp. 15–16.
50. WRH to Reid, Aug. 16, 1861, TCD notebook 123.5, pp. 215ff.
51. WRH to Reid, Aug. 26, 1861, TCD notebook 123.5, pp. 220ff.

Chapter 26

1. WRH to Lady Wilde, Dec. 29, 1855, NLD 905, pp. 44–45, and WRH to Penn Wood, Nov. 23, 1857, TCD notebook 140.5, p. 281. Hamilton was particularly pleased that his book was selling in America. WRH to Augustus De Morgan, Jan. 7, 1858, OS 968. The full citation is *Lectures on Quaternions* (Dublin: Hodges and Smith, 1853).
2. Graves, 2:478 and 3:6.
3. J.F.W. Herschel to WRH, Nov. 18, 1859, OR 1460; OS 1129; and Graves, 3:121. Herschel's letters of Jan. 8 and Mar. 1, 1860 (OS 1138, 1144), were further efforts to keep Hamilton on the right track.
4. De Morgan to WRH, Mar. 27, 1859, OS 1062; Graves, 3:557.
5. De Morgan to WRH, June 26, 1864, OS 1308; Graves, 3:609, and WRH to De Morgan, June 29, 1864, OS 1310. From the beginning De Morgan warned that "if you do not take care, you will begin to spin too much." De Morgan to WRH, Sept. 7, 1859, OS 1106.
6. WRH to Dr. Andrew Searle Hart, July 26, 1860, ORSUP 12.
7. WRH to De Morgan, Sept. 4, 1859, OS 1104. The final title was *Elements of Quaternions* (London: Longmans, Green & Co., 1866).
8. Dr. J. N. Andrews to WRH, Aug. 11, 1858, OS 1003.
9. Peter Guthrie Tait to WRH, Aug. 19, 1858, OS 1007. These were, of course, the aspects of the quaternions that would later become the core of vector analysis.
10. WRH to Tait, Dec., 1858, Graves, 3:107.
11. Tait to WRH, Aug. 20, 1858, OS 1009.
12. Tait to WRH, Oct. 20, 1860, OS 1153.
13. WRH to De Morgan, Nov. 14, 1860, OS 1158.
14. WRH to Aubrey De Vere, Nov. 30, 1860, NLD 5765, fol. 823.
15. Tait to WRH, Dec. 4, 1860, OS 1160, and Dec. 11, 1860, OS 1161.
16. WRH to De Morgan, Aug. 7, 1861, OS 1190.
17. Hamilton did, however, think that he might finish *writing* the book in 1861, for he wrote to De Morgan: "I have signed 320 octavo pages of my Elements. . . . I do trust that I am in sight of land, and that the *writing*, at least, of the Book will have been quite finished by Christmas—or before New Year's Day. . . . No attempt whatever has been made to adopt a *popular style*, such as was aimed at, near the commencement of the Lectures: but I think that to a certain class of students, the new work will be acceptable and useful" (WRH to De Morgan, Nov. 9, 1861, OS 1200; Graves, 3:568).
18. Tait to WRH, Dec. 16, 1860, OS 1164.
19. Hamilton wrote a "prodigious letter" of 216 folio pages on anharmonic coordinates to Andrew Hart, Senior Fellow of Trinity College, followed by a postscript of 64 pages (Graves, 3:123). Anharmonic coordinates is one of the subjects to be included in the projected fourth volume of Hamilton's *Mathematical Papers*.
20. Tait to WRH, Jan. 1, 1861, OS 1167.
21. Tait to WRH, Sept. 23, 1862, OS 1253.
22. WRH to Tait, Aug. 28, 1863, OR 1586; Graves, 3:150.
23. Thomson later wrote: "We [Thomson and Tait] have had a thirty-eight year's war over quaternions. He had been captivated by the originality and extraordinary beauty of Hamilton's genius in this respect, and had accepted, I believe, definitely from Hamilton to take charge of quaternions after his death, which he has most loyally executed. Times without number I offered to let quaternions into Thomson and Tait, if

he could only show that in any case our work would be helped by their use. You will see that from beginning to end they were never introduced" (Kelvin to George Chrystal, 1901, quoted from Michael J. Crowe, *A History of Vector Analysis: The Evolution of the Idea of a Vectorial System* [Notre Dame, Ind.: University of Notre Dame Press, 1967], p. 119).

24. Tait to WRH, June 25, 1863, OS 1277.
25. WRH to Tait, June 27, 1863, ORSUP 199.
26. Tait to WRH, Aug. 13, 1863, OS 1283.
27. Tait to WRH, Sept. 5, 1863, OS 1284.
28. Tait to WRH, Dec. 18, 1863, OS 1299.
29. Graves, 3:133-34.
30. Peter Guthrie Tait, "On the Intrinsic Nature of the Quaternion," Royal Society of Edinburgh, *Proceedings* (July 2, 1894), in Tait's *Scientific Papers*, 2 vols. (Cambridge: At the University Press, 1898-1900), 2:395.
31. Crowe, *History of Vector Analysis*, p. 41.
32. Tait, "Intrinsic Nature of the Quaternion," p. 396.
33. TCD notebook 168.5.
34. Ibid.
35. WRH to De Morgan, Apr. 29, 1854, OS 790; Graves, 2:478 and 3:6.
36. TCD notebook 168.5.
37. WRH to Hart, May 27, 1862, Graves, 3:140.
38. Graves, 3:140.
39. WRH to John Herschel, Apr. 26, 1847, OS 388; Graves, 2:566-67.
40. WRH to Herschel, Nov. 23, 1846, OS 375; Graves, 2:533.
41. WRH to De Morgan, May 8, 1852, Graves, 3:361.
42. WRH to De Morgan, Jan., 1852, Graves, 3:328, and Aug. 3, 1852, Graves, 3:398-99.
43. WRH to De Morgan, Dec., 1853, Graves, 3:468.
44. The riot is described in *The Book of Trinity College Dublin, 1591-1891* (Dublin: Hodges, Figgis, & Co., 1892), pp. 101-3.
45. WRH to De Morgan, Mar. 25, 1858, OS 980. Arch claimed to have been an innocent bystander.
46. WRH to Lady Campbell, June 15, 1854, Graves, 3:19.
47. WRH to De Morgan, Dec., 1853, OS 770; Graves, 3:469, and De Morgan to WRH, Jan. 10, 1854, OS 772; Graves, 3:470.
48. Hamilton stayed at a hotel the entire time. WRH to sister Sydney, Sept. 12, 1858, OR 1385.
49. Graves, 3:117-20.
50. Ibid., pp. 13-16.
51. Ibid., p. 102.
52. Ibid., p. 163.
53. WRH to George Willoughby Hemans, May 30, Aug. 29, and Sept. 6, 1856, OR 2285.
54. WRH to son William, Nov. 5, 1856, OR 2285.
55. WRH to De Morgan, June 17, 1857, OS 917.
56. Sydney Hamilton to WRH, May 10, 1862, OR 1568.
57. Edward Cullen, "French Survey of the Isthmus of Darien," *Athenaeum*, no. 1779 (Nov. 30, 1861), p. 727.
58. Sydney Hamilton to WRH, May 10, 1862, OR 1568.
59. Sydney Hamilton to WRH, Aug. 4, 1862, OR 1557.
60. Sydney Hamilton to WRH, Aug. 9, 1862, OR 1558.
61. T. Romney Robinson to WRH, May 20, 1862, OR 1568.
62. WRH to sister Sydney, Aug. 5, 1862, TCD notebook 140.5, pp. 13ff.
63. David I. Folkman, Jr., *The Nicaragua Route* (Salt Lake City: University of Utah Press, 1972), p. 14. Napoleon's pamphlet was *Canal of Nicaragua, Or a Project to Connect the Atlantic and Pacific Oceans by Means of a Canal* (London: Mills and Sons, 1846).
64. Gerstle Mack, *The Land Divided: A History of the Panama Canal and Other Isthmian Canal Projects* (New York: Alfred Knopf, 1944), p. 249.
65. Ibid., pp. 249-59.

66. Ibid., pp. 149-60.
67. Folkman, *Nicaragua Route.*
68. Sydney Hamilton to WRH, 1862, OR 1567.
69. Sydney Hamilton to Lady Helen Hamilton, Sept. 6, 1862, OR 1560.
70. Sydney had written to Hamilton that she had spoken to the consul from New Granada to Great Britain, who had promised a free grant of twenty acres to every immigrant. She also remarked on the fact that the country around Bogota was extremely healthy because of the high altitude (Sydney Hamilton to WRH, no date, OS 1375).
71. WRH to Paul Askin, Dec. 31, 1862, OR 2289.
72. William O. Scroggs, *Filibusters and Financiers: The Story of William Walker and His Associates* (New York: Macmillan and Company, 1916).
73. Encyclopedia Britannica, 9th ed. (1880), gives the population of the Greytown as 955 in 1863.
74. The region around Greytown was entirely unsuitable for agriculture, and almost all foodstuffs had to be imported.
75. William Edwin Hamilton to WRH, June 1863, ORSUP 199.
76. W. E. Hamilton to WRH, June 17, 1863, OR 1578.
77. WRH to son William, July 16, 1863, ORSUP 200.
78. WRH to son William, Aug. 15, 1863, ORSUP 200.
79. W. E. Hamilton to Helen E. A. Hamilton, June 10 [1867], OR 1523.
80. WRH to son William, Jan. 29, 1864, ORSUP 201. According to his own account William Edwin stayed with the Keatings for five months. Keating billed WRH at the rate of a dollar a day, and William Edwin calculated that to be about twice what it should be (W. E. Hamilton to WRH, June 22, 1864, OS 1305).
81. WRH to son Archibald, Apr. 14, 1864, OR 1607; ORSUP 201.
82. W. E. Hamilton to WRH, July 5, 1864, OR 1621.
83. WRH, June 8, 1864, ORSUP 8.
84. WRH to Robinson, Dec. 5, 1864, OR 1631; Graves, 3:167.
85. WRH to John O'Regan, Feb. 19, 1861, OR 1503.
86. WRH to De Morgan, Mar. 22, 1857, OS 905; Graves, 3:513.
87. Charles Graves to WRH, Apr. 19, 1861, OR 1519.
88. WRH to C. Graves, Apr. 21, 1864, OR 1608.
89. Dean Ogle Moore to J. O'Regan, May 17, 1865, ORSUP 13.
90. Moore to J. O'Regan, Tuesday in Easter Week, 1865, ORSUP 13.
91. Moore to J. O'Regan, July 7, 1865, ORSUP 13.
92. Graves, 3:203.
93. W. E. Hamilton to J. O'Regan, June 13, 1865, ORSUP 193.
94. W. E. Hamilton to J. O'Regan, June 13, 1865, ORSUP 193.
95. W. E. Hamilton to J. O'Regan, June 16, 1865, ORSUP 193.
96. Graves, 3:202n.
97. WRH income tax statement signed July 29, 1865, OR 1665.
98. H.E.A. Hamilton to her mother, Lady Hamilton, Aug. 18, 1865, ORSUP 12.
99. Graves, 3:209-11.
100. WRH memo, ORSUP 194.
101. H.E.A. Hamilton, Sept. 4, 1865, OR 1679.
102. C. Graves to Thomas Larcom, Sept. 15, 1865, RIA Ms. 3E 11.
103. Dr. G. Wyse to C. Graves, Oct. 27, 1884, ORSUP 197.
104. James Disney to Robert P. Graves, Mar. 15, 1883, OR 1793.
105. Charles Graves stated in his eulogy: "His diligence of late was even excessive— interfering with his sleep, his meals, his exercise, his social enjoyments. It was, I believe, fatally injurious to his health" (Graves, 3:224). Also, G. H. Porter (with whom Arch stayed) to J. O'Regan (Oct. 1, no year, OR 2009): "After all, did not Sir W. kill himself by over mental work ... poor Lady H.—suffered for taking over in a position for which she was not fitted."
106. Archibald Henry Hamilton to his mother, Lady Hamilton, Jan. 26, 1866, OR 1701.
107. W. E. Hamilton to H.E.A. Hamilton, June 15, 1869, ORSUP lot 13.
108. R. P. Graves to W. E. Hamilton, Dec. 12, 1865, OR 1694.
109. W. E. Hamilton to H.E.A. Hamilton, Nov. 2, 1869, ORSUP 13, the same to A. H.

Hamilton, Dec. 7, 1869, ORSUP 13, and the same to J. O'Regan, July 19, 1872, OR 1753.
110. A. H. Hamilton to J. O'Regan, Nov. 11, 1871, ORSUP 13.
111. A. H. Hamilton to J. O'Regan, June 27, 1872, ORSUP 13.
112. W. E. Hamilton to Robert S. Ball, Nov. 10, 1882, JG 117.
113. *Chatham Market Guide* 6, no. 2 (May 30, 1891), OR 1833.
114. Moore to H.E.A. Hamilton, Sept. 9, 1867, ORSUP 13, and R. P. Graves to H.E.A. Hamilton, Sept. 25, 1867, OR 1717.
115. A. H. Hamilton to J. O'Regan, June 1870, ORSUP 13.
116. Sydney Hamilton to [H.E.A. Hamilton(?)], Mar. 9, 1871, OR 1746.
117. H.E.A. Hamilton, 1860 ORSUP.
118. H.E.A. Hamilton, OR 2302.
119. R. P. Graves to J. O'Regan, June 14, 1870, OR 1908.
120. H.E.A. Hamilton, OR 2302.
121. R. P. Graves to Ellen De Vere O'Brien, June 2, 1873, TCD ms 5020.
122. Sydney Hamilton to J. O'Regan, July 21, 1875, ORSUP 13.
123. Sydney Hamilton to J. O'Regan, July 21, 1875, and obituary notice, ORSUP 13.
124. Notes by Phoebe Alice O'Regan, Jan. 2, 1966, and press cutting from the *Irish Times* or the *Fermanagh Times*, June, 1914.
125. A. H. Hamilton to R. P. Graves, Nov. 10, 1882, ORSUP 13.

Chapter 27

1. The importance of philosophy in the Scottish University curriculum is described well in George Elder Davie, *The Democratic Intellect: Scotland and Her Universities in the Nineteenth Century* (Edinburgh: At the University Press, 1961). The debate over the quantification of the predicate between De Morgan and William Hamilton of Edinburgh was an important stimulus to George Boole in his studies on the logic of mathematics, but Hamilton does not seem to have taken an active interest in it, nor did he recognize the importance of Boole's work. See Luis M. Laita, "Influences on Boole's Logic: The Controversy between William Hamilton and Augustus De Morgan," *Annals of Science* 36 (1979):46-65.
2. *Math. Papers*, 3:5.
3. Samuel Taylor Coleridge, *Aids to Reflection, and the Confessions of an Inquiring Spirit* (London: George Bell and Sons, 1901), p. 224.
4. Quoted from Geoffrey Durrant, *Wordsworth and the Great System: A Study of Wordsworth's Poetic Genius* (Cambridge: At the University Press, 1970), p. 5.
5. Graves, 1:519.
6. John Passmore, *A Hundred Years of Philosophy* (New York: Basic Books, 1966), p. 50.
7. WRH to Aubrey De Vere, Aug. 7, 1855, NLD 905, pp. 88-89.
8. Charles Graves to Thomas Larcom, Sept. 15, 1865, RIA ms. 3E 11.

Bibliographical Essay

This book is based primarily on manuscript sources and on published scientific papers. Full citations appear in the notes, and therefore the following bibliographical essay is limited to a discussion of major manuscript collections and general secondary sources.

By far the largest collection of Hamilton's manuscripts is in the Library of Trinity College Dublin (see the Introduction). MSS 1492 contains approximately 250 notebooks and 10 boxes of loose papers. The notebooks range from small pocket-sized books to very large leather-bound ledger books. The notebooks are largely mathematical and are roughly chronological. The entries are usually dated. Hamilton also made some attempt to put all of his research on a single subject in the same book, in which case entries from different years may be found side by side. Some of the notebooks are letterbooks; in some cases they contain the only extant copy of a letter. Unfortunately Hamilton followed no plan consistently, and often parts of letters are scattered among mathematical calculations. George Salmon prepared a brief catalog of the notebooks soon after Hamilton's death (MS 1492, loose papers, box 10). MS 3558 is a more detailed catalog of the notebooks that J. L. Synge prepared in 1927 while editing the first volume of the *Mathematical Papers.* A third catalog of the notebooks was prepared in 1966 by the late Kenneth O. May. I have a copy of this catalog, which I intend to send to the Library at Trinity College.

The loose papers (also in MSS 1492) contain rough calculations and early drafts of published papers. Modern readers have attempted to put these papers in order, but a great deal of confusion remains. The loose papers contain several important items, including a third book of the "Theory of Systems of Rays," which Hamilton described in his table of contents but never published.

The Hamilton letters are in four major collections at the Trinity College Library. MSS 1492 contains 1421 letters, including most of the scientific and mathematical correspondence. MSS 7762-72 and 7773-76 are two large collections of personal letters recently acquired from the O'Regan family. Xerox copies of these letters are also at Trinity College Dublin; they carry the manuscript numbers 5123-33 and 7243-46, respectively. MSS 4015 is a smaller collection of 127 items. These are copies of letters held by the Graves family. The originals plus several additional letters are now at the library of the American Philosophical Society. MSS 2172-73 are the letters between WRH and Peter Guthrie Tait.

MS 2505 is a catalog of Hamilton's papers (mostly letters) made by R. P. Graves in 1878. MS 2504 is the correspondence of Hamilton's sister Eliza. MS Q.1.3 is a poetry book owned by Elinor De Vere O'Brien, and MSS 5020 (in the papers of Mr. Florence Vere O'Brien) contains letters between Elinor De Vere and Robert P. Graves about Hamilton. Other occasional Hamilton manuscripts are MSS 2165, 2176, and 2246.

Important biographical information is contained in the Trinity College muniments, including the visitation book of the observatory (V. MUN. 99/6), records of examinations (V. MUN. 27.6), and a packet of "Letters etc respecting the Election of a Professor of Astronomy—1827" (WB X 10/22). The Main Library at Trinity College Dublin also has collections of obituaries, offprints owned by Hamilton, copies of his scientific papers, and scrapbooks of "memorials" collected by C. M. Ingleby.

The National Library of Ireland, Dublin, has one very important notebook (MS 905) that contains Hamilton's personal correspondence regarding Catherine Disney. MSS 5764-65 is the complete correspondence between Hamilton and Aubrey De Vere. MSS 11132 is the Edgeworth Correspondence, containing many letters between Hamilton and members of the Edgeworth family. Other occasional Hamilton manuscripts at the National Library are MSS 10531 and 1754.

The Royal Irish Academy has several letters (MSS 23.0.49 and 3 E 11) and an original copy of the Icosian Game (Strong Room, box 3. D. 8 [40]).

The few occasional manuscripts outside of Dublin are listed in *Manuscript Sources for the History of Irish Civilisation* (11 vols., ed. Richard J. Hayes [Boston: G. K. Hall, 1965]). Hamilton's correspondence with Wordsworth is at Cornell University.

A complete bibliography of Hamilton's published works is in Robert Perceval Graves, *Life of Sir William Rowan Hamilton* (3 vols. [Dublin: Hodges, Figgis & Co., 1882-91], 3:645-58). Since this bibliography is readily available it does not seem necessary to reproduce it here. One addition to the bibliography should be noted. Hamilton's letters to Lord Adare were published by Robert P. Graves as "Sir W. Rowan Hamilton on the Elementary Conceptions of Mathematics" (*Hermathena* 6 [1879]: 469-89). Graves's biography is composed largely of letters, many of which are not among the manuscript collections described above (see the Introduction for further information on Graves's biography). The *Mathematical Papers* (3 vols. [Cambridge: at the University Press, 1931-67]) is the other major printed source on Hamilton. The three volumes are on his optics, dynamics, and algebra, respectively, and a fourth volume is planned to cover Hamilton's geometry and miscellaneous writings. These volumes contain both published and unpublished papers, and include explanatory notes and appendixes. The main purpose of the editors was to select the papers of greatest mathematical importance, but they have also included papers that are primarily of historical interest. The *Mathematical Papers* do not contain Hamilton's *Lectures on Quaternions* (Dublin: Hodges and Smith, 1853) or his *Elements of Quaternions* (London: Longmans, Green & Co., 1866).

Because Hamilton was well known during his lifetime there are many biographical notices and obituaries from the nineteenth century. After Graves's *Life* appeared, it became the basis for all later biographical sketches. Most of these later sketches add little that is new, but they are written with a historical perspective and a knowledge of Hamilton's mathematical work that Graves did not have. The earlier sketches are Robert Perceval Graves, "Sir William R. Hamilton, Our Portrait Gallery no. XXVI" (*Dublin University Magazine* 19 [Jan.-June, 1842]: 94-110); Charles Graves, "Eloge of William Rowan Hamilton" (*Math. Papers* 1:ix-xvi); P. G. Tait, "Sir William Rowan Hamilton" (*North British Review* 45 [1866]: 37-74); Augustus De Morgan, "Sir W. R. Hamilton" (*Gentleman's Magazine* 220 [1866]: 128-34]; Charles Pritchard, "William Rowan Hamilton" (*Monthly Notices of the Astronomical Society of London* 26 [1866]: 109-19, found also in *American Journal of Science and Arts*, 2d ser. 42 [1866]: 292-302). The article in the *Dictionary of National Biography* by R. E. Anderson provoked a controversy with R. P. Graves that appeared in the *Athenaeum* (see note 7 to chapter 1). In this century biographical articles have been written by Edward Study, "Sir William Rowan Hamilton" (*Jahresbericht der deutschen Mathematiker-Vereinigung* 14 [1905]: 421-24); A. W. Conway, "The Influence of the Work of Sir William Rowan Hamilton on Modern Mathematical Thought" (*Scientific Proceedings of the Royal Dublin Society* 20 [1931]: 125-28) and "Hamilton, His Life, Work and Influence" (*Proceedings of the Second Canadian Mathematical Congress,* 1949 [Toronto: University of Toronto Press, 1951], pp. 32-41); Edmund T. Whittaker, "The Hamiltonian Revival" (*Mathematical Gazette* 24 [1940]: 153-58), followed by a controversy in the same journal (25 [1941]: 106-8, 298-300) and also by Whittaker, "William Rowan Hamilton," (*Scientific American* 190 [1954]: 82-88); David Eugene Smith, "Sir William Rowan Hamilton" (*Scripta Mathematica* 10 [1944]: 9-11); J. L. Synge, "The Life and Early Work of Sir William Rowan Hamilton" (*Scripta Mathematica* 10 [1944]: 13-24); A. J. McConnell, "William Rowan Hamilton" (*Advancement of Science* 14 [1958]: 323-32); and Cornelius Lanczos, "William Rowan Hamilton— An Appreciation" (*University Review* [National University of Ireland] 4 [1967]: 151-66 and *American Scientist* 55 [1967]: 129-43). The best-known biographical sketch is a chapter entitled "An Irish Tragedy" in Eric Temple Bell's *Men of Mathematics* (New York: Simon and Schuster, 1965). This account is unreliable, however. Bell comments further on Hamilton in his *Development of Mathematics* (2d ed. [New York: McGraw-Hill, 1945]). The most recent biographical sketch is my article in the *Dictionary of Scientific Biography* (15 vols. [New York: Charles Scribner's Sons, 1970-78], 6:85-93).

Useful articles on the cultural setting of Victorian science are W. F. Cannon, "History in Depth: The Early Victorian Period" (*History of Science* 3 [1964]: 20-38), and two articles by L. Pearce Williams, "The Historiography of Victorian Science" (*Victorian Studies* 9 [1966]: 197-204), and "The Physical Sciences in the First Half of the Nineteenth Century: Problems and Sources" (*History of Science* 1 [1962]: 1-15). On the political and social setting in Ireland the most useful general history is J. C. Beckett, *The Making of Modern Ireland, 1603-1923* (London: Faber and Faber, 1966). Also valuable are R. B. McDowell's *Public Opinion and Government Policy in Ireland, 1801-1846* (London: Faber and Faber, 1952) and *Social Life in Ireland, 1800-1845* ("Irish Life and Culture" 12, The Cultural Relations Committee of Ireland [Dublin: Three Candles, 1957]). Demographic information on Ireland is most easily found in T. W. Freeman, *Pre-Famine Ireland: A Study in Historical Geography* (Manchester: Manchester University Press, 1957), and K. H. Connell, *Population of Ireland, 1750-1845*

(Oxford: At the Clarendon Press, 1950). Identification of local towns and residences is greatly facilitated by Samuel Lewis, *A Topographical Dictionary of Ireland* (2 vols. [London: S. Lewis, 1837]). A context for Hamilton's references to rebellion in the countryside is provided by Galen Broeker, *Rural Disorder and Police Reform in Ireland, 1812-1836* (London: Routledge and Kegan Paul, 1970). The best source of information on the Church of Ireland is Walter Alison Phillips, *History of the Church of Ireland from the Earliest Times to the Present Day* (3 vols. [Oxford: Oxford University Press, 1933]). A good source for the Oxford Movement is R. W. Church, *The Oxford Movement: Twelve Years, 1833-1845* (1891; Hamden, Conn.: Archon Books, 1966).

On the group of scientists at Trinity College Dublin see A. J. McConnell's "The Dublin Mathematical School in the First Half of the Nineteenth Century" (Quaternion Centenary Celebration, Royal Irish Academy, *Proceedings* 50, sect. A, no. 6 [Feb., 1945]: 75-88). A more detailed history of this school is the Ph.D. thesis of James G. O'Hara, "Humphrey Lloyd (1800-1881) and the Dublin Mathematical School of the Nineteenth Century" (University of Manchester Institute of Science and Technology, 1979). Another thesis on the Dublin scientific community is being prepared by Patrick Cross at the University of Oklahoma, and Gordon L. H. Davies is editing a book entitled *Science in Trinity College, 1774-1920* in honor of the quater-centenary of Trinity College in 1991 and 1992. General histories of Trinity College are William Macneile Dixon, *Trinity College Dublin* (London: F. E. Robinson & Co., 1902); Constantia Elizabeth Maxwell, *A History of Trinity College Dublin, 1591-1892* (Dublin: The University Press, Trinity College, 1946); John William Stubbs, *The History of the University of Dublin from Its Foundation to the End of the 18th Century* (Dublin: Hodges, Figgis & Co., 1889); and *The Book of Trinity College Dublin, 1591-1891* (Dublin: Hodges, Figgis & Co., 1892). The report of the Commissioners on Trinity College Dublin in 1853 (Great Britain, *Parliamentary Papers*, vol. 45, no. 1637 [1852-53], p. 82) contains the transcript of a hearing in which Hamilton was subjected to a rather severe grilling about the affairs of Dunsink Observatory.

Information on Aubrey De Vere comes from Wilfrid Philip Ward, *Aubrey De Vere: A Memoir Based on His Unpublished Diaries and Correspondence* (London: Longmans, Green and Co., 1904), and S. M. Paraclita Reilly, *Aubrey De Vere:Victorian Observer* (Dublin: Clonmore and Reynolds, 1956). The best work on Maria Edgeworth is Marilyn Butler's *Maria Edgeworth: A Literary Biography* (Oxford: At the Clarendon Press, 1972). Also helpful are Michael Hurst, *Maria Edgeworth and the Public Scene: Intellect, Fine Feeling, and Landlordism in the Age of Reform* (Coral Gables, Fla.: University of Miami Press, 1969), and Christina Colvin, ed., *Maria Edgeworth: Letters from England, 1813-1844* (Oxford: At the Clarendon Press, 1971).

The relationship between romanticism and nineteenth-century science is a subject that cannot be grasped without attention to particular cases. General treatments that I have found helpful are L. Pearce Williams, "Kant, Naturphilosophie and Scientific Method" (in *Foundations of Scientific Method: The Nineteenth Century*, ed. Ronald N. Giere and Richard S. Westfall [Bloomington: Indiana University Press, 1973], pp. 3-22); Barry Gower, "Speculation in Physics: Theory and Practice of Naturphilosophie" (*Studies in the History and Philosophy of Science* 3 [1973]: 301-56); and D. M. Knight, "The Physical

Sciences and the Romantic Movement" (*History of Science* 9 [1970]: 54-75). The argument that British chemists drew upon German metaphysical idealism is made most strongly in L. Pearce Williams, *Michael Faraday: A Biography* (New York: Basic Books, 1965). More on the subject is in Trevor H. Levere, *Affinity and Matter: Elements of Chemical Philosophy, 1800-1865* (Oxford: At the Clarendon Press, 1971).

On Coleridge and Wordsworth the works of Meyer H. Abrams are helpful: *The Mirror and the Lamp: Romantic Theory and the Critical Tradition* (New York: W. W. Norton Co., 1956), *Natural Supernaturalism: Tradition and Revolution in Romantic Literature* (New York: W. W. Norton, 1971), and especially "Coleridge's 'A Light in Sound': Science, Meta-Science, and Poetic Imagination" (*Proceedings of the American Philosophical Society* 116, no. 6 [Dec., 1972]: 456-76). Also on Coleridge see E. S. Shaffer, "Coleridge and Natural Philosophy: A Review of Recent Literary and Historical Research" (*History of Science* 12 [1974]: 284-98). Charles R. Sanders, in *Coleridge and the Broad Church Movement* (Durham, N.C.: Duke University Press, 1942), explains Coleridge's religious position and his important distinction between reason and understanding. Other books that help throw light on the obscurities of Coleridge's philosophy are John H. Muirhead, *Coleridge as Philosopher* (New York: Humanities Press, 1954); G.N.G. Orsini, *Coleridge and German Idealism: A Study in the History of Philosophy with Unpublished Material from Coleridge's Manuscripts* (Carbondale: Southern Illinois University Press, 1969); Thomas McFarland, *Coleridge and the Pantheistic Tradition* (Oxford: At the Clarendon Press, 1969); and Alice D. Snyder, *Coleridge on Logic and Learning, with Selections from the Unpublished Manuscripts* (New Haven, Conn.: Yale University Press, 1929). The biography of Wordsworth by Mary T. Moorman, *William Wordsworth; A Biography* (2 vols. [Oxford: At the Clarendon Press, 1957-65]), gives a context for the events mentioned by Hamilton and Wordsworth in their correspondence. Wordsworth's visit to Ireland is described in Herbert V. Fackler, "Wordsworth in Ireland, 1829: A Survey of His Tour" (*Eire/Ireland* 6 [1971]: 53-64), and George Dodd, "Wordsworth and Hamilton" (*Nature* 228 [1970]: 1261-63). On Wordsworth's philosophy, I found the most helpful books to be Geoffrey Durrant, *Wordsworth and the Great System: A Study of Wordsworth's Poetic Genius* (Cambridge: At the University Press, 1970), and Melvin Miller Rader, *Wordsworth: A Philosophical Approach* (Oxford: At the Clarendon Press, 1967).

The early history of the British Association is currently receiving much attention, and our understanding of events leading up to its organization may soon change as a result. The best brief accounts are the articles by A. D. Orange, "The Origins of the British Association for the Advancement of Science" (*British Journal for the History of Science* 6 [1972]: 152-76), and "The Idols of the Theatre: The British Association and Its Early Critics" (*Annals of Science* 32 [1975]: 277-94). The argument that a controversial presidential election at the Royal Society led to the creation of the British Association is given by L. Pearce Williams, "The Royal Society and the Founding of the British Association for the Advancement of Science" (*Notes and Records of the Royal Society* 16 [1961]: 221-33).

A substantial amount has been written about the scientific philosophies of William Whewell and John Herschel, but not much has been written about the specific problems they discussed with Hamilton. For information on Whewell's ideas about mechanics one has to turn to I. Todhunter, *William Whewell, D.D.: An Account of His Writings* (2 vols. [London: Macmillan and Co., 1876]), and to Whewell's own numerous writings on mechanics, especially his article "On the Nature of the Truth of the Laws of Motion" (Cambridge Philosophical Society, *Transactions* 5, pt. 2 [1833-35]: 149-72). Whewell's method is explored by Robert E. Butts in *William Whewell's Theory of Scientific Method* (Pittsburgh: University of Pittsburgh Press, 1968) and in "Whewell's Logic of Induction" (in *Foundations of Scientific Method: The Nineteenth Century,* ed. Ronald N. Giere and Richard S. Westfall [Bloomington: Indiana University Press, 1973], pp. 53-85). John Herschel's philosophy is described in Walter F. Cannon, "John Herschel and the Idea of Science" (*Journal of the History of Ideas* 22 [1961]: 215-39), and Joseph Agassi, "Sir John Herschel's Philosophy of Success" (*Historical Studies in the Physical Sciences* 1 [1969]: 1-36). Most of Hamilton's correspondence with Herschel had to do with optics and is discussed in a later section.

Hamilton learned about Kant indirectly through Coleridge, Carlyle, Madame de Staël, and Dugald Stewart, but he built his metaphysics of mathematics on a direct reading of Kant's works. An appreciation of Hamilton's arguments about the foundations of algebra therefore requires a plunge into Kant's critical philosophy. My chief guide in this undertaking was Norman Kemp Smith's *A Commentary to Kant's "Critique of Pure Reason"* (2d ed. [London: Macmillan and Co., 1923]), without which I would have floundered hopelessly. Two articles were helpful in discovering Kant's intent. These are Kaarlo Jaakko Hintikka, "Kant on the Mathematical Method" (*Monist* 51 [1967]: 352-75), and Charles Parsons, "Kant's Philosophy of Arithmetic" (in *Philosophy, Science and Method,* ed. Sidney Morgenbesser, Patrick Suppes, Morton White, [New York: St. Martin's Press, 1969], pp. 568-94). René Wellek, *Immanuel Kant in England, 1793-1838* (Princeton, N.J.: Princeton University Press, 1931), gives valuable information about the availability of Kant's ideas in Great Britain during the nineteenth century.

Hamilton's work in optics is described in G. C. Steward, "On the Optical Writings of Sir William Rowan Hamilton" (*Mathematical Gazette* 16 [1932]: 179-91), and in Georg Prange, "W. R. Hamilton's Bedeutung für die geometrische Optik" (*Jahresbericht der deutschen Mathematiker-Vereinigung* 30 [1921]: 69-82). George Sarton wrote on the "Discovery of Conical Refraction by William Rowan Hamilton and Humphrey Lloyd (1833)" (*Isis* 17 [1932]: 154-170). The most reliable descriptions of Hamilton's optics are in the introduction and appendixes to the first volume of *Math. Papers,* edited by John L. Synge and A. W. Conway, and in a helpful book by Synge, *Geometrical Optics: An Introduction to Hamilton's Method* ("Cambridge Tracts in Mathematics and Mathematical Physics," no. 37 [Cambridge: At the University Press, 1937]). On the relationship between Hamilton's characteristic function and Bruns' Eikonal see Felix Klein, "Über das Brunsche Eikonal" (*Zeitschrift für Mathematik und Physik* 46 [1901]: 603-6), and two articles by Synge, "Hamilton's Method in Geometrical Optics" and "Hamilton's Characteristic Function and Bruns' Eiconal" (*Journal of the Optical Society of America* 27 [1937]: 75-82 and 138-44).

On the developments in analytical geometry that led to Hamilton's first study of rays, see René Taton, *L'oeuvre scientifique de Monge* (1st ed. [Paris: Presses Universitaires de France, 1951]), and Carl Boyer, "Cartesian Geometry from Fermat to Lacroix" (*Scripta Mathematica* 13 [1947]: 133-53). For the history of the wave theory of light I relied on three Ph.D. dissertations: Robert H. Silliman, "Augustin Fresnel (1788-1827) and the Establishment of the Wave Theory of Light" (Princeton University, 1968); Eugene Frankel, "Jean Baptiste Biot: The Career of a Physicist in Nineteenth-Century France" (Princeton University, 1972); and David B. Wilson, "The Reception of the Wave Theory of Light by Cambridge Physicists (1820-1850): A Case Study in the Nineteenth-Century Mechanical Philosophy" (The Johns Hopkins University, 1968). Charles Gillispie has a good chapter on Fresnel and Young in his *Edge of Objectivity* ([Princeton, N.J.: Princeton University Press, 1960], pp. 406-35). Henry John Steffens, *The Development of Newtonian Optics in England* (New York: Science History Publications, 1977), is good on the eighteenth century, and Geoffrey Cantor, "The Reception of the Wave Theory of Light in Britain: A Case Study Illustrating the Role of Methodology in Scientific Debate" (*Historical Studies in the Physical Sciences* 6 [1975]: 109-32), describes the controversies in which Hamilton was involved.

Field theories in the nineteenth century are described in Edmund T. Whittaker, *A History of the Theories of Aether and Electricity* (2 vols. [New York: Harper Torchbooks, 1960]); Kenneth F. Schaffner, *Nineteenth-Century Aether Theories* (Oxford: Pergamon Press, 1972); Mary B. Hesse, *Forces and Fields: The Concept of Action at a Distance in the History of Physics* (London: Thomas Nelson and Sons, 1961); and Barbara Gusti Doran, "Origin and Consolidation of Field Theory in Nineteenth-Century Britain: From the Mechanical to the Electromagnetic View of Nature" (*Historical Studies in the Physical Sciences* 6 [1975]: 133-260). A good impression of the points at issue can be obtained from reading Humphrey Lloyd's *Treatise on Light and Vision* (London: Longman, Rees, Orme, Brown, and Green, 1831) and his "Report on the Progress and Present State of Optics" (*British Association Report,* [1834]: 295-413), and the article "Light" by John F. W. Herschel in the *Encyclopaedia Metropolitana* (29 vols. [London: B. Fellowes; F. & J. Rivington, etc., 1817-45], 4:341-586). Developments in the middle of the century are described by R. T. Glazebrook, "Report on Optical Theories" (*British Association Reports* [1885], pp. 157-261). The early textbook by Thomas Preston, *The Theory of Light* (2d ed. [London: Macmillan, 1895]), describes the many unusual optical phenomena that the wave theory was required to explain.

It is difficult to find clear descriptions of Hamilton's dynamics. H. Bateman's "Hamilton's Work in Dynamics and Its Influence on Modern Thought" (*Scripta Mathematica* 10 [1944]: 51-63) gives numerous references to later theoreticians who have developed Hamilton's work. René Dugas, "Sur la pensée dynamique d'Hamilton; origines optiques et prolongements modernes" (*Revue Scientifique* 79 [1941]: 15-23), covers much of the same ground. A more thorough study is Wolfgang Buchheim, "William Rowan Hamilton und das Fortwirken seiner Gedanken in der modernen Physik" (*NTM: Schriftenreihe für Geschichte der Naturwissenschaft, Technik, und Medizin* 5, pt. 1 [1968]: 19-30; 6, pt. 2 [1969]: 43-60). A good contemporary evaluation is Arthur Cayley, "Report on the

Recent Progress of Theoretical Dynamics" (*British Association Reports* [1857], pp. 1-42). On the hodograph see J. M. Child, "Hamilton's Hodograph" (*Monist* 25 [1915]: 615-24). Descriptions in modern textbooks of Hamilton's dynamics usually do not follow Hamilton's arguments closely. The most historical account is Wolfgang Yourgrau and Stanley Mandelstam, *Variational Principles in Dynamics and Quantum Theory* (3d ed. [Philadelphia: W. B. Saunders Co., 1968]). Others are Georg Prange, "Die allgemeinen Integrationsmethoden der analytischen Mechanik" (*Encyklopädie der mathematischen Wissenschaften* 4, no. 2 [1933]: 505-804); Cornelius Lanczos, *The Variational Principles of Mechanics* (4th ed. [Toronto: University of Toronto Press, 1970]); and Herbert Goldstein, *Classical Mechanics* (Reading, Mass.: Addison-Wesley Publishing Co., 1950).

Hamilton's contributions to algebra are discussed in several general histories of mathematics. These include Felix Klein, *Vorlesungen über die Entwicklung der Mathematik im 19. Jahrhundert* (2 vols. [Berlin: Verlag von Julius Springer, 1926-27]); Morris Kline, *Mathematical Thought from Ancient to Modern Times* (New York: Oxford University Press, 1972); and Dirk Struik, *A Concise History of Mathematics* (New York: Dover, 1948). The more specialized history by Luboš Nový, *Origins of Modern Algebra* (Prague: Academia Publishing House of the Czechoslovak Academy of Sciences, 1973), gives details that are not included in other accounts and is especially good on the work of Martin Ohm. Another useful article is C. C. MacDuffee, "Algebra's Debt to Hamilton" (*Scripta Mathematica* 10 [1944]: 25-35). Ernest Nagel has a good article on the debate over negative and imaginary numbers entitled "'Impossible Numbers': A Chapter in the History of Modern Logic" (*Studies in the History of Ideas*, 3 vols., ed. the Department of Philosophy of Columbia University [New York: Columbia University Press, 1918-35], 3:429-74). Two articles by Florian Cajori also give a historical context for Hamilton's ideas on the foundations of algebra. These are "Historical Note on the Graphic Representation of Imaginaries before the Time of Wessel" (*American Mathematical Monthly* 19 [1912]: 167-71) and "History of the Exponential and Logarithmic Concepts" (*American Mathematical Monthly* 20 [1913]: 5, 35, 75, 107, 148, 173, 205). Three theses have been especially helpful: Helena Pycior, "The Role of Sir William Rowan Hamilton in the Development of British Modern Algebra" (Ph.D. diss., Cornell University, 1976), Barbara Underwood Cohen, "From Algebra to Algebras: A Study of Changing Conceptions of Mathematics" (Undergraduate thesis, Harvard University, 1966); and Stella Mills, "British Group Theory, 1850-1890" (Ph.D. diss., University of Birmingham, 1976). Also on group theory, see G. A. Miller, "Note on William R. Hamilton's Place in the History of Abstract Group Theory" (*Bibliotheca mathematica* 11-12, ser. 3 [1910-12]: 314-15).

Jerold Mathews, in "William Rowan Hamilton's Paper of 1837 on the Arithmetization of Analysis" (*Archive for History of Exact Science* 19 [1978]: 177-200), explains Hamilton's construction of number and criticizes his arguments in light of modern mathematics. A shorter article limited to Hamilton's construction of irrational numbers is H. E. Hawkes, "Note on Hamilton's Determination of Irrational Numbers" (*Bulletin of the American Mathematical Society* 7 [1900]: 306-7). A good statement of the intuitionist approach in modern mathematics is L.E.J. Brouwer, "Intuitionism and Formalism" (in *Philosophy of Mathematics: Selected Readings*, ed. Paul Benacerraf and Hilary Putnam [Englewood Cliffs,

N.J.: Prentice-Hall, 1964]). The search for a general solution to the equation of the fifth degree is described in J. Pierpont, "Zur Geschichte der Gleichung des V. Grades (bis 1858)" (*Monatshefte für Mathematik und Physik* 6 [1895]: 15-68); H. O. Foulkes, "The Algebraic Solution of Equations" (*Science Progress in the Twentieth Century* 26 [1932]: 601-8); and Felix Klein, *Lectures on the Icosahedron and the Solution of Equations of the Fifth Degree* (2d ed., trans. George Gavin Morrice [1884; New York: Dover Publishing, 1956]).

The centennial of the discovery of quaternions in 1943 produced two collections of papers on Hamilton and the quaternions. These are the "Quaternion Centenary Celebration" (Royal Irish Academy, *Proceedings* 50, sect. A, no. 6 [Feb., 1945]: 69-121) and *A Collection of Papers in Memory of Sir William Rowan Hamilton* (ed. David Eugene Smith, *Scripta Mathematica Studies,* no. 2 [New York, 1945]). Both of these collections contain valuable papers, some of which I have already mentioned. On the discovery of quaternions see E. T. Whittaker, "The Sequence of Ideas in the Discovery of Quaternions," pages 93-98 in "Quaternion Centenary Celebration," and B. L. van der Waerden, "Hamilton's Discovery of Quaternions" (*Mathematics Magazine* 49 [1976]: 227-34). Hamilton's significance for the later development of vector analysis is described in great detail by Michael J. Crowe in his *A History of Vector Analysis: The Evolution of the Idea of a Vectorial System* (Notre Dame, Ind.: University of Notre Dame Press, 1967). Articles on the same subject are Reginald J. Stephenson, "Development of Vector Analysis from Quaternions" (*American Journal of Physics* 34 [1966]: 194-201), and Alfred M. Bork, "Vectors Versus Quaternions—The Letters in *Nature*" (*American Journal of Physics* 34 [1966]: 202-11). Kenneth O. May discusses the triplets in "The Impossibility of a Division Algebra of Vectors in Three Dimensional Space" (*American Mathematical Monthly* 73 [1966]: 289-91), and the history of theorems on the sums of squares is given in Charles W. Curtis, "The Four and Eight Square Problem and Division Algebras" (in "Studies in Mathematics," vol. 2, *Studies in Modern Algebra,* ed. A. A. Albert [Buffalo: Mathematical Association of America, 1963], 1: 100-125) and L. E. Dickson, "On Quaternions and Their Generalization and the History of the Eight Square Theorem" (*Annals of Mathematics,* 2d ser. 20 [1919]: 155-71).

For the importance of quaternions in the history of linear algebra and matrices, see two articles by Thomas Hawkins, "Cauchy and the Spectral Theory of Matrices," (*Historia Mathematica* 2 [1975]: 1-29) and "Another Look at Cayley and the Theory of Matrices" (*Archives internationales d'histoire des sciences* 27 [1977]: 82-112). Peter Guthrie Tait and his relations with Hamilton are described in Cargill Gilston Knott, *Life and Scientific Works of Peter Guthrie Tait* (Cambridge: At the University Press, 1911), followed by a debate in *Nature* (87 [1911]: 35-37, 77, and 111).

Descriptions of Hamilton's Icosian Calculus are in Norman L. Biggs, E. Keith Lloyd, and Robin J. Wilson, *Graph Theory, 1736-1936* (Oxford: At the Clarendon Press, 1976); Ross Honsberger, *Mathematical Gems* (2 vols. [Washington, D.C.: Mathematical Association of America, 1973-76]); Edouard Lucas, *Recréations mathématiques* (4 vols. [Paris: Gauthier-Villars, 1882-94]); and J. Riversdale Colthurst, "The Icosian Calculus" (Royal Irish Academy, *Proceedings* 50, sect. A [1945]: 112-21). The recent confirmation of the four-color conjecture has revived interest in its history. See especially John Wilson, "New Light on the Origin of the Four-Color Conjecture" (*Historia Mathematica* 3 [1976]: 329-30), and Kenneth O. May, "The Origin of the Four-Color Conjecture" (*Isis* 56 [1965]: 346-48). A readable description of the solution is Kenneth Appel and Wolfgang Haken, "The Solution of the Four-Color-Map Problem" (*Scientific American* 237, no. 4 [Oct., 1977]: 108-21). John Graves's library, where Hamilton had the inspiration of the icosian calculus, is described in Alison R. Dorling, "The Graves Mathematical Collection in University College London" (*Annals of Science* 33 [1976]: 307-10).

There have been no recent studies on the history of the optical-mechanical analogy as it came from Hamilton, but there have been many studies of the background to Erwin Schrödinger's wave mechanics. Among others I have used V. Raman and Paul Forman, "Why Was It Schrödinger Who Developed de Broglie's Ideas?" (*Historical Studies in the Physical Sciences* 1 [1969]: 291-314); Armin Hermann, "Schrödinger" (*Dictionary of Scientific Biography*, vol. 12, pp. 217-23); William T. Scott, *Erwin Schrödinger: An Introduction to His Writtings* (Amherst: University of Massachusetts Press, 1967); J. Gerber, "Geschichte der Wellenmechanik" (*Archive for History of Exact Science* 5 [1968-69]: 349-416); Paul Hanle, "The Schrödinger-Einstein Correspondence and the Sources of Wave Mechanics" (*American Journal of Physics* 47 [1979]: 644-48); and Max Jammer, *The Conceptual Development of Quantum Mechanics* (New York: McGraw-Hill Book Co., 1966). Easy access to the manuscript material on the history of quantum mechanics is provided by Thomas Kuhn, John L. Heilbron, Paul Forman, and Lini Allen, *Sources for History of Quantum Physics: An Inventory and Report* (Philadelphia: American Philosophical Society, 1967).

Index